Contemporary Problems in Mathematical Physics

Proceedings of the Second International Workshop on

Contemporary Problems in Mathematical Physics

Cotonou, Republic of Benin 28 October – 2 November 2001

editors

Jan Govaerts
Catholic University of Louvain, Belgium

M. Norbert Hounkonnou
University of Abomey-Calavi, Republic of Benin

Alfred Z. Msezane
Clark Atlanta University, USA

World Scientific
New Jersey • London • Singapore • Hong Kong

Published by

World Scientific Publishing Co. Pte. Ltd.
P O Box 128, Farrer Road, Singapore 912805
USA office: Suite 1B, 1060 Main Street, River Edge, NJ 07661
UK office: 57 Shelton Street, Covent Garden, London WC2H 9HE

British Library Cataloguing-in-Publication Data
A catalogue record for this book is available from the British Library.

CONTEMPORARY PROBLEMS IN MATHEMATICAL PHYSICS
Proceedings of the Second International Workshop

ISBN 981-02-4935-7

Printed in Singapore by World Scientific Printers (S) Pte Ltd

FOREWORD

This volume is the Proceedings of the Second International Workshop on Contemporary Problems in Mathematical Physics, which was held in Cotonou (Republic of Benin) from October 28th to November 2nd, 2001. Being the second such scientific gathering, following the First Workshop on Contemporary Problems in Mathematical Physics also organized in Cotonou from October 31st to November 5th, 1999, it is hoped these two first events are called to launch a long and prosperous scientific tradition in Western Africa through meetings such as these, foreseen to occur on a regular basis every second year in the Republic of Benin. Indeed, one should hope that in the future these COPROMAPH conferences, as they shall henceforth be known, will be organized under the auspices of the recently institutionalized International Chair in Mathematical Physics and Applications, within the University of Abomey-Calavi, Republic of Benin.

Scientific meetings such as the COPROMAPH Workshops provide a unique opportunity for mathematical and theoretical physicists of the African continent to meet and initiate stimulating collaborations both within Africa as well as with scientists from other parts of the world also attending these events. In particular, it is hoped that the greatest benefits will go to the younger generation of African scientists, whose responsability it is to build the basis for a scientific and technological tradition and expertise in Africa extending far beyond the new century into the young millenium.

Close to 100 participants from Africa, Europe and North America attended this second Workshop. Presentations included two series of invited plenary lectures that lasted throughout the Workshop, as well as shorter research communications organized in three parallel sessions every day, according to the following general themes.

. Theoretical Methods of Modern Classical and Quantum Physics

. Coherent states, Wavelets and
Geometric Methods in Theoretical Physics

. Functional Analysis, Special Functions and
Orthogonal Polynomials

A significant fraction of these contributions is collected in this Proceedings volume, following a strict refereeing process through which not all the submitted manuscripts could be accepted for publication.

This Second International Workshop on Contemporary Problems in Mathematical Physics was made possible through the generous financial support in different forms of many Institutions. The organizers wish to express their most special thanks to the Government of the Republic of Benin through its Ministry for Higher Education and Scientific Research, the University of Abomey-Calavi and its Vice-Chancellor, and all the other Sponsors mentioned in the official opening and closing Addresses reproduced in the next pages, as well as in the list provided hereafter.

Beyond the contributions of the participants, the efforts of many people, too numerous to be all named explicitly here, helped shape this second Workshop into the successful meeting it turned out to be once again. That they may all find here the expression of our gratitude.

Let us all meet again, together with new collaborators, at the Third International Workshop on Contemporary Problems in Mathematical Physics, already foreseen for October-November 2003.

Jan Govaerts
M. Norbert Hounkonnou
Alfred Z. Msezane

Scientific Advisory Committee

Organizing Committee

E. Azatassou (Benin)
E. Baloïtcha (Benin)
J.O. Chabi (Benin)
H. Capo (Benin)
X. Chapuisat (France)
J.M. Combes (France)
C. Duqué (Belgium)
N. Fonton (Benin)
J. Govaerts (Belgium)
M.N. Hounkonnou (Benin)
A.Z. Msezane (USA)
B. Olory (Benin)

Sponsored and Supported by

. The European Physical Society (Mulhouse, France)

. The Abdus Salam International Center for Theoretical Physics (Trieste, Italy)

. Université Paris Sud XI (Paris, France)

. Université catholique de Louvain (Louvain-la-Neuve, Belgique)

. Coopération Universitaire au Développement, Commission Interuniversitaire des Universités Francophones (CUD-CIUF, Belgique)

. The Center for Theoretical Studies of Physical Systems, Clark Atlanta University (Atlanta, Georgia, USA)

. Université de Perpignan, France

. Ministère de la Coopération, France (Ambassades de France au Cameroun, au Tchad et au Bénin)

. Hôtel Croix du Sud, Cotonou, République du Bénin

. Hôtel du Port, Cotonou, République du Bénin

. Centre Béninois pour la Recherche Scientifique et Technique (CBRST), République du Bénin

CONTENTS

Part II. Coherent States, Wavelets and Geometric Methods in Theoretical Physics

Part III. Functional Analysis
Special Functions and Orthogonal Polynomials

Opening Address by the Representative of His Excellency, the Minister for Higher Education and Scientific Research

· Vice-Chancellor of the University of Abomey-Calavi,
· Deputy Vice-Chancellor of the University of Abomey-Calavi,
· Members of the cabinet of the Vice-Chancellor of the University of Abomey-Calavi,
· Director of the executive Secretariat of COPROMAPH2,
· Dean of the Faculty of Arts,
· Representative of the Director of the Benin Center for Sustainable Development,
· Members of the scientific council of COPROMAPH2,
· Members of the organizing committee,
· Dear Professors and researchers,
· Honorable guests,

First of all, on behalf of His Excellency, the Minister for Higher Education and Scientific Research, I would like to welcome you all, especially all our partners who are now here in Benin to attend the Second International Workshop in Mathematical Physics.

The present Workshop which opens today, and which is the Second International Workshop on Contemporary Problems in Mathematical Physics (COPROMAPH2) is a great honour for our country, the Republic of Benin.

It is a well-known fact that our Government shares with the scientific world the conviction that development is the aspiration towards a better well-being for the whole of humankind and that this can but only happen based on progress in the most advanced sciences. This is why, from very early on, the Benin Government has favourably welcomed the idea of holding this series of International Workshops on Contemporary Problems in Mathematical Physics. It is in this context that our country has had the great honour to host COPROMAPH1, in this very same Centre St. Jean Eudes of Atrokpocodji.

Furthermore, seeing the success achieved by COPROMAPH1, Benin has been retained to host the headquarters of the Executive Secretariat of the institution under the supervision of a high ranking researcher who, at the age of 46, has proved himself internationally and has reached in 2000 the rank of professorship through CAMES standards, I mean our fellow countryman, posted at the University of Abomey-Calavi, Professor Mahouton Norbert Hounkonnou.

Finally, during COPROMAPH1, it was agreed that the International Workshop on Contemporary Problems in Mathematical Physics will be held every two years in the Republic of Benin. This is a responsability which, thanks to the dynamism of the members of the local organizing committee, the Executive Secretariat has met in a remarkable way. As you all know, COPROMAPH1 was held from October 31st to November 5th, 1999. And today, October 28th, 2001, we are proud to open COPROMAPH2.

Ladies and gentlemen researchers,

The mention of Mathematical Physics as a main theme opens up a prosperous field to the generations of students of our Universities. Mathematical Physics is a noble instrument for researchers all over the world, which is devoted to the progress of the physical sciences, and beyond that, to the progress of almost all the domains of knowledge. The degree of abstraction contained in the formulae of Mathematical Physics does allow this science to be the basis of all development, even though its so-called abstraction actually projects into the future the potential to see practical implementations of your research results. Science establishes the basis for tomorrow's technology. Denying its appropriation is to close onto oneself the door towards the future. The Benin Government is very grateful to you all for your contribution in giving Africa a chance to get settled in the third millennium in harmony with the most advanced countries.

I would now like to take this opportunity to express my sincere gratitude to the eminent professors and researchers who have come all the way from other African countries, from America and from Europe in order to examine together the contemporary problems in Mathematical Physics.

As I wish every success to your conference, I am confident that the conclusions that you will achieve will be largely disseminated and will contribute to the progress of science.

Long live the International Workshop on Contemporary Problems in Mathematical Physics!

Long live International Scientific Cooperation!

I now formally declare open the Second International Workshop on Contemporary Problems in Mathematical Physics.

Thank you.

Dr. R. N'Tia

Adjoint Director of the Minister's Cabinet,

speaking on behalf of Minister Professor D. Sossa.

Opening Address by
the Vice-Chancellor of the University of Abomey-Calavi

· Representative of His Excellency,
  the Minister for Higher Education and Scientific Research,
· Deputy Vice-Chancellor of the University of Abomey-Calavi,
· Members of the cabinet of the Vice-Chancellor
  of the University of Abomey-Calavi,
· Director of the executive Secretariat of COPROMAPH2,
· Dean of the Faculty of Arts,
· Representative of the Director
  of the Benin Center for Sustainable Development,
· Members of the scientific council of COPROMAPH2,
· Members of the organizing committee,
· Dear Professors and researchers,
· Honorable guests,

After having successfully organized the First International Workshop on Contemporary Problems in Mathematical Physics in October-November 1999, the University of Abomey-Calavi (the National University of Benin) is happy to show its abilities in organizing effectively, two years later, the second edition of the conference, known under the name of COPROMAPH2. The positive response that you, eminent researchers and high ranking teachers from all over the world, have given to the call from Cotonou is more than an encouragement on an international scale for the developement of high level research in our African Universities. It is thanks to you that we have been able to face this double challenge of presiding the Executive Secretariat and calling this international conference on Contemporary Problems in Mathematical Physics. From here on, we can speak of a tradition that I invite you to perpetuate in taking courageous measures to support the President of the International Chair which the Executive Secretary has mentioned already. In fact, COPROMAPH gathers eminent high ranking researchers of a domain at the boundary between Mathematics and Physics and which requires the combination of scientific minds pursuing research in the specific domains of Mathematics on the one hand, and of Physics on the other. We are allowed to ask ourselves if Theoretical Physics alone does not already encompass the field of Mathematical Physics. We are also allowed to ask ourselves about possibly including Applied Mathematics in this domain as well. In fact, and as it appears from the topics of the contributed papers at this conference, Mathematical Physics proceeds from all these aspects altogether. A major difficulty encountered by our African Universities in the adoption of teaching research programmes is this possibility of a common approach to problems of Physics by physicists and mathematicians alike. The creation of a Ph. D. programme and the organization of numerous international conferences related to these doctoral studies show the commitment of our University to the development of fundamental research.

Dear Minister,
Dear Dean,
Ladies and gentlemen,

The University of Abomey-Calavi is happy to welcome you at this second edition of COPROMAPH which gathers great physicists and mathematicians from various scientific Institutions from North America, Europe and Africa. The Ph. D. students' aspiration is really to permanently relive these moments of scientific exchange with most eminent researchers. They will draw from them some ideas that will greatly benefit their training through research. Their teachers will find in this conference a privileged opportunity for the presentation of their research results.

The University of Abomey-Calavi is grateful to its sister Institutions for willingly allowing their researchers to attend this meeting. It is particularly grateful to the Executive Secretariat and to the international and local organization committees. This is also the opportunity to thank the Sponsors which have contributed to making this conference reality. To list just a few, I would like to mention,

. The European Physical Society;

. The Abdus Salam International Center for Theoretical Physics;

. The Université Paris Sud XI;

. The Université catholique de Louvain (Louvain-la-Neuve, Belgique);

. The Center for Theoretical Studies of Physical Systems, Clark Atlanta University, USA;

. The Université de Perpignan, France;

. The Ministère de la Coopération, France (Ambassades de France au Cameroun, au Tchad et au Bénin);

. The Hôtel Croix du Sud, Cotonou, Republic of Benin;

. The Hôtel du Port, Cotonou, Republic of Benin;

. The Centre Béninois pour la Recherche Scientifique et Technique (CBRST).

Dear Minister,

We are aware that without your support, such meetings have no chance to succeed. Once again, the University of Abomey-Calavi thanks you.

On behalf of the University community of Benin, I wish full success to your conference.

Thank you.

Professor I. Takpara

Closing Address by the COPROMAPH2 Executive Director

· His Excellency, the Minister for Higher Education and Scientific Research,
· Vice-Chancellor of the University of Abomey-Calavi,
· Deputy Vice-Chancellor of the University of Abomey-Calavi,
· Members of the cabinet of the Vice-Chancellor
of the University of Abomey-Calavi,
· Dean of the Faculty of Arts,
· Members of the scientific council of COPROMAPH2,
· Members of the organizing committee,
· Dear Professors and researchers,
· Honorable guests,

In a few moments, we shall be closing the Second International Workshop on Contemporary Problems in Mathematical Physics (COPROMAPH2). I would like to assess the proceedings and say that, thanks to your various contributions, our conference has been a success for a number of reasons which are worth mentioning.

First of all, attendance has been numerous and regular, meeting the expectations of the organizing committee. There were as many as 89 participants who have come from 17 countries and from three continents, namely America, Europe and Africa. Represented countries are Belgium, Benin, Burkina-Faso, Burundi, Cameroon, Canada, Congo, France, Gabon, Lesotho, Niger, Poland, Senegal, Switzerland, Tchad, Togo, United States of America and Zambia.

Secondly, the papers presented in the workshops have been of a good quality. Besides, attendance has been numerous and regular, which shows the great interest which you, participants, have found in the various presentations made at this Conference.

Thirdly, the working environment was relaxed and warm. This could be seen in the rich debates which followed presentations and also during the different breaks.

Ladies and gentlemen, Honorable guests,
Dear participants, Dear colleagues,

Generally speaking, I would like to say that COPROMAPH2 has been a great celebration of science, an exchange forum for high-ranking scientists. Yet the most significant achievement of our conference, and which will stand out in the history of our subject, is the resolution our conference has made to create an International Chair in Mathematical Physics and Applications (ICMPA) at the University of Abomey-Calavi. I am personally embarrassed by the decision of the Scientific Council of the Chair to appoint me as its President. I remain nevertheless confident that I will benefit from the support of all the Vice-Presidents and that of all participants who have contributed to the decision. I would like to assure you all that the tasks you have set us will be carried out with great care and consciousness.

Ladies and gentlemen,

Our conference has been a great success indeed, and this is why, with your permission, I would like to acknowledge the help of the following Institutions,

. The Government of Benin and most especially, His Excellency the Minister for Higher Education and Scientific Research, Professor D. Sossa, as well as the Director General of the Benin Center for Scientific and Technical Research, Professor F. Toukourou, for their financial, material and moral support;

. The University of Abomey-Calavi, and more particularly its Vice-Chancellor, Professor Issifou Takpara, together with his cabinet for their devotion and appreciable help;

. The Gbe International Laboratory for its appreciable scientific help as well as the contributions of its researchers, with a special mention to Professor H.B.C. Capo;

. The Mathematics and Physical Sciences Institute as well as the Faculty of Science and Technology of the University of Abomey-Calavi for the support of their lecturers, researchers and students;

. All our partners for their effective or forthcoming institutional, financial and material support. In this respect, we would like to mention especially all our Sponsors.

Forward with a science in the service of development!
Forward with the perpetuation of COPROMAPH as one of the activities of the International Chair in Mathematical Physics and Applications (ICMPA)!
Long live the International Scientific Community!

And see you all again by the end of October early November 2003 here in Cotonou.

Once again, thank you to all.

Professor M. Norbert Hounkonnou

Closing Address by
the Vice-Chancellor of the University of Abomey-Calavi

· His Excellency, the Minister for Higher Education and Scientific Research,
· Dean of the Faculty of Arts,
· Members of the scientific council of COPROMAPH2,
· Members of the organizing committee,
· Dear Professors and researchers,
· Honorable guests,

You have just completed a whole week of exchanges and are now about to go back to your various institutions and families. On behalf of the entire University community of Benin, I would like to express once again my sincere gratitude for having accepted to take part in the Second International Workshop on Contemporary Problems in Mathematical Physics.

You have rendered Africa and the International Scientific Community an invaluable service and we can take for granted the positive impact of your exchanges on the future of scientific research in all our institutions. The evidence lies in the high quality of your papers which cover all of Mathematical Physics. I am convinced that the younger researchers have greatly benefited from them. The holding of a high-ranking conference such as yours on the African Continent is such a rare event that the University of Abomey-Calavi is happy to have contributed to its realization.

On behalf of the whole University Community, I would like to thank His Excellency, the Minister for Higher Education and Scientific Research for the support he granted both at the preparatory stage and during the Conference.

I would like to congratulate the organizers of this Conference for their courage and enthusiasm and for the actual holding of this international gathering. Allow me to acknowledge the work done by the COPROMAPH2 Executive Secretariat who have spent numerous days and, I am sure, will have to spend more. You have done the University of Abomey-Calavi a great honour by taking the decision to create in our Institution an International Chair in Mathematical Physics and its Applications. We will take all the necessary academic and administrative steps for the implementation of the Chair, because the honour granted by this exceptional event will no doubt have repercussions on our country as well as on Africa as a whole.

I wish everyone of you a safe journey back home, hoping that you have enjoyed your stay in Benin.

Professor I. Takpara

Closing Address by

His Excellency,

the Minister for Higher Education and Scientific Research

· Vice-Chancellor of the University of Abomey-Calavi,
· Dean of the Faculty of Arts,
· Members of the scientific council of COPROMAPH2,
· Members of the organizing committee,
· Dear Professors and researchers,
· Honorable guests,

As we are about to draw the curtains on the Second International Workshop on Contemporary Problems in Mathematical Physics, I am very happy to be with you to voice my satisfaction for the quality of the work you have done during the past five days. Five days appear rather short, yet they have been enough for you to examine the main contemporary problems in Mathematical Physics. Who else could achieve such a feat if not hard-working researchers like you who know how to manage time. Judging by the news that has reached me, I am confident that the COPROMAPH conferences which started two years ago can be perpetuated. As evidence, I would like to mention the relevance of the different papers that have been presented by outstanding specialists and also the rich debates that followed. All this appears to me as the sign of you determination and willingness to contribute to the solution for the problems of development. I would like to congratulate you on your resolution to create an International Chair in Mathematical Physics and its Applications. I am very grateful to you for choosing the University of Abomey-Calavi, not only to host the Chair, but also for choosing Professor Mahouton Norbert Hounkonnou as its President.

On behalf of the Government of my country, please accept my congratulations. As from now on, you can take for granted the support of the Government of my country for the actual implementation of the International Chair in Mathematical Physics and its Applications at the University of Abomey-Calavi, in the hospitable land of Benin. Besides, you can also take for granted the support of the Government of my country for the organization of COPROMAPH3 in two years time, which will be the first one ever to be organized by the newly created Chair. So, let us meet again in 2003!

I hope you have been able to take some time out and enjoy the historical, cultural and artistic aspects of Benin.

As I wish you a safe journey back to your respective countries, I now declare closed the Second International Conference on Contemporary Problems in Mathematical Physics.

Thank you.

Professor D. Sossa

Resolution on the Creation of an

International Chair in Mathematical Physics and Applications

at the University of Abomey-Calavi, Republic of Benin

---

1. Given the efforts made by the Authorities of the University of Abomey-Calavi in general, and by the Benin Institute of Mathematical and Physical Sciences in particular, to foster the scientific and inter-university collaboration between the researchers and the Universities of the sub-region and of the world;

2. Given the necessity for Africa not to be left out from the Mathematical Physics development in the world;

3. Given the necessity of creating a permanent framework for the exchange of research projects, fostering cooperation and promoting excellence in the Universities of the South;

4. Given the necessity to encourage the young generation of Mathematical Physics practitioners to be competitive at the international level and proposing quality contributions to the technological development in Africa and elsewhere;

5. Given that the First International Workshop on Contemporary Problems in Mathematical Physics (COPROMAPH1), held in Cotonou from October 31st to November 5th, 1999, has been a tremendous success;

6. Given that the Second International Workshop on Contemporary Problems in Mathematical Physics (COPROMAPH2) has effectively been held in Cotonou in conformity with the initial planning and has been a resounding success;

7. Given the firm intention of the Scientific Council and Local and International Organizing Committees to perpetuate COPROMAPH conferences;

8. Given the firm intention of the officials of the University of Abomey-Calavi and of the Ministry for Higher Education and Scientific Research to financially and materially sustain the creation of an International Chair in accordance with the request of the COPROMAPH2 Scientific Committee;

We, the participants to the Second International Workshop on Contemporary Problems in Mathematical Physics (COPROMAPH2), have resolved to,

1. Agree on the idea of creating an International Chair in Mathematical Physics and Applications at the University of Abomey-Calavi;
2. Agree on the composition for the Chair Presidency as follows:

   Professor Mahouton Norbert Hounkonnou (Benin): President;
   Professor Alfred Z. Msezane (USA): Vice-President;
   Professor S. Twareque Ali (Canada): Vice-President;
   Professor Tudor Ratiu (Switzerland): Vice-President;
   Professor Anatole Odzijewicz (Poland): Vice-President;
   Professor Xavier Chapuisat (France): Vice-President;
   Professor Jean-Pierre Antoine (Belgium): Vice-President;
   Professor Mary Tew Niane (Senegal): Vice-President;
3. Define the objectives of the Chair as follows:
   i) The working-out and the dissemination of the knowledge and know-how in Mathematical Physics and Applications;
   ii) The promotion of young people and women in the field of Mathematical Physics and Applications;
   iii) The research of an active partnership between the Chair and public institutions and industries for concrete applications of Mathematical Physics to development problems;
4. Recommend to the current COPROMAPH Executive Secretary to take the necessary steps for the effective creation of the said Chair and to elaborate the Status and Internal Regulations in conformity with the existing academic and administrative standards;

5. Invite the Vice-Chancellor of the University of Abomey-Calavi to provide the Chair with the necessary structures and installations for its functioning;

6. Invite his Excellency the Minister for Higher education and Scientific Research to officially appoint the members of the Chair Scientific Council as they are proposed by the President of the Chair;

7. Welcome the initiative of the pioneering members of the Executive Secretariat that has led today to the birth of the young Chair.

Cotonou, November 2nd, 2001

The participants to COPROMAPH2

A Vote of Thanks

We, the participants to the *Second International Workshop on Contemporary Problems in Mathematical Physics* (COPROMAPH2) held in Cotonou (Benin) from October 28th to November 2nd, 2001, would like to express,

. Our gratitude to the Government of Benin and also to the Minister for Higher Education and Scientific Research for their efficient support and the subsequent backing promised to COPROMAPH2 at the opening ceremony;

. Our gratitude to the Vice-Chancellor of the University of Abomey-Calavi for the appreciable support he granted to the COPROMAPH2 Organizing Committee;

. Our sympathy to the St. Jean Eudes Center of Atrokpocodji for their precious assistance;

. Our thanks to the various Institutions that have funded the organisation of COPROMAPH2;

. Our congratulations to the Organizing Committee for their efficient dedication and the intelligence with which they welcomed the participants, and also for giving the participanst the opportunity to enjoy the rich cultural aspects of Benin.

Cotonou, November 2nd, 2001

The participants to COPROMAPH2

Plenary Lectures

QUANTIZATION TECHNIQUES: A QUICK OVERVIEW

S. TWAREQUE ALI

Department of Mathematics and Statistics, Concordia University, Montréal, Québec, CANADA H4B 1R6

E-mail: stali@mathstat.concordia.ca

These pages outline, in a fairly self contained manner, of some of the more well-known quantization techniques. The intimate relationship between the geometry of the configuration and phase spaces and the transition process from classical to quantum mechanics is put into focus. The crucial role played by the symplectic geometry of classical mechanics is highlighted. Simple illustrative examples are worked out each time a new quantization technique is discussed. The limits of the different procedures are also discussed.

1 Introduction

These notes are intended to provide the reader with a rapid overview of some of the more well-known quantization techniques appearing in the current literature. An attempt has been been made to make the presentation self-contained; it is hoped that the notes would be understandable to graduate students in physics and mathematics. For reasons of time, it was not possible to include a discussion of deformation quantization or quantization using Feynman path integrals.

Quantization is the method by which one makes the transition:

$$\textit{Classical Mechanics} \quad \longrightarrow \quad \textit{Quantum Mechanics}$$

There is a general belief, mostly for historical reasons, that a classical theory should contain enough information within its geometric and algebraic structure for possible quantum theories – of which the given classical theory is the limit – to be built. Of course, the understanding here is that quantum mechanics is the more refined theory, to which classical mechanics is an approximation. The opposite process

$$\textit{Quantum Mechanics} \quad \underset{\hbar \longrightarrow 0}{\lim} \quad \textit{Classical Mechanics}$$

by which, starting with the (in a sense more fundamental) quantum theory, one arrives at its classical limit, is called *dequantization* (or the *correspondence principle*). To use a medieval metaphor, one believes that the ghost of a quantum theory hovers over its classical remains. *Quantizing then amounts to incarnating this spirit, while dequantizing involves exorcising the ghost!* Of course, there are quantum systems for which no classical counterpart exists (e.g., He-II superfluidity or even a spin-1/2 particle). For these systems a quantization procedure would not make sense. However, there is a certain mathematical richness in the various theories of quantization, in cases where such a procedure does make sense, to make a study of quantization intrinsically interesting. Moreover, such an analysis also gives a deeper overview of the entire relationship between classical and quantum physics.

In the past three decades, a large number of techniques have been devised to make this transition from classical to quantum mechanics. The point of departure is always an analysis of the geometrical structure of either the classical phase space or the classical configuration space. There is, however, no general theory of quantization presently available, which is applicable in all cases, and indeed, often the technique used to quantize has to be tailored to the problem in question.

Quantum mechanics, like any other physical theory, e.g., classical mechanics, electrodynamics, relativity, thermodynamics, etc., cannot be derived. The laws of quantum mechanics, expressed in mathematical form, are the results of deep physical intuition, as indeed, are all other physical theories. Their validity can only be checked experimentally and rests ultimately in their predictive power. From this point of view, quantization is not a method for *deriving* quantum mechanics – rather it is a way to understand the deeper physical reality which underlies the structure of both classical and quantum mechanics and which unifies the two from a geometrical perspective. On the other hand, from a practical point of view, since it is often easier to write down the equations governing the classical dynamics of a system, the various available quantization techniques can help one to arrive at the corresponding quantum theory at a second stage. But then, as mentioned earlier, there do exist quantum systems for which a classical limit does not exist. Yet even here an understanding of the underlying geometry often helps in analyzing their nature.

The literature on quantization is enormous; the choice of good starting points is rich.[4,5,7,12,13,14,15,16,20,22,24,18,26]

2 Review of Classical Mechanics

This section is a quick review of Newtonian mechanics and the rudiments of the Lagrangian and Hamiltonian formulations. There exist many good references for the material in this section.[1,11,12]

2.1 The Newtonian formulation

Classical mechanics studies the dynamics of macroscopic objects at (usually) low, i.e., non-relativistic speeds. Underlying the mathematical structure of classical mechanics are the three laws of Newton:

1. Every body tends to continue in its state of rest, or of uniform motion, except in so far as it is compelled by an external impressed force to change that state.

2. The rate of change of momentum is proportional to the external impressed force, and it is effected in the direction in which the force acts.

3. To every action there is an equal and opposite reaction.

It is Newton's second law of motion, which expressed in mathematical form looks as:

$$\mathbf{F} = \frac{d\mathbf{p}}{dt}, \qquad \mathbf{p} = m\mathbf{v}, \tag{1}$$

that forms the basis for classical mechanics. If m is assumed constant (clearly not true, for example, for a rocket propelled by burning fuel) then,

$$\begin{aligned} \mathbf{F} &= m\frac{d\mathbf{v}}{dt} = m\mathbf{a}, \\ \mathbf{a} &= \frac{d\mathbf{v}}{dt} = \frac{d^2\mathbf{r}}{dt^2} = \ddot{\mathbf{r}}. \end{aligned} \tag{2}$$

Note immediately, that if $\mathbf{F} = \mathbf{0}$, then $\frac{d\mathbf{p}}{dt} = \mathbf{0}$, i.e., the linear momentum is *conserved.*

In general, a mechanical system is composed of a set of N $(< \infty)$ particles, at points $\mathbf{r}_i$, masses m_i and external forces $\mathbf{F}_i^{(e)}$, $i = 1, 2, \ldots, N$. In addition, there could be a force of interaction $\mathbf{F}_{ji}$ of the j-th particle on the i-th particle. We assume $\mathbf{F}_{ii} = \mathbf{0}$ and $\mathbf{F}_{ij} = -\mathbf{F}_{ji}$ (Newton's third law). Then, the *equations of motion* for the system are:

$$m_i\ddot{\mathbf{r}}_i = \mathbf{F}_i^{(e)} + \sum_j \mathbf{F}_{ji}. \tag{3}$$

Thus, in principle at least, once all the masses and forces are known, one should be able to obtain the $\mathbf{r}_i$ as functions of the time t, for all t, and hence solve the problem completely.

2.2 *A simple example*

We work out, as an example of the above, the most well known of physical examples, the simple pendulum. We consider a simple pendulum of mass m, and length ℓ, oscillating in the X_2X_3-plane. The forces acting on it are:

- $-mg\widehat{\mathbf{k}}$ = force due to gravity ($\widehat{\mathbf{k}}$ is the unit vector in the vertical direction),
- $\mathbf{t}$ = tension on the string,
- $\mathbf{a}$ = centrifugal force.

The equation of motion is

$$m\ddot{\mathbf{r}} = -mg\widehat{\mathbf{k}} + \mathbf{t} + \mathbf{a}, \tag{4}$$

We set up the coordinate system in a way such that the origin of coordinates coincides with the undisturbed position of the mass centre of the pendulum, the X_3-axis is aligned with the vertical direction and the motion of the pendulum takes place in the X_2X_3-plane. Then, in coordinate form, (4) becomes,

$$\begin{aligned} m\ddot{x}_1 &= 0, \\ m\ddot{x}_2 &= -\|\mathbf{t}\| \sin\theta + \|\mathbf{a}\| \sin\theta, \\ m\ddot{x}_3 &= -mg + \|\mathbf{t}\| \cos\theta - \|\mathbf{a}\| \cos\theta, \end{aligned} \tag{5}$$

where θ is the angular displacement,

$$\sin\theta = \frac{x_2}{\ell}, \qquad \cos\theta = \frac{\ell - x_3}{\ell}. \tag{6}$$

The above relations are a consequence of the *constraint equation*:

$$x_2^2 + (\ell - x_3)^2 = \ell^2, \tag{7}$$

the string being unstretchable. We are only interested in the planar motion of the pendulum; hence, solving the first equation in (5), with the initial condition, $\dot{x}_1(0) = x_1(0) = 0$, gives $x_1 = 0$, for all t. Using (6) we may now rewrite the two remaining equations in (5) as,

$$\begin{aligned} m\ell\cos\theta\dot{\theta}^2 + m\ell\sin\theta\ddot{\theta} &= -mg + (\|\mathbf{t}\| - \|\mathbf{a}\|)\cos\theta, \\ m\ell\sin\theta\dot{\theta}^2 + m\ell\cos\theta\ddot{\theta} &= -(\|\mathbf{t}\| - \|\mathbf{a}\|)\sin\theta, \end{aligned} \tag{8}$$

which then yield the single equation for the one remaining variable θ:

$$\ell\ddot{\theta} = -g\sin\theta. \tag{9}$$

Thus, by taking the constraint (7) into consideration and using the initial condition $\dot{x}_1(0) = x_1(0) = 0$, we have reduced the problem to a single second order differential equation for the "generalized coordinate" θ, corresponding to the one surviving degree of freedom. If θ is small, (9) becomes

$$\ddot{\theta} + \frac{g}{\ell}\theta = 0, \tag{10}$$

to which the general solution is

$$\theta(t) = \theta_0 \sin(\omega t + \delta), \qquad \omega = \sqrt{\frac{g}{\ell}}, \tag{11}$$

where θ_0 and δ are constants.

2.3 The Lagrangian formulation

In the case of the simple pendulum above, we started out with a single particle of mass m, having 3 degrees of freedom. But there were *constraints* on the problem, which reduced the number of degrees of freedom to one. Thus, instead of working with the three Cartesian coordinates, x_i, $i = 1, 2, 3$, we only needed to work with a single *generalized coordinate* θ.

Going back to (3), we realize that for a general mechanical system, the vectors $\mathbf{r}_i$ may not all be independent. There could be constraints acting on the system, e.g., if we have a rigid body, then

$$(\mathbf{r}_i - \mathbf{r}_j)^2 = c_{ij}^2 = \text{const.} \tag{12}$$

Constraints restrict the number of degrees of freedom – from a total of $3N$. We shall assume all our constraints to be *holonomic*, i.e., expressible as (smooth) functional relations of the type,

$$f_i(\mathbf{r}_1, \mathbf{r}_2, \ldots, \mathbf{r}_N, t) = 0, \qquad i = 1, 2, \ldots, k \leq N. \tag{13}$$

Geometrically, the configuration space goes down from being a $3N$-dimensional Euclidean space to an $n = (3N - k)$-dimensional (differential) manifold. Constraints can also be non-holonomic, e.g., given by inequalities:

$$g(\mathbf{r}_1, \mathbf{r}_2, \ldots, \mathbf{r}_N, t) \leq 0, \tag{14}$$

but we shall not consider them here.

Constraints ultimately do reflect additional forces in the problem, and the obvious way to take account of them is to move to a different set of n independent variables, assuming that this can be done. Let us denote this new independent set by $q^1, q^2, \ldots, q^n$. Then,

$$\mathbf{r}_i = \mathbf{r}_i(q^1, q^2, \ldots, q^n, t), \qquad i = 1, 2, \ldots, N \tag{15}$$

and the idea will be to write the equations of motion (3) in terms of the q^i's. This will have the effect that the constraints will have been taken into account implicitly.

In (3), denote by $\mathbf{F}_i$ the total applied force on the i-th particle,

$$\mathbf{F}_i = \mathbf{F}_i^{(e)} + \sum_{j=1}^{N} \mathbf{F}_{ji}, \qquad i = 1, 2, \ldots, N, \tag{16}$$

and assume that all these forces are derivable from a *scalar potential*:

$$\mathbf{F}_i = -\nabla_i V, \qquad i = 1, 2, \ldots, N. \tag{17}$$

Note that $V = V(\mathbf{r}_1, \mathbf{r}_2, \ldots \mathbf{r}_N)$, and if $\mathbf{r}_i = (x_{i1}, x_{i2}, x_{i3})$, then

$$\nabla_i = \left(\frac{\partial}{\partial x_{i1}}, \frac{\partial}{\partial x_{i2}}, \frac{\partial}{\partial x_{i3}}\right).$$

Next, using D'Alembert's principle of virtual work, it is possible to show that the equations of motion (3) are equivalent to the set of n uncoupled equations:

$$\frac{d}{dt}\left(\frac{\partial L}{\partial \dot{q}^j}\right) - \frac{\partial L}{\partial q^j} = 0, \qquad j = 1, 2, \ldots, n, \tag{18}$$

where the Lagrangian L is defined to be the function,

$$L = L(q^1, q^2, \ldots, q^n, \dot{q}^1, \dot{q}^2, \ldots, \dot{q}^n, t) \equiv L(q, \dot{q}, t) = T - V, \tag{19}$$

where T denotes the total kinetic energy,

$$T = \frac{1}{2}\sum_{i=1}^{N} m_i \|\dot{\mathbf{r}}_i\|^2 = \frac{1}{2}\sum_{i=1}^{N} m_i \|\mathbf{v}_i\|^2, \tag{20}$$

of the system and V is now assumed to be a function of t and the q^i's alone, obtained by substituting for the $\mathbf{r}_i$'s using (15).

We call the q^j's the *canonical coordinates* of the system. Corresponding to each q^j, let us define the *canonically conjugate momentum* by

$$p_j = \frac{\partial L}{\partial \dot{q}^j}. \tag{21}$$

It follows from (18) that if the Lagrangian L does not involve the variable q^j explicitly, i.e., if q^j is *cyclic*, then the canonically conjugate momentum p_j is conserved, i.e.,

$$\frac{dp_j}{dt} = \frac{d}{dt}\left(\frac{\partial L}{\partial \dot{q}^j}\right) = 0, \quad \text{since} \quad \frac{\partial L}{\partial q^j} = 0. \tag{22}$$

The problem of solving the original equations of motion (3), now becomes one of:

1. Obtaining a suitable set of n independent generalized (canonical) coordinates q^j.

2. Writing down the Lagrangian function L as a function of the q^j's and $\dot{q}^j$'s using (15).

3. Solving the system of n (second order) differential equations (18) and reconverting to the original set of coordinates $\mathbf{r}_i$, to get the time evolution of the mechanical system, $\mathbf{r}_i = \mathbf{r}_i(t)$.

Note that L is a function of both the q^j's and the $\dot{q}^j$'s, and both sets of variables are to be treated as independent. In other words, thinking of the q^j's as defining an n-dimensional differential manifold, L is a function defined on its tangent bundle.

2.4 The Hamiltonian formulation

The Lagrangian equations of motion (18) are second order differential equations in the time variable. Using the definition (21) of the conjugate momenta, it should be possible to obtain a set of first order equations in terms of the independent variables $q^j, p_j, \quad j = 1, 2, \ldots, n$. This is the objective of the Hamiltonian formulation of classical mechanics. In order to make this change of variables, consider first a function $f : \mathbb{R}^2 \to \mathbb{R}$, and

$$df = \frac{\partial f}{\partial x}dx + \frac{\partial f}{\partial y}dy = udx + vdy. \tag{23}$$

We now want to move to a new basis: $(x, y) \to (u, y)$. For this, let us make a *Legendre transformation* to a related function g:

$$g = f - ux. \tag{24}$$

Then,

$$\begin{aligned} dg &= df - udx - xdu = udx + vdy - udx - xdu \\ \Rightarrow dg &= vdy - xdu. \end{aligned} \tag{25}$$

Thus, writing $g = g(u, y)$, we get from the above,

$$\begin{aligned} dg &= \frac{\partial g}{\partial u}du + \frac{\partial g}{\partial y}dy = -xdu + vdy \\ \Rightarrow x &= -\frac{\partial g}{\partial u}, \qquad v = \frac{\partial g}{\partial y}. \end{aligned} \tag{26}$$

Applying this idea to L, let us write

$$H(q,p,t) = \sum_{i=1}^{n} \dot{q}^i p_i - L(q,\dot{q},t), \tag{27}$$

so that

$$\begin{aligned} dH &= \sum_{i=1}^{n} \left(\frac{\partial H}{\partial q^i} dq^i + \frac{\partial H}{\partial p_i} dp_i \right) + \frac{\partial H}{\partial t} dt \\ &= \sum_{i=1}^{n} \left(\dot{q}^i dp_i + p_i d\dot{q}^i - \frac{\partial L}{\partial q^i} dq^i - \frac{\partial L}{\partial \dot{q}^i} d\dot{q}^i \right) - \frac{\partial L}{\partial t} dt \end{aligned} \tag{28}$$

From (18), $\dot{p}_i = \dfrac{\partial L}{\partial q^i}$, and using (21) in (28) we obtain,

$$\begin{aligned} dH &= \sum_{i=1}^{n} \left(\frac{\partial H}{\partial q^i} dq^i + \frac{\partial H}{\partial p_i} dp_i \right) + \frac{\partial H}{\partial t} dt \\ &= \sum_{i=1}^{n} (\dot{q}^i dp_i - \dot{p}_i dq^i) - \frac{\partial L}{\partial t} dt. \end{aligned} \tag{29}$$

Hence,

$$\dot{q}^i = \frac{\partial H}{\partial p_i}, \quad \dot{p}_i = -\frac{\partial H}{\partial q^i}, \quad -\frac{\partial L}{\partial t} = \frac{\partial H}{\partial t}, \qquad i = 1, 2, \ldots n. \tag{30}$$

The above equations provide us with a set of $2n+1$ equations, all first order in time, which is completely equivalent to the set of Lagrange equations (18). We call the equations (30) the canonical *Hamiltonian equations of motion.* The function H is called the *Hamiltonian* of the system. In a large class of problems of physical interest (27) actually reduces to

$$H = T + V, \tag{31}$$

and if the relations (15), involving $\mathbf{r}_i$ and q^i, do not depend on t explicitly, then H is conserved whenever V is independent of time. Whenever H is expressible as (31), it denotes the total energy of the system.

A point to be made here, and to which we shall return later, is that the Hamiltonian is a function of (q^i, p_i) and t, and hence is a a function on the *cotangent bundle* of the manifold defined by the q^i's. We shall analyze this point in some detail in Section 3.

2.5 *The pendulum revisited*

We go back to the simple pendulum which we solved in Section 2.2 using Newton's laws of motion, and this time tackle the problem using first the Lagrangian and then the Hamiltonian formalisms.

Clearly,

$$L = T - V = \frac{1}{2}m\ell^2\dot{\theta}^2 - mg\ell(1 - \cos\theta), \tag{32}$$

and for small θ,

$$L = \frac{1}{2}m\ell^2\dot{\theta}^2 - \frac{1}{2}mg\ell\theta^2, \tag{33}$$

from which

$$\frac{\partial L}{\partial \dot{\theta}} = m\ell^2\dot{\theta}, \qquad \frac{\partial L}{\partial \theta} = -mg\ell\theta. \tag{34}$$

Hence, the Lagrangian equation of motion,

$$\frac{d}{dt}\left(\frac{\partial L}{\partial \dot{q}^j}\right) - \frac{\partial L}{\partial q^j} = 0,$$

with $q^j = \theta$, becomes

$$m\ell^2\ddot{\theta} + mg\ell\theta = 0,$$

which is the same as (10).

Next, using the Hamiltonian approach,

$$p_\theta = \frac{\partial L}{\partial \dot{\theta}} = m\ell^2\dot{\theta}, \tag{35}$$

so that,

$$\begin{aligned} H = T + V &= \frac{1}{2}m\ell^2\dot{\theta}^2 + \frac{1}{2}mg\ell\theta^2 \\ &= \frac{p_\theta^2}{2m\ell^2} + \frac{1}{2}mg\ell\theta^2. \end{aligned} \tag{36}$$

The Hamiltonian equations of motion are now

$$\dot{q}^j = \frac{\partial H}{\partial p_j} \quad \Rightarrow \quad \dot{\theta} = \frac{p_\theta}{m\ell^2}, \tag{37}$$

which is the same as (35), and

$$\dot{p}_j = -\frac{\partial H}{\partial q^j} \quad \Rightarrow \quad \dot{p}_\theta = -mg\ell\theta. \tag{38}$$

Computing $\ddot{\theta}$ from (37) and substituting for $\dot{p}_\theta$ using (38), we again get (10).

Note also, that the *two-form* $d\theta \wedge dp_\theta$, written in terms of θ and $\dot{\theta}$ (using (35)) becomes:

$$d\theta \wedge dp_\theta = (d\theta \wedge d\dot{\theta})m\ell^2. \tag{39}$$

Finally, let us change to the new *canonical variables*, x, p_x, where

$$x = \ell\theta = \text{arc length}, \tag{40}$$

so that from (33),

$$L = \frac{1}{2}m(\dot{x}^2 - \omega^2 x^2) \quad \Rightarrow \quad \frac{\partial L}{\partial \dot{x}} = p_x = m\dot{x}. \tag{41}$$

Thus,

$$H = \frac{p_x^2}{2m} + \frac{m^2\omega^2}{2}x^2, \qquad \omega = \sqrt{\frac{g}{\ell}}. \tag{42}$$

Once again, computing the 2-form $dx \wedge dp_x$, we find

$$dx \wedge dp_x = (d\theta \wedge d\dot{\theta})m\ell^2, \tag{43}$$

exactly as in (39). This is an explicit example of the fact that the canonical two-form is independent of the generalized (canonical) coordinates used to express it.

3 Symplectic Formulation of Classical Mechanics

The configuration space of a physical system, consisting of N particles, and subject to k holonomic constraints is a $(3N-k)$-dimensional differential manifold. We denote this manifold by $\mathfrak{Q}$, and as before write $n = 3N - k$ for its dimension. It will be necessary to study the underlying geometry of the configuration space and its tangent and cotangent bundles. We collect, in this section, a few important results and concepts. More detailed discussions are available elsewhere.[1,11,12,25]

3.1 The phase space as a cotangent bundle

The generalized coordinates q^j, $j = 1, 2, \ldots, n$, map the configuration space $\mathfrak{Q}$ locally to the Euclidean space $\mathbb{R}^n$. Denoting this map by ϕ, at any given point $q \in \mathfrak{Q}$, its local coordinates are then given by

$$q^j = \phi(q)^j, \quad q \in U, \qquad j = 1, 2, \ldots, n, \tag{44}$$

where U is some open set in $\mathfrak{Q}$, containing the point q. We shall also use the vector notation,

$$\mathbf{q} = (q^1, q^2, \ldots, q^n) = \phi(q) \in \mathbb{R}^n. \tag{45}$$

Differentiating smooth local curves, in the neighbourhood of a point $q \in U$, then gives tangent vectors at q. Let $\mathbf{e}_j,\ \ j = 1, 2, \ldots, n$, be the canonical basis of $\mathbb{R}^n$ (i.e., unit vectors with 1 in the i-th position and zeroes elsewhere), and denote by $T_q\mathfrak{Q}$ the tangent space of $\mathfrak{Q}$ at the point q. Then, writing $T_q(\phi) : T_q\mathfrak{Q} \to \mathbb{R}^n$ for the tangent (or derivative) map of ϕ at q, we define a natural basis in $T_q(\mathfrak{Q})$ using the vectors

$$\frac{\partial}{\partial q^j} = T_q(\phi)^{-1}\mathbf{e}_j. \tag{46}$$

Explicitly, if $T_q(\phi)^{-1}$ maps $\mathbf{x} \in \mathbb{R}^n$ to $X \in T_q(\mathfrak{Q})$, then (for t in some open neighbourhood of the origin),

$$X = \frac{d}{dt}\phi^{-1}(\mathbf{q} + t\mathbf{x})|_{t=0}. \tag{47}$$

Also, writing $\mathbf{x} = \sum_{i=1}^n \dot{q}^i \mathbf{e}_i$, we find that

$$X = \sum_{i=1}^n \dot{q}^i \frac{\partial}{\partial q^i}. \tag{48}$$

The cotangent space $T_q^*\mathfrak{Q}$ of $\mathfrak{Q}$ at the point q is the dual space of $T_q\mathfrak{Q}$. A convenient basis for this space are the covectors dq^j, constituting the dual basis of (46):

$$\langle dq^i\ ;\ \frac{\partial}{\partial q^j}\rangle = \delta^i_j, \qquad i, j = 1, 2, \ldots, n. \tag{49}$$

A general cotangent vector will be denoted by X^*. The tangent and cotangent bundles are then the $2n$-dimensional manifolds,

$$T\mathfrak{Q} = \bigcup_{q \in \mathfrak{Q}} T_q\mathfrak{Q}, \qquad T^*\mathfrak{Q} = \bigcup_{q \in \mathfrak{Q}} T_q^*\mathfrak{Q}. \tag{50}$$

Lagrangian mechanics, which works with the variables $q^j, \dot{q}^j$, is therefore based on the tangent bundle, while Hamiltonian mechanics is based on the cotangent bundle, as we shall presently see.

Consider next the kinetic energy T of the system. Usually, for fixed q, this is a posititive definite symmetric quadratic in the variables $\dot{q}^j$:

$$T = T(q) = \frac{1}{2}\sum_{i,j=1}^n t_{ij}(q)\dot{q}^i\dot{q}^j, \qquad t_{ij}(q) = t_{ji}(q) \in \mathbb{R},\ \ i, j = 1, 2, \ldots n. \tag{51}$$

We shall denote by $\mathbf{T}$ the symmetric, positive definite matrix with elements t_{ij}, so that (suppressing q)

$$T = \frac{1}{2}\dot{\mathbf{q}} \cdot \mathbf{T}\dot{\mathbf{q}}, \qquad \dot{\mathbf{q}} = (\dot{q}_1, \dot{q}_2, \ldots, \dot{q}_n). \tag{52}$$

It is now seen that the kinetic energy gives rise to a *Riemannian metric* g on $\mathfrak{Q}$, in the manner

$$g = \frac{1}{2}\sum_{i,j=1}^{n} t_{ij} dq^i \otimes dq^j, \tag{53}$$

so that,

$$g(X, X') = \frac{1}{2}\sum_{i,j=1}^{n} t_{ij}\dot{q}^i \dot{q}'^j \quad \text{and} \quad g(X, X) = T,$$

$$\text{where,} \qquad X = \sum_{i=1}^{n} \dot{q}^i \frac{\partial}{\partial q^i}, \quad X' = \sum_{i=1}^{n} \dot{q}'^i \frac{\partial}{\partial q^i} \in T_q\mathfrak{Q}, \tag{54}$$

and in particular,

$$g(\frac{\partial}{\partial q^i}, \frac{\partial}{\partial q^j}) = \frac{1}{2}t_{ij}. \tag{55}$$

Suppose now that the potential V depends only on the generalized coordinates q^j, and not on the velocities $\dot{q}^j$. Then writing as before, $H = T + V$, the components of the generalized momenta are,

$$p_j = \frac{\partial H}{\partial \dot{q}^j} = \frac{\partial T}{\partial \dot{q}^j} = \sum_{i=1}^{n} t_{ij}\dot{q}^j, \tag{56}$$

in terms of which

$$g(X, X') = \frac{1}{2}\sum_{j=1}^{n} p_j \dot{q}'^j. \tag{57}$$

Furthermore, since g defines a scalar product on $T_q\mathfrak{Q}$, it can be used to establish a vector space isomorphism between $T_q\mathfrak{Q}$ and $T_q^*\mathfrak{Q}$. Indeed, with X as in (54) and p_j as in (56),

$$g(\cdot\,,\, X) = \frac{1}{2}\sum_{i,j=1}^{n} t_{ij}\dot{q}^j dq^i = \frac{1}{2}\sum_{i=1}^{n} p_i dq^i. \tag{58}$$

Thus, the map,

$$I : T_q\mathfrak{Q} \to T_q^*\mathfrak{Q}, \qquad I(X) = g(\cdot\,,\, X), \tag{59}$$

is the postulated isomorphism. Furthermore, it is easily checked, that if t^{ij} denote the elements of the inverse matrix, $\mathbf{T}^{-1}$, then for an arbitrary $X^* = \sum_{i=1}^n p_i dq^i \in T_q^* \mathbf{\Omega}$, the inverse isomorphism I^{-1} acts as:

$$I^{-1}(X^*) = \sum_{i=1}^{n} \dot{q}^i \frac{\partial}{\partial q^i}, \quad \text{where,} \quad \dot{q}^i = \sum_{j=1}^{n} t^{ij} p_j. \tag{60}$$

The dual pairing between $T_q^* \mathbf{\Omega}$ and $T_q \mathbf{\Omega}$ can be written in terms of the metric g as:

$$\langle X^* \, ; \, X' \rangle = \sum_{i=1}^{n} p_i \dot{q}'^i = 2g(I^{-1}(X^*) \, , \, X'). \tag{61}$$

To check the consistency of the above scheme, note that,

$$\begin{aligned} \langle dq^i \, ; \, \frac{\partial}{\partial q^j} \rangle &= 2 \sum_{k=1}^{n} t^{ki} g(\frac{\partial}{\partial q^k} \, ; \, \frac{\partial}{\partial q^j}) \\ &= 2 \sum_{k=1}^{n} t^{ki} \, \frac{1}{2} t_{kj}, \quad \text{by (55)} \\ &= \delta^i_j, \end{aligned} \tag{62}$$

as it should be. Equations (56), (58) and (59) clearly show that the p_j are the components of covectors (i.e., elements of the cotangent bundle $T^* \mathbf{\Omega}$). Thus Hamiltonian mechanics, which works with the variables q^j, p_j, is based on the cotangent bundle of $\mathbf{\Omega}$.

The phase space of the physical system is defined to be the manifold described by the local coordinates q^j, p_j and hence has the structure of a cotangent bundle. We shall use for it the notation,

$$\mathbf{\Gamma} = T^* \mathbf{\Omega} = \text{phase space}. \tag{63}$$

3.2 Canonical transformations and symplectic invariance

Suppose that we are given a classical mechanical system of n degrees of freedom, a system of canonical coordinates of position and momentum (q^i, p_i), $i = 1, 2, \ldots, n$, the Hamiltonian function

$$H = \sum_{i=1}^{n} \dot{q}^i p_i - L(q, \dot{q}, t), \qquad L = T - V, \tag{64}$$

and the Hamiltonian equations of motion:

$$\dot{q}^i = \frac{\partial H}{\partial p_i}, \quad \dot{p}_i = -\frac{\partial H}{\partial q^i}, \quad -\frac{\partial L}{\partial t} = \frac{\partial H}{\partial t}, \qquad i = 1, 2, \ldots n. \tag{65}$$

We ask under what kinds of coordinate transformations the above equations remain invariant. The nature of the invariance group should afford us a deeper insight into the structure of classical mechanics.

In order to answer this question, let us change to a new set of coordinates (Q^i, P_i), $i = 1, 2, \ldots n$, which are functions (assume them to be smooth) of the old coordinates (q^i, p_i), $i = 1, 2, \ldots n$, and possibly also of time:

$$\left.\begin{aligned} Q^i &= Q^i(\mathbf{q}, \mathbf{p}, t) \\ P_i &= P_i(\mathbf{q}, \mathbf{p}, t) \end{aligned}\right\}, \qquad \mathbf{q} = \begin{pmatrix} q^1 \\ q^2 \\ \vdots \\ q^n \end{pmatrix}, \qquad \mathbf{p} = \begin{pmatrix} p_1 \\ p_2 \\ \vdots \\ p_n \end{pmatrix}. \tag{66}$$

But at the same time, we want these new coordinates to be also *canonical coordinates.* This means that we require the existence of a new Hamiltonian function $K(\mathbf{Q}, \mathbf{P}, t)$, such that

$$\dot{Q}^i = \frac{\partial K}{\partial P_i}, \quad \dot{P}_i = -\frac{\partial K}{\partial Q^i}, \qquad i = 1, 2, \ldots, n. \tag{67}$$

It can be shown, very generally, that such a transformation of coordinates implies a relationship between the old and the new Hamiltonians of the type:

$$\sum_{i=1}^{n} \dot{q}^i p_i - H = \sum_{i=1}^{n} \dot{Q}^i P_i - K + \frac{dF}{dt}, \tag{68}$$

where F is a function of the phase space coordinates and time, continuous up to the second derivative. It is this function F which actually determines the canonical transformation, and again, very generally,

$$K = H + \frac{\partial F}{\partial t}, \tag{69}$$

so that if the transformation equations (66) do not involve time explicitly, then $K = H$, and hence, the Hamiltonian does not change under such a canonical transformation. We shall assume, for the moment, that we are in fact working with such a restricted class of transformations:

$$\begin{aligned} Q^i &= Q^i(\mathbf{q}, \mathbf{p}), \qquad P_i = P_i(\mathbf{q}, \mathbf{p}), \\ \dot{Q}^i &= \frac{\partial H}{\partial P_i}, \qquad \dot{P}_i = -\frac{\partial H}{\partial Q^i}, \quad i = 1, 2, \ldots, n, \end{aligned} \tag{70}$$

and try to study the nature of the Jacobian of the transformation.

Since $Q^i = Q^i(\mathbf{q}, \mathbf{p})$,

$$\begin{aligned}\dot{Q}^i &= \sum_{j=1}^{n} \left(\frac{\partial Q^i}{\partial q^j} \dot{q}_j + \frac{\partial Q^j}{\partial p_j} \dot{p}_j \right) \\ &= \sum_{j=1}^{n} \left(\frac{\partial Q^i}{\partial q^j} \frac{\partial H}{\partial p_j} - \frac{\partial Q^i}{\partial p_j} \frac{\partial H}{\partial q^j} \right), \end{aligned} \tag{71}$$

the last equality following the Hamiltonian equations. On the other hand, since we are also assuming the transformation $(\mathbf{q}, \mathbf{p}) \mapsto (\mathbf{Q}, \mathbf{P})$ to be invertible, so that

$$q^i = q^i(\mathbf{Q}, \mathbf{P}), \quad p_i = p_i(\mathbf{Q}, \mathbf{P}), \quad i = 1, 2, \ldots n, \tag{72}$$

implying $H(\mathbf{q}, \mathbf{p}, t) = H(\mathbf{Q}, \mathbf{P}, t)$, we find that

$$\frac{\partial H}{\partial P_i} = \sum_{j=1}^{n} \left(\frac{\partial H}{\partial q^j} \frac{\partial q^j}{\partial P_i} + \frac{\partial H}{\partial p_j} \frac{\partial p_j}{\partial P_i} \right). \tag{73}$$

Since

$$\dot{Q}^i = \frac{\partial H}{\partial P_i},$$

comparing (71) and (73) we get,

$$\left(\frac{\partial Q^i}{\partial q_j} \right)_{\mathbf{q},\mathbf{p}} = \left(\frac{\partial p_j}{\partial P_i} \right)_{\mathbf{Q},\mathbf{P}}, \qquad \left(\frac{\partial Q^i}{\partial p_j} \right)_{\mathbf{q},\mathbf{p}} = -\left(\frac{\partial q^j}{\partial P_i} \right)_{\mathbf{Q},\mathbf{P}}. \tag{74}$$

The subscripts in the above relations indicate the independent variables.

Similarly, writing out the $\dot{P}_i$ and $\dfrac{\partial H}{\partial Q^i}$ in terms of $\mathbf{q}, \mathbf{p}$, we find

$$\left(\frac{\partial P_i}{\partial q_j} \right)_{\mathbf{q},\mathbf{p}} = -\left(\frac{\partial p_j}{\partial Q^i} \right)_{\mathbf{Q},\mathbf{P}}, \qquad \left(\frac{\partial P_i}{\partial p_j} \right)_{\mathbf{q},\mathbf{p}} = \left(\frac{\partial q^j}{\partial Q^i} \right)_{\mathbf{Q},\mathbf{P}}. \tag{75}$$

In order to simplify the manipulations, let us introduce the vectors

$$\boldsymbol{\eta} = \begin{pmatrix} \mathbf{q} \\ \mathbf{p} \end{pmatrix}, \quad \boldsymbol{\zeta} = \begin{pmatrix} \mathbf{Q} \\ \mathbf{P} \end{pmatrix} \in \mathbb{R}^{2n}$$

$$\frac{\partial H}{\partial \boldsymbol{\eta}} = \begin{pmatrix} \dfrac{\partial H}{\partial \mathbf{q}} \\ \dfrac{\partial H}{\partial \mathbf{p}} \end{pmatrix} \in \mathbb{R}^{2n}, \quad \frac{\partial H}{\partial \mathbf{q}} = \begin{pmatrix} \dfrac{\partial H}{\partial q^1} \\ \dfrac{\partial H}{\partial q^2} \\ \vdots \\ \dfrac{\partial H}{\partial q^n} \end{pmatrix}, \quad \frac{\partial H}{\partial \mathbf{p}} = \begin{pmatrix} \dfrac{\partial H}{\partial p_1} \\ \dfrac{\partial H}{\partial p_2} \\ \vdots \\ \dfrac{\partial H}{\partial p_n} \end{pmatrix} \in \mathbb{R}^n,$$

$$\frac{\partial H}{\partial \boldsymbol{\zeta}} = \begin{pmatrix} \dfrac{\partial H}{\partial \mathbf{Q}} \\ \dfrac{\partial H}{\partial \mathbf{P}} \end{pmatrix} \in \mathbb{R}^{2n}, \quad \frac{\partial H}{\partial \mathbf{Q}} = \begin{pmatrix} \dfrac{\partial H}{\partial Q^1} \\ \dfrac{\partial H}{\partial Q^2} \\ \vdots \\ \dfrac{\partial H}{\partial Q^n} \end{pmatrix}, \quad \frac{\partial H}{\partial \mathbf{P}} = \begin{pmatrix} \dfrac{\partial H}{\partial P_1} \\ \dfrac{\partial H}{\partial P_2} \\ \vdots \\ \dfrac{\partial H}{\partial P_n} \end{pmatrix} \in \mathbb{R}^n. \tag{76}$$

Also, let J be the $2n \times 2n$ matrix,

$$J = \begin{pmatrix} \mathbf{0} & \mathbb{I}_n \\ -\mathbb{I}_n & \mathbf{0} \end{pmatrix}, \qquad J^2 = -\mathbb{I}_{2n}, \tag{77}$$

where $\mathbb{I}_k$ denotes the $k \times k$ unit matrix.

In this notation, the Hamiltonian equations (65) and (67) assume the forms:

$$\dot{\boldsymbol{\eta}} = J\frac{\partial H}{\partial \boldsymbol{\eta}}, \qquad \dot{\boldsymbol{\zeta}} = J\frac{\partial H}{\partial \boldsymbol{\zeta}}, \tag{78}$$

while the transformations $(\mathbf{q}, \mathbf{p}) \mapsto (\mathbf{Q}, \mathbf{P})$ and $(\mathbf{Q}, \mathbf{P}) \mapsto (\mathbf{q}, \mathbf{p})$ are conveniently written as

$$\boldsymbol{\zeta} = \boldsymbol{\zeta}(\boldsymbol{\eta}), \qquad \boldsymbol{\eta} = \boldsymbol{\eta}(\boldsymbol{\zeta}). \tag{79}$$

Since these are inverse transformations, if M is the Jacobian of $\boldsymbol{\zeta} = \boldsymbol{\zeta}(\boldsymbol{\eta})$, that of $\boldsymbol{\eta} = \boldsymbol{\eta}(\boldsymbol{\zeta})$ is M^{-1}. Moreover, we see that (in an obvious notation):

$$M = \left[\frac{\partial \zeta^i}{\partial \eta^j}\right] = \begin{pmatrix} \left[\dfrac{\partial Q^i}{\partial q^j}\right] & \left[\dfrac{\partial Q^i}{\partial p_j}\right] \\ \left[\dfrac{\partial P_i}{\partial q^j}\right] & \left[\dfrac{\partial P_i}{\partial p_j}\right] \end{pmatrix}, \qquad M^{-1} = \left[\frac{\partial \eta^i}{\partial \zeta^j}\right] = \begin{pmatrix} \left[\dfrac{\partial q^i}{\partial Q^j}\right] & \left[\dfrac{\partial q^i}{\partial P_j}\right] \\ \left[\dfrac{\partial p_i}{\partial Q^j}\right] & \left[\dfrac{\partial p_i}{\partial P_j}\right] \end{pmatrix} \tag{80}$$

while the transposed matrix M^T is:

$$M^T = \left[\frac{\partial \zeta^j}{\partial \eta^i}\right] = \begin{pmatrix} \left[\dfrac{\partial Q^j}{\partial q^i}\right] & \left[\dfrac{\partial P_j}{\partial q^i}\right] \\ \left[\dfrac{\partial Q^j}{\partial p_i}\right] & \left[\dfrac{\partial P_j}{\partial p_i}\right] \end{pmatrix}. \tag{81}$$

Thus, differentiating (79) with respect to t (see (71)),

$$\dot{\boldsymbol{\zeta}} = M\dot{\boldsymbol{\eta}}, \tag{82}$$

while (74) and (75) together imply,

$$M^{-1} = -JM^TJ \quad \Rightarrow \quad MJM^T = J. \tag{83}$$

A matrix M, satisfying (83) is said to be *symplectic*, and thus we have shown that the Jacobian, at any point, of a canonical transformation (which does not involve time explicitly) is a symplectic matrix. Since $\det J = 1$, we see from (83) that

$$\det M = \pm 1. \tag{84}$$

The set of all real $2n \times 2n$ matrices M, satisfying (83) and (84) is clearly a group. The connected component of this group, to the identity, has $\det M = 1$, and is called the $2n$-dimensional, real *symplectic group*, denoted $\mathrm{Sp}(2n, \mathbb{R})$. This is the invariance group of the canonical transformations.

It ought to be emphasized here that, while the canonical transformations $\boldsymbol{\eta} \to \boldsymbol{\zeta}$ are transformations acting on the phase space manifold $\boldsymbol{\Gamma}$ itself, and are in general non-linear, the elements of the symplectic group $\mathrm{Sp}(2n, \mathbb{R})$ are matrices M, which are linear transformations on vectors $\dot{\boldsymbol{\eta}}$ tangent to the phase space manifold. In other words, $\mathrm{Sp}(2n, \mathbb{R})$ is the gauge group of the tangent bundle $T\boldsymbol{\Gamma}$, or of the cotangent bundle $T^*\boldsymbol{\Gamma}$. It can also be shown that if $\boldsymbol{\eta} \mapsto \boldsymbol{\zeta}(\boldsymbol{\eta}, t)$ is a canonical transformation which depends on time, then the infinitesimal transformation $\boldsymbol{\zeta}(\boldsymbol{\eta}, t_0) \mapsto \boldsymbol{\zeta}(\boldsymbol{\eta}, t_0 + dt)$ is brought about by a symplectic matrix.

Let us next look at a few quantities which are *canonical invariants*, in other words, quantities which do not change under canonical transformations. By far the most important canonical invariant is the *Poisson bracket*, $\{\ ,\ \}$, of two functions u, v of the phase space variables $\mathbf{q}, \mathbf{p}$. We define it as:

$$\{u, v\}_{\mathbf{q},\mathbf{p}} = \sum_{i=1}^{n} \left[\frac{\partial u}{\partial q^i}\frac{\partial v}{\partial p_i} - \frac{\partial u}{\partial p_i}\frac{\partial v}{\partial q^i}\right] = -\frac{\partial u}{\partial \boldsymbol{\eta}}^{\dagger} J \frac{\partial v}{\partial \boldsymbol{\eta}}, \tag{85}$$

where † denotes forming the corresponding row vector. Again the subscripts $\mathbf{q}, \mathbf{p}$ denote the independent variables. If we make a canonical transformation to the new variables $\mathbf{Q}, \mathbf{P}$, then (82) and (83) together imply

$$\{u,v\}_{\mathbf{Q},\mathbf{P}} = -\frac{\partial u}{\partial \zeta}^{\dagger} J \frac{\partial v}{\partial \zeta} = -\frac{\partial u}{\partial \boldsymbol{\eta}}^{\dagger} J \frac{\partial v}{\partial \boldsymbol{\eta}} = \{u,v\}_{\mathbf{q},\mathbf{p}}, \tag{86}$$

and hence $\{\ ,\ \}$ is a canonical invariant, so that the subscripts may be dropped.

The invariance of the Poisson bracket now implies the invariance of a certain *2-form* on the phase space, which we try to display next. Note that,

$$J^{-1} = \begin{pmatrix} \mathbf{0} & -\mathbb{I}_n \\ \mathbb{I}_n & \mathbf{0} \end{pmatrix} \quad := \quad \omega, \qquad \omega^{-1} = J, \tag{87}$$

and define the vector $\mathbf{X}_u \in \mathbb{R}^{2n}$, for any smooth function u on phase space by:

$$\mathbf{X}_u = \omega^{-1}\frac{\partial u}{\partial \boldsymbol{\eta}}. \tag{88}$$

The Poisson bracket (85) may then be re-expressed as:

$$\{u,v\}_{\mathbf{q},\mathbf{p}} = \mathbf{X}_u^{\dagger}\omega^{-1}\mathbf{X}_v. \tag{89}$$

The phase space, which we denoted by $\mathbf{\Gamma}$ is itself a differential manifold and the coordinates $(\mathbf{q}, \mathbf{p})$ refer to a specific coordinate chart. In other words, there exists an open set $V \subset \mathbf{\Gamma}$ and a homeomorphism $\psi : V \to \mathbb{R}^{2n}$, such that, for all $x \in V$, the components of $\boldsymbol{\eta} = \psi(x)$ are given in terms of the components of $\mathbf{q}, \mathbf{p}$ as:

$$\eta^i = \psi(x)^i = \begin{cases} q^i, & i = 1, 2, \ldots, n \\ p_i, & i = n+1, n+2, \ldots, 2n. \end{cases} \tag{90}$$

In terms of these coordinates, the tangent space $T_x\mathbf{\Gamma}$ has the basis vectors,

$$\frac{\partial}{\partial \eta^i} = T_x(\psi)^{-1}\mathbf{e}_i = \begin{cases} \dfrac{\partial}{\partial q^i}, & i = 1, 2, \ldots, n \\ \dfrac{\partial}{\partial p_{i-n}}, & i = n+1, n+2, \ldots, 2n, \end{cases} \tag{91}$$

where again the $\mathbf{e}_i$ are the canonical basis vectors of $\mathbb{R}^{2n}$. The cotangent space $T_x^*\mathbf{\Gamma}$ has the dual basis, $\{d\eta^i\}_{i=1}^{2n}$, such that

$$\langle d\eta^i; \frac{\partial}{\partial \eta^j}\rangle = \delta_{ij}. \tag{92}$$

In terms of the $\mathbf{q}$ and $\mathbf{p}$,

$$d\eta^i = \begin{cases} dq^i, & i = 1, 2, \ldots, n \\ dp_{i-n}, & i = n+1, n+2, \ldots, 2n, \end{cases} \tag{93}$$

so that

$$\begin{aligned} \langle dq^i; \frac{\partial}{\partial q^j} \rangle &= \langle dp_i; \frac{\partial}{\partial p_j} \rangle = \delta_{ij} \\ \langle dq^i; \frac{\partial}{\partial p_j} \rangle &= \langle dp_i; \frac{\partial}{\partial q^j} \rangle = 0, \quad i, j = 1, 2, \ldots, n. \end{aligned} \tag{94}$$

Note also, that as x varies over U, $\boldsymbol{\eta} = \psi(x)$ varies smoothly in $\mathbb{R}^{2n}$, so that the bases $\left\{ \frac{\partial}{\partial \eta^i} \right\}_{i=1}^{2n} \subset T_x\mathbf{\Gamma}$ and $\{d\eta^i\}_{i=1}^{2n} \subset T_x^*\mathbf{\Gamma}$ also vary smoothly as the tangent and cotangent spaces change.

In terms of these coordinates, let us introduce the 2-form, using $(\omega_{ij}) = \boldsymbol{\omega}$ in (87):

$$\Omega = \frac{1}{2} \sum_{i,j=1}^{2n} \omega_{ij} d\eta^i \wedge d\eta^j = \sum_{i=1}^{n} dp_i \wedge dq^i. \tag{95}$$

Also, using $\mathbf{X}_u$ in (88), let us define the *Hamiltonian vector fields*:

$$X_u = \sum_{i=1}^{2n} (\mathbf{X}_u)^i \frac{\partial}{\partial \eta^i}. \tag{96}$$

Note that, if ω^{ij} denote the matrix elements of $\boldsymbol{\omega}^{-1}$, then explicitly,

$$X_u = \sum_{i,j=1}^{2n} \omega^{ij} \frac{\partial u}{\partial \eta^j} \frac{\partial}{\partial \eta^i} = -\sum_{i=1}^{n} \left[\frac{\partial u}{\partial q^i} \frac{\partial}{\partial p_i} - \frac{\partial u}{\partial p_i} \frac{\partial}{\partial q^i} \right], \tag{97}$$

implying, for any smooth function $v : \mathbf{\Gamma} \to \mathbb{R}$, that

$$X_u(v) = X_u \rfloor dv = -\{u, v\}. \tag{98}$$

Furthermore, it is easily verified that

$$dv = -X_v \rfloor \Omega = -\Omega(X_v, \cdot\,), \tag{99}$$

and

$$\{u, v\} = -\Omega(X_u, X_v) = -X_v \rfloor (X_u \rfloor \Omega), \tag{100}$$

so that, in particular,

$$X_u(v) = \Omega(X_u, X_v). \tag{101}$$

In the above equations, $\rfloor$ denotes the contraction of a form with a vector. Our conventions are such that if V and V^* are (real) vector spaces, dual to each other, and if $\mathbf{a}^*, \mathbf{b}^* \in V^*$, then

$$\beta := \mathbf{a}^* \wedge \mathbf{b}^* = \mathbf{a}^* \otimes \mathbf{b}^* - \mathbf{b}^* \otimes \mathbf{a}^*, \tag{102}$$

while, if $\mathbf{a}, \mathbf{b} \in V$, then

$$\mathbf{a} \rfloor \beta = \mathbf{b}^* \langle \mathbf{a}^*; \mathbf{a} \rangle - \mathbf{a}^* \langle \mathbf{b}^*; \mathbf{a} \rangle, \tag{103}$$

$\langle\ ;\ \rangle$ being the dual pairing between V and V^*. Also, we write

$$\begin{aligned} \mathbf{a} \rfloor \beta &= \beta(\mathbf{a}, \cdot), \\ \mathbf{b} \rfloor (\mathbf{a} \rfloor \beta) &= \beta(\mathbf{a}, \mathbf{b}). \end{aligned} \tag{104}$$

Writing the Poisson bracket (85) in terms of the 2-form Ω, as done in (100), immediately shows that this 2-form is a canonical invariant. Indeed, since as proved in (86), the Poisson bracket itself is a canonical invariant, the same same property is shared by Ω. If $M \in \mathrm{Sp}(2n, \mathbb{R})$, then the invariance of Ω is expressed by the fact that for arbitrary vector fields,

$$X = \sum_{i=1}^{2n} a_i \frac{\partial}{\partial \eta^i}, \qquad Y = \sum_{i=1}^{2n} b_i \frac{\partial}{\partial \eta^i}, \tag{105}$$

and transformed fields,

$$X' = \sum_{i=1}^{2n} a_i' \frac{\partial}{\partial \eta^i}, \qquad Y' = \sum_{i=1}^{2n} b_i' \frac{\partial}{\partial \eta^i}, \tag{106}$$

where,

$$a_i' = \sum_{i=1}^{2n} M_{ij} a_j, \qquad b_i' = \sum_{i=1}^{2n} M_{ij} b_j, \tag{107}$$

we have

$$\Omega(X, Y) = \Omega(X', Y'). \tag{108}$$

Poisson brackets also satisfy the *Jacobi identity*. If u, v and w are three smooth functions on phase space, then

$$\{u, \{v, w\}\} + \{v, \{w, u\}\} + \{w, \{u, v\}\} = 0. \tag{109}$$

As will be seen later, this is equivalent to the property of Ω being closed, i.e.,

$$d\Omega = 0. \tag{110}$$

Finally, there is another canonical invariant on phase space, which turns out to be useful too. It is the *Liouville measure*, which in the coordinates $\mathbf{q}, \mathbf{p}$ (in which Ω has the form (95)) is :

$$d\eta = \prod_{i=1}^{n} dp_i \, dq^i. \tag{111}$$

The invariance of $d\eta$ follows trivially from the fact that the Jacobian of the transformation $\boldsymbol{\eta} = (\mathbf{q}, \mathbf{p}) \mapsto \boldsymbol{\zeta} = (\mathbf{Q}, \mathbf{P})$ is a symplectic matrix M, of unit determinant (see (84)), and hence,

$$d\boldsymbol{\zeta} = |\det M| \, d\boldsymbol{\eta} = d\boldsymbol{\eta}. \tag{112}$$

The 2-form Ω in (95) is called a *symplectic form* and it characterizes the phase space $\boldsymbol{\Gamma}$ of the classical system with which we are working. Moreover, it affords us an invariant formulation of classical mechanics, opening up the possibility of a general geometric treatment. We shall look at some aspects of such a generalization in the next section.

4 Some Symplectic Geometry

In this section we briefly review some mathematical notions related to symplectic vector spaces and manifolds. More exhaustive treatments are available elsewhere.[1,12,14,22,25,26]

4.1 Symplectic vector spaces

The considerations of the last section show that the geometry of phase space is symplectic, determined by the symplectic form Ω. The specific structure of Ω in (95) was clearly dictated by the particular choice of coordinates in which the the Hamiltonian equations (65) were expressed. Let us look a little more carefully now at the geometry of symplectic space.

Definition 4.1 *A symplectic vector space* (V, Ω) *is a real vector space* V*, together with an antisymmetric, non-degenerate bilinear form* Ω *on* V*.*

This means that $\Omega : V \times V \to \mathbb{R}$ is a bilinear map, satisfying

1. $\Omega(a, b) = -\Omega(b, a), \quad \forall a, b \in V;$
2. for fixed $a \in V$, $\Omega(a, b) = 0$, for all $a \in V$, implies that $a = \mathbf{0}$.

Let $\dim(V) = N$, and suppose that $\{e_i\}_{i=1}^N$ is a basis for V. Denote by V^* the dual space of V. Then $\dim(V^*) = N$ and denote by $\{e^{i*}\}_{i=1}^N$ the dual basis in V^*, i.e.,

$$\langle e^{i*} \,;\, e_j \rangle = \delta^i_j, \qquad i, j = 1, 2, \ldots, N, \tag{113}$$

$\langle\ ;\ \rangle$ denoting the dual pairing between V and V^*. It is then fairly straightforward to see that Ω can be written as

$$\Omega = \frac{1}{2}\sum_{i,j=1}^{N} \omega_{ij} e^{i*} \wedge e^{j*} = \sum_{i,j=1}^{N} \omega_{ij} e^{i*} \otimes e^{j*} \tag{114}$$

(see (102) - (104) for wedge product conventions), where, ω_{ij} are the elements of an $N \times N$ real, antisymmetric matrix $\boldsymbol{\omega}$, which is nonsingular:

$$\boldsymbol{\omega} = -\boldsymbol{\omega}^T, \qquad \det \boldsymbol{\omega} \neq 0. \tag{115}$$

For $a, b \in V$, such that

$$a = \sum_{i=1}^{N} a^i e_i, \qquad b = \sum_{i=1}^{N} b^i e_i. \tag{116}$$

we get

$$\Omega(a,b) := b\rfloor(a\rfloor\Omega) = \sum_{ij=1}^{N} a^i \omega_{ij} b^j. \tag{117}$$

If we introduce the vectors $\boldsymbol{\alpha}, \boldsymbol{\beta} \in \mathbb{R}^N$, such that

$$\boldsymbol{\alpha} = (a^1, a^2, \ldots, a^N), \qquad \boldsymbol{\beta} = (b^1, b^2, \ldots, b^N), \tag{118}$$

then clearly,

$$\Omega(a,b) = \boldsymbol{\alpha}^{\dagger}\boldsymbol{\omega}\boldsymbol{\beta}. \tag{119}$$

It can be shown that every symplectic vector space (V, Ω) is necessarily of even dimension, i.e., $N = 2n$, and that there exists a basis $\{u_i, v_i\}_{i=1}^n$ in V for which,

$$\Omega(u_i, v_j) = \delta_{ij}, \quad \Omega(u_i, u_j) = \Omega(v_i, v_j) = 0, \quad i,j = 1,2,\ldots,n. \tag{120}$$

Such a basis is called a *symplectic frame*. Let $\{u^{i*}, v^{i*}\}_{i=1}^n$ be the dual basis in V^*, chosen so that

$$\langle u^{i*}\ ;\ u_j\rangle = \langle v^{i*}\ ;\ v_j\rangle = \delta^i_j, \quad \langle u^{i*}\ ;\ v_j\rangle = \langle v^{i*}\ ;\ u_j\rangle = 0, \quad i,j = 1,2,\ldots,n. \tag{121}$$

Then, in this basis,

$$\Omega = \sum_{i=1}^{n} u^{i*} \wedge v^{i*}. \tag{122}$$

Symplectic spaces are easily constructed. One can start with any real, $2n$-dimensional vector space V and an antisymmetric, non-singular $2n \times 2n$

matrix $\boldsymbol{\omega}$. One can then choose a basis $\{e^{i*}\}_{i=1}^{2n}$ in the dual space V^* and use (114) to define Ω.

The bilinear form Ω establishes a linear isomorphism between V and V^* in the following way:

$$a \mapsto a^* = a \rfloor \Omega \in V^*, \quad \forall a \in V. \tag{123}$$

Also, assuming Ω has been brought into the form (122), if Q denotes the subspace of V generated by the vectors $\{v_i\}_{i=1}^{n}$ in (120), then it follows from

$$\langle v^{i*} \; ; \; v_j \rangle = \delta^i_j,$$

in (121), that the $\{v^{i*}\}_{i=1}^{n}$ generate Q^*. Moreover, writing

$$V = Q \oplus Q', \tag{124}$$

where Q' is the subspace of V generated by $\{u_i\}_{i=1}^{n}$, it follows from (123) that Q' can be identified with Q^*. Indeed, from (121) it is easily seen that

$$v^{i*} = u_i \rfloor \Omega. \tag{125}$$

Thus, we may set

$$V \simeq Q \oplus Q^*. \tag{126}$$

Writing elements $a \in V$ now as

$$a = (q, p), \qquad q \in Q, \quad p \in Q^*, \tag{127}$$

we get

$$\Omega(a, a') = \Omega((q,p) \; ; \; (q', p')) = \langle p \; ; \; q' \rangle_{Q,Q^*} - \langle p' \; ; \; q \rangle_{Q,Q^*} \tag{128}$$

where now $\langle \; ; \; \rangle_{Q,Q^*}$ denotes the dual pairing between Q and Q^*.

Let (V, Ω) be a symplectic vector space and $F \subset V$ a subspace. The *symplectic complement* of F is the subspace

$$F^\perp = \{a \in V \mid \Omega(a, b) = 0, \;\; \forall b \in F\}. \tag{129}$$

If F and G are two subspaces of V, then the following results are easy to prove:

$$\begin{aligned} F \subset G \Rightarrow G^\perp &\subset F^\perp, \\ (F^\perp)^\perp &= F, \\ (F + G)^\perp &= F^\perp \cap G^\perp, \\ (F \cap G)^\perp &= F^\perp + G^\perp, \\ \dim(V) &= \dim(F) + \dim(F^\perp). \end{aligned} \tag{130}$$

Definition 4.2 *A subspace F of a symplectic vector space (V, Ω) is said to be*

$$\begin{aligned} \text{isotropic} &\iff F \subset F^{\perp}, \\ \text{coisotropic} &\iff F^{\perp} \subset F, \\ \text{symplectic} &\iff F \cup F^{\perp} = \{0\}, \\ \text{Lagrangian} &\iff F = F^{\perp}. \end{aligned}$$

From (120) it follows that the subspaces Q and Q', generated by $\{v_i\}_{i=1}^n$ and $\{u_i\}_{i=1}^n$, respectively, are both Lagrangian. Hence Lagrangian subspaces always exist. On a symplectic space F, the bilinear form restricts to a non-degenerate bilinear form Ω_F, making (F, Ω_F) itself into a symplectic vector space.

Definition 4.3 *A linear* canonical transformation *of a symplectic vector space (V, Ω) is a linear map $S : V \to V$, such that*

$$\Omega(Sa, Sb) = \Omega(a, b), \tag{131}$$

for all $a, b \in V$. The group of all such transformations is denoted $\mathrm{SP}(V, \Omega)$.

Introducing the symplectic frame defined in (120), it is easily established that S can be identified with a symplectic matrix

$$SJS^T = J,$$

so that $\mathrm{SP}(V, \Omega)$ is isomorphic to the symplectic group $\mathrm{Sp}(2n, \mathbb{R})$.

4.2 Symplectic manifolds

Let us pass now to a consideration of a symplectic manifold, for as we have seen, the phase space of a classical mechanical system is such a manifold.

Definition 4.4 *A symplectic manifold is a pair $(\mathbf{\Gamma}, \Omega)$, where $\mathbf{\Gamma}$ is a real even dimensional, differential manifold and Ω a closed antisymmetric two-form on $\mathbf{\Gamma}$, called the symplectic form.*

Generally, if X and Y are vector fields on $\mathbf{\Gamma}$, then:

$$\Omega(X, Y) = -\Omega(Y, X), \qquad \Omega(X, Y) = 0, \text{for all } Y \Rightarrow X = 0, \tag{132}$$

and the map $T_x\mathbf{\Gamma} \to T_x^*\mathbf{\Gamma}$, given pointwise using $\Omega(x)$ (see (123)), by

$$X(x) \mapsto X(x) \rfloor \Omega(x) \quad \text{or,} \quad X \mapsto X \rfloor \Omega, \tag{133}$$

is an isomorphism.

Thus, the tangent space $T_x\mathbf{\Gamma}$ at any point $x \in \mathbf{\Gamma}$ is a symplectic vector space, with the bilinear form $\Omega(x)$ (the restriction of Ω to $T_x\mathbf{\Gamma}$), which is non-degenerate. The closedness of Ω is the statement that its exterior derivative vanishes:

$$d\Omega = 0. \tag{134}$$

We shall see later that this condition is equivalent to the existence of the Jacobi identity for the associated Poisson brackets.

The phase space of a classical mechanical system is the cotangent bundle $T^*\mathbf{\Omega}$ of the configuration space manifold $\mathbf{\Omega}$. Such a manifold $\mathbf{\Gamma} = T^*\mathbf{\Omega}$ always has a symplectic structure. Indeed, a point $x \in \mathbf{\Gamma}$ is of the form $x = (q, p)$, where $q \in \mathbf{\Omega}$ and p is a covector at q, i.e., $p \in T_q^*(\mathbf{\Omega})$. Using a coordinate chart, $\phi : U \to \mathbb{R}^n$, where U is an open set in $\mathbf{\Omega}$, we can introduce the local coordiantes,

$$q^i = \phi^i(q), \qquad i = 1, 2, \ldots, n, \tag{135}$$

on $\mathbf{\Omega}$. This gives us the basis, $\left\{\dfrac{\partial}{\partial q^i}\right\}_{i=1}^n$ in the tangent space $T_q\mathbf{\Omega}$ to the manifold $\mathbf{\Omega}$ at the point q, and the basis $\{dq^i\}_{i=1}^n$, in the cotangent space $T_q^*\mathbf{\Omega}$. The covector $p \in T_q^*\mathbf{\Omega}$ can then be written as

$$p = \sum_{i=1}^n p_i dq^i, \tag{136}$$

where the p_i are simply the coordinates of the covector p at q, in the basis $\{dq^i\}$. The symplectic structure on $\mathbf{\Gamma} = T^*\mathbf{\Omega}$ is then given by the canonical two-form

$$\Omega = \sum_{i=1}^n dp_i \wedge dq^i. \tag{137}$$

There is a coordinate independent way of arriving at Ω. Let $\pi : \mathbf{\Gamma} = T^*\mathbf{\Omega} \to \mathbf{\Omega}$ be the canonical projection map,

$$\pi(q, p) = q.$$

Its derivative π_* at the point (q, p) is the tangent map,

$$\pi_*(q, p) = T_{(q,p)}(\pi) : T_{(q,p)}\mathbf{\Gamma} \to T_q\mathbf{\Omega}.$$

Let us define a one-form θ on $\mathbf{\Gamma}$, such that for any vector field X on $\mathbf{\Gamma}$,

$$(X \rfloor \theta)(q, p) = (\pi_* X)(q, p) \rfloor p, \tag{138}$$

and we set

$$\Omega = d\theta. \tag{139}$$

To see what θ looks like in the coordinates introduced above, write X as

$$X = \sum_{i=1}^{n} [\alpha^i \frac{\partial}{\partial q^i} + \beta_i \frac{\partial}{\partial p_i}],$$

where $\alpha^i : \mathbf{\Gamma} \to \mathbb{R}$ and $\beta_i : \mathbf{\Gamma} \to \mathbb{R}$ are smooth functions. Then,

$$\pi_* X = \sum_{i=1}^{n} \alpha^i \frac{\partial}{\partial q^i},$$

so that with p as in (136), it follows that

$$\pi_* X(q,p) \rfloor p = \sum_{i=1}^{n} \alpha^i p_i.$$

Thus, since this must equal $(X \rfloor \theta)(q,p)$, we get

$$\theta(q,p) = \sum_{i=1}^{n} p_i dq^i,$$

whence (139) implies that Ω has the form postulated in (137). We also see, as a consequence of (138) that the one-form θ is globally defined. Hence, Ω is not only closed, i.e., $d\Omega = 0$, in this case it is also *exact.* This latter property, of exactness, is not however shared by the two-form coming from an arbitrary symplectic manifold. In particular, a compact symplectic manifold cannot have a global one-form θ satisfying (139), i.e., for such a manifold, Ω cannot be exact.

On any symplectic manifold, we can however, always find a local one-form θ, such that locally, $\Omega = d\theta$. Such a one-form is called a *symplectic potential.* Moreover, θ is not unique. The uniqueness of local symplectic potentials θ, and hence the fact that every symplectic manifold locally looks like a cotangent bundle, is a consequence of Darboux's theorem:

Theorem 4.5 *Let* $(\mathbf{\Gamma}, \Omega)$ *be a* $2n$*-dimensional symplectic manifold and let* $x \in \mathbf{\Gamma}$. *Then there is a neighbourhood* U *of* x, *and a coordinate system* $\{q^i, p_i\}$, $i = 1, 2, \ldots, n$, *on* U *such that*

$$\Omega = \sum_{i=1}^{n} dp_i \wedge dq^i$$

on U.

Local coordinates, in which Ω has the above form, are called *canonical or Darboux coordinates.* Of course, such coordinates are not unique, since local canonical transformations, which are diffeomorphisms $\rho : \mathbf{\Gamma} \to \mathbf{\Gamma}$, inducing linear transformations ρ^* on the tangent spaces, satisfy (see (108) and (131)):

$$\Omega(\rho^* X\ ,\ \rho^* Y) = \Omega(X\ ,\ Y), \tag{140}$$

for arbitrary vector fields X and Y. We know, of course, that at each point $x \in \mathbf{\Gamma}$, the set of all such maps $\rho^*(x)$ is isomorphic to $\mathrm{Sp}(2n, \mathbb{R})$.

From now on, by a classical mechanics we shall mean a symplectic manifold $(\Omega, \mathbf{\Gamma})$ of dimension $2n$. A local set of coordinates will be denoted by $\eta^i,\ i = 1, 2, \ldots, 2n$, in terms of which the tangent and cotangent spaces have bases:

$$\left\{\frac{\partial}{\partial \eta^i} = \partial_i\right\}_{i=1}^{2n}, \quad \text{and} \quad \{d\eta^i\}_{i=1}^{2n},$$

respectively, with

$$\langle d\eta^i\ ;\ \partial_i \rangle = \delta^i_j. \tag{141}$$

Vector fields X, Y on $\mathbf{\Gamma}$ will be denoted locally by

$$X = \sum_{i=1}^{2n} v^i \partial_i, \qquad Y = \sum_{i=1}^{2n} w^i \partial_i, \tag{142}$$

the components, v^i, w^i being smooth functions on $\mathbf{\Gamma}$. Similarly, the two-form Ω locally assumes the shape (see (95)):

$$\Omega = \frac{1}{2} \sum_{i,j=1}^{2n} \omega_{ij} d\eta^i \wedge d\eta^j, \tag{143}$$

where, fixed i, j, $\omega_{ij}(x)$ is a smooth function of $x \in \mathbf{\Gamma}$, and for fixed $x \in \mathbf{\Gamma}$,

$$\omega_{ij}(x) = -\omega_{ji}(x). \tag{144}$$

Thus,

$$\Omega(X, Y) = \sum_{i,j=1}^{2n} v^i\ \omega_{ij}\ w^j. \tag{145}$$

The condition for non-degeneracy (see (132)) is now

$$\det \omega_{ij}(x) \neq 0, \qquad \forall x. \tag{146}$$

This also implies that the inverse matrix $\big(\omega^{ij}(x)\big)$ exists for all x:

$$\sum_{k=1}^{2n} \omega^{ik}\omega_{kj} = \sum_{k=1}^{2n} \omega_{jk}\omega^{ki} = \delta^i_j.$$

In terms of the inverse matrix $\big(\omega^{ij}(x)\big)$, the Hamiltonian vector field, corresponding to the smooth function v on phase space is defined to be (see (97)),

$$X_v = \sum_{i,j=1}^{2n} \partial_j v\, \omega^{ij} \partial_i, \tag{147}$$

which then satisfies (99). Furthermore, the Poisson bracket of two functions u and v is now (see (100)),

$$\{u, v\} = \sum_{i,j=1}^{2n} \partial_i u\, \omega^{ji} \partial_j v = -X_u(v) = -\Omega(X_u, X_v), \tag{148}$$

which is manifestly antisymmetric.

Let us calculate the contraction of the 3-form $d\Omega$ with three arbitrary vector fields, X_u, X_v and X_w, corresponding to the three functions u, v and w. From the definition of the exterior derivative of a 2-form we have,

$$\begin{aligned} d\Omega(X_u, X_v, X_w) &= \sum_{j=1}^{2n} [\partial_j \Omega(X_u, X_v)\, \langle d\eta^j\ ;\ X_w\rangle + \partial_j \Omega(X_v, X_w)\, \langle d\eta^j\ ;\ X_u\rangle \\ &\quad + \partial_j \Omega(X_w, X_u)\, \langle d\eta^j\ ;\ X_v\rangle] \\ &= \sum_{i,j=1}^{2n} \omega^{ji} [\partial_j \Omega(X_u, X_v)\, \partial_i w + \partial_j \Omega(X_v, X_w)\, \partial_i u \\ &\quad + \partial_j \Omega(X_w, X_u)\, \partial_i v], \end{aligned}$$

by (147). Using (148), we obtain

$$d\Omega(X_u, X_v, X_w) = \{u, \{v, w\}\} + \{v, \{w, u\}\} + \{w, \{u, v\}\}. \tag{149}$$

Hence, the 3-form is closed, i.e., $d\Omega = 0$, if and only if the Jacobi identity (109) is satisfied.

If H is the Hamiltonian of the system, then the time evolution is obtained by solving the Hamiltonian equations (see (78)), which can be conveniently rewritten in the form:

$$\dot{\eta} = \{\eta, H\}, \tag{150}$$

where H is assumed to be independent of time. On the other hand, for a Hamiltonian vector field X_u, with components (see (147)),

$$\begin{aligned} X_u^i &= \sum_{j=1}^{2n} \partial_j u \, \omega^{ij} = \sum_{j,k=1}^{2n} \partial_j u \, \omega^{ij} \partial_k \eta^i \\ &= \{\eta^i, u\}, \qquad i = 1, 2, \ldots, 2n, \end{aligned} \tag{151}$$

its *flow* is (by definition) given by $\eta^i = \eta^i(t)$, such that,

$$X_u^i = \dot{\eta}^i, \qquad i = 1, 2, \ldots, 2n. \tag{152}$$

Combining (151) and (152),

$$\dot{\eta}^i = \{\eta^i, u\}, \qquad i = 1, 2, \ldots, 2n. \tag{153}$$

5 The Concept of Quantization

This section is devoted to a discussion of the general concept of quantization. We also have a tentative look at a method for quantizing, which while at first glance may look rather ad hoc, does nevertheless serve as an ingredient in several more formally developed techniques. Relevant references for the material in this section are of course available.[2,3,4,6,13,16,18,24]

5.1 *Some generalities*

Classical mechanics, as we have seen, is formulated on phase space. Very generally, we may call a symplectic space $(\mathbf{\Gamma}, \Omega)$, equipped with a Poisson bracket emanating from Ω a classical mechanics. *States* in classical mechanics are probability measures μ, defined on $\mathbf{\Gamma}$, i.e., Borel measures satisfying

$$\int_{\mathbf{\Gamma}} d\mu(x) = 1. \tag{154}$$

In particular, a pure state, localized at $x = x_0$ is a δ-measure concentrated at x_0:

$$\delta_{x_0}(f) = f(x_0), \qquad f \in \mathcal{C}^\infty(\mathbf{\Gamma}). \tag{155}$$

(Note that $\mathcal{C}^\infty$ denotes the space of smooth, i.e., infinitely differentiable functions). *Observables* in classical mechanics are real valued functions $f : \mathbf{\Gamma} \to \mathbb{R}$, and the value of an observable f, in a state μ, is the expectation value:

$$\langle \mu \, ; \, f \rangle = \int_{\mathbf{\Gamma}} f(x) \, d\mu(x). \tag{156}$$

Quantum mechanics, on the other hand, is based on a (complex and separable) Hilbert space $\mathfrak{H}$. The pure states of a quantum system are rays $\mathbb{C}\phi \subset \mathfrak{H}$, $\phi \neq 0$, in this Hilbert space. Alternatively, a pure state can also be represented by a one-dimensional projection operator,

$$\Phi = \frac{|\phi\rangle\langle\phi|}{||\phi||^2}, \qquad \phi \neq 0, \tag{157}$$

on $\mathfrak{H}$. We shall see later, that the space of all such projection operators, which we denote by $\mathbb{CP}(\mathfrak{H})$, also has the structure of a (generally infinite dimensional) manifold. On the other hand, a *mixed state* in quantum mechanics can be represented by a density matrix,

$$\rho = \sum_{n=1}^{N} \alpha_n |e_n\rangle\langle e_n|, \quad \alpha_n \geq 0, \quad \sum_{n=1}^{N} \alpha_n = 1, \quad N = \dim(\mathfrak{H}), \tag{158}$$

where $\{e_n\}_{n=1}^{N}$ is an orthonormal basis in $\mathfrak{H}$.

The observables of quantum mechanics are self-adjoint operators A, on the Hilbert space $\mathfrak{H}$, and the value of A in the state Φ is given by

$$\langle \Phi \,;\, A\rangle = \langle \widehat{\phi} | A\widehat{\phi}\rangle_{\mathfrak{H}} = \mathrm{tr}[\Phi A], \tag{159}$$

where $\widehat{\phi}$ is any vector in the range of the projection operator Φ, and $\langle \cdot \mid \cdot \rangle_{\mathfrak{H}}$ denotes the scalar product in $\mathfrak{H}$. For the mixed state ρ,

$$\langle \rho \,;\, A\rangle = \mathrm{tr}[\rho A] = \sum_{n=1}^{N} \alpha_n \langle e_n | A e_n\rangle. \tag{160}$$

The set of classical observables is a (non-associative) algebra with respect to the product

$$f \circ_{cl} g = \{f \,,\, g\}, \tag{161}$$

given by the Poisson bracket, while the set of quantum observables is a (non-associative) algebra with respect to the commutator bracket

$$A \circ_{qu} B = \frac{1}{i\hbar}[A, B] = \frac{1}{i\hbar}(AB - BA). \tag{162}$$

Let $\mathcal{S}(\mathfrak{H})$ denote some suitably large class of self-adjoint operators on $\mathfrak{H}$ (defined on some common dense domain), equipped with the product (162), and $\mathcal{A}(\mathbf{\Gamma})$ a suitably large subalgebra of $C^\infty(\mathbf{\Gamma})$, equipped with the product (161). Then a *quantization* of the classical mechanics $(\mathbf{\Gamma}, \Omega)$ would seem to imply finding an appropriate Hilbert space $\mathfrak{H}$ and an algebraic homomorphism between $\mathcal{A}(\mathbf{\Gamma})$ and $\mathcal{S}(\mathfrak{H})$. Of course, this rather general statement needs to be made more precise, and in fact even modified somewhat, as we shall see.

As noted earlier, in general terms, *quantization* is the method by which one makes the transition:

$$\textit{Classical Mechanics} \quad \longrightarrow \quad \textit{Quantum Mechanics}, \tag{163}$$

while the opposite process,

$$\textit{Quantum Mechanics} \quad \lim_{\hbar \longrightarrow 0} \quad \textit{Classical Mechanics}, \tag{164}$$

is called *dequantization*. Let $f \in \mathcal{A}(\mathbf{\Gamma})$ and let $\widehat{f}$ be its quantized counterpart. Then, according to the general rules of quantization (attributed to Dirac), we should require that the quantization map $f \mapsto \widehat{f}$ satisfy the properties:

1. $f \mapsto \widehat{f}$ be real linear
2. for some appropriate subset $\{f\} \subset \mathcal{A}(\mathbf{\Gamma})$, which is closed under the Poisson bracket operation and chosen for physically motivated reasons, the corresponding quantized operators act irreducibly on $\mathfrak{H}$;
3. $\mathbb{I} \mapsto I$, where $\forall x \in \mathbf{\Gamma}$, $\mathbb{I}(x) = 1$, and I is the identity operator on $\mathfrak{H}$;
4. for a chosen set of functions $f, g, \ldots$, (which in general cannot be made to include all of $\mathcal{A}(\mathbf{\Gamma})$ without relaxing condition 2. above),

$$\{f, g\} \longrightarrow \frac{1}{i\hbar}[\widehat{f}, \widehat{g}], \tag{165}$$

where $[\widehat{f}, \widehat{g}]$ is the commutator of the operators $\widehat{f}$ and $\widehat{g}$ on $\mathfrak{H}$.

Note that the *quantization condition* (165) may be expressed in the convenient notation:

$$[\widehat{f}, \widehat{g}] = i\hbar\, \widehat{\{f, g\}}. \tag{166}$$

Note also that, if a certain subset $\{f\} \subset \mathcal{C}^\infty(\mathbf{\Gamma})$ is stable under formation of Poisson brackets, then it constitutes a subalgebra of $\mathcal{C}^\infty(\mathbf{\Gamma})$, and physics often dictates that a particular *finite dimensional* subalgebra be irreducibly realized on $\mathfrak{H}$. Let $\mathcal{A}_0(\mathbf{\Gamma})$ denote such a finite dimensional subalgebra. Often this could be the Lie algebra of a physically pertinent symmetry group of the system. As an example, consider the functions $u(q,p) = q^i$ and $v(q,p) = p_i$, $i = 1, 2, \ldots, n$. Correspondingly, we have the Hamiltonian vector fields, $X_{q^i} = -\dfrac{\partial}{\partial p_i}$ and $X_{p_i} = \dfrac{\partial}{\partial q^i}$, $i = 1, 2, \ldots, n$. Then,

$$\{q^i\,,\,p_j\} = -\Omega(X_{q^i}\,,\,X_{p_j}) = \delta^i_j, \qquad i, j = 1, 2, \ldots, n, \tag{167}$$

and along with the (central) element $u = \mathbb{I}$ they constitute a closed subalgebra $\mathcal{A}_0(\mathbf{\Gamma})$, which in fact is the Lie algebra of the *Weyl-Heisenberg* group $G_{WH}(n)$, for a system with n degrees of freedom. In view of (165), the quantized operators $\widehat{q}^i$ and $\widehat{p}_i$, together with the identity operator I would have to satisfy,

$$[\widehat{q}^i \, , \, \widehat{p}_j] = i\hbar I \delta^i_j, \qquad i,j = 1,2,\ldots,n. \tag{168}$$

This is simply the Hilbert space representation of the generators of the same abstract Weyl-Heisenberg group. If we require that this set be represented irreducibly on $\mathfrak{H}$, then according to a theorem of von Neumann there is, up to unitary equivalence, only one such representation. This representation may be realized on $\mathfrak{H} = L^2(\mathbb{R}^n, d\mathbf{x})$ as:

$$(\widehat{q}^i\psi)(\mathbf{x}) = x^i\psi(\mathbf{x}), \qquad (\widehat{p}_i\psi)(\mathbf{x}) = -i\hbar\frac{\partial}{\partial x_i}\psi(\mathbf{x}), \qquad \mathbf{x} = (x^1, x^2, \ldots, x^n), \tag{169}$$

the above relations holding on some appropriate dense set of elements $\psi \in \mathfrak{H}$. The relations (168) are called the *canonical commutation relations (CCR)*, and accordingly, the Weyl-Heisenberg group is the group of the canonical commutation relations. The Poisson brackets (167) and the CCR (168) can be rewritten, in terms of the coordinates η^i, $i = 1,2,\ldots,2n$, (see (90)). We get,

$$\{\eta^i \, , \, \eta^j\} = J^{ij}, \qquad [\widehat{\eta}^i \, , \, \widehat{\eta}^j] = i\hbar I \; J^{ij} \qquad i,j = 1,2,\ldots,2n, \tag{170}$$

J^{ij} being the elements of the matrix J introduced in (77). In an obvious notation, the above relations may be re-expressed compactly as:

$$\{\boldsymbol{\eta} \, , \, \boldsymbol{\eta}\} = J, \qquad [\widehat{\boldsymbol{\eta}} \, , \, \widehat{\boldsymbol{\eta}}] = i\hbar I \; J, \tag{171}$$

and the quantization condition as:

$$[\widehat{\boldsymbol{\eta}} \, , \, \widehat{\boldsymbol{\eta}}] = i\hbar \, \widehat{\{\boldsymbol{\eta} \, , \, \boldsymbol{\eta}\}} \tag{172}$$

Now, Poisson brackets are invariant under the symplectic group. If $S \in \mathrm{Sp}(2n, \mathbb{R})$, has matrix elements S^i_j then the following must hold (as can also be verified by direct computation):

$$[\widehat{\eta}\,'^i \, , \, \widehat{\eta}\,'^j] = i\hbar I \; J^{ij}, \qquad \widehat{\eta}\,'^i = \sum_{j=1}^{2n} (S^{-1})^i_j \, \widehat{\eta}^j, \qquad i = 1,2,\ldots,n. \tag{173}$$

Furthermore, in the representation space $\mathfrak{H}$ of the CCR it is also possible to construct a unitary operator $U(S)$, corresponding to each $S \in \mathrm{Sp}(2n, \mathbb{R})$, such that

$$\widehat{\eta}\,'^i = U(S)\widehat{\eta}^i U(S)^{-1}, \tag{174}$$

in other words, this set of canonical transformations is unitarily implemented.

Unfortunately, however, if the Weyl-Heisenberg group is implemented irreducibly on $\mathfrak{H}$, then the set of other Poisson brackets for which (165) can be satisfied is rather small. We shall return to this point later Section 6.1. On the other hand, if the irreducibility condition is dropped then it is possible to (*pre-*)quantize the entire Poisson algebra $\mathcal{C}^\infty(\mathbf{\Gamma})$. For this reason, in the method of geometric quantization, one proceeds in two stages. In the first stage or *prequantization*, a map $f \mapsto \widehat{f}$, satisfying (165), for all $f \in \mathcal{C}^\infty(\mathbf{\Gamma})$ is obtained. The result is that one ends up with much too large a Hilbert space on which condition 2. is violated. Thus, in the second stage, called *polarization*, one seeks to isolate a suitable maximal subset of observables for which (165) is required to hold and the quantized versions of which then act irreducibly on a smaller Hilbert space.

In the theory of quantization, there are usually two points of departure:

1. Start with the system localized in the position space $\mathbf{\Omega}$ and then proceed to quantize the theory.

2. Start with the system localized on the phase space $\mathbf{\Gamma}$, or more generally, on an arbitrary symplectic manifold and then make the transition to a quantum theory.

In either case one exploits the *geometry* and/or the *Borel structure* of these underlying spaces. Additionally, a kinematical symmetry at the classical level is expected to manifest itself at the quantum level as well – a fact which is also to be exploited. In the next sections we shall look at several quantization techniques which illustrate both the routes mentioned above. But before that let us look at a motivating example.

5.2 A general procedure using frames

As a first example, we briefly look at a very general quantization map and a technique which has sometimes been referred to in the literature as *prime quantization* or *coherent state quantization.* The method is very similar to *Berezin quantization* and also shares many features with the method of *geometric quantization.* Let $\mathfrak{H}$ be a Hilbert space and $\mathcal{F}(\eta_x^i, A, n)$ be a *frame* in $\mathfrak{H}$, with η_x^i labelled by points x of the phase space $\mathbf{\Gamma}$. This means that $x \mapsto F(x) = \sum_{i=1}^n |\eta_x^i\rangle\langle\eta_x^i|$ is a positive operator-valued function on $\mathbf{\Gamma}$, and

$$\int_{\mathbf{\Gamma}} F(x)\, d\nu(x) = A, \tag{175}$$

ν being the Liouville measure on $\mathbf{\Gamma}$, a bounded, positive operator on $\mathfrak{H}$ with bounded inverse. There is an obvious method for associating a self-adjoint operator $\widehat{f}$ with a classical observable f, using this frame. Consider the *tight frame* $\mathcal{F}(\widetilde{\eta}^i_x, I, n)$ obtained from $\mathcal{F}(\eta^i_x, A, n)$ by setting

$$\widetilde{\eta}^i_x = A^{-\frac{1}{2}}\eta^i_x\,, \qquad \sum_{i=1}^{n}\int_{\mathbf{\Gamma}} |\widetilde{\eta}^i_x\rangle\langle\widetilde{\eta}^i_x|\; d\nu(x) = I. \tag{176}$$

A quantization scheme, $f \mapsto \widehat{f}$, can now be obtained by setting

$$\widehat{f} = \int_{\mathbf{\Gamma}} f(x)\widetilde{F}(x)\; d\nu(x), \quad \widetilde{F}(x) = \sum_{i=1}^{n} |\widetilde{\eta}^i_x\rangle\langle\widetilde{\eta}^i_x|. \tag{177}$$

Of course, the agebra of phase space functions $\mathcal{A}(\mathbf{\Gamma})$, for which (177) is well-defined, depends on both the operator-valued function $\widetilde{F}$ and the measure ν. A justification for calling the above correspondence procedure a quantization, is that it is possible to isolate a sufficiently large subset of functions $\{f\}$ for which (165) holds. Moreover, one can describe the localizability properties of the quantized system on phase space by means of the *positive operator-valued (POV) measure* a, based on $\mathfrak{H}$, and given for each Borel subset Δ of $\mathbf{\Gamma}$ by

$$a(\Delta) = \int_{\Delta} \widetilde{F}(x)\, d\nu(x) = \sum_{i=1}^{n}\int_{\Delta} |\widetilde{\eta}^i_x\rangle\langle\widetilde{\eta}^i_x|\, d\nu(x). \tag{178}$$

More precisely, the existence of the POV measure (178) allows us to introduce the following notion of quantum localization on $\mathbf{\Gamma}$: If we interpret a positive operator ρ on $\mathfrak{H}$, with unit trace, as an accessible state of a quantum system, the quantity $\mathrm{Tr}[a(\Delta)\rho]$ defines the probability of finding the system localized in the subset Δ of $\mathbf{\Gamma}$ when it is in the state ρ. Of course this probabilistic interpretation requires the normalization condition $a(\mathbf{\Gamma}) = I$, and hence necessitates the passage to the tight frame $\mathcal{F}(\widetilde{\eta}^i_x, I, n)$. If $\mathbf{\Gamma}$ is a homogeneous space of a locally compact group G and the measure ν is invariant with respect to the action of G on $\mathbf{\Gamma}$, then the POV-measure a can often be used to define a *system of covariance* for $(\mathbf{\Gamma}, G)$. This means that on $\mathfrak{H}$, there is also a unitary irreducible representation $g \mapsto U(g)$ of the group G, for which

$$U(g)a(\Delta)U(g)^* = a(g\Delta), \tag{179}$$

for all Borel sets $\Delta \subset \mathbf{\Gamma}$ and all $g \in G$. (In the above, $g\Delta$ denotes the Borel set obtained from Δ by translating it using the action of g on the homogeneous space $\mathbf{\Gamma}$). In this way one obtains a family of covariant localization operators $a(\Delta)$ on phase space. This particular quantization procedure may also be formulated in terms of completely positive maps of C^*-algebras. Of course, in order for the procedure to be effective, a systematic method has to be devised for determining the Hilbert space $\mathfrak{H}$ and the frame $\mathcal{F}(\eta^i_x, A, n)$.

Let us work out a simple example to illustrate the above procedure. Consider a classical particle of mass m, having a single degree of freedom and moving on the configuration space $\mathbf{\Omega} = \mathbb{R}$ and the phase space $\mathbf{\Gamma} = T^*\mathbb{R} = \mathbb{R}^2$. Consider the Hilbert space $\mathfrak{H} = L^2(\mathbb{R}, dx)$ and on it take the vector η (ground state of the harmonic oscillator):

$$\eta(x) = (\pi\hbar)^{-\frac{1}{4}} \exp[-\frac{x^2}{2\hbar}], \tag{180}$$

and using it define the family of vectors, $\eta_{q,p}$, $(q,p) \in \mathbf{\Gamma}$,

$$\eta_{q,p} = \exp[\frac{i}{\hbar}(x - \frac{q}{2})p]\eta(x - q). \tag{181}$$

As will be explained later, these vectors, called the *canonical coherent states*, are coherent states associated to the Weyl-Heisenberg group. At this point let us just note that they form a tight frame, $\mathcal{F}(\eta_{q,p}, I, 1)$. Indeed a straightforward computation shows that

$$\int_{\mathbb{R}^2} F(q,p)\, d\nu(q,p) = I, \qquad F(q,p) = |\eta_{q,p}\rangle\langle\eta_{q,p}|, \qquad d\nu(q,p) = \frac{dq\, dp}{h}. \tag{182}$$

If now $f \in \mathcal{C}^\infty(\mathbf{\Gamma})$, then its quantized version, according to (177), is

$$\widehat{f} = \int_{\mathbb{R}^2} f(q,p) F(q,p)\, d\nu(q,p), \tag{183}$$

and if $\phi \in \mathfrak{H}$ is chosen from the domain (assumed dense) of this operator, then

$$(\widehat{f}\phi)(x) = \frac{1}{h}\int_{\mathbb{R}^2} f(q,p)\, [e^{-\frac{i}{\hbar}(x'-x)p}\eta(x'-q)\phi(x')\, dx']\, \eta(x-q)\, dqdp. \tag{184}$$

From this it follows that if $f(q,p) = f(q)$ is a function of q alone, then $\widehat{f}$ is the operator of multiplication by the function $|\eta|^2 \circ f$ (the circle denotes a convolution):

$$|\eta|^2 \circ f(x) = \int_{\mathbb{R}^2} |\eta(x-q)|^2 f(q)\, dq, \tag{185}$$

while if $f(q,p)$ is a function of p alone, $f(q,p) = f(p)$, then $\widehat{f}$ is the (in general pseudo-)differential operator

$$\widehat{f} = f(-i\hbar\frac{\partial}{\partial x}), \tag{186}$$

(formally, if $f(p)$ is written as a power series in p, then $\widehat{f}$ is obtained by replacing p by $-i\hbar\frac{\partial}{\partial x}$). In particular, taking $f(q) = q$ and $f(p) = p$, we get:

$$(\widehat{q}\phi)(x) = x\phi(x), \qquad (\widehat{p}\phi)(x) = -i\hbar\frac{\partial\phi}{\partial x}(x). \tag{187}$$

Thus, if $f = H$ is the harmonic oscillator Hamiltonian,

$$H = \frac{p^2}{2m} + \frac{m^2\omega^2}{2}q^2, \tag{188}$$

then

$$\widehat{H} = -\frac{\hbar^2}{2m}\frac{d^2}{dx^2} + \frac{m^2\omega^2}{2}x^2 + C, \tag{189}$$

where C is the constant

$$C = \frac{\sqrt{\pi}}{4}\hbar^{\frac{3}{2}}m^2\omega^2. \tag{190}$$

which simply changes the ground state energy.

We see in this example, that this method of quantization yields the expected result, in that the Poisson bracket $\{q, p\}$ is properly mapped to the commutator bracket $[\widehat{q}, \widehat{p}]$, and the algebra generated by $\widehat{q}, \widehat{p}$ and I is irreducibly represented on $\mathfrak{H}$. On the other hand, taking for $f(q) = q^m$ and $f(p) = p^n$, with $m, n \geq 3$, it is easily seen that (165) is violated.

6 Canonical Quantization and its Generalizations

We discuss in this section the original idea of quantization, introduced in the early days of quantum mechanics – rather simple minded and ad hoc, but extremely effective – and some later refinements of it. There also exist quite some useful references in the literature.[7,8,9,10,13,17,20,21]

6.1 The early notion of quantization

The originators of quantum theory used the following simple technique for quantizing a classical system: As before, let $q^i, p_i, \quad i = 1, 2, \ldots, n$, be the canonical position and momenta, respectively, of a classical system with n degrees of freedom. Then their quantized counterparts, $\widehat{q}^i, \widehat{p}_i$, are to be realized as operators on the Hilbert space $\mathfrak{H} = L^2(\mathbb{R}^n, d\mathbf{x})$, by the prescription:

$$(\widehat{q}^i\psi)(\mathbf{x}) = x^i\psi(\mathbf{x}) \qquad (\widehat{p}_i\psi)(\mathbf{x}) = -i\hbar\frac{\partial}{\partial x^i}\psi(\mathbf{x}), \tag{191}$$

on an appropriately chosen dense set of vectors ψ in $\mathfrak{H}$. This simple procedure is known as *canonical quantization*. As already discussed in Section 5.1, von Neumann's *uniqueness theorem* now states that, up to unitary equivalence, this is the only representation which realizes the canonical commutation relations (CCR):

$$[\widehat{q}^i, \widehat{p}_j] = i\hbar I \; \delta^i_j, \qquad i, j = 1, 2, \ldots, n, \tag{192}$$

irreducibly on a separable Hilbert space. Let us examine this question of irreducibility a little more closely.

The Weyl-Heisenberg group $G_{WH}(n)$, for a system with n degrees of freedom, is topologically isomorphic to $\mathbb{R}^{2n+1}$, and consists of elements $(\theta, \boldsymbol{\eta})$, with $\theta \in \mathbb{R}$ and $\boldsymbol{\eta} \in \mathbb{R}^{2n}$ (as in (76)). The product rule is

$$(\theta, \boldsymbol{\eta})(\theta', \boldsymbol{\eta}') = (\theta + \theta' + \xi(\boldsymbol{\eta}, \boldsymbol{\eta}'), \boldsymbol{\eta} + \boldsymbol{\eta}'), \tag{193}$$

where, the *multiplier* ξ is given by

$$\xi(\boldsymbol{\eta}, \boldsymbol{\eta}') = \frac{1}{2}\, \boldsymbol{\eta}^\dagger \boldsymbol{\omega} \boldsymbol{\eta}' = \frac{1}{2}\, (\mathbf{p} \cdot \mathbf{q}' - \mathbf{q} \cdot \mathbf{p}'), \tag{194}$$

and once again, the matrix $\boldsymbol{\omega}$ is the inverse of J (see (87)). The multiplier ξ is precisely the quantity which is needed to realize $G_{WH}(n)$ as the *central extension* of the abelian group $\mathbb{R}^{2n}$. From the general theory of group representations on Hilbert space, it is known that each unitary irreducible representation (UIR) of $G_{WH}(n)$ is characterized by a non-zero real number, which we shall write as $\frac{1}{\hbar}$, and eventually identify $\hbar$ with Planck's constant (of course, for a specific value of it). Each UIR is carried by the Hilbert space $\mathfrak{H} = L^2(\mathbb{R}^n, d\mathbf{x})$ via the following unitary operators:

$$\begin{aligned}(U^\hbar(\theta, \boldsymbol{\eta})\psi)(\mathbf{x}) &= (\exp\left[\frac{i}{\hbar}\, \{\theta + \boldsymbol{\eta}^\dagger \boldsymbol{\omega} \widehat{\boldsymbol{\eta}}\}\right] \psi)(\mathbf{x}) \\ &= \exp\left[\frac{i}{\hbar}\left\{\theta + \mathbf{p} \cdot \mathbf{x} - \frac{1}{2}\mathbf{p} \cdot \mathbf{q}\right\}\right] \psi(\mathbf{x} - \mathbf{q}), \;\; \psi \in \mathfrak{H}.\end{aligned} \tag{195}$$

This shows that the $2n$ quantized (unbounded) operators, $\widehat{\eta}^i = \widehat{q}^i$ with $i = 1, 2, \ldots, n$ and $\widehat{\eta}^i = \widehat{p}_{i-n}$ with $i = n+1, n+2, \ldots, 2n$, which are the components of $\widehat{\boldsymbol{\eta}}$, along with the identity operator I on $\mathfrak{H}$, are the infinitesimal generators spanning the representation of the Lie algebra $\mathfrak{g}_{WH}(n)$ of the Weyl-Heisenberg group $G_{WH}(n)$. Since the representation (195) is irreducible, so also is the representation (192) of the Lie algebra. This is the precise mathematical sense in which we say that the algebra of Poisson brackets $\{\widehat{q}^i, \widehat{p}_j\} = \delta^i_j$ is irreducibly realized by the representation (192) of the CCR.

One could justifiably ask at this point, how many other Poisson brackets could be added to the set $\mathfrak{g}_{WH}(n)$ and the resulting enlarged algebra still be represented irreducibly on the same Hilbert space $\mathfrak{H}$. In other words, does there exist a larger algebra, containing $\mathfrak{g}_{WH}(n)$, which is also irreducibly represented on $\mathfrak{H} = L^2(\mathbb{R}^n, d\mathbf{x})$? To analyse this point further, let us look at functions $u \in \mathcal{C}^\infty(\mathbf{\Gamma})$ which are homogeneous, symmetric polynomials in the variables q^i and p_j of degree two. Any such polynomial can be written as:

$$u(\boldsymbol{\eta}) = \frac{1}{2} \sum_{i,j=1}^{2n} \eta^i U_{ij} \eta^j = \frac{1}{2} \boldsymbol{\eta}^\dagger U \boldsymbol{\eta}, \tag{196}$$

where the U_{ij} are the elements of a $2n \times 2n$ real, symmetric matrix U. Set

$$U = JX(u), \tag{197}$$

with $X(u) = -JU$, a $2n \times 2n$ real matrix satisfying

$$X(u) = JX(u)^T J. \tag{198}$$

It follows, therefore, that every such homogeneous symmetric polynomial u is characterized by a $2n \times 2n$ real matrix $X(u)$ satisfying (198), and conversely, every such matrix represents a homogeneous symmetric polynomial of degree two via

$$u(\boldsymbol{\eta}) = \frac{1}{2} \boldsymbol{\eta}^\dagger JX(u) \boldsymbol{\eta}. \tag{199}$$

Computing the Poisson bracket of two such polynomials u and v using (148), we easily see that

$$\{u, v\} = \frac{1}{2} \boldsymbol{\eta}^\dagger J[X(u), X(v)] \boldsymbol{\eta}, \quad \text{where,}$$
$$[X(u), X(v)] = X(u)X(v) - X(v)X(u). \tag{200}$$

In other words, the set of homogeneous, symmetric, quadratic polynomials constitutes a closed algebra under the Poisson bracket operation, which we denote by $\mathfrak{P}_2$, and the corresponding set of matrices $X(u)$ is closed under the bracket relation,

$$[X(u), X(v)] = X(\{u, v\}), \tag{201}$$

constituting thereby a matrix realization of the same algebra, $\mathfrak{P}_2$. Moreover, we also see that

$$\{\eta^i, u\} = (X(u)\boldsymbol{\eta})^i, \qquad i = 1, 2, \ldots, 2n, \tag{202}$$

or compactly,

$$\{\boldsymbol{\eta}, u\} = X(u)\boldsymbol{\eta}, \tag{203}$$

which can be thought of as giving the action of the Poisson algebra of quadratic polynomials on the Weyl-Heisenberg group $G_{WH}(n)$.

Going back now to the symplectic group $\mathrm{Sp}(2n, \mathbb{R})$, consider an element $S = e^{\varepsilon X}$, close to the identity, where $\varepsilon > 0$ and X is a $2n \times 2n$ real matrix. The fact that S can be written this way is guaranteed by the exponential mapping theorem for Lie groups. The defining condition $SJS^T = J$, for an element of $\mathrm{Sp}(2n, \mathbb{R})$, then implies,

$$(\mathbb{I}_{2n} + \varepsilon X)J(\mathbb{I}_{2n} + \varepsilon X)^T + \mathcal{O}(\varepsilon^3) = J.$$

Simplifying and dividing by ε,

$$XJ + JX^T + \varepsilon XJX^T + \mathcal{O}(\varepsilon^2) = 0.$$

Hence, letting $\varepsilon \to 0$, we find that

$$XJ + JX^T = 0 \quad \Rightarrow \quad X = JX^T J. \tag{204}$$

Thus, JX is a symmetric matrix and X a matrix of the type (198) with an associated second degree, homogeneous, symmetric polynomial:

$$X = X(u), \qquad u(\boldsymbol{\eta}) = \frac{1}{2}\boldsymbol{\eta}^\dagger JX\boldsymbol{\eta}. \tag{205}$$

On the other hand, the matrices X in $S = e^{\varepsilon X}$ constitute the Lie algebra $\mathrm{sp}(2n, \mathbb{R})$ of the Lie group $\mathrm{Sp}(2n, \mathbb{R})$, and thus we have established an algebraic isomorphism $\mathfrak{P}_2 \simeq \mathrm{sp}(2n, \mathbb{R})$. Moreover, the relations (200) and (203) together then constitute the Lie algebra of the *metaplectic group*, which is the semi-direct product $\mathrm{Mp}(2n, \mathbb{R}) = G_{WH}(n) \rtimes \mathrm{Sp}(2n, \mathbb{R})$. The Lie algebra, $\mathrm{mp}(2n, \mathbb{R})$, of this group consists, therefore, of all polynomials in the variables $q^i, p_i \;\; i = 1, 2, \ldots, n$ up to order two. The group $\mathrm{Mp}(2n, \mathbb{R})$ has elements $(\theta, \boldsymbol{\eta}, S)$ and the multiplication rule is:

$$(\theta, \boldsymbol{\eta}, S)(\theta', \boldsymbol{\eta}', S') = (\theta + \theta' + \xi(\boldsymbol{\eta}, S\boldsymbol{\eta}'), \boldsymbol{\eta} + S\boldsymbol{\eta}', SS'), \tag{206}$$

with the same multiplier ξ as in (194).

The metaplectic group has a UIR on the same space $\mathfrak{H}$, extending the representation of $U^\hbar$ of $G_{WH}(n)$ given in (195). We denote this representation again by $U^\hbar$ and see that since $(\theta, \boldsymbol{\eta}, S) = (\theta, \boldsymbol{\eta}, \mathbb{I}_{2n})(0, \mathbf{0}, S)$,

$$U^\hbar(\theta, \boldsymbol{\eta}, S) = U^\hbar(\theta, \boldsymbol{\eta})U^\hbar(S), \tag{207}$$

where for $S = e^{\varepsilon X(u)}$, the unitary operator $U^{\hbar}(S)$ can be shown to be

$$U^{\hbar}(S) = \exp\left[-\frac{i\varepsilon}{\hbar}\widehat{X}(u)\right], \qquad \widehat{X}(u) = -\frac{1}{2}\widehat{\boldsymbol{\eta}}^{\dagger} J X(u)\widehat{\boldsymbol{\eta}}. \tag{208}$$

Furthermore, using the unitarity of $U^{\hbar}(S)$, it is easily shown that

$$[\widehat{X}(u), \widehat{X}(v)] = i\hbar\, \widehat{X}(\{u, v\}), \tag{209}$$

that is, the quantization of $\boldsymbol{\eta}$ now extends to the second degree polynomials in the manner $u \to \widehat{X}(u) := \widehat{u}$. The self-adjoint operators $\widehat{\boldsymbol{\eta}}$ and $\widehat{X}(u)$ of the representation of the Lie algebra $\mathrm{mp}(2n, \mathbb{R})$, on the Hilbert space $\mathfrak{H}$, has the full set of commutation relations,

$$\begin{aligned} [\widehat{\boldsymbol{\eta}}, \widehat{X}(u)] &= i\hbar X(u)\widehat{\boldsymbol{\eta}}, \\ [\widehat{X}(u), \widehat{X}(v)] &= i\hbar\, \widehat{X}(\{u, v\}) = i\hbar\, [\widehat{X(u), X(v)}]. \end{aligned} \tag{210}$$

It was shown by Groenewold, and later by van Hove that this is the best one can do. In other words, it is not possible to find an algebra larger than $\mathrm{mp}(2n, \mathbb{R})$, which could also be irreducibly represented on $L^2(\mathbb{R}^n, d\mathbf{x})$. On the other hand, van Hove also showed that if we relax the irreducibility condition, then on $L^2(\boldsymbol{\Gamma}, d\nu)$, it is possible to represent the full Poisson algebra of $\mathcal{C}^{\infty}(\boldsymbol{\Gamma})$. This is the so-called *prequantization* result, to which we shall return later.

A number of other questions also arise in connection with the scheme of canonical quantization:

- Let $\mathfrak{Q}$ be the position space manifold of the classical system and q any point in it. Geometrically, the phase space of the system is the cotangent bundle $\boldsymbol{\Gamma} = T^*\mathfrak{Q}$. If $\mathfrak{Q}$ is linear, i.e., $\mathfrak{Q} \simeq \mathbb{R}^n$, then the replacement $q^i \to x^i,\ p_j \to -i\hbar \dfrac{\partial}{\partial x^j}$ works fine. But what if $\mathfrak{Q}$ is not a linear space?

- How do we quantize observables which involve higher powers of q^i, p_j, such as for example $f(q, p) = (q^i)^n (p_j)^m$, when $m + n \geq 3$?

- How should we quantize more general phase spaces, which are symplectic manifolds but not necessarily cotangent bundles?

In the next section we review two procedures which have been proposed to extend canonical quantization to provide, among others, the answers to these questions.

6.2 Segal and Borel quantization

A method for quantizing on an arbitrary configuration space manifold $\mathfrak{Q}$ was proposed by Segal, as a generalization of canonical quantization and very much within the same spirit. A group theoretical method was suggested by Mackey, within the context of the theory of *induced representations* of finite dimensional groups. A much more general method, combining the Segal and Mackey approaches, was later developed by Doebner *et al.*, under the name of *Borel quantization*, while a method using infinite dimensional diffeomorphism groups was suggested by Goldin. The relation to diffeomorphism groups was also noticed by Segal, who in fact in the same paper, lifted the theory to the cotangent bundle $T^*\mathfrak{Q}$ and thereby anticipated the theory of geometric quantization. Segal also pointed out that the number of inequivalent such quantizations was related to the first cohomology group of $\mathfrak{Q}$.

Briefly, the technique suggested by Segal is as follows: The configuration space $\mathfrak{Q}$ is an n-dimensional C^∞-manifold. Since on $\mathfrak{Q} = \mathbb{R}^n$, canonical quantization represents the classical position observables q^i as the operators $\widehat{q}^i$ of multiplication by the corresponding position variable, Segal generalized this idea and defined an entire class of observables of position using the smooth functions $f : \mathfrak{Q} \to \mathbb{R}$. Similarly, since canonical quantization on $\mathfrak{Q} = \mathbb{R}^n$ replaces the classical observables of momentum, p_i, by derivatives with respect to these variables, in Segal's scheme an entire family of quantized momentum observables is obtained by using the vector fields X of the manifold $\mathfrak{Q}$. In local coordinates,

$$X = \sum_{i=1}^{n} a^i(q) \frac{\partial}{\partial q^i},$$

for C^∞-functions $a^i : \mathfrak{Q} \to \mathbb{R}$. Quantization now involves first choosing a Borel measure μ on $\mathfrak{Q}$, locally equivalent to the Lebesgue measure on $\mathbb{R}^n$, and the Hilbert space $\mathfrak{H} = L^2(\mathfrak{Q}, d\mu)$. The quantum observables of position and momentum are then given by the mappings, $f \mapsto \widehat{q}(f)$ and $X \mapsto \widehat{p}(X)$, respectively, where on some suitable dense set of vectors $\psi \in \mathfrak{H}$,

$$(\widehat{q}(f)\psi)(q) = f(q)\psi(q), \qquad \widehat{p}(X) = -i\hbar(X + K_X), \tag{211}$$

the various quantities appearing above being defined as:

$$\begin{aligned}(X\psi)(q) &= \sum_{i=1}^{n} a^i(q) \frac{\partial}{\partial q^i}\psi(q), \\ (K_X\psi)(q) &= \frac{1}{2}(Xw)(q)\psi(q) + \frac{1}{2}\ell(X)(q)\psi(q),\end{aligned} \tag{212}$$

where,

$$\ell(X) = \sum_{i=1}^{n} \frac{\partial a^i}{\partial q^i}, \quad \text{and locally,} \quad d\mu(q) = w(q)\, dq^1\, dq^2 \ldots dq^n.$$

The local density function w for the measure $d\mu$ is also assumed to be smooth. The additional term K_X is needed to ensure self-adjointness of the operator $\widehat{p}(X)$. In terms of the Lie bracket $[X, Y] = X \circ Y - Y \circ X$ of the vector fields, one obtains for the quantized operators the following commutation relations, which clearly generalize the CCR (168):

$$\begin{aligned} [\widehat{p}(X), \widehat{p}(Y)] &= -i\hbar\widehat{p}([X, Y]) \\ [\widehat{p}(X), \widehat{q}(f)] &= -i\hbar\widehat{q}(Xf) \\ [\widehat{q}(f), \widehat{q}(g)] &= 0. \end{aligned} \tag{213}$$

As mentioned earlier, Segal actually suggested going over to the group of diffeomorphisms of $\mathfrak{Q}$ and its unitary representations, to attend to domain questions associated to $\widehat{q}(f), \widehat{p}(X)$ and then gave a classification of possible unitarily inequivalent quantizations in these terms. Note also, that the Segal quantization method is based on configuration space, rather than on phase space. As such, the primary preoccupation here is to generalize the method of canonical quantization. On the other hand, as we said before, Segal also extended the theory to phase space and in that sense, Segal's method leads to similar results as other mehods that we shall study, on the representations of the Poisson algebra on Hilbert space.

Rather than discussing further details of Segal's method, it is possibly more instructive to look at the related, and certainly more assiduously studied, method of *Borel quantization.* The method of Borel quantization focuses on both the geometric and measure theoretic properties of the configuration space manifold $\mathfrak{Q}$ as well as attempting to incorporate internal symmetries by lifting $\mathfrak{Q}$ to a complex hermitian vector bundle with connection and curvature, compatible with the hermitian structure.

In order to motivate this technique, let us go back to the Weyl-Heisenberg group $G_{WH}(3)$, for three degrees of freedom. From (195) we get the UIR, this time on $\mathfrak{H} = L^2(\mathbb{R}^3, d\mathbf{x})$:

$$(U^\hbar(\theta, \mathbf{q}, \mathbf{p})\psi)(\mathbf{x}) = \exp\left[\frac{i}{\hbar}\left\{\theta + \mathbf{p}\cdot\mathbf{x} - \frac{1}{2}\mathbf{p}\cdot\mathbf{q}\right\}\right]\psi(\mathbf{x} - \mathbf{q}), \qquad \psi \in \mathfrak{H}.$$

Note that the Hilbert space is built out of configuration space functions, $\mathfrak{H} = L^2(\mathfrak{Q}, d\mathbf{x})$, where now $\mathfrak{Q} = \mathbb{R}^3$. Let us enlarge the group to include rigid rotations of $\mathbb{R}^3$ as well. The rotation group is $SO(3)$ (3×3 real orthogonal matrices), but in order to have a theory which will include quantum particles of both integral and half-integral spins, it is necessary to work with the group $SU(2)$ (2×2 complex, unitary matrices of unit determinant), which is simply connected and is the two-fold covering group of $SO(3)$. The homomorphism $R : SU(2) \to SO(3)$ has the kernel $\{\mathbb{I}_2, -\mathbb{I}_2\}$ and is given by,

$$R(\rho)_{kj} = \frac{1}{2}\text{tr}\,[\rho\sigma_j\rho^{-1}\sigma_k], \qquad k, j = 1, 2, 3, \;\; \rho \in SU(2), \tag{214}$$

where the σ_k are the usual Pauli matrices:

$$\sigma_1 = \begin{pmatrix} 0 & 1 \\ 1 & 0 \end{pmatrix}, \quad \sigma_2 = \begin{pmatrix} 0 & -i \\ i & 0 \end{pmatrix}, \quad \sigma_3 = \begin{pmatrix} 1 & 0 \\ 0 & -1 \end{pmatrix}.$$

The group $SU(2)$ has an action on the Weyl-Heisenberg group $G_{WH}(3)$, and the extended group $\mathcal{G}_{orth}$, called the *orthochronous Galilei group*, obeys the multiplication rule:

$$\begin{aligned}(\theta, \mathbf{q}, \mathbf{p}, \rho)(\theta', \mathbf{q}', \mathbf{p}', \rho') = \; & (\theta + \theta' + \xi[(\mathbf{q}, \mathbf{p});\; (R(\rho)\mathbf{q}', R(\rho)\mathbf{p}')], \\ & \mathbf{q} + R(\rho)\mathbf{q}', \mathbf{p} + R(\rho)\mathbf{p}', \rho\rho'). \end{aligned} \tag{215}$$

It is clear that $\mathcal{G}_{orth}$ is a subgroup of the metaplectic group $\mathrm{Mp}(6, \mathbb{R})$ and is in fact the semi-direct product:

$$\mathcal{G}_{orth} = G_{WH}(3) \rtimes SU(2).$$

Note also, that the Weyl-Heisenberg group $G_{WH}(3)$ now appears as the subgroup of $\mathcal{G}_{orth}$, consisting of elements of the type $(\theta, \mathbf{q}, \mathbf{p}, \mathbb{I}_2)$, where $\mathbb{I}_2$ is the identity element of $SU(2)$.

Unitary irreducible representations of $SU(2)$ are carried by the so-called spinor spaces $\mathfrak{K}^j$, which are isomorphic to $\mathbb{C}^{2j+1}$, $\;j = 0, 1/2, 1, 3/2, \ldots$. We shall denote these representations by $\mathcal{D}^j$ and assume them to be carried by $\mathbb{C}^{2j+1}$, which is a $(2j+1)$-dimensional Hilbert space equipped with the usual scalar product

$$\langle \mathbf{z} | \mathbf{z}' \rangle = \sum_{m=-j}^{j} \overline{z}_m z'_m, \qquad \mathbf{z} = (z_{-j}, z_{-j+1}, \ldots, z_{j-1}, z_j), \;\; \text{etc.}, \;\; \in \mathbb{C}^{2j+1}.$$

Using $\mathcal{D}^j$ we may construct a UIR of $\mathcal{G}_{orth}$ on the Hilbert space $\mathfrak{H} = \mathbb{C}^{2j+1} \otimes L^2(\mathbf{\Omega}, d\mu) = \mathbb{C}^{2j+1} \otimes L^2(\mathbb{R}^3, d\mathbf{x})$. Denoting this representation by $U^{\hbar,\, j}$,

$$(U^{\hbar,\, j}(\theta, \mathbf{q}, \mathbf{p}, \rho)\psi)(\mathbf{x}) = \exp \frac{i}{\hbar}\left[\theta + \mathbf{p}\cdot\mathbf{x} - \frac{1}{2}\mathbf{p}\cdot\mathbf{q}\right] \mathcal{D}^j(\rho)\psi(R(\rho)^{-1}(\mathbf{x} - \mathbf{q})). \tag{216}$$

Note that the elements ψ of the Hilbert space $\mathfrak{H}$ are $(2j+1)$-component vector valued funcions and the $\mathcal{D}^j(\rho)$ are $(2j+1)\times(2j+1)$ matrices acting on them:

$$\psi(\mathbf{x}) = \{\psi_m(\mathbf{x})\}_{m=-j}^{j}, \qquad (\mathcal{D}^j(\rho)\psi)_m(\mathbf{x}) = \sum_{m'=-j}^{j} \mathcal{D}^j(\rho)_{mm'}\psi_{m'}(\mathbf{x}).$$

Denote the σ-algebra of the Borel sets of $\mathbf{\Omega} = \mathbb{R}^3$ by $\mathcal{B}(\mathbb{R}^3)$. This is the σ-algebra generated by the open sets of $\mathbb{R}^3$. Corresponding to each $\Delta \in \mathcal{B}(\mathbb{R}^3)$, define an operator $P(\Delta)$ on $\mathfrak{H}$:

$$(P(\Delta)\psi)(\mathbf{x}) = \chi_\Delta(\mathbf{x})\psi(\mathbf{x}), \qquad \chi_\Delta(\mathbf{x}) = \begin{cases} 1, \text{ if } \mathbf{x} \in \Delta, \\ 0, \text{ otherwise.} \end{cases} \tag{217}$$

This is a projection operator, $P(\Delta) = P(\Delta)^* = P(\Delta)^2$, and has the following measure theoretic properties:

$$\begin{aligned} P(\emptyset) &= 0, & P(\mathbb{R}^3) &= I \\ P(\cup_{i\in J}\Delta_i) &= \sum_{i\in J} P(\Delta_i), & \Delta_i \cap \Delta_j &= \emptyset, \;\; i \neq j, \end{aligned} \tag{218}$$

where J is a discrete index set and the convergence of the sum is meant in the weak sense. Such a set of projection operators $P(\Delta)$, $\Delta \in \mathcal{B}(\mathbb{R}^3)$, is called a (normalized) *projection valued (PV)-measure.* Note that, for any $\psi \in \mathfrak{H}$,

$$\begin{aligned} \mu_\psi(\Delta) &= \langle\psi|P(\Delta)\psi\rangle, \\ &= \int_\Delta \|\psi(\mathbf{x})\|^2 \, d\mathbf{x}, \qquad \Delta \in \mathcal{B}(\mathbb{R}^3), \end{aligned} \tag{219}$$

defines a real measure, absolutely continuous with respect to the Lebesgue measure. In fact, if $\{\phi_n\}_{n=1}^\infty$ is an orthonormal basis of $\mathfrak{H}$, then

$$\nu(\Delta) = \sum_{n=1}^\infty \frac{1}{2^n}\langle\phi_n|P(\Delta)\phi_n\rangle, \tag{220}$$

defines a probability measure on $\mathbb{R}^3$, which is *equivalent* to the Lebesgue measure – i.e., there exists a measurable function $f(\mathbf{x}) > 0$, almost everywhere, such that,

$$\begin{aligned} \nu(\Delta) &= \int_\Delta f(\mathbf{x})\, d\mathbf{x}, \qquad \Delta \in \mathcal{B}(\mathbb{R}^3), \\ \nu(\mathbb{R}^3) &= \int_{\mathbb{R}^3} f(\mathbf{x})\, d\mathbf{x} = 1. \end{aligned} \tag{221}$$

In fact,

$$f(\mathbf{x}) = \sum_{i=1}^{\infty} \frac{1}{2^n} ||\phi_n(\mathbf{x})||^2.$$

The PV-measure P has the following crucial property with respect to the representation $U^{\hbar,\, j}$:

$$U^{\hbar,\, j}(\theta, \mathbf{q}, \mathbf{p}, \rho) P(\Delta) U^{\hbar,\, j}(\theta, \mathbf{q}, \mathbf{p}, \rho)^* = P(R(\rho)\Delta + \mathbf{q}), \qquad \Delta \in \mathcal{B}(\mathbb{R}^3, \tag{222}$$

where $R(\rho)\Delta + \mathbf{q}$ is the *translate* of the set Δ under the action of the group element $(\theta, \mathbf{q}, \mathbf{p}, \rho)$, i.e.,

$$R(\rho)\Delta + \mathbf{q} = \{\mathbf{y} = R(\rho)\mathbf{x} + \mathbf{q} \mid \mathbf{x} \in \Delta\}.$$

Consequently, we shall write $R(\rho)\Delta + \mathbf{q}$ as $(\theta, \mathbf{q}, \mathbf{p}, \rho)\Delta$ and the relation (222) as

$$U^{\hbar,\, j}(\theta, \mathbf{q}, \mathbf{p}, \rho) P(\Delta) U^{\hbar,\, j}(\theta, \mathbf{q}, \mathbf{p}, \rho)^* = P((\theta, \mathbf{q}, \mathbf{p}, \rho)\Delta). \tag{223}$$

In this context, it should also be noted that the configuration space $\mathfrak{Q}$, on which P is defined, can be identified with the coset space,

$$\mathfrak{Q} = \mathcal{G}_{orth}/\Theta \otimes (\mathcal{V} \rtimes SU(2)), \tag{224}$$

where $\Theta \otimes (\mathcal{V} \rtimes SU(2))$ is the subgroup of $\mathcal{G}_{orth}$ consisting of elements of the type $(\theta, \mathbf{0}, \mathbf{p}, \rho)$, i.e., phase changes, velocity transformations ($\mathbf{p} = m\mathbf{v}$), and rotations. The pair $\{U^{\hbar,\, j}, P\}$, satisfying (222) is called a *system of imprimitivity* and according to Mackey's theory of induced representations of groups, the pair determines the representation $U^{\hbar,\, j}$ as an *induced representation*, and the structure of the Hilbert space $\mathfrak{H}$ as an L^2-space over $\mathfrak{Q}$. Note also, that the space $\mathfrak{H} = \mathbb{C}^{2j+1} \otimes L^2(\mathfrak{Q}, \mathbf{x})$ is suggestive of the existence of a $\mathbb{C}^{2j+1}$-complex vector bundle over $\mathfrak{Q}$ and the possible set of such bundles should correspond to UIRs of $\mathcal{G}_{orth}$. Indeed, the representation $U^{\hbar,\, j}$ is intimately related to a $\mathbb{C}^{2j+1}$-associated bundle (associated to the principal bundle with total space $\mathcal{G}_{orth}$ and structure group $\Theta \otimes (\mathcal{V} \rtimes SU(2))$.

The projection operators $P(\Delta)$ also describe localization properties of the system. If the system is in a state given by the wave function $\psi \in \mathfrak{H}$ (note this means that $||\psi|| = 1$), then

$$\langle \psi | P(\Delta) \psi \rangle = \int_{\Delta} ||\psi(\mathbf{x})||^2 \, d\mathbf{x} \tag{225}$$

is the probability of finding the system localized in the volume Δ. The imprimitivity condition (222) is then simply a covariance condition for this probability under phase changes, translations in position and velocity and under rotations.

Using the measure theoretic properties of P, we can define an entire commutative algebra of quantized observables. Note first, that the three position operators $\widehat{q}^i$ are defined as:

$$\widehat{q}^i = \int_{\mathbb{R}^3} x_i \, dP(\mathbf{x}), \qquad i = 1, 2, 3, \tag{226}$$

where the integral above is to be interpreted in the sense:

$$\langle \psi | \widehat{q}^i \psi \rangle = \int_{\mathbb{R}^3} x_i \, d\mu_\psi(\mathbf{x}) = \int_{\mathbb{R}^3} x_i \|\psi(\mathbf{x})\|^2 \, d\mathbf{x}, \tag{227}$$

μ_ψ being the measure defined in (219), and the ψ being chosen from a dense set such that

$$\int_{\mathbb{R}^3} x_i \|\psi(\mathbf{x})\|^2 \, d\mathbf{x} < \infty.$$

Next, if $f : \mathbf{\Omega} \to \mathbb{R}$ is a sufficiently well-behaved function, then the operator of multiplication $\widehat{q}(f)$:

$$\widehat{q}(f)\psi(\mathbf{x}) = f(\mathbf{x})\psi(\mathbf{x}), \qquad \psi \in \mathfrak{H}, \tag{228}$$

is in fact the operator,

$$\widehat{q}(f) = \int_{\mathbb{R}^3} f(\mathbf{x}) \, dP(\mathbf{x}), \tag{229}$$

which is the quantized version of f, given by Segal's procedure. The operators $\widehat{q}(f), \widehat{q}(g), \ldots$, all commute and constitute the *algebra of quantized position observables.*

Consider now the three momentum operators $\widehat{p}^i = -i\hbar \dfrac{\partial}{\partial x^i}$. Their exponentiated versions:

$$\exp\left[-\frac{i}{\hbar}\widehat{p}^i\right] = \exp\left[\frac{\partial}{\partial x^i}\right], \qquad i = 1, 2, 3, \tag{230}$$

generate *shifts*:

$$(\exp\left[\frac{\partial}{\partial x^i}\right]\psi)(\mathbf{x}) = \psi(\mathbf{x} + \mathbf{e}_i), \tag{231}$$

$\mathbf{e}_i$ being the i-th unit vector in the canonical basis. (The above relation is made obvious if one thinks in terms of the usual Taylor series expansion:

$$f(x_0 + h) = \sum_{n=0}^{\infty} \frac{h^n}{n!} f^{(n)}(x_0) = \sum_{n=0}^{\infty} \frac{h^n}{n!} \left(\frac{\partial}{\partial x}\right)^n f(x)|_{x=x_0} \quad).$$

The vector field X_i, such that $X_i(\mathbf{x}) = \dfrac{\partial}{\partial x^i}$, is thus seen to give a diffeomorphism of the configuration space. Denoting this diffeomorphism by ϕ^{X_i}, we have,

$$\phi^{X_i}(\mathbf{x}) = \mathbf{x} + \mathbf{e}_i, \qquad \mathbf{x} \in \mathbb{R}^3. \tag{232}$$

Next, we define a one-parameter set of such diffeomorphisms, $\phi_s^{X_i}, \quad s \in \mathbb{R}$, by

$$\phi_s^{X_i}(\mathbf{x}) = \mathbf{x} + s\mathbf{e}_i. \tag{233}$$

These form a one-parameter commutative group under composition:

$$\begin{aligned} \phi_{s_1}^{X_i} \circ \phi_{s_2}^{X_i} &= \phi_{s_1+s_2}^{X_i}, \\ \phi_0^{X_i} = \mathbb{I}, \qquad & \phi_{-s}^{X_i} = [\phi_s^{X_i}]^{-1}. \end{aligned} \tag{234}$$

Furthermore,

$$\frac{d}{ds}\phi_s^{X_i}(\mathbf{x})|_{s=0} = \mathbf{e}_i. \tag{235}$$

But now this $\mathbf{e}_i$ is to be interpreted as a tangent vector to $\mathbb{R}^3$ at $\mathbf{x}$ and hence we make the identification:

$$\frac{d}{ds}\phi_s^{X_i}(\mathbf{x})|_{s=0} = X_i(\mathbf{x}). \tag{236}$$

In other words, X_i is the infinitesimal generator of the one-parameter group of diffeomorphisms $\phi_s^{X_i}$. On the other hand, since $-i\hbar\dfrac{\partial}{\partial x^i}$ is the quantized momentum operator on $\mathfrak{H}$, the quantization condition $\widehat{p}(X_i) = -i\hbar X_i$ is seen to emerge in this case. Finally, note that in view of the imprimitivity condition (222), we obtain

$$V(\phi_s^{X_i})P(\Delta)V(\phi_{-s}^{X_i}) = P(\Delta + s\mathbf{e}_i), \tag{237}$$

where we have written

$$V(\phi_s^{X_i}) = U^{\hbar,\, j}(0, s\mathbf{e}_i, \mathbf{0}, \mathbb{I}_2) = \exp\left[-\frac{i}{\hbar}s\widehat{p}(X_i)\right]. \tag{238}$$

We now generalize from this example. Consider a one-parameter family of diffeomorphisms $s \mapsto \phi_s$ of $\mathbb{R}^3$, which are suficiently well behaved in the parameter $s \in \mathbb{R}$, in an appropriate sense. Then,

$$\frac{d}{ds}\phi_s(\mathbf{x})|_{s=0} = X(\mathbf{x}), \tag{239}$$

defines a vector field X. If we were working on the Hilbert space $\mathfrak{H} = L^2(\mathbb{R}^3, d\mathbf{x})$, its quantized form $\widehat{p}(X)$, according to Segal's procedure would be a general momentum observable acting on $\psi \in L^2(\mathbb{R}^3, d\mathbf{x})$ in the manner

$$(\widehat{p}(X)\psi)(\mathbf{x}) = -i\hbar(X\psi)(\mathbf{x}) - \frac{i\hbar}{2}\frac{\partial a^i}{\partial x^i}(\mathbf{x})\psi(\mathbf{x}), \quad \text{where,} \quad X(\mathbf{x}) = \sum_{i=1}^{n} a^i(\mathbf{x})\frac{\partial}{\partial x^i}, \tag{240}$$

and together, the set of all such momentum observables then form an algebra under the bracket operation (see (213)):

$$[\widehat{p}(X),\, \widehat{p}(Y)] = -i\hbar\widehat{p}([X,\, Y]). \tag{241}$$

However, since in general we want to work on the Hilbert space $\mathfrak{H} = \mathbb{C}^{2j+1} \otimes L^2(\mathbb{R}^3, d\mathbf{x})$, additional terms will appear in the expression for $\widehat{p}(X)$ in (240) to reflect the *internal degrees of freedom*. We write $\phi_s = \phi_s^X$, to indicate the generator, and define the transformed Borel sets

$$\phi_s^X(\Delta) = \{\phi_s^X(\mathbf{x}) \mid \mathbf{x} \in \Delta\}, \tag{242}$$

for each $\Delta \in \mathcal{B}(\mathbb{R}^3)$. It is then easily checked that for each $s \in \mathbb{R}$,

$$V(\phi_s^X) = \exp[-i\hbar s\widehat{p}(X)], \tag{243}$$

defines a unitary operator on $\mathfrak{H}$, such that $\{V, P\}$ is a system of imprimitivity in the sense:

$$V(\phi_s^X)P(\Delta)V(\phi_{-s}^X) = P(\phi_s^X(\Delta)). \tag{244}$$

Now considering *all* such one-parameter diffeomorphism groups and their associated systems of imprimitivity, we find that the collective system is certainly irreducibly realized on $\mathfrak{H} = L^2(\mathbb{R}^3, d\mathbf{x})$, and we would like the same to be true for representations on $\mathfrak{H} = \mathbb{C}^{2j+1} \otimes L^2(\mathbb{R}^3, d\mathbf{x})$.

The aim of Borel quantization is to construct such irreducible systems on arbitrary configuration space manifolds $\mathfrak{Q}$. It is clear that the problem is related to that of finding irreducible representations of the diffeomorphism group, Diff($\mathfrak{Q}$), which admit systems of imprimitivity based on the Borel sets of $\mathfrak{Q}$.

6.3 Synopsis of Borel quantization

Let us briefly outline the method of Borel quantization, along with some of its technical details. Let $\mathfrak{Q}$ be a configuration space manifold, of dimension n, μ a smooth measure on $\mathfrak{Q}$ (i.e., locally equivalent to the Lebesgue measure on $\mathbb{R}^n$) and let $\widetilde{\mathfrak{H}} = \mathbb{C}^n \otimes L^2(\mathbf{\Omega}, d\mu)$, where $k \geq 1$ is an integer. Let $\widetilde{P}$ be the PV-measure on $\widetilde{\mathfrak{H}}$:

$$(\widetilde{P}\widetilde{\psi})(x) = \chi_E(x)\widetilde{\psi}(x), \qquad \widetilde{\psi} \in \widetilde{\mathfrak{H}}, \qquad E \in \mathcal{B}(\mathbf{\Omega}), \tag{245}$$

χ_E being the characteristic function of the set E and $\mathcal{B}(\mathbf{\Omega})$ denoting the set of all Borel sets of $\mathbf{\Omega}$. Now let $\mathfrak{H}$ be another Hilbert space and P a PV-measure on it (also defined over $\mathcal{B}(\mathbf{\Omega})$).

Definition 6.1 *The pair $\{\mathfrak{H}, P\}$ is called a k-homogeneous localized quantum system if and only if it is unitarily equivalent to $\{\widetilde{\mathfrak{H}}, \widetilde{P}\}$, i.e., iff there exists a unitary map $W : \mathfrak{H} \longrightarrow \widetilde{\mathfrak{H}}$ such that*

$$WP(E)W^{-1} = \widetilde{P}(E), \qquad E \in \mathcal{B}(\mathbf{\Omega}). \tag{246}$$

Let $f \in \mathcal{C}^\infty(\mathbf{\Omega}) =$ (space of infinitely differentiable, real-valued functions on $\mathbf{\Omega}$).

Definition 6.2 *Let $\{\mathfrak{H}, P\}$ be a k-homogeneous localized quantum system. The self-adjoint operator,*

$$\widehat{q}(f) = \int_{\mathbf{\Omega}} f(x)\, dP_x, \tag{247}$$

defined on the domain,

$$\mathcal{D}(\widehat{q}(f)) = \{\psi \in \mathfrak{H} \mid \int_{\mathbf{\Omega}} |f(x)|^2\, d\langle\psi|P_x\psi\rangle < \infty\},$$

is called a generalized position operator.

Note that under the isometry (246), $\widehat{q}(f)$ becomes the operator of multiplication by f on $\widetilde{\mathfrak{H}}$. The following properties of these operators are easily verified:

1. $\widehat{q}(f)$ is a bounded operator if and only if f is a bounded function.

2. $\widehat{q}(f) = 0$ if and only if $f = 0$.

3. $\widehat{q}(\alpha f) = \alpha\widehat{q}(f)$, for $\alpha \in \mathbb{R}$.

4. $\widehat{q}(f+g) \supseteq \widehat{q}(f) + \widehat{q}(g)$ and $\mathcal{D}(\widehat{q}(f)) + \widehat{q}(g)) = \mathcal{D}(\widehat{q}(f+g)) \bigcap \mathcal{D}(\widehat{q}(f))$.

5. $\widehat{q}(f \cdot g)) \supseteq \widehat{q}(f)\, \widehat{q}(g)$ and $\mathcal{D}(\widehat{q}(f)\, \widehat{q}(g)) = \mathcal{D}(\widehat{q}(f \cdot g)) \bigcap \mathcal{D}(\widehat{q}(f))$.

We had mentioned earlier the notion of a *shift* on the manifold $\mathbf{\Omega}$. This is a one parameter group of diffeomorphisms: $\phi_s : \mathbf{\Omega} \longrightarrow \mathbf{\Omega}, \quad \phi_{s_2} \circ \phi_{s_1} =$

$\phi_{s_1+s_2}$, $s, s_1, s_2 \in \mathbb{R}$, ϕ_0 being the identity map. Each such shift defines a *complete vector field*,

$$X := \frac{d}{ds}\phi_s|_{s=0}, \tag{248}$$

and conversely, every such vector field X gives rise to a shift ϕ_s^X, called the *flow* of the vector field:

$$\phi_s^X(x) = e^{sX(x)}\phi(x). \tag{249}$$

There is a natural action of the shifts on Borel sets $E \subset \mathfrak{Q}$,

$$E \longmapsto \phi_s^X(E) = \{\phi_s^X(x) \mid x \in E\}. \tag{250}$$

Since ϕ_s^X is smooth, the resulting set $\phi_s^X(E)$ is also a Borel set. We want to represent the shifts ϕ_s^X on $\widetilde{\mathfrak{H}}$ as one-parameter unitary groups on Hilbert spaces $\mathfrak{H}$. Let $\mathcal{U}(\mathfrak{H})$ denote the set of all unitary operators on $\mathfrak{H}$ and $\mathfrak{X}_c(\mathfrak{Q})$ the set of all complete vector fields of the manifold $\mathfrak{Q}$.

Definition 6.3 *Let $\{\mathfrak{H}, P\}$ be a quantum system localized on $\mathfrak{Q}$. A map*

$$V : \phi_s^X \longmapsto V(\phi_s^X) \in \mathcal{U}(\mathfrak{H}), \tag{251}$$

is called a shift of the localized quantum system if, for all $X \in \mathfrak{X}_c(\mathfrak{Q})$, the map $s \longmapsto V(\phi_s^X)$ gives a strongly continuous representation of the additive group of $\mathbb{R}$ and $\{V(\phi_s^X), P\}$ is a system of imprimitivity with respect to the group of real numbers $\mathbb{R}$ and the Borel $\mathbb{R}$-space $\mathfrak{Q}$ with group action ϕ_s^X, i.e.,

$$V(\phi_s^X)P(E)V(\phi_{-s}^X) = P(\phi_s^X(E)). \tag{252}$$

The triple $\{\mathfrak{H}, P, V\}$ is called a localized quantum system with shifts.

Two localized quantum systems with shifts, $\{\mathfrak{H}_j, P_j, V_j\}$, $j = 1, 2$, are said to be unitarily equivalenty if there exists a unitary map $W : \mathfrak{H}_1 \longrightarrow \mathfrak{H}_2$, such that $WP_1(E)W^{-1} = P_2(E)$, $E \in \mathcal{B}(\mathfrak{Q})$ and $WV_1(\phi_s^X)W^{-1} = V_2(\phi_s^X)$, $x \in \mathfrak{X}_c(\mathfrak{Q})$, $s \in \mathbb{R}$. The map $\widehat{p} : \mathfrak{X}_c(\mathfrak{Q}) \longrightarrow \mathcal{S}(\mathfrak{H})$, where $\widehat{p}(X)$ is defined via Stone's theorem as the infinitesimal generator of

$$V(\phi_s^X) = \exp[\frac{i}{\hbar}s\widehat{p}(X)], \tag{253}$$

is called the *kinematical momentum* of $\{\mathfrak{H}, P, V\}$.

The imprimitivity relation (252) has the following important consequences.

Lemma 6.4 *Let $\{\mathfrak{H}, P, V\}$ be a k-homogeneous localized quantum system with shifts. Then*

$$V(\phi_s^X)\widehat{q}(f)V(\phi_{-s}^X) = \widehat{q}(f \circ \phi_s^X). \tag{254}$$

A k-homogeneous quantum system with shifts $\{\mathfrak{H}, P, V\}$, *is unitarily equivalent to* $\{\widetilde{\mathfrak{H}}, \widetilde{P}, \widetilde{V}\}$, *with* $\widetilde{\mathfrak{H}}$ *and* $\widetilde{P}$ *as in (245).*

The representation $\widetilde{V}$ acquires a very specific form. To understand it we need the concept of a cocycle. Let G be a locally compact group, H a standard Borel group, X a Borel G-space with group action $x \longmapsto gx$ and $[\nu]$ a G-invariant measure class on X. (This means that if ν is any measure in the class, then so also is ν_g, where $\nu_g(E) = \nu(gE)$, for all $E \in \mathcal{B}(\mathbf{\Omega})$.)

A Borel measurable map $\xi : G \times X \longrightarrow H$ is called a *cocycle* of G, relative to the measure class $[\nu]$ on X, with values in H if,

$$\xi(e, x) = 1,$$
$$\xi(g_1 g_2, x) = = \xi(g_1, g_2 x)\; \xi(g_2, x), \tag{255}$$

for $[\nu]$−almost all $x \in X$ and almost all (with respect to the Haar measure) $g_1, g_2 \in G$ (e is the identity element of G). Two cocycles ξ_1 and ξ_2 are said to be *cohomologous* or *equivalent* if there exists a Borel function $\zeta : X \longrightarrow H$, such that,

$$\xi_2(g, x) = \zeta(gx)\; \xi_1(g, x)\; \zeta(x)^{-1}$$

for almost all $g \in G$ and $x \in X$. The equivalence classes $[\xi]$ are called *cohomology classes of cocycles.* The following classification theorem for localized quantum systems then holds.

Theorem 6.5 *Any k-homogeneous quantum system* $\{\mathfrak{H}, P, V\}$ *on* $\mathbf{\Omega}$, *with shifts, is unitarily equivalent to a canonical representation on* $\{\widetilde{\mathfrak{H}}, \widetilde{P}, \widetilde{V}\}$, *with* $\widetilde{\mathfrak{H}} = \mathbb{C} \otimes L^2(\mathbf{\Omega}, d\mu)$, *for some smooth measure* μ *on* $\mathbf{\Omega}$,

$$(\widetilde{P}(E)\widetilde{\psi})(x) = \chi_E(x)\widetilde{\psi}(x),$$

for all $\widetilde{\phi} \in \widetilde{\mathfrak{H}}$ *and all* $E \in \mathcal{B}(\mathbf{\Omega})$, *and*

$$(V(\phi^X_s)\widetilde{\psi})(x) = \xi^X(s, \phi^X_{-s}(x))\, \sqrt{\lambda(\phi^X_s, \phi^X_{-s}(x))}\; \widetilde{\psi}(\phi^X_{-s}(x)), \tag{256}$$

for all $\widetilde{\psi} \in \widetilde{\mathfrak{H}}$ *and all* $X \in \mathfrak{X}_c(\mathbf{\Omega})$, *where* ξ^X *is a cocycle of the abelian group* $\mathbb{R}$ *(relative to the class of smooth measures on* $\mathbf{\Omega}$*), having values in* $\mathcal{U}(k)$ *(the group of* $k \times k$ *unitary matrices) and* λ *is the unique smooth Radon-Nikodym derivative,*

$$\lambda(\phi^X_s, x) = \frac{d\mu_{\phi^X_s}}{d\mu}(x).$$

Moreover, equivalence classes of k-homogeneous localized quantum systems are in one-to-one correspondence with equivalence classes of cocycle sets

$[\{\xi^X\}_{X\in\mathfrak{X}_c(\mathfrak{Q})}]$, *where*

$$\{\xi_1^X\}_{X\in\mathfrak{X}_c(\mathfrak{Q})} \sim \{\xi_2^X\}_{X\in\mathfrak{X}_c(\mathfrak{Q})}$$

if there exists a Borel function $\zeta : \mathfrak{Q} \longrightarrow \mathcal{U}(k)$, *such that, for all* $X \in \mathfrak{X}_c(\mathfrak{Q})$, $s \in \mathbb{R}$ *and* $x \in \mathfrak{Q}$,

$$\xi_2^X(s,x) = \zeta(\phi_s^X(x))\, \xi_1^X(s,x)\, \zeta(x)^{-1}.$$

Differentiating (256)with respect to s, using (253), and then setting $s = 0$, we obtain,

$$\widehat{p}(X)\widetilde{\psi} = -i\hbar\, \mathfrak{L}_X\widetilde{\psi} - \frac{i\hbar}{2}\,\mathrm{div}_\nu(X)\widetilde{\psi} + \omega(X)\widetilde{\psi}, \tag{257}$$

where $\mathfrak{L}_X\widetilde{\psi}$ is the Lie derivative of $\widetilde{\psi}$ along X and,

$$\frac{1}{2}\mathrm{div}_\nu(X)(x) = \frac{d}{ds}\sqrt{\lambda(\phi_s^X, \phi_{-s}^X(x)}|_{s=o}$$

$$\omega(X)(x) = -i\hbar\,\frac{d}{ds}\,\xi^X(s,\phi_{-s}^X(x))|_{s=0}. \tag{258}$$

The first two terms in (257) are linear in X. It is now possible to show that the following commutation relations hold:

$$[\widehat{q}(f),\widehat{q}(g)] = 0,$$
$$[\widehat{p}(X),\widehat{q}(f)] = -i\hbar\,\widehat{q}(\mathfrak{L}_X f),$$
$$[\widehat{p}(X),\widehat{p}(Y)] = -i\hbar\,\widehat{p}([X,Y]) - i\hbar\,\Omega(X,Y), \tag{259}$$

for all $f, g \in \mathcal{C}^\infty$, $X, Y \in \mathfrak{X}_c(\mathfrak{Q})$, and where,

$$\Omega(X,Y) = -i\hbar\,[\omega(X),\omega(Y)] + \mathfrak{L}_X\omega(Y) - \mathfrak{L}_Y\omega(X) - \omega([X,Y]). \tag{260}$$

The two-form Ω and the one-form ω on $\mathfrak{Q}$ are related in the same way as the curvature two-form $\frac{i}{\hbar}\,\Omega$ of a $\mathbb{C}^1$-bundle and its connection form $\frac{i}{\hbar}\,\omega(X)$. Indeed, one can show that if D is the covariant derivative defined by the connection, then $D\Omega = 0$, which is the *Bianchi identity*.

Definition 6.6 *Let* $\{\mathfrak{H}, P, V\}$ *be a* k*-homogeneous localized quantum system with shifts on* $\mathfrak{Q}$ *and* Ω *a differential two-form on* $\mathfrak{Q}$ *with values in* $\mathcal{S}(\mathbb{C}^k)$ *(set of all* $k \times k$ *Hermitian matrices). The kinematical momentum* $\widehat{p}$ *is called* Ω*-compatible if in a canonical representation* $\{\widetilde{\mathfrak{H}}, \widetilde{P}, \widetilde{V}\}$, *the associated kinematical momenta* $\widetilde{p}$ *satisfy*

$$[\widetilde{p}(X),\widetilde{p}(Y)]\widetilde{\psi} = -i\hbar\,(\widetilde{p}([X,Y])\widetilde{\psi} + \Omega(X,Y)\widetilde{\psi}). \tag{261}$$

In this case, the quadruple $\{\mathfrak{H}, \widehat{q}, \widehat{p}, \Omega\}$ *is called an* Ω*-compatible* k*-Borel kinematics.*

In order to arrive at a classification theory of localized quantum systems, we first impose some additional smoothness conditions. An Ω-compatible k-quantum Borel kinematics $\{\mathfrak{H}, \widehat{q}, \widehat{p}, \Omega\}$ is said to be *differentiable* if it is equivalent to $\{\widetilde{\mathfrak{H}}, \widetilde{q}, \widetilde{p}, \widetilde{\Omega}\}$, where

1. $\widetilde{\mathfrak{H}} = L^2(\mathbb{E}, \langle\cdot|\cdot\rangle, d\nu)$ for a $\mathbb{C}^k$-bundle $\mathbb{E}$ over $\mathbf{\Omega}$, with Hermitian metric $\langle\cdot|\cdot\rangle$ and a smooth measure ν on $\mathbf{\Omega}$.

2. $\widetilde{\Omega}$ is a two-form with (self-adjoint) values in the endomorphism bundle $L_{\mathbb{E}} = \mathbb{E} \otimes \mathbb{E}^*$.

3. $(\widetilde{q}(f)\sigma)(x) = f(x)\sigma(x)$, for all $f \in C^\infty(\mathbf{\Omega})$ and smooth sections $\sigma \in \Gamma_0$ (= smooth sections of compact support).

4. $\widetilde{p}(X)\Gamma_0 \subset \Gamma_0$, for all $X \in \mathfrak{X}_c(\mathbf{\Omega})$.

We then have the following canonical representation of a differentiable quantum Borel kinematics:

Theorem 6.7 *Let $\{\mathfrak{H}, \widehat{q}, \widehat{p}, \Omega\}$ be a local differentiable quantum Borel kinematics on $\mathbf{\Omega}$ in standard form. Then there is a hermitian connection ∇ with curvature $\frac{i}{\hbar}\,\Omega$ on $\mathbb{E}$ and a covariantly constant self-adjoint section Φ of $L_{\mathbb{E}} = \mathbb{E} \otimes \mathbb{E}^*$, the bundle endomorphisms of $\mathbb{E}$, such that for all $X \in \mathfrak{X}_c(\mathbf{\Omega})$ and all $\sigma \in \Gamma_0$,*

$$\widehat{p}(X)\sigma = -i\hbar\,\nabla_X\sigma + (-\frac{i\hbar}{2}\,\mathbb{I} + \Phi)\mathrm{div}_\nu(X)\sigma. \tag{262}$$

For an elementary quantum Borel kinematics, i.e., when the $\mathbb{C}^k$-bundle is a line bundle, one can give a complete classification of the possible equivalence classes of quantum Borel kinematics. Indeed, for hermitian line bundles, one has the classification theorem:

Theorem 6.8 *Let $\mathbf{\Omega}$ be a connected differentiable manifold and B a closed two-form on $\mathbf{\Omega}$ (i.e., $dB = 0$). Then there exists a hermitian complex line bundle $(\mathbb{E}, \langle\cdot|\cdot\rangle, \nabla)$, with compatible connection and curvature $\frac{i}{\hbar}\,B$, if and only if B satisfies the integrality condition*

$$\frac{1}{2\pi\hbar}\int_\Sigma B \in \mathbb{Z}, \tag{263}$$

for all closed two-surfaces Σ in $\mathbf{\Omega}$. Furthermore, the various equivalence classes of $(\mathbb{E}, \langle\cdot|\cdot\rangle, \nabla)$ (for fixed curvature $\frac{i}{\hbar}\,B$) are parametrized by $H^1(\mathbf{\Omega}, \mathcal{U}(1)) \simeq \pi_1(\mathbf{\Omega})^*$, where $\pi_1(\mathbf{\Omega})^*$ denotes the group of characters of the first fundamental group of $\mathbf{\Omega}$.

The classification of the associated elementary quantum Borel kinematics is then spelled out in the following theorem:

Theorem 6.9 *The equivalence classes of elementary local differentiable quantum Borel kinematics are in one-to-one correspondence with $I^2(\mathfrak{Q}) \times \pi_1(\mathfrak{Q})^* \times \mathbb{R}$, where $I^2(\mathfrak{Q})$ denotes the set of closed real two-forms on $\mathfrak{Q}$, satisfying the integrality condition (263).*

For $\mathbb{C}^k$-bundles only a weaker result, for $\Omega = 0$, is known:

Theorem 6.10 *The equivalence classes of ($\Omega = 0$)-compatible differentiable and local k-quantum Borel kinematics are in one-to-one correspondence with the equivalence classes $\{(D, A)\}$ of pairs of unitary representations $D \in \mathrm{Hom}(\pi_1(\mathfrak{Q}), \mathcal{U}(k))$ and self-adjoint matrices $A \in \mathcal{S}(\mathbb{C}^k) \bigcap D'$, where D' is the commutant of the representation D, i.e., $D' = \{M \in \mathcal{L}(\mathbb{C}^k) \mid [M, D(g)] = 0, \ \forall g \in \pi_1(\mathfrak{Q})\}$. Here two pairs (D_1, A_1) and (D_2, A_2) are equivalent if there is a unitary matrix U such that $D_2 = U D_1 U^{-1}$ and $A_2 = U A_1 U^{-1}$.*

7 Geometric Quantization

A much discussed method for quantizing classical systems is the Kostant-Souriau technique for geometric quantization. In contradistinction to canonical or Segal-Borel quantization, this procedure exploits the symplectic geometry of the classical phase space Γ and consists of two stages, provided a technical condition of *integrality* is satisfied by the associated 2-form. In the first stage or *prequantization*, a map $f \mapsto \widehat{f}$, satisfying (165), is obtained for all classical observables f. In the second stage, called *polarization*, a maximal family of commuting observables is identified and the quantum algebra is represented irreducibly on a reduced Hilbert space, the relation (165) holding only for a restricted set of observables. Exhaustive treatments of geometric quantization, and related mathematical tools are available in the literature.[14,15,22,26] In order to discuss the method of geometric quantization, we need a few mathematical facts about vector and line bundles.

7.1 Vector bundles and line bundles

A *vector bundle* over a manifold M is a manifold B (called the *total space*), together with a smooth map $\pi : B \longrightarrow M$ (called the *projection map*), such that:

1. For each $m \in M$, the set $B_m = \pi^{-1}(m)$ is a vector space over $\mathbb{K}$ (i.e., $\mathbb{R}$ or $\mathbb{C}$), with constant dimension r. The vector space B_m is called the *fibre* over m and r is called the *fibre dimension*.

2. For each $m_0 \in M$, there exists a neighbourhood $U \subset M$ of m_0, together with a diffeomorphism $\psi : U \times \mathbb{K}^r \longrightarrow \pi^{-1}(U)$, such that the restricted map , $\psi|_m : \{m\} \times \mathbb{K}^r \longrightarrow B_m$ is a linear isomorphism of vector spaces, for each $m \in M$. The pair $\{U, \psi)\}$ is called a *local trivialization* of the bundle.

A vector bundle with one-dimensional fibres is called a *line bundle*. We shall mostly be concerned with complex line bundles, for which each fibre is linearly isomorphic to $\mathbb{C}$.

A map $s : U \subset M \longrightarrow B$, such that $\pi(s(m)) = m$, for every $m \in U$, is called a *section* over U. The set of all sections over U is denoted $\Gamma_B(U)$ and the subset of smooth sections is denoted $\mathcal{C}_B^\infty(U)$. There is a natural vector space structure (over $\mathbb{K}$) on $\Gamma_B(U)$ and $\mathcal{C}_B^\infty(U)$ and elements in these spaces can be multiplied by $\mathbb{K}$-valued functions on U. Let B be a complex vector bundle (i.e., $\mathbb{K} = \mathbb{C}$) and denote by $\mathfrak{X}_\mathbb{C}(M)$ the space of *complexified* vector fields on the base space.

A connection ∇ on B is a map which assigns to each $X \in \mathfrak{X}_\mathbb{C}(M)$, an operator ∇_X on $\mathcal{C}_B^\infty(M)$ such that,

1. $\nabla_{fX+gY} = \nabla_{fX} + \nabla_{gY}, \quad f, g \in \mathcal{C}_\mathbb{C}^\infty(M), \quad X, Y \in \mathfrak{X}_\mathbb{C}(M),$
2. $\nabla_X(fs) = X(f)s + f\nabla_X s, \quad X \in \mathfrak{X}_\mathbb{C}(M), \quad s \in \mathcal{C}_B^\infty(M), \quad f \in \mathcal{C}_\mathbb{C}^\infty(M),$
3. $\nabla_X(s + s') = \nabla_X\, s + \nabla_X\, s', \quad X \in \mathfrak{X}_\mathbb{C}(M), \quad s, s' \in \mathcal{C}_B^\infty(M).$

One has the same definition in the real case, except that X, Y and f are real valued. Note, that a connection is essentially a way to differentiate sections. For a flat connection (i.e., $\Gamma_{ij}^k = 0$ for all i, j, k), this is just the directional derivative of Y along X. (In classical differential geometry one defines,

$$\nabla_X Y = \sum_{i,k} X^i\left(\frac{\partial Y^k}{\partial u^i} + \sum_j \Gamma_{ij}^k Y^j\right)\mathbf{x}_k,$$

where X^i, Y^k are components, in the basis $\{\mathbf{x}_k\}$, of the vector fields X, Y, respectively, u^i are local coordinates and the $\Gamma_{ij}^k = \sum_\ell \langle \mathbf{x}_{ij}, \mathbf{x}_\ell\rangle\, g^{\ell k}$ are Christoffel symbols. The connection so defined, has the above properties.)

A *hermitian structure* on a complex vector bundle B is an inner product $\langle\cdot|\cdot\rangle_m$ on each fibre $\pi^{-1}(m)$ that is smooth, in the sense that for any two smooth sections s, s', the map $m \longmapsto \langle s(m)|s'(m)\rangle_m$ is smooth. We denote a hermitian structure by H, so that pointwise, $H(s, s')(m) = \langle s(m)|s'(m)\rangle_m$. A connection ∇ and a Hermitian structure H are said to be *compatible* if

$$X(H(s, s')) = H(\nabla_X\, s, s') + H(s, \nabla_X\, s'), \tag{264}$$

for every $X \in \mathfrak{X}_{\mathbb{C}}(M)$ and all $s, s' \in C_B^\infty(M)$.

Let $\mathbb{L}$ be a complex hermitian line bundle over M with connection ∇. The *curvature two-form*, $\operatorname{curv}\nabla = R$ on $\mathbb{L}$ is defined as:

$$R(X,Y)s = ([\nabla_X, \nabla_Y] - \nabla_{[X,Y]})s, \qquad X, Y \in \mathfrak{X}(M), \quad s \in C_{\mathbb{L}}^\infty(M). \tag{265}$$

Note that R is bilinear in the variables X, Y over $C^\infty(M)$ and linear in the variable s over $C_{\mathbb{C}}^\infty(M)$. Next let (U, ψ) be a local trivialization of $\mathbb{L}$. We may assume that U is dense in M. Let $\widehat{s} : U \longrightarrow \mathbb{L}$ denote a *reference section*, for example, we may take it to be the *unit section*,

$$\widehat{s}(m) = \psi(m, 1), \quad \text{and} \quad \pi \circ \widehat{s}(m) = m. \tag{266}$$

Then, any other section $s : U \longrightarrow \mathbb{L}$ is of the form

$$s = f\widehat{s}, \quad \text{i.e.,} \quad s(m) = f(m)\widehat{s}(m), \qquad m \in U,$$

where $f : U \longrightarrow \mathbb{C}$ is a smooth function. From the general theory of line bundles it then follows that there exists a (local) one-form θ on U such that

$$\nabla_X\, s = X(f)\widehat{s} - i(X \rfloor \theta)s. \tag{267}$$

Alternatively, we may also write $\nabla\widehat{s} = -i\theta \otimes \widehat{s}$. The one-form θ is then called a *connection form.* A straightforward computation now shows that

$$iR = i\operatorname{curv}\nabla = d\theta. \tag{268}$$

Given a hermitian line bundle, there is essentially only one connection compatible with the hermitian structure. Furthermore, compatibility forces the connection form to be real.

7.2 Summary of geometric quantization

As before, let (Γ, Ω) be a classical mechanics, i.e., Γ is a symplectic manifold and Ω the associated (closed non-degenerate) two-form, defining the structure of a Poisson bracket (see (148)). In order to pass to a quantum mechanics, we need to assume that (Γ, Ω) satisfies an important *integrality condition*, namely, that the integral of Ω over any closed 2-surface be an integral multiple (including zero) of 2π (see Theorem 6.8). This assumption then implies that there exists a hermitian line bundle $\mathbb{L}$ over Γ, with a compatible connection ∇ and connection form θ such that,

$$\Omega = d\theta = i\operatorname{curv}\nabla. \tag{269}$$

Using the Liouville measure ν of Γ, which in the Darboux coordinates has the form (111), let us construct the Hilbert space $\mathfrak{H}$, of all sections s of $\mathbb{L}$, which are square-integrable with respect to the given Hermitian form and this measure, i.e.,

$$\int_\Gamma H(s,s)(x)\, d\nu(x) < \infty. \tag{270}$$

Note that the scalar product on $\mathfrak{H}$, of two sections, $s_i = \widehat{s} f_i,\ \ i = 1, 2$, is given by,

$$\langle s_1, | s_2 \rangle = \int_\Gamma H(s_1, s_2)(x)\, d\nu(x) = \int_\Gamma \overline{f_1(x)} f_2(x) H(\widehat{s}, \widehat{s})(x)\, d\nu(x).$$

Thus $\mathfrak{H}$ is naturally (unitarily) isomorphic to $L^2(\Gamma, d\nu)$.

A *prequantization* of the algebra $\mathcal{C}^\infty(\Gamma)$ is then given by the operators,

$$\widehat{f} = -i\hbar\nabla_{X_f} + f, \tag{271}$$

acting on $\mathfrak{H}$. These operators are generally unbounded, but self-adjoint for appropriately chosen f. It can then be checked that

$$[\widehat{f}, \widehat{g}] = \widehat{f}\widehat{g} - \widehat{g}\widehat{f} = i\hbar\widehat{\{f, g\}}, \tag{272}$$

which is the quantization condition. In this prequantization phase, any two functions $f, g \in \mathcal{C}^\infty(\Gamma)$ are represented by operators on $\mathfrak{H}$ whose commutator is the quantized version of the corresponding Poisson bracket. But this also means that the prequantization representation, $f \longmapsto \widehat{f}$ is highly reducible. It is necessary therefore to reduce the size of the Hilbert space $\mathfrak{H}$ in order to obtain an irreducible representation. This is now achieved by introducing a *polarization.*

The concept of a polarization is linked to that of a *Lagrangian submanifold* and a *commuting set of observables*. The Lagrangian submanifold has half the dimension of Γ:

Definition 7.1 *A Lagrangian submanifold of* Γ *is a submanifold, the tangent spaces of which are generated by vector fields,* $X_j,\ \ j = 1, 2, \ldots n$, *such that*

1. $[X_i, X_j] = \sum_{k=1}^n c_{ijk} X_k$, *where the* c_{ijk} *are real numbers.*

2. $\Omega(X_i, X_j) = 0$.

3. *The dimension of the submanifold is* n, *which is half the dimension of* Γ.

The set of vector fields $\{X_j\}_{j=1}^n$ *is then said to constitute a polarization.*

A polarization of the above type determines a foliation of the manifold. We shall assume that the space of leaves of this foliation is itself a smooth manifold. Such a polarization is called a *real polarization* (as opposed to a *complex polarization*). As an example, if $\Gamma = T^*\mathbf{\Omega}$ is a cotangent bundle, then the vector fields $\dfrac{\partial}{\partial p_i}$ define the vertical polarization and the space of leaves can be identified with $\mathbf{\Omega}$ itself. In order to obtain a quantization, one looks at sections which are covariant constants along the leaves of the polarization, i.e., sections s which satisfy,

$$\nabla_{X_i} s = 0, \qquad i = 1, 2, \ldots n, \tag{273}$$

where the X_i are the vector fields defining the polarization. But now it may so happen that such sections are no longer square-integrable in the sense of (270) and hence, no longer in the Hilbert space $\mathfrak{H}$. It is therefore necessary, at this point, to change the measure on $\mathfrak{H}$. We illustrate these by an example.

7.3 An easy example

The simplest example of the theory outlined above is the case where the symplectic manifold is $\Gamma = T^*\mathbb{R}$, the cotangent bundle of the real line. The two-form Ω is just $dp \wedge dq$ and the connection form may be taken to be $\theta = pdq$. The line bundle is trival and identifiable with $\mathbb{C} \times \Gamma$. For the reference section we simply take $\widehat{s}(q,p) = 1$, so that the Hilbert space of prequantization is $L^2(\mathbb{R}^2, dp\; dq)$. Applying (271) to the functions $f(q,p) = q$ and $f(q,p) = p$, we get the operators,

$$\widehat{q} = i\hbar\frac{\partial}{\partial p} + q, \qquad \widehat{p} = -i\hbar\frac{\partial}{\partial q}, \tag{274}$$

which obviously satisfy the canonical commutation relation $[\widehat{q}, \widehat{p}] = i\hbar\, I$. We next take the vertical polarization, defined by, $X = \dfrac{\partial}{\partial p_i}$. Then, for any section $s = \psi\widehat{s}$ we obtain,

$$\nabla_X s = \frac{\partial \psi}{\partial p}\widehat{s}, \tag{275}$$

and in order for this to vanish, ψ must be a function of q alone. We therefore truncate the measure on $\mathfrak{H}$ accordingly, and take for the Hilbert space of our quantum mechanics the space $L^2(\mathbb{R}, dq)$. In general, it might be necessary to use *half densities*[26] as a systematic way to get the truncated measure. Furthermore, for the above choice of polarization, it is only operators involving at most linear terms in p which leave the polarization invariant.

8 Berezin Quantization

A method of quantization, also based on a phase space raher than on a configuration space, and exploiting its symplectic structure was developed by Berezin.[5] While this technique shares certain features with the canonical or Segal-Borel method, it is closer in spirit to the geometric and coherent state quantization techniques. However, it works best when the symplectic manifold also has a complex structure. In order to understand Berezin's method, let us go back to classical mechanics on a symplectic manifold and assume now, that this manifold is also endowed with a complex structure. Let (Γ, Ω) be such a manifold and $\mathcal{A}(\Gamma)$ an algebra of differentiable, complex-valued functions on Γ with Poisson bracket defined via Ω (i.e., in the local coordinates in which Ω is of the form (143), the Poisson bracket is given by (148)). We take this as the classical algebra.

8.1 Definition of Berezin quantization

Berezin then proposes the following definition of a quantization of the classical mechanics: An algebra with involution $\mathcal{A}$ is called a *quantization* of the classical algebra $\mathcal{A}(\Gamma)$ if the following conditions are satisfied:

1. There exists a family of associative algebras $\mathcal{A}_h$ with involution such that:
   (a) $h \in E \subset \mathbb{R}^+$; 0 is a limit point of E, but $0 \notin E$;
   (b) $\mathcal{A} \subset \bigoplus_{h\in E} \mathcal{A}_h$, as a subalgebra.
2. There is a homomorphism of algebras, $\varphi : \mathcal{A} \to \mathcal{A}(\Gamma)$, with the properties:
   (a) for any two distinct points $x_1, x_2 \in \Gamma$, there exists a function $f \in \varphi(\mathcal{A})$ such that $f(x_1) \neq f(x_2)$ (in other words, $\varphi(\mathcal{A}) \subset \mathcal{A}(\Gamma)$ is separating for the points of Γ);
   (b) for any two elements $A, B \in \mathcal{A}$,
   $$\varphi(A \circ B - B \circ A) = \frac{ih}{2\pi}\{\varphi(A), \varphi(B)\}, \tag{276}$$
   where $\circ$ denotes multiplication in $\mathcal{A}$ and $\{\,,\,\}$ the Poisson bracket in $\mathcal{A}(\Gamma)$;
   (c) for all $A \in \mathcal{A}$, $\varphi(A^*) = \overline{\varphi(A)}$, $*$ denoting involution in $\mathcal{A}$

In the above, h plays the role of Planck's constant. It is convenient to represent the elements of $\mathcal{A}$ as $\mathcal{A}_h$-valued functions, i.e., $A(h) \in \mathcal{A}_h$, $\forall h \in E$.

Denoting multiplication and involution in $\mathcal{A}_h$ by $\circ_h$ and $*_h$, respectively, we then have, for all $A, B \in \mathcal{A}$,

$$A^*(h) = [A(h)]^{*_h}, \quad (A \circ B)(h) = A(h) \circ_h B(h). \tag{277}$$

In fact, we shall drop the subscript h in $\circ_h$ and $*_h$ and use the same symbols $\circ$ and $*$ for multiplication and involution, respectively, in both $\mathcal{A}$ and $\mathcal{A}_h$.

Properties 2.(a) and 2.(b) are called the *correspondence principle.* A quantization is said to satisfy a *weak correspondence principle* if conditions 2, 2.(a), 2.(b) and 2.(c) are replaced by

2$'$. There exists a linear manifold $\mathcal{A}' \subset \mathcal{A}$, on which there is defined a mapping, $\varphi : \mathcal{A}' \to \mathcal{A}(\Gamma)$, satisfying:

   (a) for any two distinct points $x_1, x_2 \in \Gamma$, there exists a function $f \in \varphi(\mathcal{A}')$ such that $f(x_1) \neq f(x_2)$;

   (b) for $A, B, A \circ B$ and $B \circ A \in \mathcal{A}'$,

$$\varphi(A \circ B - B \circ A) = \frac{ih}{2\pi}\{\varphi(A), \varphi(B)\}; \tag{278}$$

   (c) $\varphi(A^*) = \overline{\varphi(A)}$, for all $A \in \mathcal{A}'$.

A quantization is said to be *special* if it satisfies the additional properties:

3. For each $h \in E$, the algebra $\mathcal{A}_h$ consists of functions $A : \Gamma \to \mathbb{C}$.

4. The algebra $\mathcal{A}$ consists of functions $A : E \times \Gamma \to \mathbb{C}$, such that for fixed $h \in E$, $A(h, \cdot) \in \mathcal{A}_h$.

5. The homomorphism $\varphi : \mathcal{A} \to \mathcal{A}(\Gamma)$ is given by

$$\varphi(A)(x) = \lim_{h \to 0} A(h, x). \tag{279}$$

We should point out here that the Berezin technique does not hinge so much on finding a map $f \mapsto \hat{f}$, from a classical observable to its quantized counterpart, as on obtaining an entire quantum algebra $\mathcal{A}$, with appropriate *dequantization* properties leading to the classical algebra $\mathcal{A}(\Gamma)$. In this sense the technique is better adapted to a study of dequantization.

The Berezin quantization technique is most effective when Γ is a Kähler manifold, i.e., when it comes equipped with an integrable complex structure. This means, that for all $x \in \Gamma$, the tangent space $T_x(\Gamma)$ has a complex structure which varies smoothly with x. In order to see how the Berezin method applies to such manifolds, it will first be necessary to define a certain algebra, related to a frame.

8.2 Covariant and contravariant symbols and complex structures

Let ν denote the Liouville measure on the symplectic manifold Γ, i.e., in the local coordinates $q^i, p_j, \quad i,j = 1,2,\ldots,n$, in which Ω has the form $d\nu = \frac{1}{(2\pi)^n}\prod_{i=1}^n dp_i \wedge dq^i$. Let $\mathcal{F}(\eta_x, I, 1)$ be a rank-1 tight frame for the Hilbert space $\mathfrak{H}$, labelled by points $x \in \Gamma$:

$$\int_\Gamma |\eta_x\rangle\langle\eta_x|\, d\nu(x) = I, \tag{280}$$

and define the second measure μ on Γ:

$$d\mu(x) = \|\eta_x\|^2 d\nu(x). \tag{281}$$

Let $P(x)$ be the one-dimensional projection operator corresponding to the vector η_x, and suppose that A is a linear operator on $\mathfrak{H}$, the domain of which includes all these vectors. Then the function $a : \Gamma \to \mathbb{C}$,

$$a(x) = \mathrm{Tr}[P(x)A] = \frac{\langle\eta_x|A\eta_x\rangle}{\|\eta_x\|^2}. \tag{282}$$

is called the *covariant symbol* of A. An extension of this symbol, to a function on $\Gamma \times \Gamma$, is defined by

$$a(x,y) = \frac{\langle\eta_x|A\eta_y\rangle}{\langle\eta_x|\eta_y\rangle}, \tag{283}$$

and if $\phi \in \mathfrak{H}$ is mapped to $\Phi \in L^2(\Gamma,\nu)$, under the isometry $W : \mathfrak{H} \to L^2(\Gamma,\nu)$, where $(W\phi)(x) = \Phi(x) = \langle\eta_x|\phi\rangle$, then,

$$(W(A\phi))(x) = \int_\Gamma a(x,y)\frac{\langle\eta_x|\eta_y\rangle}{\|\eta_y\|^2}\Phi(y)\, d\mu(y). \tag{284}$$

Let A, B and $AB = C$ be operators in $\mathfrak{H}$, with covariant symbols a, b and c, respectively. Then, writing the symbol product as $c = a \circ b$, we have

$$c(x) = a \circ b(x) = \frac{\langle\eta_x|AB\eta_x\rangle}{\|\eta_x\|^2} = \int_\Gamma a(x,y)b(y,x)\frac{|\langle\eta_x|\eta_y\rangle|^2}{\|\eta_x\|^2\|\eta_y\|^2}d\mu(y). \tag{285}$$

Evidently, if a is the covariant symbol of A, then its complex conjugate $\overline{a}$ is the covariant symbol of the adjoint operator A^*. In general, while the extended function $a(x,y)$ determines A uniquely, the covariant symbol $a(x)$ may not always do so. However, as we shall see, for certain special frames arising from Kähler manifolds, every bounded operator will be uniquely determined by its covariant symbol, and consequently, the algebraic structure of the set of bounded operators will go over to an algebraic structure, with respect to the product (285), on the space of their covariant symbols.

A function $\tilde{a} : \Gamma \to \mathbb{C}$ is called the *contravariant symbol* of the operator A on $\mathfrak{H}$, if

$$A = \int_\Gamma \tilde{a}(x)P(x)\, d\mu(x) = \int_\Gamma \tilde{a}(x)|\eta_x\rangle\langle\eta_x|\, d\nu(x). \tag{286}$$

(see (177)). The covariant and contravariant symbols, $a, \tilde{a}$, of an operator A are related by

$$a(x) = \int_\Gamma \tilde{a}(y)\frac{|\langle\eta_x|\eta_y\rangle|^2}{||\eta_x||^2||\eta_y||^2}d\mu(y). \tag{287}$$

In the sequel we shall also require a certain relationship, between unitary and hermitian matrices on the one hand and symplectic and antisymmetric matrices on the other. Let U be a complex, $n \times n$ unitary matrix:

$$UU^* = U^*U = \mathbb{I}_n, \qquad U^*_{ij} = \overline{U}_{ji},$$

and let H be a complex $n \times n$ Hermitian matrix:

$$H = H^*.$$

Let us write U and H in terms of their real and imaginary parts,

$$U = U_r + iU_i, \qquad H = H_r + iH_i.$$

Then, the real $2n \times 2n$ matrix

$$S = \begin{pmatrix} U_r & -U_i \\ U_i & U_r \end{pmatrix} \tag{288}$$

is symplectic, i.e., $SJS^T = J$ (see (83)). This follows from the fact that the unitarity of U implies that

$$U_rU_r^T + U_iU_i^T = \mathbb{I}_n, \qquad U_rU_i^T - U_iU_r^T = 0.$$

Since a hermitian matrix can always be diagonalized by a unitary transformation, suppose that U diagonalizes H, i.e.,

$$UHU^* = D = \begin{pmatrix} \lambda_1 & 0 & \cdots & 0 \\ & & \ddots & \\ 0 & \cdots & 0 & \lambda_n \end{pmatrix}, \qquad \lambda_n \in \mathbb{R}.$$

Then, since D is a real $n \times n$ matrix, using the decomposition of U and H into real and imaginary parts, we find:

$$\begin{aligned} U_rH_rU_r^T + U_rH_iU_i^T - U_iH_iU_r^T + U_iH_rU_i^T &= D, \\ U_rH_iU_r^T - U_rH_rU_i^T + U_iH_rU_r^T + U_iH_iU_i^T &= 0. \end{aligned} \tag{289}$$

Next consider the real $2n \times 2n$ matrix ω,

$$\omega = \begin{pmatrix} H_i & -H_r \\ H_r & H_i \end{pmatrix} = -\omega^T. \tag{290}$$

This matrix is antisymmetric, since $H_i^T = -H_i$, $H_r^T = H_r$ (as follows from $H = H^*$). Then it is easily verified, using (289) that,

$$S\omega S^T = \begin{pmatrix} 0 & -D \\ D & 0 \end{pmatrix} := \omega_S,$$

in other words, any antisymmetric matrix of the above type can be brought into canonical form by a symplectic transformation.

Note that if ω_{ij} are the matrix elements of ω, then under the symplectic map S the antisymmetric bilinear form (see (114)),

$$\frac{1}{2}\sum_{i,j=1}^{2n} \omega_{ij}\, e^{i\,*} \wedge e^{j\,*} := \Omega$$

changes to

$$\Omega = \frac{1}{2}\sum_{i=1}^{n} \lambda_i a^{i\,*} \wedge a^{n+i\,*}, \tag{291}$$

where,

$$a^{i\,*} = \sum_{i=1}^{n} (S^{-1})_{ki} e^{k\,*}.$$

As we mentioned earlier, we shall study Berezin quantization mainly on complex, Kähler manifolds. Thus, our phase space Γ is a complex manifold, equipped with a Kähler structure. In other words, we can introduce local complex coordinates on Γ,

$$z^k = x^k + iy^k := \frac{1}{\sqrt{2}}(q^k - ip^k), \qquad k = 1, 2, \ldots, n. \tag{292}$$

We shall also use the vector notation, $\mathbf{z} = (z^1, z^2, \ldots z^n)$ and its complex conjugate (we are assuming that the complex dimension of Γ is n). In these coordinates, the 2-form (143) may be rewritten as

$$\Omega = \frac{1}{i}\sum_{i,j=1}^{n} H_{ij}\, d\overline{z}^i \wedge dz^j, \tag{293}$$

where the H_{ij} are the elements of an $n \times n$ complex, hermitian matrix H, which is related to ω via (290). We are suppressing the dependence of H and H_{ij} on $\mathbf{z}, \overline{\mathbf{z}}$. Since Ω is non-degerate,

$$\det H = \det[H_{ij}] \neq 0.$$

Note also, that if $\mathbf{w}$, $\mathbf{w}' \in \mathbb{C}^n$ and if

$$X = \sum_{i=1}^{n} w^i \frac{\partial}{\partial z^i}, \qquad \overline{X}' = \sum_{i=1}^{n} \overline{w}'^i \frac{\partial}{\partial \overline{z}^i}$$

are holomorphic and anti-holomorphic tangent vectors, respectively, then

$$\Omega(\overline{X}', X) = 2\, \Im \langle \mathbf{w}' | H \mathbf{w} \rangle_{\mathbb{C}^n}, \tag{294}$$

$\langle | \rangle_{\mathbb{C}^n}$ denoting the standard Euclidean scalar product in $\mathbb{C}^n$.

The Poisson bracket of two functions $f_1, f_2 \in \mathcal{A}(\Gamma)$, in complex coordinates, is now given by

$$\{f_1,\, f_2\} = 2i \sum_{j,k=1}^{n} H^{jk} \Big(\frac{\partial f_1}{\partial z^k} \frac{\partial f_2}{\partial \overline{z}^j} - \frac{\partial f_1}{\partial \overline{z}^j} \frac{\partial f_2}{\partial z^k}\Big), \tag{295}$$

where (H^{jk}) is the inverse of the matrix (H_{jk}).

8.3 Summary of Berezin quantization

As stated earlier, Berezin quantization works best on complex manifolds which are also Kähler. The Kähler property of Γ implies that it is a complex manifold equipped with a Hermitian structure, given locally by a metric $\sum_{j,k=1}^{n} H_{jk}\, d\overline{z}^j \otimes dz^k$ and which has the additional property that the associated *real* 2-form (293) is closed and non-degenerate. The Kähler metric H_{jk} is derivable from a *real* potential function $\Phi(z, \overline{z})$:

$$\frac{\partial^2 \Phi}{\partial \overline{z}^j \partial z^k} = H_{jk}. \tag{296}$$

It will be assumed that Φ exists globally on Γ and that it possesses an analytic extension $\Phi(z, \overline{v})$ to some neighbourhood of the diagonal in $\Gamma \times \Gamma$. Furthermore, the local coordinates z^k will be attached to an open dense set $U \subset \Gamma$, via an analytic map $\mathbf{h} : U \to \mathbb{C}^n$, $z^k = \mathbf{h}(x)^k$, and we shall write $\mathfrak{M} = \mathbf{h}(U) \subset \mathbb{C}^n$. Consider the Hilbert space $\widetilde{\mathfrak{H}}_h$ of all measurable functions $f : \mathfrak{M} \to \mathbb{C}$, which are square integrable with respect to the measure,

$$d\nu_h(z, \overline{z}) = \Big(\frac{i}{\pi}\Big)^n \exp\Big[-\frac{2\pi}{h}\Phi(z, \overline{z})\Big] \det[H_{jk}] \prod_{i=1}^{n} dz^i \wedge d\overline{z}^i. \tag{297}$$

In general, when expressed in local coordinates, f is a function of both the holomorphic variables z^k and the anti-holomorphic variables $\bar{z}^k$, so that the associated scalar product in $\widetilde{\mathfrak{H}}_h$ can be written as

$$\langle f|g\rangle_h = c(h) \int_{\mathfrak{M}} \overline{f(z,\bar{z})} g(z,\bar{z})\, d\nu_h(z,\bar{z}), \tag{298}$$

where $c(h)$ is a normalization constant to be fixed later. It turns out that, for the type of Kähler manifolds which will interest us, *viz* symmetric homogeneous spaces, the subset of all *holomorphic functions* $f(z)$ in $\widetilde{\mathfrak{H}}_h$ is a closed Hilbert subspace $\mathfrak{H}_h \subset \widetilde{\mathfrak{H}}_h$. Furthermore, $\mathfrak{H}_h$ admits a reproducing kernel K_h, which can be constructed by taking any orthonormal basis $\{f_k\}_{k=1}^{\infty}$ in $\mathfrak{H}_h$, and setting

$$K_h(z,\bar{v}) = \sum_{k=1}^{\infty} f_k(z)\overline{f_k(v)}, \quad z,\, v \in \mathfrak{M}. \tag{299}$$

(The convergence of the above sum is absolute and uniform.) It can be shown that the kernel K_h is independent of the choice of the orthonormal basis used to construct it. Moreover, it has the (reproducing) property that for any function $f \in \mathfrak{H}_h$,

$$c(h) \int_{\mathfrak{M}} K(z,\bar{v}) f(v)\, d\nu_h(v,\bar{v}) = f(z). \tag{300}$$

From the general theory of reproducing kernels it then also follows that the set of vectors,

$$\mathfrak{S}_h = \{\eta_{\bar{v}} \mid \eta_{\bar{v}}(z) = K_h(z,\bar{v}),\ v \in \mathfrak{M}\}, \tag{301}$$

called a family of *generalized coherent states*, is overcomplete in $\mathfrak{H}_h$ and

$$c(h) \int_{\mathfrak{M}} |\eta_{\bar{v}}\rangle\langle\eta_{\bar{v}}|\, d\nu_h(v,\bar{v}) = I_h, \tag{302}$$

where I_h is the identity operator on $\mathfrak{H}_h$.

We now restrict ourselves to a class of Kähler manifolds, which includes most physically interesting cases, and for which the potential Φ is the logarithm of the kernel K_h. Thus we assume that there exists a set $E \subset \mathbb{R}^+$, having 0 as a limit point, but $0 \notin E$ (see quantization condition 1(a) above), and such that, $\forall h \in E$,

$$K_h(z,\bar{z}) = \lambda \exp\left[\frac{2\pi}{h}\Phi(z,\bar{z})\right], \tag{303}$$

(λ = constant). With some additional technical assumptions it can then be shown, that choosing

$$c(h) = \left[\int_{\mathfrak{M}} \frac{d\nu_h(v, \bar{v})}{K_h(v, \bar{v})}\right]^{-1}, \tag{304}$$

one obtains

$$K_h(z, \bar{z}) = \exp\left[\frac{2\pi}{h}\Phi(z, \bar{z})\right]. \tag{305}$$

Let $\mathcal{A}_h$ denote the algebra of symbols, under the product (285), of all bounded operators on $\mathfrak{H}_h$ (using the coherent states (301)). Define $\mathcal{A}$ to be the algebra of all functions $A(h, \cdot) \in \mathcal{A}_h$, $h \in E$. Let $\mathcal{A}'$ be the set of all functions in $\mathcal{A}$ which can be represented in the form

$$A(h, z, \bar{z}) = A(0, z, \bar{z}) + \frac{h}{2\pi} A_1(z, \bar{z}) + \frac{h^2}{4\pi^2} A_2(z, \bar{z}), \tag{306}$$

where $A(0, z, \bar{z})$, $A_1(z, \bar{z})$ and $A_2(z, \bar{z})$ are functions which have analytic continuations to $\Gamma \times \Gamma$. It can then be shown that $\mathcal{A}$ is a *special quantization* of $\mathcal{A}(\Gamma)$, with functions in $\mathcal{A}'$ satisfying the *weak correspondence principle.* More precisely, for arbitrary $A, B \in \mathcal{A}'$, as $h \to 0$,

$$A \circ B(h, z, \bar{z}) = A(0, z, \bar{z}) B(0, z, \bar{z}) + \mathcal{O}(h), \tag{307}$$

and

$$(A \circ B - B \circ A)(h, z, \bar{z}) = \frac{ih}{2\pi} \{A, B\}(0, z, \bar{z}) + \mathcal{O}(h), \tag{308}$$

$\{\cdot, \cdot\}$ denoting the Poisson bracket (295).

The simplest application of the above technique is to $\Gamma = \mathbb{C}$, when $\Phi(z', \bar{z}) = z'\bar{z}$. In this case,

$$\Omega = \frac{1}{i} \frac{\partial^2 \Phi(z, \bar{z})}{\partial z \, \partial \bar{z}} \, d\bar{z} \wedge dz = \frac{1}{i} d\bar{z} \wedge dz,$$

so that the Poisson bracket becomes,

$$\{f_1, f_2\} = 2i \left(\frac{\partial f_1}{\partial z} \frac{\partial f_2}{\partial \bar{z}} - \frac{\partial f_2}{\partial z} \frac{\partial f_1}{\partial \bar{z}}\right).$$

Also,

$$d\nu_h(z, \bar{z}) = \frac{1}{\pi} \exp\left[-\frac{2\pi}{h} |z|^2\right] dz \wedge d\bar{z}.$$

The set of all analytic functions in $\tilde{\mathfrak{H}}_h = L^2(\mathbb{C}, \frac{2\pi}{h}\, d\nu_h)$ is now a closed Hilbert space $\mathfrak{H}_h$, with reproducing kernel,

$$K_h(z', \bar{z}) = \exp[\frac{2\pi}{h} z'\bar{z}],$$

associated coherent states,

$$\eta_{\bar{v}}(z) = \exp[\frac{2\pi}{h} z\bar{v}],$$

and resolution of the identity,

$$\frac{2\pi}{h} \int_{\mathbb{C}} |\eta_{\bar{z}}\rangle\langle\eta_{\bar{z}}|\, d\nu_h(z, \bar{z}) = I.$$

(Note that, in this case, $c(h) = \dfrac{2\pi}{h}$). To any bounded operator A on $\mathfrak{H}$, there corresponds its covariant symbol, $A(h, z, \bar{z}) = \langle\eta_{\bar{z}}|A\eta_{\bar{z}}\rangle$, and the operator is completely determined by its symbol. There is also the quantization rule associated to the frame defined by the coherent states,

$$f(z, \bar{z}) \longrightarrow \widehat{f} = \frac{2\pi}{h} \int_{\mathbb{C}} f(z, \bar{z}) |\eta_{\bar{z}}\rangle\langle\eta_{\bar{z}}|\, d\nu_h(z, \bar{z}), \tag{309}$$

according to which $z \longrightarrow a$, the quantum mechanical annihilation operator on $\mathfrak{H}_h$ and $\bar{z} \longrightarrow a^\dagger$, the creation operator. For polynomials in $z, \bar{z}$, the quantization procedure (309) yields the well-known Wick ordering and the the Fock representation of quantum mechanics.

A second interesting example of Berezin quantization is when $\Gamma = \{z \in \mathbb{C} \mid |z|^2 < 1\} = \mathcal{D}$, the open unit disc. This time,

$$\Phi(z, \bar{z}) = -\log(1 - z'\bar{z}), \qquad \Omega = \frac{1}{i} \frac{d\bar{z} \wedge dz}{(1 - |z|^2)^2}.$$

Also, the role of h is played by the number $1/j$, where, $j = 1, 3/2, 2, 5/2, \ldots$, and

$$d\nu^j(z, \bar{z}) = \frac{(1 - |z|^2)^{2j-2}}{2\pi i}\, d\bar{z} \wedge dz.$$

Furthermore, $c(\frac{1}{j}) = 2j - 1$. The associated Hilbert space is $\tilde{\mathfrak{H}}^j = L^2(\mathcal{D}, (2j - 1)\, d\nu^j)$ and again, the set of all holomorphic functions in it is a closed subspace $\mathfrak{H}^j$, with reproducing kernel and associated coherent states,

$$K^j(z', \bar{z}) = (1 - z'\bar{z})^{-2j}, \qquad \eta_{\bar{v}}(z) = (1 - z\bar{v})^{-2j},$$

respectively. Finally, one has the resolution of the identity on $\mathfrak{H}^j$,

$$(2j-1)\int_{\mathcal{D}} |\eta_{\bar{z}}\rangle\langle\eta_{\bar{z}}|\, d\nu^j(z,\bar{z}) = I.$$

Once again, the covariant symbol $A(1/j,z,\bar{z}) = \langle\eta_{\bar{z}}|A\eta_{\bar{z}}\rangle$ determines the operator A completely.

Both of the above two examples of Kähler manifolds are homogeneous spaces of groups – the Weyl-Heisenberg group in the first case and the group $SU(1,1)$ in the second. The situation is generic for any hermitian symmetric homogenous space, in the sense that any such space is a Kähler manifold of the above type and on which the same quantization procedure applies.

9 Coherent State Quantization

Given a rank-1 tight frame $\mathcal{F}(\eta_x, I, 1)$, $x \in \Gamma$, on phase space, it is possible to demonstrate under appropriate technical conditions, the equivalence of the coherent state map $x \mapsto \eta_x$ with a prequantization map $f \mapsto \hat{f}$. This equivalence, as we explain below, is obtained by exploiting the symplectic structure of $\mathbb{CP}(\mathfrak{H})$, the (complex) projective space of the Hilbert space $\mathfrak{H}$ for which $\mathcal{F}(\eta_x, I, 1)$ is a frame.

9.1 The projective Hilbert space

Let $N = \dim\mathfrak{H}$, which could be (countably) infinite or finite. As a set, $\mathbb{CP}(\mathfrak{H})$ will be identified with the collection of all orthogonal projections onto one-dimensional subspaces of $\mathfrak{H}$ and for each non-zero vector $\psi \in \mathfrak{H}$ let $\Psi = \frac{1}{\|\psi\|^2}|\psi\rangle\langle\psi|$ denote the corresponding projector. There is a natural Kähler structure on $\mathbb{CP}(\mathfrak{H})$ as we now demonstrate. An analytic atlas of $\mathbb{CP}(\mathfrak{H})$ is given by the coordinate charts

$$\{(V_\Phi, \mathbf{h}_\Phi, \mathfrak{H}_\Phi) \mid \Phi \in \mathbb{CP}(\mathfrak{H})\}, \tag{310}$$

where

$$V_\Phi = \{\Psi \in \mathbb{CP}(\mathfrak{H}) \mid \langle\phi|\psi\rangle \neq 0\} \tag{311}$$

is an open, dense set in $\mathbb{CP}(\mathfrak{H})$;

$$\mathfrak{H}_\Phi = (I - \Phi)\mathfrak{H} = <\phi>^\perp \tag{312}$$

is the subspace of $\mathfrak{H}$ orthogonal to the range of Φ and $\mathbf{h}_\Phi : V_\Phi \to \mathfrak{H}_\Phi$ is the diffeomorphism

$$\mathbf{h}_\Phi(\Psi) = \frac{1}{\langle\hat{\phi}|\psi\rangle}(I-\Phi)\psi, \quad \hat{\phi} = \frac{\phi}{\|\phi\|}, \tag{313}$$

where $\hat{\phi}$ is a fixed unit vector in the range of Φ. Since V_Φ is dense in $\mathfrak{H}$, it is often enough to consider only one coordinate chart. Thus, we set $e_0 = \hat{\phi}$ and choose an orthonormal basis $\{e_i\}_{i=1}^{N-1}$ of $\mathfrak{H}_\Phi$ to obtain a basis of $\mathfrak{H}$ which will be fixed from now on. We may then identify $\mathbb{CP}(\mathfrak{H})$ with $\mathbb{CP}^N$: For arbitrary $\Psi \in \mathfrak{H}$ we have

$$z_i = \langle e_i|\psi\rangle, \quad i = 1, 2, \ldots, N-1, \tag{314}$$

and the coordinates of $\Psi \in \mathbb{CP}(\mathfrak{H})$ are the standard homogeneous coordinates

$$Z_i = \frac{z_i}{z_0}, \quad i = 1, 2, \ldots, N-1, \tag{315}$$

of projective geometry.

The projection map $\pi : \mathfrak{H} \setminus \{0\} \to \mathbb{CP}(\mathfrak{H})$ that assigns to each $\phi \in \mathfrak{H} \setminus \{0\}$ the corresponding projector $\Phi \in \mathbb{CP}(\mathfrak{H})$ is holomorphic in these coordinates. For $\Phi \in \mathbb{CP}(\mathfrak{H})$ we have $\pi^{-1}(\Phi) = \mathbb{C}^* \phi$ where $\mathbb{C}^* = \mathbb{C} \setminus \{0\}$, and so $\pi : \mathfrak{H} \setminus \{0\} \to \mathbb{CP}(\mathfrak{H})$ is a $GL(1, \mathbb{C})$ principal bundle, sometimes called the *canonical line bundle* over $\mathbb{CP}(\mathfrak{H})$. We will denote the associated holomorphic line bundle by $\Xi(\mathfrak{H})$ and write elements in it as (Ψ, ψ), where $\psi \in \Psi(\mathfrak{H})$. (We again write π for the canonical projection.) A local trivialization of $\Xi(\mathfrak{H})$ over V_Φ is given by the (holomorphic) reference section $\hat{s}$ of $\Xi(\mathfrak{H})$:

$$\hat{s}(\Psi) = (\Psi, \frac{\psi}{\langle\hat{\phi}|\psi\rangle}), \tag{316}$$

and any other section $s : V_\Phi \to \Xi(\mathfrak{H})$ is given by

$$s(\Psi) = (\Psi, \kappa(\Psi)) \tag{317}$$

where $\kappa : \mathbb{CP}(\mathfrak{H}) \to \mathfrak{H} \setminus \{0\}$ is a holomorphic map with $\frac{|\kappa(\Psi)\rangle\langle\kappa(\Psi)|}{\|\kappa(\Psi)\|^2} = \Psi$. Denote by s_0 the zero-section of $\Xi(\mathfrak{H})$. The identification map $\imath_\Xi : \Xi(\mathfrak{H}) \setminus s_0(\mathbb{CP}(\mathfrak{H})) \to \mathfrak{H}$ given as

$$\imath_\Xi(\Psi, \psi) = \psi, \tag{318}$$

yields a global coordinatization of $\Xi(\mathfrak{H}) \setminus s_0(\mathbb{CP}(\mathfrak{H}))$. For any $\psi \in \mathfrak{H}$ let $\langle\psi|$ be its dual element. The restriction of $\langle\psi|$ to the fibre $\pi^{-1}(\Psi')$ in $\Xi(\mathfrak{H})$, for arbitrary $\Psi' \in \mathbb{CP}(\mathfrak{H})$, then yields a section $s^*_{\Psi'}$ of the dual bundle $\Xi(\mathfrak{H})^*$ of $\Xi(\mathfrak{H})$. Moreover, the map $\Psi' \mapsto s^*_{\Psi'}$ is antilinear between $\mathfrak{H}$ and $\Gamma(\Xi(\mathfrak{H})^*)$. We may hence realize $\mathfrak{H}$ as a space of holomorphic sections.

The tangent space $T_\Psi\mathbb{CP}(\mathfrak{H})$ to $\mathbb{CP}(\mathfrak{H})$ at the point Ψ has a natural identification with $\mathfrak{H}_\Psi$ (obtainable, for example, by differentiating curves in $\mathbb{CP}(\mathfrak{H})$ passing through Ψ). The complex structure of $\mathfrak{H}_\Psi$ then endows the tangent space $T_\Psi\mathbb{CP}(\mathfrak{H})$ with an integrable complex structure J_Ψ, making $\mathbb{CP}(\mathfrak{H})$

into a Kähler manifold. The corresponding canonical 2-form Ω_{FS}, called the *Fubini-Study 2-form*, is given pointwise by

$$\Omega_{FS}(X_\Psi, Y_\Psi) = \frac{1}{2i}(\langle\xi|\zeta\rangle - \langle\zeta|\xi\rangle), \tag{319}$$

where $\xi, \zeta \in \mathfrak{H}_\Psi$ correspond to the tangent vectors X_Ψ, Y_Ψ respectively. The associated Riemannian metric g_{FS} is given by

$$g_{FS}((X_\Psi, Y_\Psi) = \frac{1}{2}(\langle\xi|\zeta\rangle + \langle\zeta|\xi\rangle) = \Omega_{FS}(X_\Psi, J_\Psi Y_\Psi). \tag{320}$$

In the local coordinates Z_i, defined in (315), Ω_{FS} assumes the form

$$\Omega_{FS} = \frac{1}{1+\|\mathbf{Z}\|^2} \sum_{i,j=1}^{N-1} \left[\delta_{ij} - \frac{Z_i \overline{Z}_j}{1+\|\mathbf{Z}\|^2}\right] d\overline{Z}_i \wedge dZ_j, \quad \mathbf{Z} = (Z_1, Z_2, \ldots, Z_{N-1}). \tag{321}$$

Thus, clearly, $d\Omega_{FS} = 0$, implying that Ω_{FS} is a *closed* 2-form, derivable from the real Kähler potential (see (296))

$$\Phi(\mathbf{Z}, \overline{\mathbf{Z}}) = \log[1 + \|\mathbf{Z}\|^2]. \tag{322}$$

A hermitian metric H_{FS} and a connection ∇_{FS} on $\Xi(\mathfrak{H})$ can be defined using the inner product of $\mathfrak{H}$: Indeed, since $\pi^{-1}(\Psi) = \{\Psi\} \times \mathbb{C}\psi$, the hermitian structure H_{FS} is given pointwise by

$$H_{FS}((\Psi, \psi), (\Psi, \psi')) = \langle\psi|\psi'\rangle \tag{323}$$

for all $(\Psi, \psi), (\Psi, \psi') \in \pi^{-1}(\Psi)$. We will use the identification map $\imath_\Xi$ defined in (318) to construct a connection on $\Xi(\mathfrak{H})$. Define the 1-form α on $\mathfrak{H}$ by

$$\alpha(\psi) = \frac{\langle d\psi|\psi\rangle}{\|\psi\|^2}. \tag{324}$$

Then the pullback

$$\alpha_{FS} = \imath_\Xi^* \alpha \tag{325}$$

defines a $\mathbb{C}^*$-invariant 1-form on $\Xi(\mathfrak{H})$ whose horizontal space at $(\Psi, \psi) \in \Xi(\mathfrak{H})$ is $\mathfrak{H}_\Psi$. For an arbitrary section $s : V_\Phi \to \Xi(\mathfrak{H}) \setminus s_0(\mathbb{CP}(\mathfrak{H}))$, as in (317), the pullback

$$-i\theta_{FS} = s^* \alpha_{FS} \tag{326}$$

defines a local 1-form on $\mathbb{CP}(\mathfrak{H})$. Pointwise,

$$\theta_{FS}(\Psi) = i\frac{\langle d\kappa(\Psi)|\kappa(\Psi)\rangle}{\|\kappa(\Psi)\|^2} = i\overline{\partial}\log\|\kappa(\Psi)\|^2, \tag{327}$$

where $\overline{\partial}$ denotes exterior differentiation with respect to the *anti-holomorphic* variables. In terms of the coordinatization introduced in (314), with f as the holomorphic function representing κ, we have

$$\theta_{FS}(\mathbf{Z}) = i\frac{\overline{df(\mathbf{Z})}}{\overline{f(\mathbf{Z})}} + i\frac{\sum_i Z_i d\overline{Z}_i}{1+||\mathbf{Z}||^2}. \tag{328}$$

Furthermore, θ_{FS} locally defines a compatible connection ∇_{FS}

$$\nabla_{FS}\, s = -i\theta_{FS} \otimes s, \tag{329}$$

and it is easy to verify that

$$\Omega_{FS} = \partial\theta_{FS} = i\,\mathrm{curv}\nabla_{FS}, \tag{330}$$

where ∂ denotes exterior differentiation with respect to the holomorphic variables and $\mathrm{curv}\nabla_{FS}$ is the curvature of the line bundle $\Xi(\mathfrak{H})$.

Thus the hermitian line bundle $(\Xi(\mathfrak{H}), H_{FS}, \nabla_{FS})$ is a prequantization of $(\mathbb{CP}(\mathfrak{H}), \Omega_{FS})$ in the sense of Section 7.

9.2 Summary of coherent state quantization

The prequantization of $(\mathbb{CP}(\mathfrak{H}), \Omega_{FS})$ can be exploited to obtain a prequantization of an arbitrary symplectic manifold (Γ, Ω) whenever there exists a *symplectomorphism* $\mathrm{Coh} : \Gamma \to \mathbb{CP}(\mathfrak{H})$. In this case, $\Omega = \mathrm{Coh}^*\Omega_{FS}$ and $\mathbb{L} = (\mathrm{Coh}^*\Xi(\mathfrak{H}_\sigma), \mathrm{Coh}^* H_{FS}, \mathrm{Coh}^*\nabla_{FS})$ is a prequantization of Γ, i.e. in particular, $\Omega = i\,\mathrm{curv}(\mathrm{Coh}^*\nabla_{FS})$. To make the connection with coherent states, we note that the elements of $\mathbb{L}$ are pairs (x, ψ) with $\psi \in \mathfrak{H}$ and $\frac{|\psi\rangle\langle\psi|}{||\psi||^2} = \Psi = \mathrm{Coh}(x)$. Let $U \subset \Gamma$ be an open dense set such that the restriction of $\mathbb{L}$ to U is trivial. Let $\mathrm{coh} : U \to \mathfrak{H}$ be a smooth map satisfying

$$\mathrm{Coh}(x) = \frac{|\mathrm{coh}(x)\rangle\langle\mathrm{coh}(x)|}{||\mathrm{coh}(x)||^2} \tag{331}$$

(such maps can always be found). Let us also write $\eta_x = \mathrm{coh}(x), \quad \forall x \in U$. Assume furthermore that the condition

$$\int_\Gamma |\eta_x\rangle\langle\eta_x| d\nu(x) = I_{\mathfrak{H}} \tag{332}$$

is satisfied, where $I_{\mathfrak{H}}$ is the identity operator on $\mathfrak{H}$ and ν is the Liouville measure on Γ, arising from Ω. We call the vectors η_x the *coherent states* of the prequantization.

In terms of the reproducing kernel $K(x,y) = \langle\eta_x|\eta_y\rangle$ and locally on U, the expression

$$\theta_K(x) = d_1 \log K(x_1, x_2)|_{x_1=x_2=x} = i(\mathrm{Coh}^*\theta_{FS})(x), \tag{333}$$

(d_1 denoting exterior differentiation with respect to x_1) defines a 1-form on $\mathbb{L}$, for which $\Omega = d\theta_K$. The hermitian metric $H_K = \text{Coh}^* H_{FS}$ and the compatible connection ∇_K, are given by

$$H_K((x,\psi),(x,\psi')) = \langle \psi, \psi' \rangle \tag{334}$$

$$\nabla_K \, \text{coh} = -i\theta_K \otimes \text{coh}, \tag{335}$$

and $i \, \text{curv} \nabla_K = \Omega$.

More generally, if $\text{Coh} : \Gamma \to \mathbb{CP}(\mathfrak{H})$ is only assumed to be a smooth map, not necessarily a symplectomorphism, and the associated coherent states satisfy

$$\int_\Gamma |\eta_x\rangle\langle\eta_x| d\nu(x) = A, \tag{336}$$

where A is a positive, bounded invertible operator with bounded inverse, the above scheme gives us a prequantization of the symplectic manifold (Γ, Ω_K) where $\Omega_K = \text{Coh}^* \Omega_{FS}$. In fact one has

Theorem 9.1 *The triple* $(\pi : \mathbb{L} \to \Gamma, H_K, \nabla_K)$, *where* $\nabla_K \text{coh} = -i\theta_K \otimes \text{coh}$, *is a hermitian line bundle with compatible connection, and* $i \, \text{curv} \nabla_K = \Omega_K$.

As a consequence of Theorem 9.1 we see that Ω_K so constructed has integral cohomology. Thus the pair (Γ, Ω_K) satisfies the *integrality condition.* We have thus obtained a geometric prequantization on (Γ, Ω_K) from the natural geometric prequantization of $(\mathbb{CP}(\mathfrak{H}), \Omega_{FS})$ via the family of coherent states $\{\eta_x\}$. While the new two-form Ω_K on Γ is integral, this is not necessarily the case for the original form Ω. If it is, then there exists a geometric prequantization on (Γ, Ω) which we may compare with the prequantization obtained using the coherent states. The original prequantization is said to be *projectively induced,* if $\Omega = \Omega_K$; if furthermore, Γ has a complex structure which is preserved by Coh, the symplectic manifold (Γ, Ω) turns out to be a Kähler manifold. For the Berezin quantization, discussed in the last section, the coherent states (301) can be shown to give rise to a projectively induced prequantization.

It ought to be pointed out that while the map $\text{Coh} : \Gamma \to \mathbb{CP}(\mathfrak{H})$ yields a prequantization of (Γ, Ω), the method outlined above does not give an explicit way to determine $\mathfrak{H}$ itself. However, starting with the Hilbert space $L^2(\Gamma, \nu)$, one can try to obtain subspaces $\mathfrak{H}_K \subset L^2(\Gamma, \nu)$, for which there are associated coherent states.

9.3 An example

Let $\Gamma = \mathbb{R}^6$, the phase space of a classical free particle. This is a symplectic manifold with two-form, $\Omega = \sum_{i=1}^3 dp_i \wedge dq^i$. Let $\mathfrak{H} = L^2(\Gamma, d\mathbf{p} \; d\mathbf{q})$ and we look for convenient subspaces of it which admit reproducing kernels. Let $e : \mathbb{R}^3 \longrightarrow \mathbb{C}$ be a measurable function satisfying

$$\int_{\mathbb{R}^3} |e(\mathbf{k})|^2 \, d\mathbf{k} = 1.$$

for $\ell = 0, 1, 2, \ldots$, denote by $\mathcal{P}_\ell$ the Legendre polynomial of order ℓ,

$$\mathcal{P}_\ell(x) = \frac{1}{2^\ell \ell!} \frac{d^\ell}{dx^\ell} (x^2 - 1)^\ell.$$

Define

$$\begin{aligned} K_{e,\ell}(\mathbf{q}, \mathbf{p}; \mathbf{q}', \mathbf{p}') = \frac{2\ell + 1}{(2\pi)^3} \int_{\mathbb{R}^3} e^{i\mathbf{k}\cdot(\mathbf{q}-\mathbf{q}')} \, \mathcal{P}_\ell \left(\frac{(\mathbf{k} - \mathbf{p}) \cdot (\mathbf{k} - \mathbf{p}')}{\|\mathbf{k} - \mathbf{p}\| \, \|\mathbf{k} - \mathbf{p}'\|} \right) \\ \times \overline{e(\mathbf{k} - \mathbf{p})} \, e(\mathbf{k} - \mathbf{p}) \, d\mathbf{k}. \end{aligned} \tag{337}$$

It is then straightforward to verify that $K_{e,\ell}$ is a reproducing kernel with the usual properties,

$$\begin{aligned} K_{e,\ell}(\mathbf{q}, \mathbf{p}; \mathbf{q}, \mathbf{p}) &> 0, \qquad (\mathbf{q}, \mathbf{p}) \in \Gamma, \\ K_{e,\ell}(\mathbf{q}, \mathbf{p}; \mathbf{q}', \mathbf{p}') &= \overline{K_{e,\ell}(\mathbf{q}', \mathbf{p}'; \mathbf{q}, \mathbf{p})}, \\ K_{e,\ell}(\mathbf{q}, \mathbf{p}; \mathbf{q}', \mathbf{p}') &= \int_{\mathbb{R}^6} K_{e,\ell}(\mathbf{q}, \mathbf{p}; \mathbf{q}'', \mathbf{p}'') \\ &\quad \times K_{e,\ell}(\mathbf{q}'', \mathbf{p}''; \mathbf{q}', \mathbf{p}') \, d\mathbf{p}'' \, d\mathbf{q}'', \end{aligned} \tag{338}$$

and we have the associated family of *coherent states*,

$$\mathfrak{S} = \{\xi_{\mathbf{q},\mathbf{p}} \in \mathfrak{H} \mid \xi_{\mathbf{q},\mathbf{p}}(\mathbf{q}', \mathbf{p}') = K_{e,\ell}(\mathbf{q}', \mathbf{p}'; \mathbf{q}, \mathbf{p}), \;\; (\mathbf{q}, \mathbf{p}), (\mathbf{q}', \mathbf{p}') \in \Gamma\} \tag{339}$$

which span the Hilbert subspace $\mathfrak{H}_{e,\ell} \subset \mathfrak{H}$ and satisfy the resolution of the identity on it:

$$\int_{\mathbb{R}^6} |\xi_{\mathbf{q},\mathbf{p}}\rangle\langle\xi_{\mathbf{q},\mathbf{p}}| \, d\mathbf{p} \, d\mathbf{q} = I_{e,\ell}. \tag{340}$$

Using these coherent states we can do a quantization using (177), i.e.,

$$f \longmapsto \widehat{f} = \int_{\mathbb{R}^6} f(\mathbf{q}, \mathbf{p}) |\xi_{\mathbf{q},\mathbf{p}}\rangle\langle\xi_{\mathbf{q},\mathbf{p}}| \, d\mathbf{p} \, d\mathbf{q}. \tag{341}$$

In particular, we get for the position and momentum observable the operators,

$$\widehat{q^j} = q^j - i\hbar \frac{\partial}{\partial p^j}, \quad \widehat{p^j} = -i\hbar \frac{\partial}{\partial q^j}, \qquad j = 1, 2, 3, \tag{342}$$

so that

$$[\widehat{q^j}, \widehat{p^j}] = i\delta_{ij} \, I_{e,\ell}.$$

This illustrates how identifying appropriate reproducing kernel Hilbert spaces can lead to a physically meaningful quantization of the classical system.

Let us next try to bring out the connection between this quantization and the natural prequantization on $\mathbb{CP}(\mathfrak{H}_{e,\ell})$. Consider the map,

$$\text{Coh} : \Gamma = \mathbb{R}^6 \longrightarrow \mathbb{CP}(\mathfrak{H}_{e,\ell}), \qquad \text{Coh}(\mathbf{q}, \mathbf{p}) = \frac{|\xi_{\mathbf{q},\mathbf{p}}\rangle\langle\xi_{\mathbf{q},\mathbf{p}}|}{||\xi_{\mathbf{q},\mathbf{p}}||^2}. \tag{343}$$

It is straightforward, though tedious, to verify that

$$\text{Coh}^*\Omega_{FS} = \Omega = \sum_{i=1}^{3} dp_i \wedge dq^i. \tag{344}$$

The pullback of the canonical line bundle $\Xi(\mathfrak{H}_{e,\ell})$ (over $\mathbb{CP}(\mathfrak{H}_{e,\ell})$ under Coh, gives us a line bundle over $\Gamma = \mathbb{R}^6$. (The integrality condition is trivially satisfied). Hence a geometric quantization is possible.

Take the reference section $\widehat{s}(\mathbf{q}, \mathbf{p}) = \xi_{\mathbf{q},\mathbf{p}}$ in the pulled back bundle,

$$\mathbb{L} = \text{Coh}^*\Xi(\mathfrak{H}_{e,\ell}).$$

Square-integrable sections of this bundle form a Hilbert space $\mathfrak{H}_{\mathbb{L}}$, with scalar product,

$$\langle s_1|s_2\rangle = \int_{\mathbb{R}^6} \overline{\Psi_1(\mathbf{q}, \mathbf{p})}\; \Psi_2(\mathbf{q}, \mathbf{p}) K_{e,\ell}(\mathbf{q}, \mathbf{p}; \mathbf{q}, \mathbf{p})\; d\mathbf{q}\; d\mathbf{p},$$
$$s_i(\mathbf{q}, \mathbf{p}) = \widehat{s}\Psi_i, \qquad i = 1, 2,$$

and again, $\mathfrak{H}_{\mathbb{L}}$ is naturally (unitarily) isomorphic to $L^2(\Gamma, d\mathbf{q}\; d\mathbf{p})$. We take the symplectic potential,

$$\theta = \sum_{i=1}^{3} p_i\; dq^i,$$

so that $\Omega = d\theta$ and once again we obtain a prequantization, yielding the position and momentum operators

$$\widehat{q^j} = -i\hbar\frac{\partial}{\partial p^j} + q^j, \qquad \widehat{p^j} = -i\hbar\frac{\partial}{\partial q^j},$$

which are the same as in (342).

Acknowledgments

The author would like to thank the organizers of the Second International Conference on Contemporary Problems in Mathematical Physics in Cotonou, and especially Prof. Norbert Hounkonnou, for inviting him to the meeting to give this set of lectures.

References

1. R. Abraham and J.E. Marsden, *Foundations of Mechanics*, Second Edition (Benjamin/Cummings, Reading, Massachusetts, 1978).
2. S.T. Ali, *Stochastic localisation, quantum mechanics on phase space and quantum space-time*, *Riv. Nuovo Cim.* **8**, 1–128 (1985).
3. S.T. Ali and H-D. Doebner, *Ordering problem in quantum mechanics: Prime quantization and a physical interpretation*, *Phys. Rev.* **A41**, 1199–1210 (1990).
4. S.T. Ali, *Survey of quantization methods*, in *Classical and Quantum Systems – Foundations and Symmetries*: Proceedings of The Second International Wigner Symposium, Goslar, Germany, July 16-20, Eds., H.-D. Doebner, W. Scherer and F. Schroeck, Jr. (World Scientific, Singapore, 1993).
5. F.A. Berezin, *Quantization*, *Math. USSR Izvestija* **8**, 1109–1165 (1974); *General concept of quantization*, *Commun. Math. Phys.* **40**, 153–174 (1975).
6. P.A.M. Dirac, *The Principles of Quantum Mechanics* (Clarendon Press, Oxford, 1930).
7. H.D. Doebner, P. Stŏvíček and J. Tolar, *Quantization of kinematics on configuration manifolds*, preprint `arXiv:math-ph/0104013 v2 12 Sept.` 2001, p. 41.
8. M. Englis, *Berezin quantization and reproducing kernels on complex domains*, *Trans. Amer. Math. Soc.* **348**, 411–479 (1996).
9. G.A. Goldin, R. Menikoff and D.H. Sharp, *Particle statistics from induced representations of a local current group*, *J. Math. Phys.* **21**, 650–664 (1980).
10. G.A. Goldin, R. Menikoff and D.H. Sharp, *Induced representations of the group of diffeomorphisms of* $\mathbb{R}^{3n}$, *J. Phys.* **A16**, 1827–1833 (1983).
11. H. Goldstein, *Classical Mechanics*, Second Edition (Addison-Wesley, Reading, Massachusetts, 1980).
12. V. Guillemin and S. Sternberg, *Symplectic Techniques in Physics*, (Cambridge University Press, Cambridge, 1984).
13. L. van Hove, *Sur certaines représentations unitaires d'un groupe infini de transformations*, *Mem. Acad. Roy. de Belgique, Classe des Sci.* **26**, 1–102 (1951).
14. N.E. Hurt, *Geometric Quantization in Action* (Reidel, Dordrecht, 1983).
15. B. Kostant, *Quantization and unitary representations*, in *Lecture Notes in Math.* **170** (Springer-Verlag, Berlin, 1970), pp. 85–208.
16. N.P. Landsman, *Mathematical Topics Between Classical and Quantum*

Mechanics (Springer-Verlag, New York, 1998).

17. P. Nattermann, *Dynamics in Borel Quantization: Nonlinear Schrödinger Equations vs. Master Equations*, Ph.D. Dissertation, Technische Universität Clausthal, Germany (1997).
18. J. von Neumann, *Mathematical Foundations of Quantum Mechanics*, English transl. by R.T. Byer (Princeton University Press, Princeton, NJ, 1955).
19. A. Odzijewicz, *Coherent states and geometric quantization*, *Commun. Math. Phys.* **150**, 385–413 (1992).
20. I.E. Segal, *Quantization of nonlinear systems*, *J. Math. Phys.* **1**, 468–488 (1960).
21. R. Simon, E.C.G. Sudarshan and N. Mukunda, *Gaussian pure states in quantum mechanics and the symplectic group*, *Phys. Rev.* **A37**, 3028–3038 (1988).
22. J. Sniatycki, *Geometric Quantization and Quantum Mechanics* (Springer-Verlag, Berlin, 1979).
23. G.M. Tuynman *Studies in Geometric Quantization*, Ph.D. Dissertation, Amsterdam, The Netherlands (1987).
24. V.S. Varadarajan, *Geometry of Quantum Theory*, Second Edition (Springer-Verlag, New York, 1985).
25. C. von Westenholz, *Differential Forms in Mathematical Physics* (North-Holland Publishing Co., Amsterdam, 1986).
26. N.J.M. Woodhouse, *Geometric Quantization*, Second Edition (Clarendon Press, Oxford, 1992).

THE QUANTUM GEOMETER'S UNIVERSE: PARTICLES, INTERACTIONS AND TOPOLOGY

JAN GOVAERTS

Institute of Nuclear Physics, Catholic University of Louvain
2, Chemin du Cyclotron, B-1348 Louvain-la-Neuve, Belgium
E-mail: jan.govaerts@fynu.ucl.ac.be

With the two most profound conceptual revolutions of $\mathrm{XX}^{\mathrm{th}}$ century physics, quantum mechanics and relativity, which have culminated into relativistic spacetime geometry and quantum gauge field theory as the principles for gravity and the three other known fundamental interactions, the physicist of the $\mathrm{XXI}^{\mathrm{st}}$ century has inherited an unfinished symphony: the unification of the quantum and the continuum. As an invitation to tomorrow's quantum geometers who must design the new rulers by which to size up the Universe at those scales where the smallest meets the largest, these lectures review the basic principles of today's conceptual framework, and highlight by way of simple examples the interplay that presently exists between the quantum world of particle interactions and the classical world of geometry and topology.

1 Introduction

It is often said that the profound conceptual revolutions of $\mathrm{XX}^{\mathrm{th}}$ century physics may be ascribed to three fundamental physical constants, namely Newton's constant G_N characteristic of the gravitational interaction, light's velocity in vacuum c displaying the relativistic character of physical reality, and Planck's constant $\hbar = h/2\pi$ as the hallmark for the quantum character of the physical universe. All of these constants have incessantly been used much like light beacons with which to probe the as yet unexplored territories beyond the known physical laws of our material world, grasping for this ever unfulfilled dream of the ultimate unification of all of matter, radiation and their interactions.

Each of these three constants on its own has led to its separate conceptual revolution, even beyond the confines of the scientific methods of physics, in ways that shall not be recalled here. However, when considered in combination, these constants imply still further profound conceptual revisions in our understanding of the physical world, which themselves stand out as the genuine unfinished revolutions of $\mathrm{XX}^{\mathrm{th}}$ century physics. Indeed, even though the combinations of G_N with c on the one hand, and of c and $\hbar$ on the other hand, have each led to a profound new vision onto the material universe through the physicist's eye, the formulation of a conceptual framework in which all three constants play an equally important role is the wide open problem that confronts physics in this XXIst century.

As is well known, the marriage of G_N and c leads to a curved spacetime whose geometry is dynamical and is governed by the energy-matter distribution within it, a framework within which the gravitational interaction is the physical manifestation of any curvature in space and in spacetime. The most fascinating offsprings of this union are undoubtedly, on the one hand, the cosmological theory of the history of our universe from its birth to its ultimate demise if ever, and on the other hand, the prediction for regions of spacetime to be so much curled up by their energy-matter content that even light can no longer escape from such black holes. For instance, the value

$$r_0 = 2\frac{G_N M}{c^2} \tag{1}$$

for the horizon of a neutral nonrotating black hole of mass M displays the combined contribution of gravity and relativity. These examples are but two specific outcomes of classical general relativity, a relativistic invariant theory of gravity whose construction is based on a simple geometrical thus physical principle: the description of physical processes should be independent of the local spacetime observer, namely, it should be independent of the choice of local spacetime coordinate parametrization. The theory should be invariant under arbitrary local coordinate transformations in spacetime.[a] In other words, a gauge invariance principle is at work, leading to a description of the gravitational interaction based on a simple but powerful symmetry and thus geometry principle.

On the other hand, the marriage of c and $\hbar$ leads naturally to the quantum field theory description of the elementary particles and their interactions, at the most intimate presently accessible scales of space and energy, a fact made manifest by the value for their product,

$$\hbar c \simeq 197 \text{ MeV} \cdot \text{fm} . \tag{2}$$

In fact, one offspring of this second union is the unification of matter and radiation, namely of particles with their corpuscular propagating properties and fields with their wavelike propagating properties. Particles, characterized through their energy, momentum and spin values in correspondence with the Poincaré symmetries of Minkowski spacetime in the absence of gravity, are nothing but the relativistic energy-momentum quanta of a field, thereby implying a tremendous economy in the description of the physical universe,

---

[a]Einstein's theory of general relativity has furthermore inscribed into it the equivalence principle between inertial and gravitational mass.

accounting for instance at once in terms of a single field filling all of spacetime for the indistinguishability of identical particles and their statistics. Furthermore, quantum relativistic interactions are then understood simply as couplings between the various quantum fields locally in spacetime, which translate in terms of particles as diverse exchanges of the associated quanta. Such a picture lends itself most ideally to a pertubative understanding of the fundamental interactions, which has proved to be so powerful beginning with quantum electrodynamics, up to the modern $SU(3)_C \times SU(2)_L \times U(1)_Y$ Standard Model of the strong and electroweak interactions. Such a perturbative representation of processes requires a renormalization procedure of the basic field parameters—their normalizations, masses and couplings—, and one has had to learn how to identify theories for which this renormalization programme is feasable. In the course of time, a general class of renormalizable field theories has been identified, all falling again under the general spell of the gauge symmetry principle as did the gravitational interaction!

Even though the physical meaning ascribed to the renormalizability criterion has evolved in such a manner that these theories are nowadays viewed rather as effective theories for some as yet unknown more fundamental description becoming manifest and relevant at still higher energies,[1] the fact remains that the gauge symmetry principle is again at work at the most intimate level of the unification of the relativistic quantum. But this time, this invariance under local transformations in spacetime applies to some "internal" space of degrees of freedom, that fields and their quanta carry along and which are made physically manifest through the different charges and quantum numbers that particles possess. Hence, through countless experiments performed at ever increasing energies and with ever increased technical sophistication, three generations of quarks and leptons, the basic building blocks of matter, each such generation being comprised of two quarks, one charged lepton and its associated neutrino, have been identified, and their reality inscribed into the construction of the Standard Model. All interactions among these six quarks and six leptons are governed by the gauge symmetry principle, with $SU(3)_C \times SU(2)_L \times U(1)_Y$ symmetries acting within internal space and independently, though in a continuous fashion, at each point of spacetime. This local realization of the symmetry requires the existence of gauge bosons, as the carriers of the symmetry and thus of the interactions from one spacetime point to the next. There are thus eight gluons for the strong interaction, the charged and neutral massive electroweak gauge bosons $W^{\pm}$ and Z_0 for the charged and neutral current weak interactions, and finally the photon for the electromagnetic interaction. Only one member of the Standard Model family has yet to be discovered experimentally, namely the so-called higgs particle which should be responsible for a mechanism at the origin of the masses for all quarks, leptons and massive gauge bosons. The higgs hunt is on at the most powerful particle accelerators in the world, the last missing offspring of the union of c and $\hbar$.

Given the fundamental role played by symmetries, hence also geometry, in the unifications of fundamental physics concepts achieved throughout the last century, it is fair to characterize XX$^{\rm th}$ century physics as the reign supreme of the symmetry principle, this principle being pushed into its most extreme realizations possible through the gauge symmetry principle. This includes the possibility of supersymmetry, a symmetry that relates bosonic and fermionic particles which, when rendered local in spacetime, leads to theories of supergravity that must necessarily include a quantum gravitational sector. But it also appears that this symmetry principle has finally unveiled all its hidden physical secrets in the embodiement it has acquired within a field theory description of the universe, of its matter content and of its fundamental interactions. Even though the symmetry principle seems to have yielded all its potential, it proves not to be potent enough to bring order to a *ménage à trois* in which all three fundamental constants G_N, c and $\hbar$ would be living peacefully and happily together on equal terms, to bear many news fruits of their ultimate union. As is well known, there does not yet exist a commonly accepted theoretical formulation for a quantum theory of relativistic gravity which would also include the other fundamental interactions and their matter fields, all consistently expressed within a quantum framework.

Looking back at the brief and superficial highlights recalled above, one realizes that the nonquantum relativistic description available for the gravitational interaction is in fact the ideal realm of the "relativistic continuum" reigning supreme, the utmost physical application as of today of the notion of differentiable structures in geometry. Likewise, the other component of the same story, namely the relativistic quantum field theory description of the elementary particles and their other fundamental interactions, is in fact the ideal realm of the "relativistic quantum" reigning supreme, the utmost physical outcome of the ideas of quantization and its associated abstract algebraic structures. The fundamental problem that XXI$^{\rm st}$ century physics is to confront is that of the final marriage of the "continuum" and the "quantum", namely of identifying a mathematical formulation of what is referred to as "quantum geometry", the new conceptualization of what the geometry of spacetime ought to be when explored at the most extreme and smallest scales.

In terms of the three fundamental constants G_N, c and $\hbar$, it is well known how the quantum regime for relativistic gravity is characterized by Planck's mass, length and time scales,

$$M_{\rm Pl} = \sqrt{\frac{\hbar c}{G_N}} \simeq 10^{19}\ {\rm GeV}/c^2\ \ ,\ \ L_{\rm Pl} = \frac{\hbar c}{M_{\rm Pl}c^2} \simeq 2\times 10^{-35}\ {\rm m}\ \ ,$$

$$\tau_{\rm Pl} = \frac{L_{\rm Pl}}{c} \simeq 6\times 10^{-44}\ {\rm s}\ . \qquad (3)$$

Even though these values lie way beyond the reach of present day accelerators, as well as of present day theories, processes at such scales must have taken place in the early universe, while from the conceptual point of view, the fundamental conflit between the classical relativistic realm of the "continuum" for gravity with the quantum relativistic realm of the "quantum" for particles and their other interactions, cries out to the XXI$^{\rm st}$ physicist for a new conceptual revolution that ought to resolve this basic mutual inconsistency of present day physics principles. From that point of view, XXI$^{\rm st}$ century physics will be the search for the Quantum Geometry Principle, the inherited unfinished physics symphony of the XX$^{\rm th}$ century composed so far according to the rules of the Symmetry Principle.

With the advent of M-theory,[2] the nonperturbative embodiement of superstring theories,[3] and possibly also with the loop gravity programme,[4] we are most probably already getting the first glimpses of this quantum geometry waiting to be discovered by tomorrow's bright young minds. Such a pursuit in search of the possible ultimate unification, all at the same time, of matter and its interactions, and of geometry and the quantum, belongs to the best of scientific traditions finding its roots back in the earliest days of the human intellectual adventure. It should only be just that within all peoples of the world, as much from developing as from developed countries, those whose calling lies towards such an avenue should find an environment within which to contribute on equal terms to this ultimate understanding of our physical universe and its history. A workshop of this type is an opportunity to highlight some of the issues surrounding this unfulfilled quest, and hopefully entice bright new minds to dedicate themselves to this adventure at the frontiers of physical concepts. The education to critical and scientific thinking that such a research activity requires can only benefit any society within which it is pursued, both in its human and intellectual aspirations as well as in its educational, technological and economic development, bringing man always a little closer to the stars, the eternal yearning of his soul. Countless examples over human history bear witness to this fact, and many of us today benefit in so many ways from the fruits of this unswaying quest at the most abstract level as it has been pursued over centuries past.

These lecture notes do not, of course, have any pretence to outline what quantum geometry ought to be, which, after all, is the XXI$^{\rm st}$ century quantum geometer's task! Rather, these lectures wish to present sort of a guided tour of the general principles of symmetry and quantum physics that have led to the relativistic quantum field theory description of the elementary particles and their fundamental interactions, aiming at the end towards illustrations of the fact that beyond the gauge symmetry principle which seems to govern all interactions, when it comes to geometry—namely the "continuum" and gravity—and the "quantum", topology is also called to play a vital role. In fact, one is very much led to suggest that the problem of quantum gravity should find a resolution only when considered together with all the other quantum matter and interacting fields, while pure quantum gravity is oblivious actually to any geometry, and would be governed only by the rules of quantum topology. Indeed, this is the programme that was launched[5,6] with the discovery of topological quantum field theories.[5,6,7] Finally, these notes concentrate on the quantum field theory side of the above story, assuming that the reader is most familiar already with the views of classical continuum geometry as applied within the physical context of the gravitational interaction and general relativity. This is thus the spirit with which these notes are offered to the aspiring quantum geometers of the XXI$^{\rm st}$ century who are attending this Workshop.

Contents are organized as follows. Section 2 discusses the general rules of abstract canonical quantization, based on the Hamiltonian formulation of a given dynamical system. These rules are then applied to relativistic field theories in Section 3, to establish that such quantized theories provide a natural description of quantum relativistic particles in Minkowski spacetime. Section 4 introduces then to interacting quantum field theories and, as a general class of renormalizable theories in four dimensions, to general Yang-Mills theories, possibly subjected to the Higgs mechanism of spontaneous symmetry breaking. This discussion thus also serves as a motivation for Section 5 which addresses the general problem of the quantization of systems subjected to constraints in phase space, which include any gauge invariant system, following Dirac's general analysis of this issue.[8] Rather than introducing then the general methods of BRST quantization, the recent and most efficient approach towards the quantization of constrained systems based on the physical projector[9] is also discussed. As an example of its possible use, the quantization of 2+1-dimensional Chern-Simons theory is briefly described in Section 6, which in fact is one of the simplest examples of a topological quantum field theory. Finally, Sections 7 and 8 introduce to bosonic string theory and its toroidal compactification. These last three sections serve as first witnesses to the necessity to develop a new mathematical framework for quantum theories of gravity, whether they include matter degrees of freedom or not, that should define the sought-for "quantum geometry" of the fundamental unification. Finally, further comments are presented in the Conclusions.

Our conventions will be stated where appropriate. Notice also that all the discussion will be confined to bosonic degrees of freedom only, but that similar developments exist of course for systems combining both bosonic and fermionic degrees of freedom. Suggestions for some exercises are also provided, some of which could in fact become PhD research topics on their own. Finally, no attempt has been made at providing an exhaustive bibliography, for which we apologize to anyone who might feel her/his work is being overlooked. Rather, we hope that references given would suffice to quickly identify further relevant sources to any particular topic of interest.

2 Abstract Canonical Quantization

This section briefly reviews[10] the general rules of abstract canonical quantization, starting from some action principle defining the actual dynamics of a given physical system. This discussion is elaborated upon much further in Prof. S.T. Ali's lectures in these Proceedings, dedicated to the general problem of quantization. Prof. J.R. Klauder's contribution is also directly related to some of the issues addressed in this section. Furthermore, to illustrate explicitly the general discussion through the simplest of nontrivial examples, most points of relevance are also discussed in the context of the one-dimensional harmonic oscillator.

2.1 Dynamics

Let us consider a given physical system of N real degrees of freedom $q^n(t)$ ($n = 1, 2, \cdots, N$), whose dynamics derives from the variational principle based on an action which is local in time and of the form

$$S[q^n] = \int dt\, L\left(q^n, \dot{q}^n\right) \ , \tag{4}$$

defined in terms of the Lagrange function $L(q^n, \dot{q}^n)$, where $\dot{q}^n = dq^n/dt$. In the case of the one-dimensional harmonic oscillator, one simply has $N = 1$ with

$$L(q, \dot{q}) = \frac{1}{2}m\dot{q}^2 - \frac{1}{2}m\omega^2 q^2 \ , \quad \omega, m > 0 \ . \tag{5}$$

It is well known how the variational principle leads to the following Euler-Lagrange equations of motion,

$$\frac{d}{dt}\frac{\partial L}{\partial \dot{q}^n} - \frac{\partial L}{\partial q^n} = 0 \quad , \quad n = 1, 2, \cdots, N \; . \tag{6}$$

In general, any given variation of the action also includes surface term contributions—namely, total time derivative contributions for a mechanical system—which one may also want to set to zero in order to enforce the variational principle in a strong sense, namely such that the action is an exact extremum for classical solutions. Depending on one's choice of boundary conditions, as motivated by specific physics considerations in order to solve the above equations of motion, such a requirement may be too restrictive. Whatever the case, a complete set of boundary conditions must be specified in order to determine a unique solution to the equations of motion of the system.

In the case of the simple harmonic oscillator, the Euler-Lagrange equation of motion reads

$$m\ddot{q} = -m\omega^2 \, q \; , \tag{7}$$

whose general solution may be expressed as follows for later convenience,

$$q(t) = \frac{1}{\sqrt{2m\omega}} \left[\alpha \, e^{-i\omega(t-t_0)} + \alpha^* \, e^{i\omega(t-t_0)} \right] \; . \tag{8}$$

Here, α and α^* are complex integration constants (complex conjugates of one another) which may be expressed in terms of whatever choice of boundary conditions that might be contemplated, and t_0 is some specific time reference possibly associated to the choice of boundary conditions.

2.2 Hamiltonian formulation

Since canonical quantization proceeds from the Hamiltonian formulation of a dynamical system, let us now introduce the phase space description of a general system as represented in Section 2.1. Phase space is thus spanned by the functions $q^n(t)$ and their conjugate momenta $p_n(t)$ defined as

$$p_n = \frac{\partial L}{\partial \dot{q}^n} \; . \tag{9}$$

As a matter of fact, phase space provides, within the Hamiltonian formulation, the space of states which are accessible to the system throughout its history. In addition to its parametrization in terms of the local coordinates (q^n, p_n), phase space also comes equipped with a geometric symplectic structure which is related to the Poisson bracket algebra of observables or functions defined over phase space. This Poisson bracket structure is defined at a fixed reference time $t = t_0$, and Poisson brackets are to be evaluated at equal time, namely for two observables defined at the same instant in time, $t = t_0$. Given the algebraic properties of the Poisson bracket structure, its values are derived from those for the elementary phase space degrees of freedom (q^n, p_n) of the system, given by

$$\{q^n(t_0), p_m(t_0)\} = \delta^n_m. \tag{10}$$

More generally, the Poisson bracket of any two observables $F(q^n, p_n)$ and $G(q^n, p_n)$ is given by[b]

$$\{F, G\} = \frac{\partial F}{\partial q^n}\frac{\partial G}{\partial p_n} - \frac{\partial F}{\partial p_n}\frac{\partial G}{\partial q^n} . \tag{11}$$

As is well known, Poisson brackets obey a series of important algebraic properties by which they are characterized, among which their antisymmetry as well as the Jacobi identity. Furthermore, according to Darboux's theorem, there always exists a local choice of phase space parametrization for which the Poisson brackets take the canonical form in (10), in which the associated coordinates are also referred to as being canonical.

Finally, besides this kinematics information encoding specific properties of the system, its dynamics is generated through the Poisson bracket from a specific observable, namely the canonical Hamiltonian defined through the Legendre transform of the Lagrange function,

$$H_0(q^n, p_n) = \dot{q}^n p_n - L(q^n, \dot{q}^n) . \tag{12}$$

Usually at this point, it is said that this definition applies to those systems for which the relations $p_n(q^n, \dot{q}^n)$ defining the conjugate momenta may all be inverted in a unique fashion in terms of the generalized velocities $\dot{q}^n(q^n, p_n)$, namely under the condition of a regular Hessian for L,

$$\det \frac{\partial^2 L}{\partial \dot{q}^{n_1} \partial \dot{q}^{n_2}} \neq 0 . \tag{13}$$

Systems for which this is feasible are called "regular", or rather described by a "regular Lagrangian", and otherwise "singular" in the case of a singular Hessian.[c] However, whether this latter condition is met or not does not affect the fact that the quantity H_0 introduced above is always[10] a function of q^n and p_n, namely a function defined over phase space. Actually, the difference between a regular and a singular Lagrangian is that in the latter case, there

---

[b]Throughout these notes, implicit summation is understood for repeated indices.
[c]Singular systems are the topic of Section 4.

exist specific constraints among the phase space degrees of freedom, which imply that time evolution of the system may be generated by a Hamiltonian $H(q^n, p_n)$ more general than simply the canonical one $H_0(q^n, p_n)$. In the present section, we assume the system to be regular, and thus to be free of any further constraint on the phase space degrees of freedom.

Consequently, time evolution of any phase space observable $F(q^n, p_n; t)$ is determined from the Hamiltonian equation of motion

$$\dot{F} = \frac{dF}{dt} = \frac{\partial F}{\partial t} + \{F, H_0\} , \tag{14}$$

while in particular for the phase space degrees of freedom one obtains the first-order Hamiltonian equations of motion

$$\dot{q}^n = \frac{\partial H_0}{\partial p_n} \quad , \quad \dot{p}_n = -\frac{\partial H_0}{\partial q^n} . \tag{15}$$

When solving for the conjugate momenta p_n in terms of the velocities $\dot{q}^n$ through the first set of these equations, and then substituting these expressions for p_n in the second set of equations, one recovers the original Euler-Lagrange equations of motion of the Lagrangian formulation of the same dynamical system. Note that the Hamiltonian equations of motion also follow through the variational principle from a first-order phase space action given by

$$S[q^n, p_n] = \int dt \, [\dot{q}^n p_n - H_0(q^n, p_n)] , \tag{16}$$

or some other expression differing from the present one by a total time derivative of an arbitrary function of phase space.

In the particular case of the simple one-dimensional harmonic oscillator described by the Lagrange function (5), one finds for the conjugate momentum

$$p = m\dot{q} , \tag{17}$$

with the canonical Poisson bracket structure

$$\{q(t_0), p(t_0)\} = 1 . \tag{18}$$

The canonical Hamiltonian reads

$$H_0 = \frac{p^2}{2m} + \frac{1}{2} m\omega^2 q^2 , \tag{19}$$

leading to the Hamiltonian equations of motion

$$\dot{q} = \frac{1}{m} p \quad , \quad \dot{p} = -m\omega^2 q . \tag{20}$$

Their general solution may again be given in the general parametrization used previously,

$$q(t) = \frac{1}{\sqrt{2m\omega}} \left[\alpha\, e^{-i\omega(t-t_0)} + \alpha^*\, e^{i\omega(t-t_0)}\right] \ ,$$
$$p(t) = -\frac{im\omega}{\sqrt{2m\omega}} \left[\alpha\, e^{-i\omega(t-t_0)} - \alpha^*\, e^{i\omega(t-t_0)}\right] \ . \qquad (21)$$

From this particular example, we may also extend a general remark. Clearly, there is a one-to-one correspondence between specific solutions to the Hamiltonian equations of motion and the associated integration constants, for instance in the present case between the solutions for $q(t)$ and $p(t)$ and the complex integration constant α. Hence, rather than viewing the Poisson bracket structure as being defined over phase space, namely the set of all functions $(q^n(t), p_n(t))$, one may equivalently view the Poisson bracket structure as being defined on the space of integration constants for the associated equations of motion. This alternative but equivalent point of view may in fact be quite relevant and physically meaningfull. Indeed, as we shall see in the case of quantum field theories, energy-momentum quanta that realize the quantum particle content of the theory correspond precisely to the operators associated to the field integration constants, whose commutation relations are nothing but the operator realization of the corresponding classical Poisson bracket structure defined on the space of integration constants. In the simpler case of the harmonic oscillator, the same point of view implies the following nontrivial Poisson bracket for the complex integration constants α and α^*,

$$\{\alpha, \alpha^*\} = -i \ , \qquad (22)$$

while the canonical Hamiltonian then acquires the expression,

$$H_0 = \frac{1}{2}\omega\left[\alpha^*\alpha + \alpha\alpha^*\right] \ , \qquad (23)$$

an expression in which care has been exercised not to commute quantities which at the quantum level will correspond to noncommuting operators. Consequently, the Hamiltonian equations of motion for these integration constants, or rather now variables, are

$$\dot{\alpha} = -i\omega\,\alpha \quad , \quad \dot{\alpha}^* = i\omega\,\alpha^* \ , \qquad (24)$$

with the general solution

$$\alpha(t) = \alpha_0 e^{-i\omega(t-t_0)} \quad , \quad \alpha^*(t) = \alpha_0^* e^{i\omega(t-t_0)} \ , \qquad (25)$$

which thus corresponds to the following representation for the general phase space solution in (21),

$$q(t) = \frac{1}{\sqrt{2m\omega}}\left[\alpha(t) + \alpha^*(t)\right] \quad , \quad p(t) = -\frac{im\omega}{\sqrt{2m\omega}}\left[\alpha(t) - \alpha^*(t)\right] \ . \tag{26}$$

Here, the initial values α_0 and α_0^* at $t = t_0$ stand for the values α and α^* of the integration constants as they appear in the general phase space solution (21). At the quantum level, this correspondence between the solutions to the equations of motion in phase space and the solutions on the space of integration constants will translate into the Schrödinger and Heisenberg pictures for the quantized system, in which in the first case states are time dependent while operators are time independent, and vice-versa in the second case.

2.3 Canonical quantization

Canonical quantization simply proceeds through the correspondence principle. In the same manner that the classical Hamiltonian formulation of any dynamical system is characterized by three structures, namely the space of states or phase space, the kinematical structure of that space as embodied algebraically through its Poisson bracket structure, and finally its dynamics generated by the Hamiltonian, likewise any quantum system is characterized by three such structures, namely the space of quantum states or Hilbert space, equipped with a series of algebraic structures providing a linear representation space of the algebraic structure of commutation relations which are in direct correspondence with the classical Poisson brackets, and finally a quantum Hamiltonian which generates time evolution of states through the Schrödinger equation.

More specifically, the space of quantum states must be some complex vector space equipped with an hermitean inner product denoted using Dirac's bra-ket notation as $<\varphi|\psi>$ for two states $|\varphi>$ and $|\psi>$. In the best of cases, this complex space ought to be a Hilbert space in the strict mathematical sense, but often it is difficult to meet that stringent requirement, and one has to extend somewhat the relevant complex vector space. Such issues are addressed in Prof. S.T. Ali's lectures in these Proceedings.

Furthermore, this "Hilbert" space must provide a linear representation space of the quantum operator algebra of observables of the quantized system. Ideally, what one would wish is that associated to any two classical observables A and B defined over phase space, there should exist two linear quantum operators $\hat{A}$ and $\hat{B}$ whose commutator algebra is in correspondence with the

Poisson bracket of the classical observables through the rule

$$[\hat{A}, \hat{B}] = i\hbar \hat{C} \ , \tag{27}$$

where $\hat{C}$ stands for the operator to be associated to the result of the classical Poisson bracket $\{A, B\} = C$. It is understood that this correspondence rule is to be considered at a fixed reference time t_0, which has not been displayed in the above relation. Furthermore, the choice of hermitean inner product ought to be such that the operators $\hat{A}$ and $\hat{B}$ associated to classical observables which are real under complex conjugation, $A^* = A$ and $B^* = B$, are themselves self-adjoint operators, $\hat{A}^\dagger = \hat{A}$ and $\hat{B}^\dagger = \hat{B}$.

These requirements, which actually define what is precisely meant by "quantization", are by no means easily met, as is discussed in detail in Prof. S.T. Ali's lectures. To understand part of the difficulty, let us consider this requirement already for the elementary phase space degrees of freedom (q^n, p_n), which ought thus to correspond to operators $(\hat{q}^n, \hat{p}_n)$ obeying the Heisenberg algebra,

$$[\, \hat{q}^n(t_0), \hat{p}_m(t_0)\,] = i\hbar\, \delta^n_m \ , \tag{28}$$

as well as the following hermiticity properties on the space of quantum states,

$$(\hat{q}^n(t_0))^\dagger = \hat{q}^n(t_0) \ \ , \ \ (\hat{p}_n(t_0))^\dagger = \hat{p}_n(t_0) \ . \tag{29}$$

Since all other observables of the system are defined as composite operators built from these $\hat{q}^n$'s and $\hat{p}_n$'s, clearly the fact that they no longer commute at the quantum level implies that a specific choice has to be made as to the order in which they are multiplied with one another. *A priori* the only restriction that should apply to this choice of "operator ordering" is that a given composite observable real under complex conjugation should be associated to a self-adjoint operator. For some particular observables related to symmetries that a given system may possess, namely the associated Noether charges, further restrictions would follow from the requirement that these operators still generate the corresponding symmetry for the quantized system. However, it is far from trivial that such a construction is at all possible, and the fact of the matter is that it is indeed problematic, leading to "the problem of quantization" as discussed in Prof. S.T. Ali's lectures. Not all classical observables may be put into correspondence with a self-adjoint quantum operator. When this is impossible for a symmetry generator while at the same time preserving the symmetry algebra now at the quantum level, one says that the symmetry has become anomalous or that it suffers a quantum anomaly, even though a better term would be "quantum symmetry breaking" given that the symmetry is explicitly broken at the quantum level, being no longer a symmetry of the quantized system.

In practical terms, given a classical system, there may thus exist more than one unitarily inequivalent quantum system that corresponds to it, since there may exist more than one choice of operator ordering consistent with the above correspondence rules for a certain subset of all classical observables[11]. It thus seems more appropriate to advocate the point of view that one has to define some quantum system, having in the back of one's mind some classical system, in terms of some algebra of self-adjoint elementary and composite operators, which in the limit $\hbar \to 0$ is in correspondence with some classical system. After all, the physical world is quantum mechanical, rather than being that of classical mechanics whether relativistic or not. The fact that a given classical system may correspond to more than one quantum system then becomes an issue to be resolved only through experiment. For instance, there exists a discrete infinity of rotationally invariant quantum systems labelled by the SU(2) spin value, which all correspond to the same classical system of a nonrelativistic point-particle invariant under spatial rotations. Only an experiment can determine which of these quantum representations is in fact realized in a specific physical system, whether the electron or a heavy nucleus in some excited state, for example.

Finally, the last structure which defines the abstract canonical quantization of a classical system is the Hamiltonian operator $\hat{H}_0$, which should be in correspondence with the classical Hamiltonian H_0. Since in general this observable is a composite quantity, the definition of the quantum Hamiltonian $\hat{H}_0$ involves a specific choice of operator ordering such that $\hat{H}_0$ be self-adjoint given the hermitean inner product of which the space of quantum states is equipped. This is a necessary requirement for the quantum unitarity, hence the physical consistency, of the quantum system. Indeed, time evolution of quantum states $|\psi, t>$ is generated precisely through the quantum Hamiltonian and Schrödinger's equation,

$$i\hbar\frac{d}{dt}|\psi, t>= \hat{H}_0\ |\psi, t> \ , \tag{30}$$

whose formal solution is of the form

$$|\psi, t>= \hat{U}(t, t_0)\,|\psi, t_0> \ , \tag{31}$$

where $|\psi, t_0>$ is some boundary state at the initial time t_0, while the quantum evolution operator or "propagator" of the system is formally defined by the exponential operator

$$\hat{U}(t_2, t_1) = e^{-\frac{i}{\hbar}(t_2 - t_1)\hat{H}_0} \ . \tag{32}$$

Quantum unitarity of time evolution thus requires $\hat{U}(t_2, t_1)$ to be unitary and to obey the involution property

$$\hat{U}^\dagger(t_2,t_1) = \hat{U}^{-1}(t_2,t_1) = \hat{U}(t_1,t_2) \quad , \quad \hat{U}(t_3,t_2)\,\hat{U}(t_2,t_1) = \hat{U}(t_3,t_1) \ , \quad (33)$$

which certainly requires $\hat{H}_0$ to be self-adjoint on the space of quantum states. The choice of operator ordering should therefore be consistent with this basic property for $\hat{H}_0$.

Note that the entire discussion concerning the definition of the algebraic operator structure defining the quantization of a given classical system is performed at a specific reference time t_0, namely all operators are regarded as having been defined at that reference time. Time dependency of the dynamical quantum system is totally accounted for through the time dependency of quantum states $|\psi, t>$ as solutions to the Schrödinger equation. This formulation of a quantum system is referred to as "the Schrödinger picture", which is in correspondence with the classical phase space description of the system in which the Poisson bracket structure is carried by the time dependent degrees of freedom $(q^n(t), p_n(t))$. The fact that the quantum operator algebra is defined at a fixed reference time t_0 raises the question of whether the quantized system is dependent on that choice. However, two quantizations defined at different reference times in the Schrödinger picture are unitarily equivalent through the evolution operator $\hat{U}(t_2, t_1)$ associated to a specific choice of self-adjoint quantum Hamiltonian $\hat{H}_0$. In other words, the quantum evolution operator defines the isomorphism between all unitarily equivalent representations of the algebraic structures associated to different reference times. For this isomorphism to be a unitary one, it is essential that the Hamiltonian $\hat{H}_0$ be self-adjoint. Later on, we shall discuss briefly the "Heisenberg picture" for a quantum system, as an alternative to the Schrödinger one, which will be seen to correspond at the classical level to the Poisson bracket structure being defined on the space of integration constants for the Hamiltonian equations of motion as discussed previously.

As a specific simple illustration of the above general discussion, let us consider again the one-dimensional harmonic oscillator. As a quantum system, its space of states should thus provide a linear representation space of the Heisenberg algebra

$$[\hat{q}, \hat{p}] = i\hbar \ , \quad (34)$$

equipped with an hermitean inner product for which these two operators be self-adjoint,

$$\hat{q}^\dagger = \hat{q} \quad , \quad \hat{p}^\dagger = \hat{p} \ . \quad (35)$$

Alternatively, given the physical parameters m and ω carrying physical dimensions, it is possible to represent in an equivalent way the same algebra in terms of the following combinations,

$$\hat{q} = \sqrt{\frac{\hbar}{2m\omega}}\,\left[a + a^\dagger\right] \quad , \quad \hat{p} = -i\sqrt{\frac{\hbar m\omega}{2}}\,\left[a - a^\dagger\right] \ , \tag{36}$$

which are seen to be in direct correspondence with the classical solutions in (21), or equivalently,

$$a = \sqrt{\frac{m\omega}{2\hbar}}\left[\hat{q} + \frac{i}{m\omega}\hat{p}\right] \quad , \quad a^\dagger = \sqrt{\frac{m\omega}{2\hbar}}\left[\hat{q} - \frac{i}{m\omega}\hat{p}\right] \ . \tag{37}$$

The operators a and $a^\dagger$ are simply the well known annihilation and creation operators of the harmonic oscillator, obeying the Fock space algebra,

$$[a, a^\dagger] = \mathbb{1} \quad , \tag{38}$$

with a and $a^\dagger$ being the adjoint of one another. In other words, given a physical parameter with the dimension of $m\omega$, the Heisenberg and Fock space algebras are equivalent. The quantum space of states should thus provide a representation space of either algebra.

Finally, given the classical Hamiltonian $H_0 = p^2/(2m) + m\omega^2 q^2/2$, the quantum Hamiltonian may be chosen to be

$$\hat{H}_0 = \frac{\hat{p}^2}{2m} + \frac{1}{2}m\omega^2\hat{q}^2 = \frac{1}{2}\hbar\omega[a^\dagger a + a a^\dagger] = \hbar\omega\left[a^\dagger a + \frac{1}{2}\right] \ . \tag{39}$$

Note that in terms of $\hat{q}$ and $\hat{p}$, $\hat{H}_0$ does not require some operator ordering prescription, which would not be the case had one chosen to consider its definition in terms of the creation and annihilation operators. Of course in the present case, the difference between the two possibilities amounts to a constant shift in the energy eigenvalues without any physical consequence, corresponding to the vacuum quantum fluctuation contribution. For more general systems however, such a difference may not be as innocuous.

2.4 Representations of the Heisenberg algebra

Canonical quantization thus raises generally the issue of the classification of the representations of the Heisenberg algebra. Such representations will be discussed here only in the simplest of cases, namely a set of cartesian degrees of freedom q^n, restricting the presentation to a single such degree of freedom, $N = 1$, the general case being obtained through a straightforward tensor product over $n = 1, 2, \cdots, N$. A generalization to spaces of nontrivial geometry and/or topology, for curvilinear coordinates and curved configurations spaces q^n whose topology may be compact or not, is also possible, but shall not be addressed here.[12,13] In the case of cartesian phase space canonical coordinates (q,p), it is well known that up to unitary equivalence, there exists only one representation of the Heisenberg algebra defined in (34) and (35). We refrain here from discussing how this conclusion may be reached in terms familiar to most physicists,[10,12,13] and only present the description of this representation in its different realizations, namely the configuration space, the momentum space, the Fock space, and the coherent state representations.

The configuration space representation

Given that the classical variable q may take all real values, one assumes that there exists a basis of position $\hat{q}$-eigenstates $|q>$ with eigenvalues q taking all possible real values, and normalized to unity on the real line,

$$\hat{q}|q>= q|q> \ , \quad <q|q'>=\delta(q-q') \ , \quad \mathbb{1}=\int_{-\infty}^{\infty} dq\,|q><q| \ . \tag{40}$$

This representation is known as "the configuration space representation" of the Heisenberg algebra, leading to the configuration space wave function representation $\psi(q)$ of any state $|\psi>$ in the associated space of quantum states,

$$\psi(q)=<q|\psi> \quad , \quad |\psi>=\int_{-\infty}^{\infty} dq\,|q>\psi(q) \ . \tag{41}$$

This wave function $\psi(q)$ thus provides the components of the state $|\psi>$ in the configuration space basis $|q>$. In particular, inner products of states are simply given by

$$<\varphi|\psi>=\int_{-\infty}^{\infty} dq\,\varphi^*(q)\psi(q) \ , \tag{42}$$

as follows directly from the above spectral decomposition of the unit operator $\mathbb{1}$ and the configuration space decomposition of the states $|\varphi>$ and $|\psi>$. Given this construction, it then follows that the abstract position and momentum operators $\hat{q}$ and $\hat{p}$ possess the following configuration space representations,

$$<q|\hat{q}|\psi>=q\,\psi(q) \quad , \quad <q|\hat{p}|\psi>=-i\hbar\frac{d}{dq}\psi(q) \ . \tag{43}$$

This, of course, is nothing but the most familiar wave function quantization of a single cartesian degree of freedom system.

The momentum space representation

Likewise for the conjugate phase space degree of freedom, one has the momentum space representation which is spanned by the momentum eigenstate basis $|p>$, which in the case of the real line is associated to all possible real momentum eigenvalues p, and is normalized to unity,

$$\hat{p}|p>=p|p> \quad , \quad <p|p'>=\delta(p-p') \quad , \quad \mathbb{1}=\int_{-\infty}^{\infty} dp\,|p><p| \; . \tag{44}$$

Correspondingly, arbitrary states are decomposed in that basis in terms of their momentum space wave function,

$$\psi(p)=<p|\psi> \quad , \quad |\psi>=\int_{-\infty}^{\infty} dp\,|p>\psi(p) \quad , \quad <\varphi|\psi>=\int_{-\infty}^{\infty} dp\,\varphi^*(p)\psi(p) \; . \tag{45}$$

Finally, the elementary operators are represented as

$$<p|\hat{q}|\psi>=i\hbar\frac{d}{dp}\psi(p) \quad , \quad <p|\hat{p}|\psi>=p\,\psi(p) \; . \tag{46}$$

These results are of course most familiar.

The Fock space representation

Given the fact that the Heisenberg algebra may alternatively be represented through the Fock space algebra (38) (provided a parameter with the physical dimension of $m\omega$ is available), one may also consider the Fock space representation of the Heisenberg algebra in the case of the harmonic oscillator. As is well known, a basis of this representation is constructed as follows. There exists a vacuum state $|0>$ (not to be confused with either of the states $|q=0>$ or $|p=0>$), normalized to unity, and annihilated by the annihilation operator a,

$$a|0>=0 \quad , \quad <0|0>=1 \; . \tag{47}$$

Applying then in succession powers of the creation operator $a^\dagger$ onto the Fock vacuum $|0>$, one obtains a discrete infinite set of linearly independent orthonormalized states which span the representation space, defined by

$$|n>=\frac{1}{\sqrt{n!}}\left(a^\dagger\right)^n|0> \quad , \quad <n|m>=\delta_{n,m} \; , \tag{48}$$

thus leading to the spectral representation of the unit operator

$$\mathbb{1}=\sum_{n=0}^{\infty}|n><n| \; . \tag{49}$$

Correspondingly, arbitrary states may be decomposed in that basis according to

$$\psi_n = < n|\psi > \quad , \quad |\psi >= \sum_{n=0}^{\infty} |n > \psi_n \quad , \quad < \varphi|\psi >= \sum_{n=0}^{\infty} \varphi_n^* \psi_n \; . \tag{50}$$

From the definition of the states $|n >$, one readily establishes that the annihilation and creation operators possess the following representations,

$$a|n >= \sqrt{n}\,|n-1> \quad , \quad a^\dagger|n >= \sqrt{n+1}\,|n> \quad , \quad a^\dagger a|n >= n|n > \; , \tag{51}$$

from which the Heisenberg matrix representation for the position and momentum operators $\hat{q}$ and $\hat{p}$ may be derived in terms of semi-infinite matrices.

The coherent state representation

Finally, an overcomplete set of basis vectors is provided by the holomorphic or phase space coherent states defined by, respectively,[14]

$$|z >= e^{-\frac{1}{2}|z|^2}\, e^{z a^\dagger}\, |0> \quad , \quad |q,p >= e^{-\frac{i}{\hbar} q\hat{p}}\, e^{\frac{i}{\hbar} p\hat{q}}\, |0 > \; , \tag{52}$$

z being an arbitrary complex number and (q,p) arbitrary real quantities in correspondence with the classical phase space canonical degrees of freedom. These two sets of states are related as

$$|z >= e^{\frac{i}{2\hbar} qp}\, |q,p > \; , \tag{53}$$

with

$$z = \sqrt{\frac{m\omega}{2\hbar}} \left[q + \frac{i}{m\omega} p \right] \; , \tag{54}$$

to be compared to the definitions for a and $a^\dagger$ in (37). The spectral resolution of the unit operator is then given by

$$\mathbb{1} = \int \frac{dz\, d\bar{z}}{\pi}\, |z >< z| \; = \int_{(\infty)} \frac{dq\, dp}{2\pi\hbar}\, |q,p >< q,p| \; , \tag{55}$$

while for matrix elements one finds for instance for the holomorphic coherent states (a similar expression may easily be established for the matrix element $< q_1, p_1|q_2, p_2 >$)

$$< z_1|z_2 >= e^{-\frac{1}{2}|z_1|^2}\, e^{-\frac{1}{2}|z_2|^2}\, e^{\bar{z}_1 z_2} \; . \tag{56}$$

The interest of coherent states stems, among other reasons, from the fact that they provide quantum states whose properties are the closest possible to classical states. This is made manifest for example by the following action of the annihilation operator on holomorphic coherent states,

$$a^n|z>= z^n|z> \ , \tag{57}$$

making these states an ideal choice of basis to compute matrix elements of the creation and annihilation operators of the Fock space algebra representation for the Heisenberg algebra.

Changes of bases

The fact that all these different realizations of the same basic algebraic structure, namely that of the Heisenberg algebra, are unitarily equivalent, may be made explicit in different ways, such as for example by specifying the matrix elements for the different changes of bases related to the different representations described above. Thus, the relation between the configuration and momentum space representations is provided by the matrix elements

$$<q|p>= \frac{1}{\sqrt{2\pi\hbar}}\, e^{\frac{i}{\hbar}qp} \quad , \quad <p|q>= \frac{1}{\sqrt{2\pi\hbar}}\, e^{-\frac{i}{\hbar}qp} \ , \tag{58}$$

from which it should be clear that the relation between the configuration and momentum wave functions $\psi(q)$ and $\psi(p)$ for a given state $|\psi>$ is simply that of an ordinary Fourier transformation,

$$\begin{aligned} \psi(p) &= <p|\psi>= \int_{-\infty}^{\infty} dq <p|q><q|\psi>= \int_{-\infty}^{\infty} \frac{dq}{\sqrt{2\pi\hbar}}\, e^{-\frac{i}{\hbar}qp}\, \psi(q) \ , \\ \psi(q) &= <q|\psi>= \int_{-\infty}^{\infty} dp <q|p><p|\psi>= \int_{-\infty}^{\infty} \frac{dp}{\sqrt{2\pi\hbar}}\, e^{\frac{i}{\hbar}qp}\, \psi(p) \ . \end{aligned} \tag{59}$$

Likewise, the change of basis between the configuration space representation, say, and the Fock space one, is provided by the following matrix elements,

$$<q|n>= \left(\frac{m\omega}{\pi\hbar}\right)^{1/4} \frac{1}{\sqrt{2^n\, n!}}\, e^{-\frac{m\omega}{2\hbar}q^2}\, H_n\left(q\sqrt{\frac{m\omega}{\hbar}}\right) \ , \tag{60}$$

where $H_n(x)$ is the usual Hermite polynomial of order n. Given this result, any quantity obtained within any one of these representations may be transformed into its expression in any of the other representations.

Finally, the change of basis to coherent states is best given in terms of the Fock space states, in which the relevant matrix elements reduce to simple monomials of the complex variable z,

$$<n|z>= \frac{1}{\sqrt{n!}}\, z^n\, e^{-\frac{1}{2}|z|^2} \ . \tag{61}$$

Given these different matrix elements expressing the changes of bases between all four representations of the Heisenberg algebra, it is in principle possible to establish some result using one representation in which this is most convenient, and then convert to any other representation in which to determine some physical quantity.

For instance, in the case of the one-dimensional harmonic oscillator, the Hamiltonian $\hat{H}_0$ is most readily diagonalized in the Fock space representation, since one has

$$\hat{H}_0 = \hbar\omega\left[a^\dagger a + \frac{1}{2}\right] \ , \ \ \hat{H}_0|n> = E_n|n> \ , \ \ E_n = \hbar\omega(n+\frac{1}{2}) \ , \ n = 0, 1, 2, \cdots \ . \tag{62}$$

Consequently, orthonormalized configuration space wave functions for energy eigenstates are simply given as $\psi_n(q) =< q|n >$ whose expressions are known, while the configuration space propagator of the system, namely the matrix elements of the quantum evolution operator $\hat{U}(t_2, t_1)$ for external position eigenstates $|q_1 >$ and $|q_2 >$, possesses the following simple representation,

$$< q_2|\hat{U}(t_2, t_1)|q_1 >= \sum_{n=0}^{\infty} \psi_n^*(q_2)\, e^{-\frac{i}{\hbar}(t_2-t_1)\hbar\omega(n+1/2)}\, \psi_n(q_1) \ , \tag{63}$$

given the spectral resolution of the evolution operator

$$\hat{U}(t_2, t_1) = \sum_{n=0}^{\infty} |n > \ e^{-\frac{i}{\hbar}(t_2-t_1)E_n} \ < n| \ . \tag{64}$$

The Heisenberg picture

Finally, let us reconsider the issue of the Schrödinger and Heisenberg pictures for a quantum system. Given that in the former picture, the matrix element of any operator $\hat{\mathcal{O}}(t_0)$ defined at the reference time t_0 for arbitrary time dependent external states is given by $< \varphi, t|\hat{\mathcal{O}}(t_0)|\psi, t >$, using the evolution operator as expressed in general terms in (32), and obeying the properties (33), the same matrix element may be expressed as

$$< \varphi, t|\hat{\mathcal{O}}(t_0)|\psi, t >=< \varphi, t_0|\hat{\mathcal{O}}(t)|\psi, t_0 > \ , \tag{65}$$

where the time dependent operator $\hat{\mathcal{O}}(t)$ is simply defined by

$$\hat{\mathcal{O}}(t) = \hat{U}^\dagger(t, t_0)\, \hat{\mathcal{O}}(t_0)\ \hat{U}(t, t_0) \ . \tag{66}$$

In other words, within the so-called Heisenberg picture, the same physical content as that of the Schrödinger picture is provided by time independent quantum states $|\psi, t_0>$ defined at the reference time t_0, while the whole time dependency of the system is now carried by the quantum operators $\mathcal{O}(t)$ related to their definition in the Schrödinger picture at the reference time t_0 by the above definition in terms of the quantum evolution operator $\hat{U}(t_2, t_1)$. Thus, the Schrödinger equation that governs time evolution of the system now applies to the operators rather than to the states, and is given by

$$i\hbar \frac{d}{dt}\hat{\mathcal{O}}(t) = \left[\hat{\mathcal{O}}(t), \hat{H}_0(t)\right] , \tag{67}$$

as follows directly from the definition of the Heisenberg picture in (66). Note how this quantum operator equation of motion is again in direct correspondence with the associated classical one in Hamiltonian form, namely

$$\dot{\mathcal{O}} = \{\mathcal{O}, H_0\} . \tag{68}$$

Furthermore, since the quantum Hamiltonian $\hat{H}_0(t)$ commutes with itself, $[\hat{H}_0(t), \hat{H}_0(t)] = 0$, in fact this operator is totally time independent, hence $\hat{H}_0(t) = \hat{H}_0(t_0)$, expressing a conservation law for the system, which is that of the energy in the case that t parametrizes physical time, and not some other possible evolution parameter for the system's dynamics.

Once again turning to the one-dimensional harmonic oscillator in its Fock space representation, the Heisenberg picture is thus defined in terms of time dependent creation and annihilation operators $a^\dagger(t)$ and $a(t)$, whose solution to the operator Schrödinger equation $i\hbar\dot{a}(t) = [a(t), \hat{H}_0] = \hbar\omega a(t)$ is

$$a(t) = a(t_0)\, e^{-i\omega(t-t_0)} \quad , \quad a^\dagger(t) = a^\dagger(t_0)\, e^{i\omega(t-t_0)} , \tag{69}$$

thus leading to the following representation for the position operator $\hat{q}(t)$, say, in the Heisenberg picture,

$$\hat{q}(t) = \sqrt{\frac{\hbar}{2m\omega}}\left[a(t) + a^\dagger(t)\right] = \sqrt{\frac{\hbar}{2m\omega}}\left[a(t_0)\, e^{-i\omega(t-t_0)} + a^\dagger(t_0)\, e^{i\omega(t-t_0)}\right] , \tag{70}$$

an expression which ought to be compared to its classical counterparts in (21) and (26), and thus justifying the comments made already at that stage of our discussion with respect to the Schrödinger and Heisenberg pictures for a quantum system and the corresponding characterization of the Poisson bracket structure either on phase space or on the space of integration constants for the classical Hamiltonian equations of motion.

2.5 Path integral quantization

It is well known that besides the canonical quantization path, there is another royal avenue towards the quantization of a classical system whose dynamics is defined through some action and the variational principle, namely the so-called path integral or functional integral formulation of quantum mechanics.[15] Here we shall discuss how, starting from the canonical quantization of any such system following the approach outlined in the previous sections, it is possible to set up integral representations for matrix elements of quantum operators, which acquire the interpretation of functional integrals over phase space. When reducing from these integrals the conjugate momentum degrees of freedom, one recovers a functional integral over configuration space in which the original classical action expressed in terms of the Lagrange function plays again a central role. Further remarks as to quantization directly through the functional integral are made at the end of this discussion. It should already be clear that these two approaches are complementary, each with its own advantages and difficulties both with respect to an intuitive understanding of the physics that they both encode as well as to the calculational advantages of one compared to the other. However, when properly implemented, they represent in complementary terms an identical physical content.

The procedure to construct an integral representation for matrix elements of operators, starting from canonical quantization, follows essentially always the same avenue, based on the insertion of complete sets of states in terms of which the unit operator possess a spectral resolution. Here, we shall illustrate this feature for the configuration space representation of the Heisenberg algebra, even though more general cases may be envisaged as well, for instance in terms of coherent states. Furthermore, we shall consider configuration space matrix elements of the evolution operator for a given quantum system, namely the propagator $< q_f|\hat{U}(t_f,t_i)|q_i >$ of the system (in configuration space). This operator thus writes as

$$\hat{U}(t_f,t_i) = e^{-\frac{i}{\hbar}(t_f-t_i)\hat{H}_0} = \left[e^{-\frac{i}{\hbar}\epsilon\hat{H}_0}\right]^N = \lim_{N\to\infty}\left[1-\frac{i}{\hbar}\epsilon\hat{H}_0\right]^N \quad , \tag{71}$$

with $\epsilon = (t_f - t_i)/N$, while N is some arbitrary positive integer specifying an equally spaced slicing of the finite time interval $(t_f - t_i)$. In what follows, the n index for the degrees of freedom (q^n, p_n) is suppressed, to keep expressions as transparent as possible. Given this time-sliced form of the evolution operator, the idea now is to insert twice the spectral resolution of the unit operator $\mathbb{1}$, once in terms of the position eigenstates, and once in terms of the momentum eigenstates, and this in between each of the N factors that appear in the above N-factorized form for $\hat{U}(t_f,t_i)$, as follows,

$$\mathbb{1} = \int_{-\infty}^{\infty} dp_\alpha \int_{-\infty}^{\infty} dq_{\alpha+1}\, |q_{\alpha+1}>< q_{\alpha+1}|p_\alpha>< p_\alpha| \quad , \quad \alpha = 0,1,2,\cdots,N-2\,. \tag{72}$$

Setting then $q_f = q_{\alpha=N}$ and $q_i = q_{\alpha=0}$, a straightforward substitution into the considered matrix element leads to the expression (a substitution of the unit operator as $\mathbb{1} = \int_{-\infty}^{\infty} dp|p><p|$ is also performed to the right of the external final state $<q_f|$, leading to one more integration over the p_α's than over the q_α's)

$$<q_f|\hat{U}(t_f,t_i)|q_i>=$$

$$= \int_{-\infty}^{\infty} \prod_{\alpha=1}^{N-1} dq_\alpha \prod_{\alpha=0}^{N-1} dp_\alpha \prod_{\alpha=0}^{N-1} \left[<q_{\alpha+1}|p_\alpha><p_\alpha|e^{-\frac{i}{\hbar}\epsilon\hat{H}_0}|q_\alpha>\right] . \tag{73}$$

Using then the value for the matrix element $<q|p>$ given previously, this quantity finally reduces to

$$<q_f|\hat{U}(t_f,t_i)|q_i>=$$

$$= \lim_{N\to\infty} \int_{-\infty}^{\infty} \prod_{\alpha=1}^{N-1} dq_\alpha \prod_{\alpha=0}^{N-1} \frac{dp_\alpha}{2\pi\hbar} \exp\left\{\frac{i}{\hbar}\sum_{\alpha=0}^{N-1} \epsilon\left[\frac{q_{\alpha+1}-q_\alpha}{\epsilon}p_\alpha - h_\alpha\right]\right\} , \tag{74}$$

with the Hamiltonian matrix elements

$$h_\alpha = \frac{<p_\alpha|\hat{H}_0|q_\alpha>}{<p_\alpha|q_\alpha>} . \tag{75}$$

Clearly, the discretized integral representation (74) of the configuration space propagator corresponds to a specific construction of the otherwise formal expression for the phase space path integral or functional integral corresponding to that quantity, namely

$$<q_f|\hat{U}(t_f,t_i)|q_i>= \int_{q(t_i)=q_i}^{q(t_f)=q_f} \left[\mathcal{D}q\frac{\mathcal{D}p}{2\pi\hbar}\right] e^{\frac{i}{\hbar}S[q,p]} , \tag{76}$$

in which the phase space action is that of the first-order Hamiltonian formulation of the system given in (16), namely

$$S[q,p] = \int_{t_i}^{t_f} dt \left[\dot{q}p - H_0(q,p)\right] , \tag{77}$$

which is that associated to the choice of boundary conditions corresponding to the configuration space propagator when imposing the variational principle in a strong sense, namely with the induced boundary terms also required to vanish through the boundary conditions $q(t_{i,f}) = q_{i,f}$. Note that contrary to what the formal expression (76) may lead one to believe, the integration measure is not quite the phase space Liouville measure, since in fact there is always one more p_α integration than the number of q_α integrations. One should always keep this remark in mind when developing formal arguments based on the formal expression (76) of the functional integral.

Considering the momentum space matrix elements of the same operator, a similar analysis leads to an analogous specific discretized expression corresponding to the formal quantity

$$< p_f|\hat{U}(t_f,t_i)|p_i >= \int_{p(t_i)=p_i}^{p(t_f)=p_f} \left[\frac{\mathcal{D}q}{2\pi\hbar}\mathcal{D}p\right] e^{\frac{i}{\hbar}S[q,p]} , \tag{78}$$

where the appropriate Hamiltonian first-order action now reads

$$S[q,p] = \int_{t_i}^{t_f} dt\,[-q\dot{p} - H_0(q,p)] , \tag{79}$$

being this time associated to the choice of boundary conditions $p(t_{i,f}) = p_{i,f}$ as opposed to $q(t_{i,f}) = q_{i,f}$ for the propagator in configuration space. Note that the same remark as above concerning the phase space Liouville measure applies here as well.

In the particular situation that the Hamiltonian is such that the matrix elements h_α are quadratic in the momenta,

$$h_\alpha = \frac{p_\alpha^2}{2m} + V(q_\alpha) , \tag{80}$$

the integration over momentum space may be completed explicitly in the above discretized expressions, thereby leading to the configuration space functional integral representation,

$$\begin{aligned} < q_f|\hat{U}(t_f,t_i)|q_i > &= \lim_{N\to\infty} \left(\tfrac{m}{2i\pi\hbar\epsilon}\right)^{N/2} \textstyle\int_{-\infty}^{\infty}\prod_{\alpha=1}^{N-1} dq_\alpha \times \\ &\times\exp\left\{\tfrac{i}{\hbar}\textstyle\sum_{\alpha=0}^{N-1}\epsilon\left[\tfrac{1}{2}m\left(\tfrac{q_{\alpha+1}-q_\alpha}{\epsilon}\right)^2 - V(q_\alpha)\right]\right\} , \end{aligned} \tag{81}$$

or at the formal level,

$$< q_f|\hat{U}(t_f,t_i)|q_i >= \int_{q(t_i)=q_i}^{q(t_f)=q_f} [\mathcal{D}q]\, e^{\frac{i}{\hbar}S[q]} , \tag{82}$$

with

$$S[q] = \int_{t_i}^{t_f} dt\, L(q,\dot{q}) \quad , \quad L(q,\dot{q}) = \frac{1}{2}m\dot{q}^2 - V(q) . \tag{83}$$

The above explicit discretized representation of this latter formal functional integral coincides exactly with the explicit construction performed by Feynman.[15]

Hence, we have come back full circle. Starting from the action principle defined within the Lagrangian formulation of dynamics, the canonical Hamiltonian formulation of the same dynamics on phase space has been constructed, allowing for the canonical operator quantization of the associated algebraic and geometric structures, for which operator matrix elements may be given a functional integral representation on phase space or configuration space, in which the classical Hamiltonian or Lagrangian action functionals reappear on equal terms. The concept which is central to this whole construction is that of the action, through one of the many forms by which it contributes whether for the classical or the quantum dynamics.

Having chosen to follow the operator quantization path, once a specific choice of operator ordering has been made, in principle the functional integral representations acquires a totally unambiguous and well defined discretized expression, which defines in an exact manner otherwise ill defined formal path integral expressions whose actual meaning always still needs to be specified properly. Nonetheless, as we have indicated, difficulties lie at the operator level precisely in the choice of operator ordering in order to obtain a consistent unitary quantum theory.

Had one taken the functional integral path towards quantization, whether from the Lagrangian or Hamiltonian classical actions, the difficulty of a proper construction of the quantized system then lies hidden in the necessity to give a precise definition and meaning, through some discretization procedure or otherwise, to the formal and thus ill defined functional integrals such as those in (76), (78) and (82). As a matter of fact, the arbitrariness which exists at this level in the choice of discretization procedure and functional integration measure (whether over configuration, momentum or phase space) is in direct correspondence with the arbitrariness which exists on the operator side of this relationship in terms of the choice of operator ordering. Taking either path towards quantization, for appropriate choices on both sides which are in correspondence, the same dynamical quantum system is being represented in a complementary manner. It is extremely fruitful to constantly keep in one's mind these equivalent representations of a quantum dynamics when properly implemented, in particular in a manner that should ensure its quantum unitarity.

3 Relativistic Quantum Particles and Field Theories

Starting with this section, we shall explicitly work in four-dimensional Minkowski spacetime with coordinates x^μ ($\mu = 0,1,2,3$) and a metric $\eta_{\mu\nu}$ of signature $(+---)$. Furthermore as is customary in quantum field theory, units such that $\hbar = 1 = c$ are also being used throughout, so that mass and energy on the one hand, as well as time and space on the other, are each measured in the same units, while energy and time, for instance, are of inverse dimensions. Hence, any mechanical quantity may always be expressed in units of mass to some power.

3.1 Motivation

It is an experimental fact that there exist particles in nature, which behave both with relativistic and quantum properties, have definite energy, momentum and thus invariant mass values, and may be created or annihilated through different physical processes. Which type of mathematical framework would be able to account for all these physical properties all at once?

As we have recalled above, the quantization of the harmonic oscillator leads to such a framework. Indeed, the operators a and $a^\dagger$, which obey the Fock algebra $[a, a^\dagger] = \mathbb{1}$, provide for the annihilation and creation of energy quanta, each carrying an identical amount $\hbar\omega$ of energy. Furthermore, we also know that associated to these operators, there exists some configuration space operator $\hat{q}$ which in the Heisenberg picture has a time dependency defined by (from now on, the choice of reference time will be $t_0 = 0$)

$$\hat{q}(t) = \sqrt{\frac{\hbar}{2m\omega}} \left[a\, e^{-i\omega t} + a^\dagger\, e^{i\omega t}\right] , \tag{84}$$

which, in the classical limit, thus defines the entire real line as the space of classical configurations of the system. Hence, the configuration space quantum operator $\hat{q}(t)$ in the Heisenberg picture obeys the following equation

$$\left[\frac{d^2}{dt^2} + \omega^2\right] \hat{q}(t) = 0 , \tag{85}$$

which also coincides with the classical equation of motion for the system, which derives from the Lagrangian action

$$S[q] = m \int dt \left[\frac{1}{2}\left(\frac{dq}{dt}\right)^2 - \frac{1}{2}\omega^2 q^2\right] . \tag{86}$$

Let us now try to extend this mathematical framework to spinless relativistic quantum particles of definite energy-momentum $k^\mu = (k^0, \vec{k})$ and mass m such that $k^0 = (\vec{k}^{\,2} + m^2)^{1/2} = \omega(\vec{k})$, and which may be created or annihilated in specific physical processes. Thus, for each of the possible momentum values $\vec{k}$, one should introduce a pair of creation and annihilation operators $a^\dagger(\vec{k})$ and $a(\vec{k})$ obeying the Fock space algebra

$$\left[a(\vec{k}), a^\dagger(\vec{k}')\right] = (2\pi)^3\, 2\omega(\vec{k})\, \delta^3(\vec{k} - \vec{k}')\ , \tag{87}$$

where, compared to the Fock algebra for the harmonic oscillator, the normalization of the operators has been modified for a reason to be specified presently. Thus in particular, 1-particle quantum states are obtained from the normalized Fock vacuum $|0>$ as

$$|\vec{k}>= a^\dagger(\vec{k})\,|0> \quad , \quad <0|0>= 1\ . \tag{88}$$

Proceeding by analogy with the harmonic oscillator case, in order to identify the configuration space for such a quantum system, let us also consider superpositions of these operators such as in (84). However, since we wish to develop a formalism which is manifestly spacetime covariant under Lorentz transformations, the product ωt appearing in the imaginary exponentials that multiply the operators and which thus corresponds to the product of the energy value of a quantum by the time interval, must be extended into the Minkowski invariant product $\omega(\vec{k})t - \vec{k}\cdot\vec{x} = k \cdot x$, where the last expression denotes the inner product of four-vectors with the four-dimensional Minkowski metric. Furthermore, since in the present case we have an infinity of quantum operators labelled by the vector values $\vec{k}$ and which are all on an equal footing, one should consider a general superposition of all such linear combinations of the creation and annihilation operators with a $\vec{k}$-independent weight. Hence finally, one is led to consider the following operator, again in the Heisenberg picture, as the relativistic invariant extension of (84),

$$\hat{\phi}(x^\mu) = \int_{(\infty)} \frac{d^3\vec{k}}{(2\pi)^3 2\omega(\vec{k})} \left[a(\vec{k})e^{-ik\cdot x} + a^\dagger(\vec{k})e^{ik\cdot x}\right]\ . \tag{89}$$

Note that having rescaled the creation and annihilation operators by a factor $(\omega(\vec{k}))^{1/2}$, the $d^3\vec{k}$ integration measure includes the same dimensionful normalization factor as in (84) for the harmonic oscillator. The choice of numerical factor $(2\pi)^3$ is made for later convenience. As a matter of fact, the reason for the specific choice of normalization in (87) is that the integration measure in (89), namely $d^3\vec{k}/2\omega(\vec{k})$, is invariant under Lorentz transformations, as may easily be checked. In other words, this parametrization of the operator $\hat{\phi}(x^\mu)$ is manifestly Lorentz covariant.

Hence, associated to the algebra (87), one expects that the actual configurations of the corresponding system is that of a real scalar field in spacetime! Indeed, in the classical limit, the combination (89) defines a real number $\phi(x^\mu)$ attached at each spacetime point. In other words, an arbitrary collection of identical relativistic free quantum point-particles with causal and unitary propagation corresponds to quanta of a single relativistic quantum field in Minkowski spacetime. Furthermore, even though these particles display corpuscular properties by having definite energy-momentum values, their spacetime dynamical propagation also displays wavelike properties, since the field obeys the following equation of motion,

$$\left[\frac{\partial^2}{\partial t^2} - \vec{\nabla}^2 + m^2\right] \hat{\phi}(x^\mu) = 0 \ , \tag{90}$$

which is indeed a wave equation, known as the Klein-Gordon equation, and is nothing but the straightforward relativistic invariant extension of the equation of motion for the harmonic oscillator. Likewise, the corresponding classical action principle thus reads, in a manifestly Lorentz invariant form,

$$S[\phi] = \int dt \int_{(\infty)} d^3\vec{x} \left[\frac{1}{2}\left(\frac{\partial}{\partial t}\phi\right)^2 - \frac{1}{2}\left(\vec{\nabla}\phi\right)^2 - \frac{1}{2}m^2\phi^2\right] \ . \tag{91}$$

From this point of view, the configuration space that has been identified corresponds to an infinite set of harmonic oscillators sitting all adjacent next to one another in the three dimensions of space, and while they each oscillate away from their equilibrium position, the gradient term $\vec{\nabla}\phi$ in the action or in the equation of motion induces a coupling between adjacent oscillators, thereby leading to a propagating wave behaviour of the system in space as a function of time. This term in $\vec{\nabla}\phi$ is required by Lorentz invariance from the similar term in $\partial\phi/\partial t$ which is necessary for the time dependent dynamics of the system.

In conclusion, having considered the possibility to describe an arbitrary collection of identical relativistic free quantum spinless point-particles of definite energy-momentum and mass which may be created and annihilated locally in Minkowski spacetime, we are naturally led to consider a formulation which is that of a local real relativistic scalar field in spacetime with its dynamical wave properties, whose action is real under complex conjugation (which guarantees quantum unitarity), Poincaré invariant (necessary for causality, and also leading to states of definite energy-momentum and angular momentum, which are the conserved Noether charges for the Poincaré invariance group of Minkowski spacetime), and finally local in spacetime (thus guaranteeing spacetime causality and locality of particle propagation, and later on also for their interactions). At this stage, given the algebra (87), one is only describing interactionless particles, since the complete space of energy eigenstates is the simple tensor product over all $\vec{k}$ values of a Fock space representation, without any nonvanishing matrix element of the Hamiltonian between different factors of this tensor product, which would otherwise indeed represent energy-momentum exchange, namely interactions.

3.2 The classical free relativistic real scalar field

Let us thus consider as a classical system a real scalar field $\phi(x)$ over spacetime, whose dynamics is governed by the spacetime local action

$$S[\phi] = \int d^4x^\mu\, \mathcal{L}_0(\phi, \partial_\mu\phi) \ , \tag{92}$$

with the Lagrangian density

$$\mathcal{L}_0(\phi, \partial_\mu\phi) = \frac{1}{2}\eta^{\mu\nu}\partial_\mu\phi\partial_\nu\phi - \frac{1}{2}m^2\phi^2 = \frac{1}{2}\left(\partial_\mu\phi\right)^2 - \frac{1}{2}m^2\phi^2 \ . \tag{93}$$

We shall apply to this system exactly the same procedure of canonical quantization as has been described in Section 2, and establish that we have indeed a formulation of free relativistic quantum spinless particles of mass m. The infinite number of degrees of freedom is parametrized by $\phi(x^0, \vec{x}\,)$, and is thus labelled by the values of the space vector $\vec{x}$. Note that there is an abuse in our notation for the parameter m in the above Lagrangian density. At the classical level, only a length scale κ may be introduced, leading to a quadratic term of the form ϕ^2/κ^2 rather than $m^2\phi^2$ above. However, at the quantum level, it will found that the field quanta possess an invariant mass given by $m = \hbar c/\kappa$, which explains our abuse of notation at the classical level already.

In their manifestly Lorentz covariant form, the Euler-Lagrange equations read

$$\partial_\mu \frac{\partial\mathcal{L}_0}{\partial(\partial_\mu\phi)} - \frac{\partial\mathcal{L}_0}{\partial\phi} = 0 \ , \tag{94}$$

or in the present case

$$\left[\partial_\mu\partial^\mu + m^2\right]\phi = 0 \ , \tag{95}$$

which is the Klein-Gordon equation. Through Fourier analysis, the general solution is readily established, and may be expressed as

$$\phi(x^\mu) = \int_{(\infty)} \frac{d^3\vec{k}}{(2\pi)^3 2\omega(\vec{k}\,)} \left[a(\vec{k}\,)e^{-ik\cdot x} + a^*(\vec{k}\,)e^{ik\cdot x}\right] \ , \tag{96}$$

$a(\vec{k}\,)$ and $a^*(\vec{k}\,)$ being complex integration constants, while in the plane wave contributions $e^{\mp ik\cdot x}$ the value $k^0 = \omega(\vec{k}\,)$ is to be used.

In order to quantize the system, let us first consider its Hamiltonian formulation. By definition, the momentum conjugate to the field $\phi(x^0,\vec{x})$ at each point $\vec{x}$ in space is

$$\pi(x^0,\vec{x}) = \frac{\partial \mathcal{L}_0}{\partial(\partial_0\phi(x^0,\vec{x}))} = \partial_0\phi(x^0,\vec{x})\ , \tag{97}$$

while the phase space degrees of freedom $(\phi(x^0,\vec{x}),\pi(x^0,\vec{x}))$ possess a Poisson bracket structure defined by the canonical brackets at equal time x^0

$$\{\phi(x^0,\vec{x}),\pi(x^0,\vec{y})\} = \delta^{(3)}(\vec{x}-\vec{y})\ . \tag{98}$$

The Hamiltonian density is

$$\mathcal{H}_0 = \partial_0\phi\,\pi - \mathcal{L}_0 = \frac{1}{2}\pi^2 + \frac{1}{2}\left(\vec{\nabla}\phi\right)^2 + \frac{1}{2}m^2\phi^2\ , \tag{99}$$

while the Hamiltonian equations of motion follow as usual from the Hamiltonian $H_0 = \int_{(\infty)} d^3\vec{x}\,\mathcal{H}_0$ (namely the sum of $\mathcal{H}_0$ over all degrees of freedom labelled by $\vec{x}$) through the Poisson brackets. For the elementary phase space degrees of freedom, one has,

$$\partial_0\phi = \pi\quad ,\quad \partial_0\pi = \left(\vec{\nabla}^2 - m^2\right)\phi\ , \tag{100}$$

clearly leading back to the Klein-Gordon equation upon reduction of the conjugate momentum π. Hence, given the solution (96) for the field $\phi(x^\mu)$, that for the conjugate momentum is

$$\pi(x^\mu) = \int_{(\infty)} \frac{d^3\vec{k}}{(2\pi)^3 2\omega(\vec{k})}\left(-i\omega(\vec{k})\right)\left[a(\vec{k})e^{-ik\cdot x} - a^*(\vec{k})e^{ik\cdot x}\right]\ . \tag{101}$$

On basis of these expressions, it is possible to also determine the Poisson bracket structure on the space of integration constants $a(\vec{k})$ and $a^*(\vec{k})$, rather than on the phase space $(\phi(x^0,\vec{x}),\pi(x^0,\vec{x}))$. A straightforward calculation finds for the only nonvanishing bracket,

$$\{a(\vec{k}),a^*(\vec{k}')\} = -i(2\pi)^3 2\omega(\vec{k})\delta^{(3)}\left(\vec{k}-\vec{k}'\right)\ , \tag{102}$$

while the Hamiltonian then reads

$$H_0 = \int_{(\infty)} \frac{d^3\vec{k}}{(2\pi)^3 2\omega(\vec{k})}\frac{1}{2}\omega(\vec{k})\left[a^*(\vec{k})a(\vec{k}) + a(\vec{k})a^*(\vec{k})\right]\ , \tag{103}$$

hence leading to the Hamiltonian equations of motion

$$\dot{a}(\vec{k}) = -i\omega(\vec{k})a(\vec{k})\quad ,\quad \dot{a}^*(\vec{k}) = i\omega(\vec{k})a^*(\vec{k})\ , \tag{104}$$

whose solutions are of course consistent with the explicit expressions already constructed above for $\phi(x^\mu)$ and $\pi(x^\mu)$.

3.3 The quantum free relativistic real scalar field

Canonical quantization of the system in the Schrödinger picture, at the reference time $t_0 = x_0^0 = 0$, is straightforward. The space of quantum states $|\psi>$, with hermitean inner product $<\chi|\psi>$, provides a representation of the Heisenberg algebra

$$\left[\hat{\phi}(\vec{x}), \hat{\pi}(\vec{y})\right] = i\delta^{(3)}\left(\vec{x} - \vec{y}\right) . \tag{105}$$

In terms of the following representation for the quantum field operators in the Schrödinger picture at $x_0^0 = 0$,

$$\begin{aligned} \hat{\phi}(\vec{x}) &= \int_{(\infty)} \frac{d^3\vec{k}}{(2\pi)^3 2\omega(\vec{k})} \left[a(\vec{k})e^{i\vec{k}\cdot\vec{x}} + a^*(\vec{k})e^{-i\vec{k}\cdot\vec{x}}\right] , \\ \hat{\pi}(\vec{x}) &= \int_{(\infty)} \frac{d^3\vec{k}}{(2\pi)^3 2\omega(\vec{k})} \left(-i\omega(\vec{k})\right) \left[a(\vec{k})e^{i\vec{k}\cdot\vec{x}} - a^*(\vec{k})e^{-i\vec{k}\cdot\vec{x}}\right] , \end{aligned} \tag{106}$$

alternatively one has the Fock space algebra

$$\left[a(\vec{k}), a^\dagger(\vec{k}')\right] = (2\pi)^3 2\omega(\vec{k})\delta^{(3)}\left(\vec{k} - \vec{k}'\right) . \tag{107}$$

The Schrödinger equation for the time evolution of quantum states in the Schrödinger picture also reads

$$i\hbar\frac{d}{dt}|\psi, t>= \hat{H}_0|\psi, t> , \tag{108}$$

with the quantum Hamiltonian given by

$$\hat{H}_0 = \int_{(\infty)} d^3\vec{x} \left[\frac{1}{2}\hat{\pi}^2 + \frac{1}{2}\left(\vec{\nabla}\hat{\phi}\right)^2 + \frac{1}{2}m^2\hat{\phi}^2\right] . \tag{109}$$

Note that this operator does not suffer any operator ordering ambiguity. On the other hand, in terms of the Fock space operators, the same quantum Hamiltonian reads

$$\hat{H}_0 = \int_{(\infty)} \frac{d^3\vec{k}}{(2\pi)^3 2\omega(\vec{k})} \frac{1}{2}\omega(\vec{k}) \left[a^\dagger(\vec{k})a(\vec{k}) + a(\vec{k})a^\dagger(\vec{k})\right] , \tag{110}$$

which leads to finite matrix elements only after normal ordering of the creation and annihilation operators, a procedure which is denoted by double dots on both sides of a quantity and is defined by commuting all operators so that all creation operators are to the left of all annihilation operators, such as for example

$$: a(\vec{k})a^\dagger(\vec{\ell}) : = a^\dagger(\vec{\ell})a(\vec{k}) \quad , \quad : a^\dagger(\vec{k})a(\vec{\ell}) : = a^\dagger(\vec{k})a(\vec{\ell}) . \tag{111}$$

Applying this operator ordering prescription to the above expression for $\hat{H}_0$, one thus finds in the Fock space representation the normal ordered Hamiltonian

$$\hat{H}_0 = \int_{(\infty)} \frac{d^3\vec{k}}{(2\pi)^3 2\omega(\vec{k}\,)} \omega(\vec{k}\,)\, a^\dagger(\vec{k}\,) a(\vec{k}\,) \ , \tag{112}$$

while an infinite normal ordering constant contribution is then subtracted away, namely

$$\int_{(\infty)} \frac{d^3\vec{k}}{(2\pi)^3 2\omega(\vec{k}\,)} \frac{1}{2}\omega(\vec{k}\,)(2\pi)^3 2\omega(\vec{0}\,)\delta^{(3)}(\vec{0}\,) \ . \tag{113}$$

This contribution corresponds to the sum of all vacuum quantum fluctuations of all the $\vec{k}$-modes of the scalar field. Provided the system is not coupled to gravity, such a renormalization of the energy eigenvalues is without physical consequence. Nonetheless, it should imply that the two representations of the quantized system, namely that achieved through the Heisenberg algebra for the fields, or that achieved through the Fock algebra for its modes, need no longer be unitarily equivalent for such a system with an infinite set of degrees of freedom,[16] in contradistinction to the situation for a system with a finite number of degrees of freedom such as the one-dimensional harmonic oscillator.

It thus appears that one might have available two possibly physically inequivalent approaches to the quantization of this system, the first based on the representations of the field Heisenberg algebra (105), and the second based on the representations of the field Fock space algebra (107). Let us first consider the Heisenberg algebra realization, say in its configuration space representation. In the Schrödinger picture, the basis of states is then spanned by states $|\phi>$ which are associated to specific classical field configurations $\phi(\vec{x}\,)$ defined over space at the reference time $x^0_0 = 0$, and which are eigenstates of the quantum field operator $\hat{\phi}(\vec{x}\,)$,

$$\hat{\phi}(\vec{x}\,)\,|\phi> = \phi(\vec{x}\,)\,|\phi> \ . \tag{114}$$

The values for the vector $\vec{x}$ being the label for degrees of freedom, at least formally one has the following normalization of these states, together with the associated spectral resolution of the unit operator,

$$<\phi|\phi'> = \prod_{\vec{x}} \delta\left(\phi(\vec{x}\,) - \phi'(\vec{x}\,)\right) \quad , \quad \mathbb{1} = \int_{-\infty}^{\infty} \prod_{\vec{x}} d\phi(\vec{x}\,)\,|\phi><\phi| \ , \tag{115}$$

in direct analogy with the situation for a system with a finite number of degrees of freedom. Hence, arbitrary quantum states $|\psi>$ possess now a configuration space wave functional representation $\Psi[\phi]$ defined by

$$\Psi[\phi] = < \phi|\psi > \quad , \quad |\psi > = \int_{-\infty}^{\infty} \prod_{\vec{x}} d\phi(\vec{x}\,)\,|\phi > \,\Psi[\phi]\,, \tag{116}$$

which thus represents the probability amplitude for observing the given quantum state $|\psi >$ in the classical field configuration $\phi(\vec{x}\,)$, again in direct analogy with the meaning of the configuration space wave function for a finite dimensional system.

Furthermore, since the field operators $\hat{\phi}(\vec{x}\,)$ and $\hat{\pi}(\vec{x}\,)$ possess the following configuration space representations,

$$< \phi|\hat{\phi}(\vec{x}\,)|\psi > = \phi(\vec{x}\,)\,\Psi[\phi] \quad , \quad < \phi|\hat{\pi}(\vec{x}\,)|\psi > = -i\hbar\frac{\delta}{\delta\phi(\vec{x}\,)}\,\Psi[\phi]\,, \tag{117}$$

the action of the quantum Hamiltonian on quantum states in their configuration space wave functional representation is

$$< \phi|\hat{H}_0|\psi > = \int_{(\infty)} d^3\vec{x}\,\frac{1}{2}\left[-\hbar^2\left(\frac{\delta}{\delta\phi(\vec{x}\,)}\right)^2 + \left(\vec{\nabla}\phi(\vec{x}\,)\right)^2 + m^2\phi^2(\vec{x}\,)\right]\,\Psi[\phi]\,. \tag{118}$$

This Schrödinger functional representation of a quantum field theory could prove to be an appropriate framework in which to attempt a nonperturbative quantization. Even though it may well be that for a noninteracting field, which is the above situation, this approach would be unitarily equivalent to the Fock space one to be discussed presently, it is far from clear that such an equivalence should survive the introduction of nonlinear interactions. Given the wide success of the perturbative treatment of particle interactions, based on the Fock space quantization of a field theory briefly described hereafter, such nonperturbative functional quantizations have not been developed to the same extent, making this issue a worthwhile topic of further investigation,[16] especially when it comes to nonlinear field theories whose space of classical solutions includes topological configurations such as solitons and higher dimensional monopole-like configurations.

Turning now to the field Fock space algebra (107) and its representations, it is clear that the space of states is spanned by all possible n-particle states ($n = 0, 1, 2, \cdots$) of arbitrary momentum values $\vec{k}_i$ ($i = 1, 2, \cdots, n$), which are built through the action of the creation operators $a^\dagger(\vec{k}\,)$ from the normalized Fock vacuum $|0 >$, itself annihilated by the $a(\vec{k}\,)$ operators, $a(\vec{k}\,)|0 > = 0$,

$$|\vec{k}_1, \vec{k}_2, \cdots, \vec{k}_n > = N(\vec{k}_1, \vec{k}_2, \cdots, \vec{k}_n)\,a^\dagger(\vec{k}_1)\,a^\dagger(\vec{k}_2)\cdots a^\dagger(\vec{k}_n)|0 >\,, \tag{119}$$

where $N(\vec{k}_1, \vec{k}_2, \cdots, \vec{k}_n)$ denotes some normalization factor. In particular, the 1-particle quantum states correspond to

$$|\vec{k}>= a^\dagger(\vec{k})|0> \quad , \quad <\vec{k}|\vec{k}'>= (2\pi)^3 2\omega(\vec{k})\, \delta^{(3)}\left(\vec{k}-\vec{k}'\right) \, . \tag{120}$$

In addition, given the manifest spacetime invariance of the system under the Poincaré group, the quantum operators $\hat{P}^\mu$ and $\hat{M}^{\mu\nu}$ associated to the conserved Poincaré Noether charges generate the Poincaré algebra on the space of quantum states, the latter thus getting organized into irreducible representations of that symmetry. The eigenstates of these operators, thus of definite energy-momentum, angular-momentum and invariant mass, define the 1-particle states of the quantized field. Clearly, these eigenstates must correspond to the 1-particle quantum states $|\vec{k}>$ constructed above, which is indeed the case. For instance, the energy-momentum operator in Fock space is given by

$$\hat{P}^\mu = \int_{(\infty)} \frac{d^3\vec{k}}{(2\pi)^3 2\omega(\vec{k})}\, k^\mu\, a^\dagger(\vec{k})a(\vec{k}) \, , \tag{121}$$

so that the 1-particle states $|\vec{k}>$ are eigenstates of this operator, namely $\hat{P}^\mu|\vec{k}>= k^\mu|\vec{k}>$, with the eigenvalues

$$\hat{P}^0 \ : \ k^0 = \omega(\vec{k}) \quad ; \quad \hat{\vec{P}} \ : \ \vec{k} \, . \tag{122}$$

In particular, the relativistic invariant mass eigenvalue of these states is m^2, showing that the parameter m indeed measures the mass of the quanta of the quantized field. Likewise for the generalized angular-momentum operator $\hat{M}^{\mu\nu}$, the 1-particle states $|\vec{k}>$ possess an eigenvalue which measures their orbital angular-momentum, thus expressing the fact that the quanta associated to the scalar field $\phi(x^\mu)$ are indeed spinless particles. In order to obtain 1-particle states with a nontrivial spin value, one has to use fields which transform nontrivially under the Lorentz group SO(3,1), such as a vector field leading then to particles of unit spin or helicity (the latter in the massless case), or a spinor field (whether a Weyl, a Dirac or a Majorana spinor) leading to particles of 1/2 spin or helicity values (Grassmann odd variables must be used to parametrize spinor field degrees of freedom, leading, at the classical level, to Grassmann graded Poisson bracket structures and, at the quantum level, to anticommutation rather than commutation rules for fermionic quantum operators).

Hence, as expected on basis of the heuristic construction of Section 3.1, the Fock space representation of a relativistic quantum field theory (whose action is quadratic in the field) shows that the physical content of such a system is that of an arbitrary ensemble of identical free relativistic quantum point-particles of definite mass, energy- and angular-momentum. The interpretation of the field quanta as being such relativistic particles is made consistent by the manifest Poincaré invariance of the action principle.

The above Fock space construction of the quantized field is performed within the Schrödinger picture at the reference time $x_0^0 = 0$. Within the corresponding Heisenberg picture, states are time independent whilst the quantum operators, among which the basic field $\hat{\phi}(\vec{x})$, are rather now explicitly time dependent and carry the whole dynamics of the system. Given the quantum Hamiltonian (112), it is straightforward to show, based exactly on the definition (66), that in the Heisenberg picture the relativistic quantum scalar field is given precisely by the expression (89) which was constructed heuristically in Section 3.1. Hence, it is precisely the ordinary rules of canonical quantization, and only these, which, when applied to the classical system describing the dynamics of a relativistic field theory, lead to a framework which readily accounts for all the observed physical spacetime properties of relativistic quantum particles including the possibility of their creation and annihilation, which is possible only within a formalism which includes both special relativity and quantum mechanics.

In particular, acting with the quantum field $\hat{\phi}(x^\mu)$ in the Heisenberg picture on the Fock vacuum, one obtains a plane wave superposition of 1-particle states of definite momentum,

$$\hat{\phi}(x^\mu)|0>=\int_{(\infty)}\frac{d^3\vec{k}}{(2\pi)^3 2\omega(\vec{k})}\,e^{ik\cdot x}\,|\vec{k}>\ . \tag{123}$$

Such a state may thus be viewed as the quantum configuration of the field such that one particle has been created exactly at the spacetime point x^μ, which, as a consequence of Heisenberg's uncertainty principle, thus possesses a totally undertermined energy-momentum value with its characteristic plane wave probability amplitude. More generally, this interpretation also enables one to construct the probability amplitude for the process in which one particle is created at a given initial spacetime point x_i^μ and then annihilated at the final point x_f, while it propagates in a causal manner between these two positions. This quantity is thus defined by the time-ordered two-point function of the field operator,

$$\begin{aligned}<0|T\left(\hat{\phi}(x_f)\hat{\phi}(x_i)\right)|0> &= \theta(x_f^0-x_i^0)\ <0|\hat{\phi}(x_f)\hat{\phi}(x_i)|0> \\ &\quad + \theta(x_i^0-x_f^0)\ <0|\hat{\phi}(x_i)\hat{\phi}(x_f)|0>\ ,\end{aligned} \tag{124}$$

($\theta(x)$ is the usual step function such that $\theta(x>0)=1$ and $\theta(x<0)=0$) and corresponds to what is called the Feynman propagator for single field quanta. Using the explicit expansion (89) of the field operator in the Heisenberg picture in terms of the creation and annihilation operators, it is straightforward to establish that the Feynman propagator is given by the manifestly spacetime invariant expression

$$< 0|T\left(\hat{\phi}(x_f)\hat{\phi}(x_i)\right)|0>=\int_{(\infty)}\frac{d^4k^\mu}{(2\pi)^4}\,e^{-ik\cdot(x_f-x_i)}\,\frac{i}{k^2-m^2+i\epsilon}\,, \tag{125}$$

where the infinitesimal parameter $\epsilon>0$ is introduced in order to specify the contour integration in the complex plane for the energy contribution k^0, so that the correct causal structure of this propagator is recovered. This quantity is also one of the Green functions for the Klein-Gordon operator $[\partial_\mu\partial^\mu+m^2]$.

Hence, the marriage of special relativity and of quantum mechanics, namely of the constants c and $\hbar$, leads in a most natural way to a fundamental convergence and unification of concepts: relativistic quantum particles are nothing but the quanta of relativistic quantum fields, displaying at the same time the corpuscular properties of particles and the wavelike properties of the spacetime dynamics of fields. This is indeed a most powerful and all encompassing outcome of the unification of relativity and quantum mechanics. Among other consequences, it explains at once why identical particles are necessarily indistinguishable, since they simply correspond to actual physical quantum fluctuations of a single physical entity filling all of spacetime, namely the corresponding relativistic quantum field, and which may be excited or absorbed, namely created or annihilated, by acting on the system through some interaction with another field. In fact, and as shall become clear in Section 4, even interactions, namely changes in the total energy-momentum content of given quantum field states, are understood in terms of exchanges of such 1-particle quanta between given quantum states. The notion of a force acting on a relativistic particle, or of a potential energy contributing to the Hamiltonian of a quantum system, is also superseded by that of fields filling all of spacetime, and interacting with one another through local spacetime couplings, thereby leading to the exchanges of 1-particle quanta. Other profound consequences of the relativistic quantum field picture of physical reality are the spin-statistics connection (namely the fact that integer spin particles obey the Bose-Einstein statistics while half-integer spin particles the Fermi-Dirac statistics), the invariance of any relativistic quantum field theory under the combined product of the parity, time reversal and charge conjugation transformations (the so-called CPT theorem), and the particle/antiparticle duality (only this latter point is discussed explicitly hereafter).

It is clear that the Fock space quantization of field theory is ideally suited for a perturbative description of interactions, namely by starting with a situation with only free quanta, corresponding to an action which is quadratic in the fields, and then adding as perturbations to be summed through a series expansion further corrections involving locally in spacetime higher order products of the fields and their couplings, thus leading to successive perturbative corrections to quantum matrix elements of specific observables which may be viewed in terms of specific 1-particle exchanges among quantum states. This procedure will briefly be outlined in Section 4. On the other hand, the Schrödinger functional quantization of a field theory is from the outset nonperturbative in character, and may thus be better suited to study nonperturbative issues in quantum field theory, in ways that have not been explored to the same extent as the perturbative picture of quantum field theory.

A final remark may also be in order concerning some vocabulary. Note that exactly the same methods of canonical quantization are applied whether for a finite or an infinite dimensional dynamical system. Often in the literature, one finds written that the first situation is that of "first quantization", while the second that of "second quantization". Furthermore, there is also quite often mention of "negative energy states" and "negative probabilities", which must be circumvented through "second quantization". The fact of the matter is that this vocabulary is due to an historical accident. Initially, one wished to develop a relativistic extension of the nonrelativistic Schrödinger equation for, say, the harmonic oscillator and its configuration space wave function. Doing so, one unavoidably encounters diverse problems of negative energy and/or probability states, which defy a consistent physical interpretation. Considering then that the "relativistic wave function" itself needs to be quantized, one discovered that these issues are evaded altogether, leading in fact to the quantum field theory representations that were described above. In other words, the correct physical point of view is that, rather than quantizing some relativistic wave function, from the outset one is in fact (first!) quantizing a classical field theory which obeys some relativistic invariant wave equation, and at no point whatsoever do issues of "negative energy or probability states" arise.[1,17,18] In the same way that quantum mechanics, whether relativistic or not, is the quantization of finite dimensional systems whose configurations represent as a function of time, say, the positions in space of a finite collection of particles, quantum field theory is the quantization of infinite dimensional systems whose configurations are, say, the values taken by a finite collection of fields in space as a function of time, all in a spacetime invariant manner in the case of a relativistic field theory.

4 Interactions and the Gauge Symmetry Principle

Having understood how the dynamics of a relativistic quantum field whose Lagrangian density is quadratic in the field in fact describes a system whose quantum states correspond to an arbitrary number of identical free relativistic quantum particles of definite energy-momentum, spin and invariant mass, it becomes possible to envisage an extension of this formalism in order to account for interactions among such particles, namely the exchange of energy and momentum between such quantum states through the creation and annihilation of the associated quanta. Clearly, such a formulation is perturbative in character, since the free particle picture provides the starting point for a perturbative expansion in which an increasing number of interaction points are included for a given physical process. The purpose of the present section is to briefly outline how this point of view, which has proved to be so powerful and relevant to high energy particle physics and their fundamental interactions except for the gravitational one, has led, on the one hand, to the local gauge symmetry principle as an essential requirement for any theory of the fundamental interactions, and on the other hand, to the Feynman diagrammatic representation of physical processes through a perturbative expansion of the associated probability amplitudes order by order in the exchanges of interacting particles.

4.1 Field coupling and interactions

For definitiness, the discussion to be presented uses the simplest of examples, namely that of an interacting real scalar field $\phi(x^\mu)$ whose Lagrangian density now includes also a quartic term in the potential contribution, in addition to the quadratic contribution considered so far,

$$\mathcal{L}(\phi, \partial_\mu\phi) = \frac{1}{2}\left(\partial_\mu\phi\right)^2 - \frac{1}{2}m^2\phi^2 - \frac{1}{4}\lambda\phi^4 , \tag{126}$$

$\lambda > 0$ being a real positive parameter which turns out to correspond to a coupling constant measuring the strength of a spacetime local interaction in which four quanta of the field $\phi(x)$ are involved in a perturbative expansion. Compared to the free field case, we thus have

$$\mathcal{L} = \mathcal{L}_0 + \mathcal{L}_{\rm int} \ , \ \ \mathcal{L}_{\rm int} = -\frac{1}{4}\lambda\phi^4 , \tag{127}$$

$\mathcal{L}_0$ being the free field Lagrangien density whose quantization has been discussed above, while $\mathcal{L}_{\rm int}$ corresponds to an additional contribution associated to some specific interaction. The canonical quantization of such a system follows the same rules as those applied in the free field case, with in particular the fundamental Poisson brackets

$$\{\phi(x^0,\vec{x}),\pi(x^0,\vec{y})\} = \delta^{(3)}(\vec{x}-\vec{y}) \ , \tag{128}$$

which remain those of the free field case. Note that the conjugate momentum is still given by the relation $\pi(x) = \partial_0\phi(x)$ (had the interacting Lagrangian $\mathcal{L}_{\rm int}$ included some derivative coupling of the field ϕ, the conjugate momentum would have been different). However, the canonical Hamiltonian density acquires an additional contribution directly related to and determined by $\mathcal{L}_{\rm int}$, namely

$$\begin{aligned} \mathcal{H} &= \frac{1}{2}\pi^2 + \frac{1}{2}\left(\vec{\nabla}\phi\right)^2 + \frac{1}{2}m^2\phi^2 + \frac{1}{4}\lambda\phi^4 \\ &= \mathcal{H}_0 \ + \ \mathcal{H}_{\rm int} \ \ , \ \ \mathcal{H}_{\rm int} = -\mathcal{L}_{\rm int} = \frac{1}{4}\lambda\phi^4 \ . \end{aligned} \tag{129}$$

The restriction on the coupling constant $\lambda > 0$ stems from the requirement that the energy spectrum of the system be bounded below, since otherwise no stable ground state may exist. The same requirement also explains why a purely ϕ^3 term, without the quartic contribution in $\mathcal{L}_{\rm int}$, is not considered in the above discussion, even though the perturbative expansion to be described presently is then somewhat simpler to implement in actual calculations.

Consequently, the canonical quantization of the system, even in the presence of the interaction contribution, may still be performed, say, in the Fock space representation in terms of the creation and annihilation operators of free particle quanta, with a specific definition of a self-adjoint Hamiltonian operator $\hat{H} = \int_{(\infty)} d^3\vec{x}\,\hat{\mathcal{H}}$ through normal ordering in these operators. However, what then becomes a nontrivial issue is the actual diagonalization of this Hamiltonian, namely the identification of the actual spectrum of the quantized interacting field theory. A perturbative approach in the parameter λ enables an order by order identification of the quantum physical content of such a system and of its physical properties, starting from the free field quanta.

The scattering matrix

In practical terms, an extremely important method for the experimental investigation of the quantum relativistic properties of physical systems is that of scattering measurements. Different components of a given system are prepared in a given initial configuration in causally separated regions of space, and are then made to scatter within a given local neighbourdhood of an interaction point, from where interaction products emerge whose properties are then measured and analyzed, in order to infer the specific characteristics of the interactions at work and responsible for the observed process. In other words, all the physical information related to these interactions is encoded into the corresponding scattering probability amplitude.

Given such a general scheme, the basic implicit idea is that the interaction takes place over a region of space whose extent is so small that for all practical purposes the interactions are only short-ranged, so that beyond that interaction region the separated components of the system are free from interactions. In a classical picture, such components may be viewed as independent free particles each following asymptotically a straight trajectory. When the interactions are "turned off", these trajectories are not modified as they pass one another, and are thus not scattered. However, when the interaction is "turned on", the more the particles approach one another, the more their trajectories deviate from a straight path, leading in the asymptotic final state to a scattered configuration of straight trajectories as the final state components which emerge from the spatial interaction region. In other words, the characterization of a nontrivial scattering process proceeds by extrapolating to both the infinite past and the infinite future the time dependent dynamics of a given configuration of the system, and by comparing the asymptotic states to what they would have been had there not been any interaction.

Clearly, the same heuristic understanding of the characterization of the scattering process applies at the quantum level, by comparing the time dependency of given in- and out-states in the presence or absence of some given interaction, provided the initial asymptotic states are identical. The characterization of the scattering process, and of the interaction responsible for it, is then obtained by identifying the operator in Hilbert space which leads to this transition between the in- and out-asymptotic states. This is the scattering operator S whose matrix elements are thus the quantities of interest, which represent the probability amplitude for a given physical scattering process to occur.

Let us translate this reasoning in mathematical terms. Concentrating first on the initial state, let us represent the free Hamiltonian by $\hat{H}_0$, the total Hamiltonian including interactions by $\hat{H}$, and assume to be working in the Schrödinger picture at some reference time t_0. A given state $|\psi_{\rm in}, t_0>$ of the free theory is then evolved backwards in time into the asymptotic in-state

$$|\psi_{\rm in}, -\infty >= \lim_{t\rightarrow-\infty} e^{-i(t-t_0)\hat{H}_0}\,|\psi_{\rm in}, t_0 >= \lim_{t\rightarrow-\infty} |\psi_{\rm in}, t > \ , \qquad (130)$$

while a given state $|\psi, t_0>$ of the interacting theory is likewise propagated back in the infinite past according to

$$|\psi, -\infty> = \lim_{t\to-\infty} e^{-i(t-t_0)\hat{H}}\,|\psi, t_0> = \lim_{t\to-\infty} |\psi, t> \ . \qquad (131)$$

However, these two asymptotic states should correspond to an identical asymptotic quantum in-state, so that the asymptotic correspondence is defined by the relation

$$|\psi, -\infty> = |\psi_{\rm in}, -\infty> \ . \qquad (132)$$

Likewise for the asymptotic quantum out-state, one has the identification

$$|\chi, +\infty> = |\chi_{\rm out}, +\infty> \ , \qquad (133)$$

where

$$|\chi_{\rm out}, +\infty> = \lim_{t\to+\infty} e^{-i(t-t_0)\hat{H}_0}\,|\chi_{\rm out}, t_0> = \lim_{t\to+\infty} |\chi_{\rm out}, t> \ , \qquad (134)$$

$$|\chi, +\infty> = \lim_{t\to+\infty} e^{-i(t-t_0)\hat{H}}\,|\chi, t_0> = \lim_{t\to+\infty} |\chi, t> \ . \qquad (135)$$

Note that behind this construction lies the fact that the quantum theories based on $\hat{H}_0$ and $\hat{H}$ share a common space of quantum states, namely an identical representation space of a common algebraic structure of commutation relations for the fundamental degrees of freedom. The scattering operator, whose matrix elements we are about to characterize, is thus an operator acting withing this common space of quantum states, which must reduce to the identity operator in the absence of any interaction, $\hat{H} = \hat{H}_0$.

Given the above formulation, it is clear that the transition probability amplitude between the asymptotic in- and out-states of the interacting theory is simply given by

$$< \chi, t|\psi, t> = < \chi, t_0|\psi, t_0> \ , \qquad (136)$$

the value of this matrix element being independent of the time t at which it is evaluated since the evolution operator $e^{-i(t-t_0)\hat{H}}$ for the interacting theory defines a unitary isomorphism between all Schrödinger pictures for all values of t. However, this matrix element may also be expressed in terms of the in- and out-states of the free theory, since the asymptotic in-states for either theory are identical. A direct substitution of the above relations then finds

$$< \chi, t|\psi, t> = < \chi, t_0|\psi, t_0> = < \chi_{\rm out}, t_0|S|\psi_{\rm in}, t_0> \ , \qquad (137)$$

where the scattering operator S is defined by the asymptotic limits

$$S = \lim_{t_\mp\to\mp\infty} M(t_+, t_0)\, M^\dagger(t_-, t_0) \ , \qquad (138)$$

with

$$M(t,t_0) = e^{i(t-t_0)\hat{H}_0}\, e^{-i(t-t_0)\hat{H}} \ . \tag{139}$$

Note that in the absence of any interaction, $\hat{H} = \hat{H}_0$, the scattering operator S indeeds reduces to the identity operator. Since the operator $M(t,t_0)$ plays such a central role in the construction of the scattering operator S, it is important to obtain alternative expressions for it. In particular, one readily establishes the differential equation

$$\begin{aligned} i\partial_t\, M(t,t_0) &= e^{i(t-t_0)\hat{H}_0}\left[\hat{H} - \hat{H}_0\right] e^{-i(t-t_0)\hat{H}} \\ &= e^{i(t-t_0)\hat{H}_0}\, \hat{H}_{\rm int}(t_0)\, e^{-i(t-t_0)\hat{H}_0}\, M(t,t_0) \\ &= \hat{H}^{(I)}_{\rm int}(t)\, M(t,t_0) \ , \end{aligned} \tag{140}$$

having introduced

$$\hat{H}^{(I)}_{\rm int}(t) = e^{i(t-t_0)\hat{H}_0}\, \hat{H}_{\rm int}(t_0)\, e^{-i(t-t_0)\hat{H}_0} \quad , \quad \hat{H}_{\rm int}(t_0) = \hat{H} - \hat{H}_0 \ . \tag{141}$$

Note that this latter definition coincides with that of the Heisenberg picture associated to the free Hamiltonian $\hat{H}_0$. Since in the interacting theory the Heisenberg picture should be defined in a likewise manner but in terms of the full Hamiltonian $\hat{H}$ rather than the free Hamiltonian $\hat{H}_0$, one refers to the "interaction picture" as being associated to the general definition of time dependent operators $\mathcal{O}_{(I)}$ given by

$$\mathcal{O}_{(I)}(t) = e^{i(t-t_0)\hat{H}_0}\, \mathcal{O}(t_0)\, e^{-i(t-t_0)\hat{H}_0} \ , \tag{142}$$

where $\mathcal{O}(t_0)$ is the operator as constructed through canonical quantization of the interacting theory in its Schrödinger picture.

In other words, in the interaction picture, quantum states as well as operators carry a split time dependency, such that the one carried by the quantum states is solely induced by the interactions and the interacting Hamiltonian $\hat{H}_{\rm int}$, while the one carried by the quantum operators is solely induced by the time dependency related to the free field dynamics and the free Hamiltonian $\hat{H}_0$. In the interaction picture, any time dependency in the quantum states is totally ascribed to the interactions only.

Returning to the equation (140) characterizing the operator $M(t,t_0)$, one sees that its solution may also be expressed in the form

$$M(t,t_0) = T\, e^{-i\int_{t_0}^{t} dt'\, \hat{H}^{(I)}_{\rm int}(t')} \ , \tag{143}$$

where the symbol T in front of the exponential in the r.h.s. of this expression stands for the time-ordered product and exponential in which products of time-dependent operators are integrated from left to right in decreasing order of their time arguments (this is indeed required given that the operator $M(t,t_0)$ is to the right of $\hat{H}^{(I)}_{\rm int}(t)$ in (140)).

Hence, using this solution for the operator $M(t,t_0)$, the scattering operator acquires the expression

$$S = T\,e^{-i\int_{t_0}^{\infty} dt\hat{H}^{(I)}_{\rm int}(t)}\,T\,e^{-i\int_{-\infty}^{t_0} dt\hat{H}^{(I)}_{\rm int}(t)} = T\,e^{-i\int_{-\infty}^{\infty} dt\int_{(\infty)} d^3\vec{x}\,\hat{\mathcal{H}}^{(I)}_{\rm int}}\ , \qquad (144)$$

which, in the absence of any derivative coupling in the interacting Lagrangian density, so that $\mathcal{L}_{\rm int} = -\mathcal{H}_{\rm int}$, finally reduces to

$$S = T\,e^{-i\int_{(\infty)} d^4x^\mu\,\hat{\mathcal{H}}^{(I)}_{\rm int}} = T\,e^{i\int_{(\infty)} d^4x^\mu\,\hat{\mathcal{L}}^{(I)}_{\rm int}}\ . \qquad (145)$$

In this form, it should be clear why this formulation of any scattering process is ideally suited for a perturbative treatment. Since scattering matrix elements are given by matrix elements of the operator S for free field external states, see (137), it suffices to consider the creation and annihilation mode expansions of the field and its conjugate momentum in the interaction picture, and substitute these in the expressions for the interacting Lagrangian and Hamiltonian densities in the interaction picture. In particular, these fields in the interaction picture retain their expressions valid for the Heisenberg picture of the free field theory. One has

$$\begin{aligned} \hat{\phi}_{(I)}(t,\vec{x}\,) &= e^{i(t-t_0)\hat{H}_0}\,\hat{\phi}(t_0,\vec{x}\,)\,e^{-i(t-t_0)\hat{H}_0}\ , \\ \hat{\pi}_{(I)}(t,\vec{x}\,) &= e^{i(t-t_0)\hat{H}_0}\,\hat{\pi}(t_0,\vec{x}\,)\,e^{-i(t-t_0)\hat{H}_0}\ , \end{aligned} \qquad (146)$$

with the mode expansions

$$\hat{\phi}_{(I)}(x) = \int_{(\infty)} \frac{d^3\vec{k}}{(2\pi)^3 2\omega(\vec{k}\,)}\left[a(\vec{k}\,)e^{-ik\cdot x} + a^\dagger(\vec{k}\,)e^{ik\cdot x}\right]\ , \qquad (147)$$

$$\hat{\pi}_{(I)}(x) = \int_{(\infty)} \frac{d^3\vec{k}}{(2\pi)^3 2\omega(\vec{k}\,)}\left(-i\omega(\vec{k}\,)\right)\left[a(\vec{k}\,)e^{-ik\cdot x} - a^\dagger(\vec{k}\,)e^{ik\cdot x}\right]\ , \qquad (148)$$

while the creation and annihilation operators still obey the usual algebra

$$\left[a(\vec{k}\,), a^\dagger(\vec{\ell}\,)\right] = (2\pi)^3 2\omega(\vec{k}\,)\delta^{(3)}\left(\vec{k}-\vec{\ell}\,\right)\ , \qquad (149)$$

since canonical quantization in the Schrödinger picture of the interacting theory still requires the commutation relations

$$\left[\hat{\phi}(t_0,\vec{x}\,),\hat{\pi}(t_0,\vec{y}\,)\right] = i\delta^{(3)}\left(\vec{x}-\vec{y}\,\right)\ . \tag{150}$$

Furthermore, once such a substitution has been effected, a straightforward expansion of the time-ordered exponential (144) defining the scattering operator in terms of the interacting Hamiltonian in the interaction picture leads to an expansion in powers of the coupling coefficient λ, namely a perturbative representation of the probability amplitude associated to a given set of external states in terms of successive exchanges of free particle quanta being created and annihilated through the interaction couplings of the fields as they contribute to the interacting Hamiltonian.

In particular, it should be clear that successive contractions of these creation and annihilation operators as they are commuted past one another in the evaluation of the matrix elements, all in a manner consistent with the causal time ordering implied by the solution (144), always lead precisely to the time-ordered two-point function of the field operator in the interaction picture, namely the Feynman propagator computed previously for the free field theory,

$$< 0|T\left(\hat{\phi}_{(I)}(x)\hat{\phi}_{(I)}(y)\right)|0>=\int_{(\infty)}\frac{d^4k^\mu}{(2\pi)^4}\frac{i}{k^2-m^2+i\epsilon}\,e^{-ik\cdot(x-y)}\ , \tag{151}$$

where $|0>$ still denotes the perturbative normalized Fock vacuum annihilated by the operators $a(\vec{k}\,)$, $a(\vec{k}\,)|0>=0$.

Even though we cannot consider here a discussion of perturbation theory in any detail whatsoever, once put within such a framework, it takes little effort of imagination to understand how a systematic set of rules for such a perturbative expansion and evaluation of scattering matrix elements may be identified, thus providing an efficient approach towards the determination of scattering cross sections of direct relevance to experimental results. Such a discussion would consist in a whole set of lectures on their own, which is not the purpose of the present notes and may be found exposed in great detail in any quantum field theory textbook.[17,18,19,20] Nonetheless, from the above description, it should be clear that Fock space quantization of relativistic quantum field theory is ideally suited for a perturbative representation of interacting relativistic quantum particles, and that this perturbation theory approach is directly based on the interacting Hamiltonian and Lagrangian contribution to the total Lagrangian density, namely all those contributions which are not purely quadratic in the fields, the latter on their own being relevant to the description of free relativistic quantum particles.

Perturbation theory

In spite of the fact that this is not the place for a detailed presentation of perturbative quantum field theory, let us nevertheless highlight some points of relevance to the discussion hereafter, particularizing again to the simplest $\mathcal{L}_{\rm int} = -\lambda\phi^4/4$ interacting Lagrangian. As far as scattering processes are concerned, all possible results are encoded into the scattering operator

$$S = \mathbb{1} + \sum_{n=1}^{\infty} \frac{1}{n!} T \left(-i \int_{(\infty)} d^4x^\mu \, \hat{\mathcal{H}}^{(I)}_{\rm int}(x) \right)^n , \tag{152}$$

where now the interacting Hamiltonian density in the interaction picture is defined according to the usual normal ordering prescription for the creation and annihilation operators $a^\dagger(\vec{k})$ and $a(\vec{k})$,

$$\hat{\mathcal{H}}^{(I)}_{\rm int}(x) = \frac{1}{4}\lambda : \hat{\phi}^4_{(I)}(x) : . \tag{153}$$

Clearly, when considering the scattering operator in this series expanded form and the evaluation of its matrix elements for external states associated to definite numbers of incoming and outgoing particles, time ordering of operator products commuted with one another implies the contribution of the Feynman propagator which, in momentum space, simply leads to the following contribution for any internal propagating line connecting two interaction vertices at which particle quanta are created or annihilated,

$$\frac{i}{k^2 - m^2 + i\epsilon} . \tag{154}$$

Likewise, whenever the operator $\hat{\mathcal{H}}^{(I)}_{\rm int}(x)$ contributes at a given order of the perturbative expansion, it implies a spacetime local interaction in which four particle quanta are either created or annihilated, with an amplitude given by the factor

$$-\frac{1}{4} i\lambda , \tag{155}$$

up to some combinatorics factor depending on the topology of the associated diagram.

In other words, it is possible to translate the mathematical expression for the relevant matrix element evaluation into a diagrammatic representation in which internal lines are connected to interaction vertices, and for which the above contributions are then multiplied with one another, and integrated over internal momenta in a manner such as to obey the rules of energy-momentum conservation at each vertex, in order to determine the associated probability amplitude. These rules relating such Feynman diagrams to the required mathematical quantity are the Feynman rules of perturbative quantum field theory. In the specific case of the $\lambda\phi^4/4$ scalar field theory, the above discussion thus establishes that these rules consist only of the single interaction vertex accompanying the scalar Feynman propagator. In principle, given such rules, any scattering amplitude for whatever physical process may be computed to an arbitrary order in the perturbative expansion in the coupling constant λ.

As far as we are concerned, the main conclusion to be drawn from the above is that once relativistic quantum fields are coupled to one another through local spacetime couplings, such as $\mathcal{L}_{\rm int} = -\lambda\phi^4/4$, one in facts has made available within a perturbative picture a formalism in which local and causal quantum interactions are directly understood in terms of exchanges of quantum particles free to propagate between interaction vertices that occur locally in spacetime but at arbitrary positions which are integrated over when they are not observed. The marriage of $\hbar$ and c leading to quantum field theory as the natural framework for the description of relativistic quantum point-particles also implies a physical understanding of the physical origin of forces and interactions simply as following from the spacetime local couplings of fields, which also translate in the dual corpuscular picture into a process in which particles are being created, annihilated and exchanged, thereby leading to changes in their energy-momentum, hence to their interactions. The mysterious action at a distance of classical mechanics is forever gone, superseded by relativistic quantum fields which provide a natural framework not only for a unified description both of the corpuscular properties of matter and of the wavelike properties of their spacetime dynamics, but also a unified understanding of the fundamental quantum interactions in terms of both spacetime local couplings of fields and causal exchanges of particle quanta, all in a manner consistent with the principles of special relativity, of unitary quantum mechanics, and of causality.

However, this amazing convergence of physical concepts based on a few general basic principles comes with a price. When considering the perturbative expansion of scattering matrix elements, one soon comes across loop diagram contributions in which one must integrate over the internal momenta running around closed loops. For instance when considering the propagation of a single particle quantum, the first order correction to the propagator is obtained by inserting into it the four-point vertex $\lambda\phi^4/4$ and then contracting two of its four external lines with one another, leading to a 1-loop contribution with the factor

$$\left(-\frac{1}{4}i\lambda\right)\int_{(\infty)}\frac{d^4p^\mu}{(2\pi)^4}\left(\frac{i}{p^2-m^2+i\epsilon}\right)\ , \tag{156}$$

the origin of each of the factors in parentheses being obvious, while the closed loop propagator must be integrated over the associated energy-momentum. Likewise, when considering a 2 → 2 scattering process with two initial and two final particles, beyond the nonscattering and one interaction vertex contributions, there appears a 1-loop correction in which two 4-point vertices are inserted with two lines of each being contracted in pairs with two lines of the other. The corresponding contribution is given by

$$\left(-\frac{1}{4}i\lambda\right)^2\int_{(\infty)}\frac{d^4p^\mu}{(2\pi)^4}\left(\frac{i}{p^2-m^2+i\epsilon}\right)\left(\frac{i}{(p+k)^2-m^2+i\epsilon}\right)\ , \tag{157}$$

where p^μ is again the energy-momentum running around the closed loop (say, that running through one of the two internal contracted lines), while k^μ is the total external energy-momentum of the two initial or final particles ($k+p$ being then the energy-momentum running through the other internal line).

The characteristic feature of such contributions, which arise whenever closed loops appear in a diagram, is their divergence for large values of the internal momentum, namely in the ultra-violet at small distances. The fundamental reason for this feature is that interactions occur locally in spacetime at given points where the fields are multiplied with one another. In order to perform calculations nonetheless, one has to introduce some regularization procedure to tame such divergencies, and hope that at the very end, when all contributions are summed up again, all the divergent contributions would combine is such a manner that physical observables remain nevertheless finite, even if affected by finite renormalization. Many different regularization procedures have been developed, and this is not the place to discuss such issues.[17,18,19,20] The most straightforward one is to introduce an upper cut-off value Λ_c in the momentum integration, to keep track of the different types of divergencies that may arise. For instance, the 1-loop correction to the scalar field propagator given above leads to a quadratic divergence proportional to Λ_c^2, while that to the 2 → 2 scattering process is only logarithmically divergent and proportional to $\ln\Lambda_c$, as may easily be seen through simple power counting and dimensional analysis of the relevant expressions.

The crucial issue thus arises as to which are the interacting quantum field theories which, in a perturbative quantization, lead to physically meaningful and thus finite predictions for scattering processes, in spite of the existence of these ultraviolet short distance higher-loop divergencies. In practical terms, and to put it into just a few words, here is how the procedure works. Given any specific regularization procedure, order after order in perturbation theory, one needs to add further and further corrections ("counterterms") to the initial Lagrangian density, in order to introduce additional contributions to scattering amplitudes such that the perturbative series summed up to the given order remains finite when the regulator is removed, thereby leading to a finite physical result, even though the basic quantities appearing separately in the renormalized Lagrangian may be divergent. However, if the number of the required countertems grows with the order in perturbation theory, no specific prediction remains possible, since each new counterterm requires the specification of a new coupling constant whose value may be inferred only from experiment. Hence, a quantum field theory possesses any predictive power provided only a finite number of counterterms is required to render the renormalized scattering amplitudes to whatever order of perturbation theory finite and thus physical. Field theories for which this programme is feasable are called renormalizable. In fact, all such renormalizable field theories are such that all counterterms belong to a finite class of local quantum operators such that the renormalization of the theory amounts to a redefinition of the field normalizations, masses and couplings (the "bare" quantities of the classical Lagrangian density) in terms of renormalized and finite physical observables directly related to the physical external states, their masses and couplings. The "bare" quantities are obtained in terms of the renormalized ones through factors multiplying the latter, these factors being given as power series expansions in the coupling constants whose coefficient are divergent as the regulator is removed. Theories for which finite renormalization is achieved in this manner are called "multiplicatively renormalizable". These are the only perturbative quantum field theories of possible relevance to relativistic quantum particle physics and their fundamental quantum interactions. Under such circumstances, one thus obtains a predictive framework for the representation and evaluation of these processes.

The above $\lambda\phi^4/4$ scalar field theory is the simplest example of such a renormalizable quantum field theory. All the required counterterms to all orders of perturbation theory simply amount to a redefinition, through a multiplicative factor, of the field normalization, its mass m^2 and its self-coupling λ, each of these renormalization factors being given as power series expansions in the coupling λ whose coefficients include both finite and infinite contributions as the regulator is removed. Nevertheless, all physical quantities remain finite in that same limit, and may be predicted in terms only of the renormalized mass and coupling of asymptotic quanta.

Renormalizable relativistic quantum field theories

Among all possible Lagrangian densities for collections of fields of a variety of spin values, how does one characterize those that define a renormalizable quantum field theory? Through power counting and dimensional analysis of loop amplitudes, a necessary condition, though not a sufficient one, for renormalizability may be established. Namely, when working in units such that $\hbar = 1 = c$ so that all dimensionful quantities may be measured in units of mass, whenever the Lagrangian density contains a specific contribution whose coupling coefficient, say λ, has a mass dimension to some strictly negative power, $\lambda = \alpha_0/\Lambda^\kappa$ with α_0 dimensionless, Λ some mass scale and $\kappa > 0$, then the associated interactions are not renormalizable.

For example, let us consider a real scalar field ϕ whose dynamics derives from the Lagrangian density

$$\mathcal{L} = \frac{1}{2}\left(\partial_\mu \phi\right)^2 - \frac{1}{2} m^2 \phi^2 - V(\phi) \ . \qquad (158)$$

Since in units such that $\hbar = 1 = c$ the quantum action must be dimensionless, in a four-dimensional spacetime the scalar field must have a mass dimension of unity, as well as the mass parameter m. Consequently, any trilinear coupling $g\phi^3$ contribution to the potential density $V(\phi)$ must have a coupling strength g of mass dimension unity, while a quartic interaction $\lambda\phi^4$ a dimensionless strength coupling λ. In other words, in four-dimensional spacetime, any quartic potential $V(\phi)$ leads to a renormalizable quantum scalar field theory (in the absence of a quartic coupling, a cubic coupling is excluded on physical grounds, since otherwise the energy is not bounded below). However, any coupling of higher order, $\lambda\phi^n$ with $n > 4$, requires a strength coupling of mass dimension $[\lambda] = 4 - n$, and thus represents a nonrenormalizable interaction in a four-dimensional spacetime.

A similar analysis may be developed for any other field theory of higher spin content. Incidentally, in the case of general relativity, the fact that Newton's constant, which then defines the coupling strength for gravity, has a strictly negative mass dimension is one of the reasons why the perturbative quantization of that classical metric field theory of spacetime geometry is nonrenormalizable.

Historically, the requirement of renormalizability was viewed as defining, albeit for physical arguments not thoroughly convincing, a basic restriction on the construction of realistic quantum field theories for the fundamental interactions of the elementary particles. Nowadays, this point of view has considerably shifted, and renormalizable quantum field theories are rather considered to define effective low energy approximations to some more fundamental underlying description of the basic physical phenomena, which need not be given even in terms of a quantum field theory.[1] By integrating out from a given theory its high energy modes above its characteristic energy scale Λ, one recovers a low energy effective description in terms of a field theory in which the effects of the underlying theory relevant to the higher energy scales contribute only through nonrenormalizable effective coupling coefficients of the form $\lambda = \alpha_0/\Lambda^\kappa$. Hence, as the energy scale of the underlying theory becomes arbitrary large, only renormalizable couplings survive in its low energy effective field theory approximation, thereby leading to a decoupling of energy scales as one passes from one level of effective description to the next. From that point of view, the principle of renormalizability for the construction of physical quantum field theories is nothing but a principle for the decoupling of energy scales when formulating a theory capable of describing phenomena up to some characteristic energy scale, without the knowledge and independently of the physics lying beyond that energy scale. The procedure of renormalization described above is then also seen to correspond to a renormalization of the low energy observables through the resummation of all the known contributions up to some cut-off energy scale, beyond which there may lie some unknown territory, and then at the same time make sure that the low energy observables remain independent of this unknown physics, and thus remain finite as well, as this cut-off scale possibly characteristic of some unknown interactions and particles is pushed to arbitrary large values. In effect, this indeed corresponds to a decoupling of scales for the effective low energy approximate quantum field theory description.

Nonetheless, this rationale for the decoupling of scales translated into the requirement of renormalizability still leaves us with the general issue of the construction of such theories. The necessary condition mentioned above in terms of the mass dimension of interaction coupling constants, even when met, is not sufficient to ensure renormalizability of the corresponding coupling. The answer to this issue has been given above in the case of scalar fields, but not yet for spinor nor vector fields in interactions, which are certainly required for a description for the fundamental interactions of quarks and leptons. It turns out this is far from a trivial matter, and throughout the 1960's and early 1970's, it has been established[21] that the only renormalizable interactions of vector fields, massive or not, with matter are those governed by the general gauge symmetry principle of Yang-Mills theories based on some internal symmetry whose algebraic group is a compact Lie group. The stringent and elegant symmetry constraints brought about by the local gauge symmetry principle on the structure of such interactions are just powerful enough to guarantee renormalizability.

Hence, in conclusion, the general principles of special relativity, quantum mechanics and decoupling of scales for effective field theory descriptions of the fundamental interactions and particles has led to the general gauge symmetry principle, and its actual realization in terms of internal symmetries, as the guiding principle for the construction of renormalizable interacting relativistic quantum unitary local field theories as the appropriate framework for the description of the causal interactions of relativistic quantum point-particles and their wavelike spacetime dynamics. Quite an achievement for the marriage of $\hbar$ and c, the genuine third conceptual revolution of XX$^{\rm th}$ century physics following general relativity and quantum mechanics!

4.2 Global internal symmetries

Hence, it is time now to turn to the meaning of internal symmetries, namely symmetries acting on a system but which are not associated to transformations in spacetime. In technical terms, a symmetry is a transformation of a system such that it leaves its equations of motion form invariant. Or in other words, a symmetry transforms a given solution to the dynamics of a system into another solution to the same dynamics. Note that a symmetry is not necessarily an invariance property of configurations of the system, but rather it is an invariance property of the set of its dynamical configurations. In particular, it may be that even for the lowest energy configuration of a system, this solution may or may not be invariant under the action of a symmetry of the equations of motion. As we shall see, this possibility has profound consequences in the context of field theory, especially when it comes to symmetries that are realized locally at each point in spacetime, so-called local gauge symmetries.

Given the character of these notes, only the simplest examples of these different issues are presented here. However, the reader should be aware that many generalizations have been developed, that are available in the literature as well as standard quantum field theory textbooks.

The simplest example

So far, we have considered only the case of a single real scalar field of mass m. Let us now extend the discussion to a system composed of two such fields $\phi_1(x)$ and $\phi_2(x)$ sharing identical masses m and interaction couplings. Consequently, such a system possesses a continuous symmetry whose transformations mix these two fields by an arbitrary amount while preserving their normalization, namely a rotation of arbitrary angle in the two-dimensional space (ϕ_1, ϕ_2) in which they take their values. Specifically, combining the two fields into a single complex valued scalar field,

$$\phi(x) = \frac{1}{\sqrt{2}}\left[\phi_1(x) + i\phi_2(x)\right] , \qquad (159)$$

the corresponding total Lagrangian density, which then reads

$$\begin{aligned}\mathcal{L} &= \partial_\mu\phi^\dagger\partial^\mu\phi - m^2\phi^\dagger\phi - V(|\phi|) \\ &= \tfrac{1}{2}(\partial_\mu\phi_1)^2 - \tfrac{1}{2}m^2\phi_1^2 + \tfrac{1}{2}(\partial_\mu\phi_2)^2 - \tfrac{1}{2}m^2\phi_2^2 - V\left(\sqrt{\phi_1^2+\phi_2^2}\right) ,\end{aligned} \qquad (160)$$

$V(|\phi|)$ being an arbitrary renormalizable interaction potential, is clearly invariant under the class of continuous transformations

$$\phi'(x) = e^{i\alpha}\,\phi(x) , \qquad (161)$$

α being an arbitrary constant real parameter representing the rotation angle of this SO(2)=U(1) symmetry.

This symmetry, which leaves the action and thus also the equations of motion invariant, is a global symmetry, since it acts in an identical fashion on the field $\phi(x)$ irrespective of the spacetime point labelled by x^μ. The symmetry shifts the phase of the complex field by an identical amount globally throughout the whole of spacetime, namely not only instantaneously through all of space but also identically throughout the whole time history of the system. Furthermore, the action of the symmetry is not on the spacetime points at which the field is evaluated, but rather within the "internal two-dimensional space" in which the complex field takes its values. From that point of view, these values for $\phi(x)$ define a two-dimensional space associated to each of the spacetime points, the "internal" space of the system. Consequently, one says that the symmetry is a global internal one.

By virtue of Noether's theorem, associated to such a continuous symmetry, there exists a current and its charge which are locally conserved for solutions to the equations of motion. In the present instance, these conserved Noether current and charge are given by

$$J^\mu = -i\left[\phi^\dagger\partial^\mu\phi - \partial^\mu\phi^\dagger\phi\right] \quad , \quad Q = \int_{(\infty)} d^3\vec{x}\,J^0(x^0,\vec{x}) , \qquad (162)$$

while for solutions to the dynamics of the system, these quantities obey the conservation conditions,

$$\partial_\mu J^\mu = 0 \quad , \quad \frac{dQ}{dt} = 0 . \qquad (163)$$

These Noether current and charge thus characterize the specific properties of the system that follow from its U(1) continuous global internal symmetry. In particular, in its Hamiltonian formulation, the charge Q generates the algebra of the symmetry group, in the present case that of the abelian group U(1), through the Poisson bracket structure. Acting on phase space through these brackets, the Noether charge also generates, in linearized form, the associated symmetry transformations of the phase space degrees of freedom. Through the correspondence principle, the same properties should remain valid at the quantum level in terms of commutation relations. However, because of possible operator ordering ambiguities for composite quantities such as Noether charges and currents, it may be that the quantum consistency requirements for the definition of quantum physical observables clash with the symmetry properties, namely that the symmetry algebra is no longer realized in terms of the commutation relations of the Noether charges. In such a case, the symmetry is said to be anomalous, by which is meant in fact that the symmetry is explicitly broken for the quantized system.

In the present case, it may be checked by straightforward construction that the U(1) internal symmetry is not anomalous. Associated to the creation and annihilation mode expansions of the real fields ϕ_1 and ϕ_2, the complex field $\phi(x)$ acquires of course also such an expansion, but in terms of creation and annihilation operators which are superpositions of those of the initial fields. Having initially two independent fields, one still obtains two independent sets of creation and annihilations operators, given by

$$\begin{aligned} a(\vec{k}) &= \frac{1}{\sqrt{2}}\left[a_1(\vec{k}) + i a_2(\vec{k})\right] \ , \quad b^\dagger(\vec{k}) = \frac{1}{\sqrt{2}}\left[a_1^\dagger(\vec{k}) + i a_2^\dagger(\vec{k})\right] \ , \\ b(\vec{k}) &= \frac{1}{\sqrt{2}}\left[a_1(\vec{k}) - i a_2(\vec{k})\right] \ , \quad a^\dagger(\vec{k}) = \frac{1}{\sqrt{2}}\left[a_1^\dagger(\vec{k}) - i a_2^\dagger(\vec{k})\right] \ , \end{aligned} \tag{164}$$

and obeying the appropriate Fock space algebras

$$[a(\vec{k}), a^\dagger(\vec{\ell})] = (2\pi)^3 2\omega(\vec{k})\delta^{(3)}\left(\vec{k} - \vec{\ell}\right) = [b(\vec{k}), b^\dagger(\vec{\ell})] \ . \tag{165}$$

The mode expansion of the complex field in the interacting picture is then

$$\hat{\phi}_{(I)}(x) = \int_{(\infty)} \frac{d^3\vec{k}}{(2\pi)^3 2\omega(\vec{k})}\left[a(\vec{k})e^{-ik\cdot x} + b^\dagger(\vec{k})e^{ik\cdot x}\right] \ , \tag{166}$$

while a direct substitution in the normal ordered expression for the quantum Noether charge $\hat{Q}$ finds

$$\hat{Q} = -\int_{(\infty)} \frac{d^3\vec{k}}{(2\pi)^3 2\omega(\vec{k})}\left[a^\dagger(\vec{k})a(\vec{k}) \ - \ b^\dagger(\vec{k})b(\vec{k})\right] \ . \tag{167}$$

In comparison with the mode expansion for a real scalar field, one notices a common structure with, however, the role played by the creation operator component now taken over by that of the independent mode $b^\dagger(\vec{k}\,)$ rather than $a^\dagger(\vec{k}\,)$, since the field need no longer be real under complex conjugation. Furthermore, it is precisely this complex character of the field which makes possible the existence of the U(1) symmetry, whose Noether charge should thus distinguish the two types of modes present in the system. Indeed, a direct calculation finds, for instance for the creation operators,

$$[\hat{Q}, a^\dagger(\vec{k}\,)] = -a^\dagger(\vec{k}\,) \quad , \quad [\hat{Q}, b^\dagger(\vec{k}\,)] = +b^\dagger(\vec{k}\,) \ , \tag{168}$$

with in particular

$$\hat{Q}\, a^\dagger(\vec{k}\,)\,|0> = -a^\dagger(\vec{k}\,)\,|0> \quad , \quad \hat{Q}\, b^\dagger(\vec{k}\,)\,|0> = +b^\dagger(\vec{k}\,)\,|0> \ . \tag{169}$$

In other words, the conserved quantum number Q associated to this Noether quantum charge which generates the U(1) symmetry of the system, is an additive quantum number for quantum states, and takes opposite values for the field quanta created by either the operators $a^\dagger(\vec{k}\,)$ or $b^\dagger(\vec{k}\,)$. To put it still differently, these two types of field quanta are distinguished by an opposite U(1) charge under the U(1) global internal symmetry. Fields neutral under complex conjugation are associated to neutral particles under some given continuous symmetry, while fields complex under complex conjugation lead to charged particles for the associated U(1) global internal symmetry. Hence, these two types of quanta correspond to particles and their antiparticles, since except for the opposite values for the U(1) conserved charge, they otherwise share identical physical properties under the spacetime Lorentz symmetry, namely their mass and spin values.

Consequently, this is yet one more outcome of the marriage of $\hbar$ and c: the existence of particles and antiparticles of identical mass and spin, but opposite charge under internal continuous symmetries, such as their electric charge. Even for electrically neutral particles, it could be that the particle and antiparticle species are still distinct due to some other conserved quantum number than the electric charge taking opposite values. Of course, a particle which coincides with its antiparticle, and whose field is thus necessarily real under complex conjugation, is necessarily electrically neutral.

The Noether charge operator $\hat{Q}$ being the generator of the U(1) global symmetry, finite transformations of parameter α are induced through the exponentiated form

$$e^{i\alpha\hat{Q}} \tag{170}$$

acting on the space of quantum states of the system. In particular, note that the perturbative vacuum $|0>$ carries a vanishing U(1) charge, $\hat{Q}|0>= 0$, hence is also invariant under the action of the symmetry group,

$$e^{i\alpha\hat{Q}}\,|0> \,=\, |0> \ . \tag{171}$$

When the ground state or vacuum of the system is left invariant under the action of the symmetry, one says that the symmetry is realized in its Wigner mode.

It is straightforward to extend the above considerations to any internal compact Lie symmetry group. Assume that a given system of fields is invariant under a continuous group G whose algebra is spanned by a set of generators T^a such that

$$[T^a, T^b] = i f^{abc}\, T^c \ , \tag{172}$$

f^{abc} being its structure constants, and for which the collection of fields spans some linear representation of that algebra. Hence, if $\phi(x)$ denotes this collection of fields (with the representation index suppressed), and T^a now stand for the G-generators in that specific representation, the action of the symmetry on the fields may be represented as

$$\phi'(x) = e^{i\theta^a T^a}\, \phi(x) \ , \tag{173}$$

θ^a being arbitrary constant but continuous parameters for G-transformations. These quantities being constant and acting independently of the value of x^μ, such transformations define a global internal symmetry, assuming of course that the Lagrangian density $\mathcal{L}(\phi, \partial_\mu \phi)$ is invariant under these transformations. Consequently, because of Noether's theorem, there exists conserved currents $J^a_\mu(x)$ and charges $Q^a = \int_{(\infty)} d^3\vec{x}\, J^{a0}$ generating the symmetry algebra and its transformations on the space of classical as well as quantum states of the system. In particular, if the ground state of the system is invariant under all G-transformations, namely if the symmetry is realized in the Wigner mode, the quantum space of states gets organized into irreducible representations of G, with in particular the one-particle states falling into the same G-representations as the original fields $\phi(x)$, since the creation and annihilation operators also carry that same representation index. All the latter properties are clearly met in the simple U(1) example above, and it should be straightforward to understand why they should remain valid for an arbitrary nonabelian symmetry group as well.

Spontaneous global symmetry breaking

The above results still leave open the case of a symmetry which is not realized in the Wigner mode, namely when the vacuum or ground state of the system is not invariant under the action of the symmetry. It is well known that specific physical systems may possess such a property, as is the case for instance for spontaneous magnetization in a ferromagnetic material below the transition temperature. Let us recall the point made already previously, namely that what is meant by a symmetry is not the invariance of any of its configurations in particular, but rather the invariance of its equations of motion, hence also of the set of its configurations solving these equations viewed as a whole. If a given solution is not invariant, the existence of the continuous symmetry simply implies that there exists an infinite degeneracy of distinct solutions of identical energy all related through the action of the symmetry transformations. For example, imagine a simple linear stick standing along the vertical direction, onto which a certain pressure is applied along that axis. This system is obviously invariant for all rotations around the vertical axis. As long as the applied pressure is mild enough, the stick does not bend, and the lowest energy configuration of the system is indeed invariant under the axial symmetry. However, as soon as the applied pressure exceeds a specific critical value, the stick does bend until it reaches some equilibrium configuration. The horizontal direction in which this bending occurs is arbitrary, but it clearly spontaneously breaks the axial symmetry. Nevertheless, all the configurations of the system associated to all possible horizontal bending directions are degenerate in energy, and are related to one another precisely by the action of the axial symmetry group. The specific solution to the equations of motion singled out by the bending process is no longer invariant, but the set of all these solutions remains invariant, all the degenerate solutions being related through the axial symmetry group. When a symmetry is realized in such a manner, namely when the ground state of the system is not invariant under the symmetry, one says that the symmetry is spontaneously broken, or that it is realized in the Goldstone mode.

Whether a symmetry is realized in the Wigner or in the Goldstone mode is governed by the details of the dynamics of the system, whether in a perturbative or a nonperturbative regime. Once again for the purpose of simplicity, here we only discuss the simplest example, namely that of the spontaneous symmetry breaking already at the level of the classical theory of a single complex scalar field $\phi(x)$ possessing the U(1) global symmetry

$$\phi'(x) = e^{i\alpha}\,\phi(x)\ , \tag{174}$$

with the real constant angular parameter α.

Let us consider again the Lagrangian density

$$\mathcal{L}\left(\phi, \partial_\mu \phi\right) = |\partial_\mu \phi|^2 - V(|\phi|)\ , \tag{175}$$

where the potential contribution is given by

$$V(|\phi|) = \mu^2|\phi|^2 + \lambda|\phi|^4 \tag{176}$$

with $\lambda > 0$. In our previous considerations, the quantity μ^2 was taken to be positive, in which case it defined the mass-squared of the particle quanta associated to the field, describing the quantum excitations of this field above its ground state, namely the perturbative vacuum $|0>$ associated to the classical value $\phi = 0$ up to the vacuum quantum fluctuations subtracted away through normal ordering, which is invariant under the U(1) symmetry.

Presently however, we shall consider the situation when $\mu^2 < 0$, corresponding to the so-called mexican hat potential, which very much looks like the bottom of a wine bottle. In such a case, the configuration $\phi = 0$ no longer defines the lowest energy configuration of the system, since the potential $V(|\phi|)$ now reaches its lowest value for

$$|\phi(x)| = \frac{1}{\sqrt{2}}v \quad , \quad v = \sqrt{\frac{-\mu^2}{\lambda}} \; . \tag{177}$$

Such a configuration also defines the lowest energy state of the field, since all field gradient contributions to the energy then vanish identically, the field being constant throughout spacetime. Such a configuration however, is no longer invariant under the U(1) symmetry, which is thus realized in the Goldstone mode. What are then the physical consequences of this spontaneous symmetry breaking in the vacuum?

In order to properly identify the physical quanta of the field, it is necessary to consider the field fluctuations about its vacuum configuration. Note that the two independent degrees of freedom per spacetime point defined by the complex scalar field may also be represented through a polar decomposition around a given choice of vacuum configuration,

$$\phi(x) = \frac{1}{\sqrt{2}}e^{i\xi(x)/v}\,[\rho(x) + v] \; , \tag{178}$$

where $\xi(x)$ and $\rho(x)$ are two real scalar fields with a mass dimension of unity. Note that the vacuum about which this expansion is performed is

$$\phi_0 = \frac{1}{\sqrt{2}}\,v \; , \tag{179}$$

but that choice may easily be modified by adding to the mode $\xi(x)$ an arbitrary real constant quantity. This remark also shows that the U(1) symmetry now leaves the radial field $\rho(x)$ invariant, while it simply shifts the field $\xi(x)$ by the product αv. All the minimal energy configurations correspond the constant field ϕ lying at the bottom of the potential, with the norm $|\phi| = v/\sqrt{2}$ but an arbitrary phase. The U(1) symmetry simply induces a transformation of any such vacuum into any another such vacuum, the difference in their phases being set by the value of the U(1) angle α (note the perfect analogy with the above example of a bent stick). Hence, one should expect that the fluctations associated to the field $\xi(x)$ are massless, since they may be excited at zero-momentum at no extra energy cost. On the other hand, the radial fluctuation $\rho(x)$, moving the field out from its lowest energy configuration, must correspond to massive quanta of the field. Furthermore, this physical conclusion does not depend on the choice of complex phase for the reference constant vacuum configuration ϕ_0, since this amounts to a simple constant shift in the massless field $\xi(x)$.

More explicitly, a direct substitution of the mode expansion (178) gives

$$\mathcal{L} = \frac{1}{2}\partial_\mu\rho\partial^\mu\rho + \frac{1}{2}\left(1+\frac{1}{v}\rho\right)^2 \partial_\mu\xi\partial^\mu\xi - \frac{1}{2}\mu^2(\rho+v)^2 - \frac{1}{4}\lambda(\rho+v)^4 \ . \quad (180)$$

Isolating then the terms quadratic in $\xi(x)$ and $\rho(x)$ indeed confirms that the mode $\xi(x)$ is massless, while the $\rho(x)$ field is massive, with the values

$$m_\rho^2 = -2\mu^2 > 0 \quad , \quad m_\xi^2 = 0 \ . \quad (181)$$

Hence, we reach the conclusion that since the vacuum is not invariant under the action of transformations which nevertheless define a symmetry of the system and its equations of motion, necessarily in the Goldstone mode realization of the symmetry there exist massless modes, namely massless quanta for a quantized field, which in the zero momentum limit correspond to the excitation of one vacuum state into another one, all these vacuum states being degenerate in energy and infinite in number. Hence, rather than being explicitly realized in the space of states as is the case for the Wigner mode, the symmetry is now hidden through the existence of Golstone bosons. Nonetheless, the symmetry is still active within the system, even though it is no longer realized in a linear fashion. Indeed, within the field basis which diagonalizes its fluctuations, the symmetry acts as

$$\rho'(x) = \rho(x) \quad , \quad \xi'(x) = \xi(x) + \alpha v \ , \quad (182)$$

which, among other consequences, implies that the Goldstone modes may only possess derivative or gradient couplings with other fields. The symmetry thus restricts to some extent the form of interactions of Goldstone fields.

In fact, it should be quite clear that this is a conclusion valid in full generality, which is known as Goldstone's theorem. Whenever a continuous global symmetry is spontaneously broken in the vacuum, associated to each of its broken generators, there exist massless quanta carrying the corresponding quantum numbers, known as the Goldstone bosons of the symmetry. This conclusion is valid whether the spontaneous symmetry breaking mechanism is perturbative or nonperturbative, and whether the symmetry is abelian or nonabelian. The only specific requirement is that the symmetry be a continuous one (in the case of a fermionic or spacetime symmetry, the Goldstone mode need not be bosonic, as is the case for instance for spontaneous supersymmetry breaking leading to a spin 1/2 goldstino massless mode).

4.3 Local or gauged internal symmetries

So far, we have briefly discussed the meaning of a global internal symmetry, and described some of its physical consequences, whether in the Wigner or the Goldstone mode. However, the existence of a global symmetry is not very appealing, at least from some theoretical aesthetic point of view. Indeed, any global internal symmetry defines transformations on the set of fields which act in an identical manner irrespective of the spacetime point at which the field values are being considered. For instance in the case of the U(1) symmetry associated to the electric charge and the electromagnetic interactions, this would mean that in order to render the transformation unobservable, one is required to change the phases of all the electrons of the Universe by exactly the same amount instantaneously throughout all of infinite space and troughout the whole of spacetime history! Although there is no technical or mathematical inconsistency that arises with such a relativistic quantum field theory, certainly it is a property of such symmetry transformations which runs counter to our belief that causality ought to be a stringent requirement on the construction of any physical theory.

Hence, one should rather prefer to develop a formalism in which internal symmetries are still possible, but such that now transformations may be realized locally in spacetime, though in a continuous fashion as to their spacetime dependency, while they would remain nevertheless unobservable to any conceivable experiment. Namely, is it possible to locally change the quantum phase of some electron while not at the same time by the same amount that of all the other electrons of the Universe, and nevertheless keep such a change hidden from any experimentalist? Clearly, this would require some information to be sent to all the other electrons in the Universe to tell them how to adjust their quantum phases accordingly, and this at the speed of light so that no experimentalist may catch up with this signal and measure the phase of some electron before it would have had the opportunity to adjust itself to the action of the symmetry transformation. In other words, by making the symmetry local, or by gauging the symmetry, one must introduce some additional propagating field coupling with equal strength to all other matter carrying the same symmetry charge, and whose quanta are necessarily massless.

This is the heuristic idea of the local gauge symmetry principle. As we shall explicitly see through the simplest examples, such a principle in fact provides a unifying principle for the existence of fundamental interactions, whose quantum carriers are massless and couple with identical strength to all other quanta with which they interact. These gauge bosons are necessarily vector fields for internal symmetries, and as stated previously, such Yang-Mills gauge theories based on compact Lie groups are the only possible renormalizable field theories including spin 0 and 1/2 matter fields interacting with vector fields.

The simplest example

As the simplest illustration of the above description, let us consider once again the theory of a single complex scalar field ϕ whose Lagrangian density is U(1) invariant under global phase transformations of the field, see (160) and (161). Clearly, if one wishes to gauge this symmetry, namely to construct a system which remains invariant under the local phase transformations

$$\phi'(x) = e^{i\alpha(x)}\,\phi(x)\ , \tag{183}$$

$\alpha(x)$ now being an arbitrary spacetime dependent parameter rather than a constant angle as in the case of a global symmetry, a problem arises with the original Lagrangian. Indeed, this Lagrangian is no longer invariant, since the gradient contribution does not transform in the same covariant manner as the original field does,

$$\partial_\mu\phi'(x) = e^{i\alpha(x)}\left[\partial_\mu\phi(x) + i\partial_\mu\alpha(x)\,\phi(x)\right]\ . \tag{184}$$

However, this expression suggests a modification of the ordinary derivative or gradient of the field of the form

$$\partial_\mu \longrightarrow D_\mu(x) = \partial_\mu + igA_\mu(x)\ , \tag{185}$$

where g is some dimensionless real quantity, which turns out to represent the coupling strength of the U(1) gauge interaction, and $A_\mu(x)$ the vector field for the gauge boson associated to the gauging of the U(1) symmetry. Indeed, it now suffices to assume that this vector field transforms under the local U(1) symmetry according to

$$A'_\mu(x) = A_\mu(x) - \frac{1}{g}\partial_\mu\,\alpha(x)\ , \tag{186}$$

to check that the modified gradient does possess the same covariant transformation as the field does under the symmetry,

$$D'_\mu(x)\phi'(x) = \left[\partial_\mu + igA'_\mu(x)\right] e^{i\alpha(x)}\phi(x) = e^{i\alpha(x)}\, D_\mu(x)\phi(x)\ , \tag{187}$$

hence the name "covariant derivative" for the differential operator $D_\mu(x)$. Clearly, a simple substitution of the ordinary derivative by the covariant one in the original Lagrangian density invariant under the global U(1) symmetry leads to an expression invariant now under any local U(1) symmetry transformation. The U(1) symmetry has been gauged.

However, we still need to provide the vector field $A_\mu(x)$ with some dynamics, which is done by adding the pure gauge Lagrangian density to that of the matter field,

$$\mathcal{L}_A = -\frac{1}{4}F_{\mu\nu}F^{\mu\nu}\quad ,\quad F_{\mu\nu} = \partial_\mu A_\nu - \partial_\nu A_\mu\ , \tag{188}$$

$F_{\mu\nu}$ being the gauge field strength, indeed the sole gauge invariant quantity that may be constructed out of the gauge field A_μ and its first-order gradients, in order to obtain a Lagrangian density which is of second-order in spacetime gradients, and thus represents a causal propagation of the gauge field (for the same reason, the absolute sign and normalization of this Lagrangian density are fixed as given). This field being real under complex conjugation, its mode expansion is of the form, in the interacting picture,

$$A_\mu(x) = \int_{(\infty)} \frac{d^3\vec{k}}{(2\pi)^3 2|\vec{k}|} \sum_{\lambda=\pm} \left[e^{-ik\cdot x}\epsilon_\mu(\vec{k},\lambda)a(\vec{k},\lambda)\ +\ e^{ik\cdot x}\epsilon^*_\mu(\vec{k},\lambda)a^\dagger(\vec{k},\lambda)\right]\ , \tag{189}$$

$a(\vec{k},\lambda)$ and $a^\dagger(\vec{k},\lambda)$ being annihilation and creation operators with the Fock space algebra normalized in the usual manner for massless quanta, and λ denotes the different polarization states possible associated to the polarization vectors $\epsilon_\mu(\vec{k},\lambda)$. These polarization tensors are subjected to some restrictions which stem from the gauge invariance properties of the field, and shall not be discussed here (even though the issue of the quantization of gauge invariant systems is discussed hereafter, but not explicitly for such abelian and non-abelian Yang-Mills theories). Note that the mass dimension of the gauge field indeed needs to be unity, hence leading to a dimensionless gauge coupling constant g.

In conclusion, the gauging of the simplest U(1) invariant scalar field theory is defined by the total Lagrangian density

$$\mathcal{L}_{\rm total} = \mathcal{L}_A + \mathcal{L}_\phi\ , \tag{190}$$

with the pure gauge Lagrangian $\mathcal{L}_A$ given above, and the matter one by

$$\begin{aligned}\mathcal{L}_\phi &= \mathcal{L}\left(\phi, D_\mu\phi\right) = |(\partial_\mu + igA_\mu)\phi|^2 - m^2|\phi|^2 - V(|\phi|) \\ &= |\partial_\mu\phi|^2 - m^2|\phi|^2 - V(|\phi|) - igA_\mu\left[\phi^\dagger\partial^\mu\phi - \partial^\mu\phi^\dagger\phi\right] + g^2A_\mu A^\mu\ .\end{aligned} \tag{191}$$

In the case that the U(1) symmetry is that associated to the electromagnetic interaction, this system is simply that of scalar electrodynamics, namely that describing the interactions of a massive charged spin 0 particle with the photon.

From the latter expression, we immediately read off the different interaction terms coupling the matter and gauge fields. The term linear in A_μ is in fact $gA_\mu J^\mu$, namely the coupling of gauge field to the U(1) Noether current, and represents the coupling of one gauge quantum to two scalar field quanta of opposite U(1) charges. Such a feature is generic for all Yang-Mills theories: gauge fields always couple linearly to the associated Noether currents. The term quadratic in A_μ describes the coupling of two gauge quanta to two scalar quanta, also of opposite U(1) charges, in order for the total U(1) charge to be conserved in the interactions. Note that the single gauge boson interaction is proportional to ig, while the quadratic interaction is proportional to ig^2. In other words, the gauge symmetry principle not only explains, on the basis of a given internal symmetry, the appearance of local interactions, but it also sets specific restrictions on the properties of these interactions by predicting particular relations between the coupling strengths of different interactions, such restrictions being a consequence of the symmetry.

Among the interactions, the gauge boson $A_\mu(x)$ does not couple to itself, but only to the charged matter field with the universal coupling strength g. The reason for the fact that the gauge boson lacks such a self-coupling is that it is neutral under the U(1) symmetry, and does not carry any U(1) charge. Indeed, under a global symmetry transformation $\alpha(x) = \alpha$, we simply have for the transformed field $A'_\mu = A_\mu$. Furthermore, it is also the U(1) symmetry, but this time in its gauged embodiement, which explains why the gauge boson quanta are massless particles. Indeed, any mass term of the form $M_A^2 A_\mu A^\mu$ is clearly not gauge invariant under the local gauge transformations of the vector field. Hence, it is the local gauge symmetry which protects the gauge boson from acquiring any mass. In particular, this implies that physical (gauge invariant) quanta of that field may possess only two transverse polarization states, such that $k^\mu\epsilon_\mu(\vec{k}, \lambda) = 0$, $\lambda = \pm$, a fact related to the issue of the quantization of such Yang-Mills fields.

All the above considerations are readily extended to other matter fields, including fermionic ones not addressed in these notes. Furthermore, even though our discussion concentrates on the abelian U(1) case, the same developments apply to a nonabelian internal symmetry group G, leading then to Yang-Mills gauges theories. In such a case, for a collection of fields transforming in a G-representation whose generators are T^a, the covariant derivative, which now is Lie-algebra valued, reads

$$D_\mu = \partial_\mu + igA^a_\mu T^a \ , \tag{192}$$

g being the real gauge coupling constant, and A^a_μ the real gauge vector fields, which, for infinitesimal local gauge transformations of parameters $\theta^a(x)$, transform according to

$$A'^a_\mu = A^a_\mu - \frac{1}{g}\partial_\mu\theta^a - f^{abc}\theta^b A^c_\mu \ , \tag{193}$$

f^{abc} being the structure constants of the Lie algebra of G (it is also straightforward to establish the transformations of the gauge bosons for finite gauge transformations). The total Lagrangian of such a system is again given by the sum of the original G-invariant Lagrangian of the matter fields in which the ordinary derivative is substituted by the covariant derivative D_μ, to which one simply adds the pure Yang-Mills Lagrangian density

$$\mathcal{L}_A = -\frac{1}{4}F^a_{\mu\nu}F^{a\mu\nu} \quad , \quad F^a_{\mu\nu} = \partial_\mu A^a_\nu - \partial_\nu A^a_\mu - gf^{abc}A^b_\mu A^c_\nu \ . \tag{194}$$

Once again, any mass term for the gauge bosons A^a_μ is forbidden by local gauge invariance, while gauged matter interactions are directly read off from the matter Lagrangian, leading again to linear and quadratic interactions of scalar fields with the gauge bosons. However, for a nonabelian symmetry, given the nonvanishing structure constants f^{abc}, the gauge bosons themselves possess now G-charges, actually those of the adjoint representation as may be seen from their gauge transformations for constant parameters $\theta^a(x) = \theta^a$. Consequently, from the expansion of the pure Yang-Mills Lagrangien, one identifies cubic and quartic terms representing gauge boson trilinear and quadrilinear couplings, whose strengths are directly proportional to g and g^2, respectively. Hence once again, the symmetry governs the details of all the gauge interactions, namely their strengths and their symmetry properties as well. Such predictions are specific to Yang-Mills theories, and provide important signatures for high energy experiments as to the relevance of the gauge symmetry principle for the physics of the fundamental interactions and the elementary particles. Note also that it is precisely these nonlinear gauge boson self-couplings which must be, in ways still to be thoroughly understood, at the origin of the specific nonperturbative phenomena of nonabelian theories, such as the property of confinement for the theory of the strong interactions among quarks, namely quantum chromodynamics (QCD) based on the local gauge symmetry $SU(3)_C$ for colour degree of freedom of quarks.

Spontaneous gauge symmetry breaking

The above discussion of the construction of abelian and nonabelian internal gauge symmetries implicitly assumed the symmetry to be realized in the Wigner mode. Hence, it is also important to consider the situation when the symmetry is rather realized in the Goldstone mode. For the purpose of illustration in the simplest case, let consider once again the U(1) gauged single scalar field theory, but this time with a potential leading to spontaneous symmetry breaking. The associated Lagrangian density is thus

$$\mathcal{L} = -\frac{1}{4}F_{\mu\nu}F^{\mu\nu} + |(\partial_\mu + igA_\mu)\phi|^2 - V(|\phi|) \ , \tag{195}$$

with

$$V(|\phi|) = \mu^2|\phi|^2 + \lambda|\phi|^4 \quad , \quad \mu^2 < 0 \ , \ \lambda > 0 \ . \tag{196}$$

This time however, because of the U(1) local symmetry transformation properties of the fields,

$$\phi'(x) = e^{i\alpha(x)}\,\phi(x) \quad , \quad A'_\mu(x) = A_\mu(x) - \frac{1}{g}\partial_\mu\alpha(x) \ , \tag{197}$$

when expanding any scalar field configuration about one of its vacuum configurations,

$$\phi(x) = \frac{1}{\sqrt{2}}e^{i\xi(x)/v}\,[\rho(x) + v] \quad , \quad \phi_0(x) = \frac{1}{\sqrt{2}}v \quad , \quad v = \sqrt{\frac{-\mu^2}{\lambda}} \ , \tag{198}$$

it is always possible to effect a local U(1) gauge transformation, with parameter

$$\alpha(x) = -\frac{1}{v}\xi(x) \ , \tag{199}$$

(note that since in general $\xi(x)$ is spacetime dependent, such a procedure is possible only when the internal symmetry is gauged), such that the Goldstone mode is completely gauged away from the scalar field, but lies hidden now in the transformed gauge field A'_μ,

$$\phi'(x) = \frac{1}{\sqrt{2}}[\rho(x) + v] \quad , \quad A'_\mu(x) = A_\mu(x) + \frac{1}{gv}\partial_\mu\xi(x) \ . \tag{200}$$

Upon substitution of the transformed fields in the Lagrangian density, which is physically equivalent to the original expression for the Lagrangian on account of local gauge invariance, one then finds

$$\mathcal{L} = -\frac{1}{4}F'_{\mu\nu}F'^{\mu\nu} + \frac{1}{2}\left[\partial_\mu\rho + igA'_\mu(\rho+v)\right]^2 - \frac{1}{2}\mu^2(\rho+v)^2 - \frac{1}{4}\lambda(\rho+v)^4 \ . \quad (201)$$

Isolating now all quadratic terms in the fields, one immediately notices that the radial field ρ still possesses the mass $m_\rho^2 = -2\mu^2 > 0$, but that in place of a massless Goldstone mode $\xi(x)$ which no longer appears in this Lagrangian by having been gauged away, there now appears an explicit mass term for the gauge boson field, with value

$$m_A^2 = g^2v^2 \ . \quad (202)$$

Hence, even though a local symmetry when realized in the Wigner mode forbids any mass for its gauge bosons, when spontaneously broken in the vacuum and realized in the Goldstone mode, gauge bosons do acquire a mass! Nevertheless, their mass is then not just any parameter in the Lagrangian, but is in fact governed by the symmetry properties and takes a very specific value proportional both to the gauge coupling constant g and the scalar field vacuum expectation value v which spontaneously breaks those symmetry generators whose gauge bosons are massive. The counting of degrees of freedom is also in order. In the Wigner phase, one has two real scalar modes (one massless and one massive, the Goldstone and the radial ones, ξ and ρ) and two massless gauge modes (the two transverse modes of the gauge field). In the Goldstone phase, one has one real massive scalar mode (the radial field ρ) and three massive gauge boson polarization modes. Note that the longitudinal massive gauge boson component is nothing but the would-be Goldstone mode ξ which has been gauged away and turned into the longitudinal component of the gauge field A'_μ, see (200).

These general features of the spontaneous symmetry breaking of a local gauge symmetry remain valid in general, and characterize the so-called Higgs mechanism. Whenever a local internal symmetry is spontaneously broken in the vacuum, those gauge bosons associated to the generators which do not leave invariant the vacuum acquire a mass proportional to the product of the gauge coupling and the scalar vacuum expectation value. Moreover, the Goldstone modes in the case of a global symmetry then provide the longitudinal polarization states of the massive gauge bosons, leaving over the massive scalar modes, referred to as higgs scalars, as the only remnants of the spontaneously broken scalar matter sector. The gauge transformation which gauges away the Goldstone modes from the scalar sector to hide them in the gauge fields is known as the unitary gauge. It is in the unitary gauge that the physical content of such a theory is most readily identified. In the simplest example above, we thus conclude that the physical field content is that of a neutral massive spin 0 particle of mass $\sqrt{-2\mu^2}$, the higgs particle, interacting with itself and with a neutral massive spin 1 particle of mass $|gv|$.

Remarks

As already mentioned, it turns out that the gauge symmetry principle uniquely singles out among all possible quantum field theories of interacting spin 0, 1/2 and 1 particles, all those that are renormalizable, whether the gauge bosons are massive or not, provided however that in the former case their mass arises through the Higgs mechanism.[21] This is quite a remarkable result, since such a local internal symmetry principle also implies the existence of specific interactions between matter particles and gauge bosons, whose detailed properties are totally governed by the underlying symmetry, whether abelian or nonabelian. In other words, all the relativistic and quantum dynamics of fundamental interactions among elementary point-particles, through the marriage of $\hbar$ and c, appears to follow simply from the very elegant and powerful idea of a fundamental symmetry based on a compact Lie group.

Thus in order to describe all the known strong, electromagnetic and weak interactions observed to act between all known quarks and leptons, a gauge group as simple as $\mathrm{SU}(3)_c\times\mathrm{SU}(2)_L\times\mathrm{U}(1)_Y$ suffices, with a specific choice of representations for the quark and leptons fermionic fields, as well as for the scalar sector required for the Higgs mechanism leading to massive electroweak gauge bosons but nonetheless a massless photon. If not yet totally unified within this Standard Model of these interactions, at least all these interactions are brought within the unified framework of relativistic quantum Yang-Mills theories, leading to predictions whose precision is without precedent and which are confirmed through remarkable particle physics experiments. Nevertheless, this raises the issue of the rationale behind such a principle, as well as for the choice of internal symmetry and matter content.

From another perspective, with such Yang-Mills theories we are encountering dynamical theories whose quantization requires an approach more general than that which was briefly reviewed in Sect.2. Indeed, considering the issue for example from within the Hamiltonian approach, when identifying the momentum conjugate to the U(1) gauge field A_μ coupled to the single scalar field through the Lagrangian density discussed above, one finds

$$\pi^\mu = \frac{\partial\mathcal{L}_{\rm total}}{\partial(\partial_0 A_\mu)} = -F^{0\mu}\ , \tag{203}$$

thus leading to the following constraint for its time component

$$\pi^0 = 0 \ . \tag{204}$$

In other words, all phase space degrees of freedom of the system are not independent. Some are in fact constrained, and as we shall see in the forthcoming section, this is a generic feature for any system possessing a local symmetry whose parameters are not constant. How is one then to quantize such systems, since their physical dynamics is not contained within all of phase space, but only within some subspace of it? Clearly, gauge invariance implies that all degrees of freedom are not physical and relevant to the dynamics. How does one then account consistenly for such redundant features of a gauge invariant system in its quantization? In the above example, it would be possible to solve for these gauge degrees of freedom, but at the cost of loosing a manifestly spacetime covariant description of such systems, which is also not welcome in itself. Hence, it is time now to turn to the discussion of the quantization of constrained dynamics.

5 Dirac's Quantization of Constrained Dynamics

5.1 Classical Hamiltonian formulation of singular systems

The system of constraints

The Hamiltonian approach towards canonical quantization discussed in Sect.2 explicitly assumed that the Lagrange function be "regular", see (13). Now, we have to develop the Hamiltonian formulation associated to a "singular" Lagrangian,[8,10] namely one for which the Hessian possesses some local zero modes,

$$\det \frac{\partial^2 L}{\partial \dot{q}^{n_1} \partial \dot{q}^{n_2}} = 0 \ . \tag{205}$$

When considered in terms of the conjugate momenta, $p_n = \partial L / \partial q^n$, for which the canonical Poisson bracket structure is still in effect,

$$\{q^{n_1}(t), p_{n_2}(t)\} = \delta^{n_1}_{n_2} \ , \tag{206}$$

since the condition (205) also writes as

$$\det \frac{\partial p_{n_1}}{\partial \dot{q}^{n_2}} = 0 \ , \tag{207}$$

it follows that singular systems are characterized by the existence of a series of primary constraints on phase space, of the form

$$\phi_m(q^n, p_n) = 0 \ , \tag{208}$$

where in this section, the index m will be reserved for the primary constraints. To be precise, a technical restriction on the choice of expression for these primary constraints, known as the regularity condition, is assumed, such that the phase space gradient of these constraints is a matrix of constant maximal rank on the constraint hypersurface (as the discussion proceeds, further constraints beyond the primary ones appear; the subset of phase space defined by the entire set of constraints is known as the constraint hypersurface). Indeed, whether a given constraint $\phi = 0$ is expressed as $\phi^2 = 0$ or $\phi^{1/2} = 0$ may seem *a priori* equally acceptable, but further developments in fact require this regularity condition for the constraints. In practice, it may be established that under the regularity condition, any quantity that vanishes on the constraint hypersurface may be written as a local linear combination of the constraints.

The canonical Hamiltonian

$$H_0(q^n, p_n) = \dot{q}^n p_n - L(q^n, \dot{q}^n) \tag{209}$$

is still defined in the usual way, in spite of the existence of constraints. It may be shown that whether the Lagrange function is regular or singular, this function $H_0(q^n, p_n)$ is indeed always a function defined over phase space (q^n, p_n), even in the presence of primary constraints. However, in contradistinction to the regular situation, time evolution of the system may *a priori* be generated by a Hamiltonian more general than the canonical one, since one may add to the latter some combination of the primary constraints. Hence, let us consider the primary Hamiltonian

$$H_*(q^n, p_n; t) = H_0(q^n, p_n) + \mathcal{U}^m(q^n, p_n; t)\phi_m(q^n, p_n) \ , \tag{210}$$

$\mathcal{U}^m(q^n, p_n; t)$ being *a priori* some arbitrary functions of phase space, possibly time dependent as well. These functions parametrize the arbitrariness which exists as to the time dependent dynamics of the system restricted to the constraint hypersurface. Some of these functions may be restricted or be totally determined by consistency conditions on this time development, as we shall discuss, whereas others may remain totally undetermined, a possibility that should be expected in the case of gauge theories since the solutions to such systems always depend on some arbitrary time dependent functions related to the gauge degrees of freedom.

The functions $\mathcal{U}^m(q^n, p_n; t)$ could indeed be restricted, since the primary Hamiltonian, which should generate a consistent time evolution, must be such that given initial data within the constraint hypersurface, the evolved configuration must always belong to that same hypersurface. In other words, whenever time evolution pulls some initial configuration away from the constraint surface, some projection mechanism must push it back onto it. This requirement is certainly met if time evolution of the primary constraints is such that their Hamiltonian equations of motion vanish identically on the constraint hypersurface, namely

$$\dot{\phi}_m = \{\phi_m, H_*\}_{|\phi_m=0} = 0 \ , \tag{211}$$

a condition also expressed as

$$\dot{\phi}_m \approx 0 \ , \tag{212}$$

where the *weak equality* sign "≈" stands for a relation which is valid when restricted to the constraint hypersurface. Since the Hamiltonian equation of motion for an arbitrary phase space observable $f(q^n, p_n; t)$ is

$$\begin{aligned} \frac{df}{dt} &= \frac{\partial f}{\partial t} + \{f, H_0\} + \mathcal{U}^m\{f, \phi_m\} + \{f, \mathcal{U}^m\}\phi_m \\ &\approx \frac{\partial f}{\partial t} + \{f, H_0\} + \mathcal{U}^m\{f, \phi_m\} \ , \end{aligned} \tag{213}$$

it follows that a consistent time evolution of the constraints requires

$$\{\phi_m, H_0\} + \mathcal{U}^{m'}\{\phi_m, \phi_{m'}\} \approx 0 \ . \tag{214}$$

This set of equations for each of the values for the index m labelling the primary constraints either implies a trivial identity, $0 = 0$, or else a nontrivial phase space restriction $\chi(q^n, p_n) = 0$ independent of the functions $\mathcal{U}^m$, or finally a genuine linear equation for the functions $\mathcal{U}^m$. Whenever restrictions $\chi(q^n, p_n) = 0$ appear at this stage, they in fact represent new or secondary constraints on phase space, for which conditions of a consistent time evolution have again to be considered. Consequently, a whole hierarchy of constraints may appear generation after generation of such secondary constraints, through the requirement of a consistent time evolution, until the whole set of constraints is exhausted. Let us denote the whole set of constraints as $\phi_j(q^n, p_n) = 0$, the index j being reserved for that purpose. Note that whenever a new constraint $\phi_j = 0$ is uncovered, there is no reason not to include it as well in the Hamiltonian through some combination $\mathcal{U}^j(q^n, p_n)\phi_j$ to be added to the linear combination of constraints, hence leading *in fine* to the total Hamiltonian

$$H_T = H_0 + \mathcal{U}^j \phi_j \ . \tag{215}$$

Hence finally, a consistent time evolution generated by this Hamiltonian requires that all constraints $\phi_j(q^n, p_n) = 0$ be preserved through the dynamical equations of motion, $\dot{\phi}_j \approx 0$, leading to the set of equations

$$\{\phi_j, H_0\} + \mathcal{U}^{j'}\{\phi_j, \phi_{j'}\} \approx 0 \ . \tag{216}$$

The general solution to these equations is

$$\mathcal{U}^j(q^n,p_n) = U^j(q^n,p_n) + \lambda^\alpha(t)V^j_\alpha(q^n,p_n) \ , \tag{217}$$

where $U^j(q^n,p_n)$ are a specific set of solutions to the inhomogeneous linear equations

$$\{\phi_j,\phi_{j'}\}U^{j'} = -\{\phi_j,H_0\} \ , \tag{218}$$

while $V^j_\alpha(q^n,p_n)$ provide a basis for the space of solutions to the homogeneous equations

$$\{\phi_j,\phi_{j'}\}V^{j'}_\alpha = 0 \ . \tag{219}$$

Note that these solutions induce the space of zero modes of the matrix $\{\phi_j,\phi_{j'}\}$. Finally, the quantities $\lambda^\alpha(t)$ are arbitrary time dependent functions which define arbitrary linear combinations of the zero mode solutions V^j_α. *A priori*, these functions could also depend on the phase space variables, $\lambda^\alpha(q^n,p_n;t)$, with no added advantage to the description of any such constrained system however. As a matter of fact, it suffices to consider them to be solely functions of time. Indeed, as discussed hereafter, the freedom in the choice of these functions is directly related to the existence of local gauge symmetries, and as is well known, gauge symmetries imply the appearance of arbitrary (space)time dependent functions in the general solutions to the equations of motion. The quantities $\lambda^\alpha(t)$ are related to nothing but these arbitrary (space)time dependent functions defining general solutions.

In terms of the above results, the dynamics of a constrained system is generated by a total Hamiltonian of the form

$$H_T = H + \lambda^\alpha(t)\phi_\alpha \ , \tag{220}$$

with

$$H = H_0 + U^j\phi_j \quad , \quad \phi_\alpha = V^j_\alpha\phi_j \ . \tag{221}$$

As we shall now discuss, the role and meaning of the quantities U^j, H, ϕ_α and λ^α are essential to the understanding of constrained dynamics.

First- and second-class quantities and constraints

As a matter of fact, constrained dynamical systems may possess physically equivalent but nonetheless different Lagrangian formulations. Whether a given constraint then appears as a primary or a secondary constraint is then function of the chosen Lagrangian formulation used as the starting point for the Hamiltonian analysis of constraints. However, there exists a characterization of constraints which is not dependent on that choice, in a way which is already suggested by the above conclusions regarding the quantities H and ϕ_α. This characterization leads to a classification of constraints which is due to Dirac.[8]

Consider any phase space quantity $R(q^n, p_n)$. By definition, this quantity is said to be a *first-class* quantity if and only if its Poisson brackets with all the constraints ϕ_j vanish weakly,

$$\{R, \phi_j\} \approx 0 \iff \{R, \phi_j\} = R_j{}^{j'} \phi_{j'} \ , \tag{222}$$

the equivalence being valid provided the constraints are regular. Otherwise, if at least one of these Poisson brackets does not vanish on the constraint hypersurface, the quantity R is said to be *second-class*.

From the Jacobi identity obeyed by Poisson brackets, it follows that the Poisson bracket $\{R_1, R_2\}$ of any two first-class quantities R_1 and R_2 is itself a first-class quantity. Furthermore, these definitions are not void of content. Indeed, from the definition of the coefficients U^j and the quantity H, it follows that

$$\{H, \phi_j\} \approx 0 \ , \tag{223}$$

showing that the Hamiltonian H is in fact a first-class quantity. Likewise, from the definition of the coefficients V^j_α, it follows that

$$\{\phi_\alpha, \phi_j\} \approx 0 \ , \tag{224}$$

showing also that all the linear combinations of constraints ϕ_α are themselves first-class quantities. Consequently, the total Hamiltonian is itself a first-class quantity, being given by the first-class Hamiltonian H summed with an arbitrary linear combination of the first-class constraints ϕ_α.

This characterization as being first- or second-class quantities applies in particular to the constraints ϕ_j. This classification of constraints in terms of first- and second-class constraints is in fact an invariant one, independent of the starting Lagrangian used in the analysis of constraints, in contradistinction to their primary or secondary character. Hence in the following, we shall assume that through appropriate linear combinations, a subset as large as possible of the original constraints ϕ_j is brought into the first-class subset, while the remaining linearly independent set of constraints is such that none of its linear combinations could be first-class. Given the above definitions of the coefficients U^j and V^j_α, the whole set of first-class constraints corresponds to the constraints ϕ_α, while the remaining linearly independent second-class constraints shall be denoted χ_s. Note that whenever the whole set of constraints ϕ_j is reducible, namely not linearly independent, the set of reducible constraints is necessarily included among the first-class one.

Second-class constraints and Dirac brackets

Having identified this invariant characterization of constraints, let us now address its intrinsic meaning, first in the case of second-class constraints. The simplest example of such a situation is provided by the constraints

$$q^1 \approx 0 \quad , \quad p_1 \approx 0 \; , \tag{225}$$

since $\{q^1, p_1\} = 1$. Clearly, the meaning of such constraints is that the degree of freedom $n = 1$ does not partake in no manner whatsoever in the dynamics of the system. One might as well remove that sector of the system altogether from its inception, namely subtract away from the definition of Poisson brackets all those contributions stemming for the degree of freedom $n = 1$, without any consequence as to the actual and genuine dynamics of the system.

From this simple example, it thus appears that the meaning of second-class constraints is that they are associated to the appearance of degrees of freedom which are totally redundant and irrelevant to the dynamics of the system, and may thus be suppressed altogether from the dynamics. This is achieved through a redefinition of the bracket structure on phase space, leading to so-called Dirac brackets in the general case, which in effect subtracts away any second-class constraint contributions to the dynamics.

For this purpose, let us consider the matrix of Poisson brackets of the second-class constraints χ_s,

$$\Delta_{ss'} = \{\chi_s, \chi_{s'}\} \; . \tag{226}$$

Given the property that none of the combinations of the second-class constraints χ_s is first-class, it follows necessarily that this matrix is regular, even on the constraint hypersurface,

$$\det \Delta_{ss'} \not\approx 0 \; . \tag{227}$$

Indeed, there would exist otherwise a nontrivial combination $\chi_s C^s$ of the constraints χ_s which would be first-class,

$$\det \Delta_{ss'} \approx 0 \quad \Longleftrightarrow \quad \{\chi_s, \chi_{s'} C^{s'}\} \approx 0 \; , \tag{228}$$

C^s being a zero mode of the matrix $\Delta_{ss'}$ (note that this result also establishes that the combination $U^j \phi_j$ appearing in the first-class Hamiltonian H belongs to the set of combinations of the second-class constraints χ_s). Hence, the matrix of Poisson brackets of all second-class constraints χ_s is invertible, even on the constraint hypersurface, leading to the definition of the Dirac bracket $\{f, g\}_D$ of any two quantities $f(q^n, p_n)$ and $g(q^n, p_n)$ on phase space,

$$\{f, g\}_D = \{f, g\} - \{f, \chi_s\} \left(\Delta^{-1}\right)^{ss'} \{\chi_{s'}, g\} \; . \tag{229}$$

This definition is such that the Dirac bracket of any phase space quantity f with any of the second-class constraints χ_s vanishes exactly as an identity valid throughout all of phase space,

$$\{f,\chi_s\}_D = \{f,\chi_s\} - \{f,\chi_{s'}\}\left(\Delta^{-1}\right)^{s's''}\{\chi_{s''},\chi_s\} = 0\ . \tag{230}$$

Consequently, provided one uses Dirac rather than Poisson brackets, the second-class constraints $\chi_s = 0$ may be imposed even before any such calculation, which is certainly not the case for Poisson brackets in which case it is essential that constraints never be enforced before any bracket evaluation. Furthermore, when considering the Hamiltonian equations of system, their expression in terms of Dirac rather than Poisson brackets does not modifiy the dynamics on the constraint hypersurface either, since one has for an arbitrary phase space quantity $f(q^n,p_n;t)$,

$$\begin{aligned}\frac{df}{dt} &= \frac{\partial f}{\partial t} + \{f,H_T\}\\ &= \frac{\partial f}{\partial t} + \{f,H_T\}_D + \{f,\chi_s\}\left(\Delta^{-1}\right)^{ss'}\{\chi_{s'},H_T\} \approx \frac{\partial f}{\partial t} + \{f,H_T\}_D\ .\end{aligned} \tag{231}$$

Consequently, through the use of Dirac rather than Poisson brackets, it becomes possible to subtract away all those contributions of the redundant degrees of freedom which are irrelevant to the time evolution of the system because of the second-class constraints $\chi_s = 0$. One may then as well explicitly solve for these constraints, leading to a set of coordinates z_A parametrizing the associated reduced phase space equipped with the bracket structure induced by the Dirac brackets,

$$\{z_A, z_B\}_D = C_{AB}(z_A)\ . \tag{232}$$

The total Hamiltonian is then still given by an expression of the form

$$H_T = H + \lambda^\alpha(t)\phi_\alpha\ . \tag{233}$$

Hence, the actual and genuine time dependent dynamics of the constrained system is now reduced into a constrained dynamics characterized by the sole first-class constraints ϕ_α restricted to the reduced phase space $\{z_A\}$. Henceforth, we shall assume to have effected such a reduction of the second-class constraints through Dirac brackets, and no longer display the "D" subindex on bracket evaluations.

It should be remarked though, that generally such a reduction of second-class constraints often entails a loss of manifest spacetime Poincaré invariance in the case of relativistic field theories, not a welcome feature. Furthermore, quite often the Dirac bracket structure that is obtained is not canonical, with in particular phase space dependent bracket values $C_{AB}(z_A)$ which sometimes are not even spacetime local functions in the case of field theories. Such circumstances then render canonical quantization of Dirac brackets problematic. In principle, by Darboux's theorem, it is always possible to locally bring the phase space coordinate system to canonical form, but again in practice this is often no small feat. However such issues may only be addressed on a case by case basis. Let us only point out here that the physical projector approach to be discussed in Sect.5.5 readily circumvents all these issues.

As an example, the reader is invited to consider the following Lagrange function,

$$L(q^n, \dot{q}^n) = \dot{q}^n K_n(q^n) - V(q^n) \ , \tag{234}$$

where $K_n(q^n)$ and $V(q^n)$ are arbitrary functions such that the matrix

$$K_{nm} = \frac{\partial K_m}{\partial q^n} - \frac{\partial K_n}{\partial q^m} \tag{235}$$

is regular. Clearly, this system is already in its Hamiltonian form,[10,22] with phase space degrees of freedom q^n, Hamiltonian $H(q^n) = V(q^n)$ and a Poisson bracket structure $\{q^n, q^m\}$ encoded in the functions $K_n(q^n)$. To identify the latter, it suffices to consider the Euler-Lagrange equations following from the above first-order Lagrangian,

$$K_{nm}\dot{q}^m = \frac{\partial V}{\partial q^n} \ , \tag{236}$$

and require these equations to be equivalent to the Hamiltonian ones,

$$\dot{q}^n = \{q^n, q^m\} \frac{\partial H}{\partial q^m} \ . \tag{237}$$

Hence, we must have for the Poisson brackets of the fundamental phase space degrees of freedom

$$\{q^n, q^m\} = \left(K^{-1}\right)^{nm} \ . \tag{238}$$

Of course, all this follows provided one notices that the above Lagrange function is already that of the Hamiltonian formulation of the system, since it is linear in the first-order time derivatives of the degrees of freedom. However, in practical instances such a feature is not necessarily so obvious, in which case one would embark onto the constraint analysis path following the general discussion of the present section. Indeed, one immediately notices the primary constraints for the conjugate momenta

$$p_n = \frac{\partial L}{\partial \dot{q}^n} = K_n(q^n) \quad , \quad \phi_n(q^n, p_n) = p_n - K_n(q^n) \ , \tag{239}$$

whose brackets $\{q^n, p_m\} = \delta^n_m$ are now canonical. The reader is thus invited to pursue the constraint analysis of this system, to conclude that these primary constraints $\phi_n = 0$ already exhaust all the constraints of the system, that these constraints are all second-class and are solved precisely by reducing the conjugate momenta $p_n = K_n(q^n)$, and that finally the reduced description based on the relevant Dirac brackets is nothing else than the Hamiltonian formulation identified above in terms of the phase space $\{q^n\}$, the Hamiltonian $H(q^n) = V(q^n)$ and the brackets $\{q^n, q^m\} = (K^{-1})^{nm}$. Note that in the case of bosonic degrees of freedom q^n, the regularity of the antisymmetric tensor K_{nm} requires an even number of coordinates q^n, as befits indeed any bosonic phase space.

Among possible examples of such systems of great interest, the most obvious one is certainly the Dirac Lagrangian density for a Dirac spinor in whatever spacetime dimension,

$$\mathcal{L} = \bar{\psi}\left(i\gamma^\mu\partial_\mu - m\right)\psi \ , \tag{240}$$

ψ being the Dirac spinor with $\bar{\psi} = \psi^\dagger\gamma^0$, m the mass of its quanta, and γ^μ the usual Dirac matrices obeying the Clifford-Dirac algebra $\{\gamma^\mu, \gamma^\nu\} = 2\eta^{\mu\nu}$. This system is thus already in its Hamiltonian form. Note however that its degrees of freedom $\psi(x^\mu)$ are now Grassmann odd quantities, hence leading to a Grassmann odd graded bracket structure, whose canonical quantization requires now anticommutation relations rather commutation ones.

First-class constraints and gauge invariance

After explicit resolution of any second-class constraints, the constrained Hamiltonian dynamics is characterized in terms of a phase space of coordinates z_A whose bracket structure is generally of the form

$$\{z_A, z_B\} = C_{AB}(z_A) \ , \tag{241}$$

and whose time evolution is generated by a total Hamiltonian

$$H_T = H + \lambda^\alpha\phi_\alpha \ , \tag{242}$$

where the first-class Hamiltonian H and constraints ϕ_α thus obey the bracket algebra

$$\{H, \phi_\alpha\} = C_\alpha{}^\beta\phi_\beta \quad , \quad \{\phi_\alpha, \phi_\beta\} = C_{\alpha\beta}{}^\gamma\phi_\gamma \ , \tag{243}$$

$C_\alpha{}^\beta$ and $C_{\alpha\beta}{}^\gamma$ being specific quantities which, in a general situation, may even be functions of phase space. The functions $\lambda^\alpha(t)$ are totally arbitrary, and thus parametrize an intrinsic freedom active within the system and directly related to the existence of the first-class constraints.

Given the above simple example of second-class constraints suggested as an exercise, it should be clear now that this whole information may also be encoded into the specification of a Hamiltonian variational principle based on the following first-order action

$$S[z_A;\lambda^\alpha] = \int dt\,\left[\dot{z}_A K^A(z_A) - H(z_A) - \lambda^\alpha(t)\phi_\alpha(z_A)\right]\ , \tag{244}$$

where the functions $K^A(z_A)$ are such that

$$\frac{\partial K^B}{\partial z_A} - \frac{\partial K^A}{\partial z_B} = \left(C^{-1}\right)^{AB}\ . \tag{245}$$

Hence, from that point of view, the functions $\lambda^\alpha(t)$ are nothing but Lagrange multipliers for the first-class constraints $\phi_\alpha = 0$. But then what is the meaning of the existence of these first-class constraints?

We shall now argue to establish that first-class constraints are the generators of local gauge symmetries of such a system, namely transformations of the phase space degrees of freedom z_A and the functions λ^α leaving the equations of motion invariant and whose parameters are local functions of time (or spacetime in the case of local field theories). The most direct way to establish this fact is by considering the following infinitesimal variations generated by the constraints

$$\delta_\zeta z_A = \{z_A, \zeta^\alpha\phi_\alpha(z_A)\}\quad ,\quad \delta_\zeta\lambda^\alpha = \dot{\lambda}^\alpha + \lambda^\gamma\zeta^\beta C_{\beta\gamma}{}^\alpha - \zeta^\beta C_\beta{}^\alpha\ , \tag{246}$$

where $\zeta^\alpha(t)$ are arbitrary time dependent infinitesimal parameters. By direct substitution into the first-order action (244), one then finds that indeed this action is invariant up to a surface term,

$$\delta_\zeta S = \int dt\,\frac{d}{dt}\left[\zeta^\alpha\left(K^A C_{AB}\frac{\partial\phi_\alpha}{\partial z_B} - \phi_\alpha\right)\right]\ , \tag{247}$$

(which may vanish for an appropriate choice of boundary conditions, though this is by no means a necessary requirement), so that the Hamiltonian equations of motion are invariant. Hence indeed, the transformations (246) do define local gauge symmetries of the system.

Alternatively, let us consider a set of initial data lying within the constraint hypersurface and let them evolve in time given two different choices $\lambda_1^\alpha(t)$ and $\lambda_2^\alpha(t)$ for the Lagrange multipliers. Accordingly, the change in the phase space variables z_A associated to an infinitesimal time interval δt for each choice is

$$\delta_1 z_A = \{z_A, H + \lambda_1^\alpha\phi_\alpha\}\delta t\quad ,\quad \delta_2 z_A = \{z_A, H + \lambda_2^\alpha\phi_\alpha\}\delta t\ , \tag{248}$$

so that the corresponding phase space trajectories differ in such a way that

$$\delta_2 z_A - \delta_1 z_A = \{z_A, (\lambda_2^\alpha - \lambda_1^\alpha)\delta t \phi_\alpha\} = (\lambda_2^\alpha - \lambda_1^\alpha)\delta t\, \{z_A, \phi_\alpha\} \,. \qquad (249)$$

However, since the choice for the Lagrange multipliers $\lambda^\alpha(t)$, which directly partake in the time dependency of the system dynamics, is totally arbitrary, the physical content and interpretation of the associated description should be equivalent irrespective of that choice. In other words, different choices of $\lambda^\alpha(t)$ are to be viewed as defining transformations between different phase space trajectories within the constraint hypersurface which describe one and the same physical configuration of the system. Namely, the freedom related to the choice in λ^α is nothing but a local gauge symmetry freedom. As the transformations (246) establish, this gauge freedom is also the one which is generated by all first-class constraints within the Hamiltonian formulation of the system. In particular, it thus appears, from (249), that the freedom in the choice of Lagrange multipliers $\lambda^\alpha(t)$ is nothing but the freedom that the system affords to include within its time evolution the possibility to also effect arbitrary gauge transformations as the system proceeds along one of its physically equivalent phase space trajectories within the constraint hypersurface. The Lagrange multipliers simply parametrize the freedom in local gauge transformations available throughout the time evolution of the system.

Having established the consistency of the gauge symmetry interpretation of the action of the first-class constraints ϕ_α, it now appears that the algebra (243) of these quantities is nothing but the local Hamiltonian gauge symmetry algebra, with structure coefficients $C_{\alpha\beta}{}^\gamma$,

$$\{\phi_\alpha, \phi_\beta\} = C_{\alpha\beta}{}^\gamma\, \phi_\gamma \,. \qquad (250)$$

In case these coefficients are in fact constant, one says that the algebra is closed, whereas otherwise it is open. In the latter case, this implies that one is in fact dealing with an algebraic structure in a strict sense provided only gauge transformed quantities are restricted onto the constraint hypersurface. Indeed, given gauge transformations associated to two independent sets of gauge parameters $\zeta_1^\alpha(t)$ and $\zeta_2^\alpha(t)$, the commutator of the bracket induced transformations of any phase space quantity f is such that, using Jacobi's identity,

$$[\delta_{\zeta_1}, \delta_{\zeta_2}]\, f = \{f, \{\zeta_1^\alpha \phi_\alpha, \zeta_2^\beta \phi_\beta\}\} = \{f, \zeta_1^\alpha \zeta_2^\beta C_{\alpha\beta}{}^\gamma \phi_\gamma\} \approx \zeta_1^\alpha \zeta_2^\beta C_{\alpha\beta}{}^\gamma \{f, \phi_\gamma\} \,. \qquad (251)$$

Finally, not only do phase space configurations and trajectories of the system fall into gauge equivalence classes in this manner, each such class being associated to a distinct physical configuration of the system, but a likewise classification of physical phase space observables as the gauge equivalence classes of first-class quantities is also relevant. Indeed, given any first-class quantity f such that

$$\{f, \phi_\alpha\} = f_\alpha{}^\beta\, \phi_\beta \ , \tag{252}$$

clearly this property is preserved through time evolution since the total Hamiltonian $H_T = H + \lambda^\alpha \phi_\alpha$ is first-class, while it also implies that all its gauge equivalent representations are all of the form $f + \chi^\alpha \phi_\alpha$ for some coefficient functions $\chi^\alpha(z_A)$. Hence, gauge invariant physical observables of the system are nothing but the gauge equivalence classes of first-class quantities, whose time evolution is well defined and independent of the choice of Lagrange multipliers $\lambda^\alpha(t)$, as it should since the latter parametrize gauge transformations throughout the time history of the system. Obvious examples of such gauge invariant physical observables are the first-class Hamiltonian H, the first-class constraints ϕ_α which must vanish for physical configurations, and thus also the total Hamiltonian H_T, which takes values independent of $\lambda^\alpha(t)$ for physical configurations.

A few final remarks are in order.[10] First, it should be stressed that even though there is in general a correspondence between Lagrangian and Hamiltonian local gauge invariances, there is by no means any necessity that the corresponding algebraic structures should be identical. The example of the scalar relativistic particle to be discussed hereafter provides an illustration. Furthermore, it may be that given the complete Hamiltonian formulation of the system, a specific choice for some of the Lagrange multipliers is implicitly made before the reduction of some conjugate momenta is effected in order to obtain a particular Lagrangian formulation of the same dynamics. In such a case, this Lagrangian shares only part of the original gauge freedom of the Hamiltonian formulation, with a specific correspondence between these gauge symmetries, rather than an identity, given the effected partial reduction of phase space. Likewise, if one fails to notice that some of the original configuration space degrees of freedom are in fact Lagrange multipliers for some constraints, and thus applies an analysis of constraints for the associated trivial conjugate momenta, one obtains further first-class constraints expressing the Lagrange multiplier character of that sector of the system. As a matter of fact, such redundant features may be gauged away without compromising the genuine dynamics of the system, leading *in fine* to the *fundamental* or *basic* Hamiltonian formulation[10] of a gauge invariant system. The above considerations also indicate how a same gauge invariant system may in fact possess quite a number of distinct Hamiltonian and Lagrangian formulations. Only a complete analysis of its constraints can uncover its actual basic Hamiltonian formulation.

Finally, let us also stress that those gauge symmetries generated by the first-class constraints are *small* Hamiltonian gauge transformations, namely are transformations which belong to the same homotopy class as the identity transformation, being continuously connected to the latter. Systems may also be invariant under *large* gauge transformations, namely gauge symmetries whose parameters are (space)time dependent functions but such that nonetheless the associated transformations are not continuously connected to the identity and thus belong to a homotopy class of the gauge symmetry group different from the identity class. In considering the Hamiltonian formulation of gauge invariant systems, whereas its small gauge transformations are directly accounted for through the first-class constraints, invariance under large gauge transformations, if relevant, has to be enforced separately.

To conclude this section, let us invite the reader to develop the analysis of constraints of a pure nonabelian Yang-Mills theory, whose Lagrangian density is

$$\mathcal{L} = -\frac{1}{4} F^a_{\mu\nu} F^{a\mu\nu} \quad , \quad F^a_{\mu\nu} = \partial_\mu A^a_\nu - \partial_\nu A^a_\mu - g f^{abc} A^b_\mu A^c_\nu \ , \tag{253}$$

$A^a_\mu(x^\mu)$ being the gauge field vector associated to the nonabelian compact Lie algebra of structure constants f^{abc} such that $[T^a, T^b] = i f^{abc} T^c$, and g the gauge coupling constant. Such an analysis is quite instructive, for instance in what concerns the Lagrange multiplier status of the time component A^a_0 of the gauge vector potentials and the identification of the basic Hamiltonian formulation of such a system.

As a particularization of pure Yang-Mills theory, it is also instructive to consider its dimensional reduction to 0+1 dimensions,[23] namely to a gauge invariant mechanical system, leading to a Lagrangian of the form

$$L(q^a_i, \dot{q}^a_i) = \frac{1}{2g^2} \left[\dot{q}^a_i + f^{abc} \lambda^b q^c_i\right]^2 - V(q^a_i) \quad , \quad V(q^a_i) = \frac{1}{2} \omega^2 \left(q^a_i\right)^2 \ , \tag{254}$$

$V(q^a_i)$ being a gauge invariant potential, such as the quadratic one indicated. In fact, upon dimensional reduction the pure Yang-Mills Lagrangian density (253) leads to quartic terms in the potential, which may be ignored without spoiling gauge invariance in 0+1 dimensions. The advantage of the quadratic choice is that usual harmonic oscillator techniques enable an explicit resolution, even at the quantum level, of this gauge invariant system, in particular with the identification of its gauge invariant physical spectrum.

5.2 The relativistic scalar particle

As a useful guide illustrating the discussion which is to follow later on, let us present now in detail the analysis of constraints for a interesting though simple enough system, namely the relativistic scalar massive particle.[10] When discussing that system, we shall take for the Minkowski spacetime metric the signature $\eta_{\mu\nu} = \mathrm{diag}\,(- + +... + +)$ in a D-dimensional spacetime, $\mu, \nu = 0, 1, 2 \cdots, D-1$. Furthermore, this system possesses two well known action principle formulations, one leading to linear equations of motion, the other to nonlinear equations. We shall consider here the linear formulation and indicate its relation to the nonlinear one where appropriate.

The action principle

Wishing to construct a manifestly spacetime Poincaré covariant formulation of the particle's trajectories, one has to consider its spacetime history in terms of a parametrized world-line $x^\mu(\tau)$ spanning some initial and final spacetime positions x^μ_i and x^μ_f. Nonetheless, the physics of the system should be independent not only of the spacetime reference frame used, namely Poincaré invariant, but it should also be independent of the world-line τ parametrization used, namely invariant under arbitrary world-line reparametrizations or diffeomorphisms. The latter include transformations

$$\tau \to \tilde{\tau} = \tilde{\tau}(\tau) \quad , \quad x^\mu(\tau) \to \tilde{x}^\mu(\tilde{\tau}) = x^\mu(\tau) \tag{255}$$

which either preserve the world-line orientation, or reverse it. Clearly, the former class of transformations defines an ensemble of small gauge transformations, whereas the latter one an ensemble of large gauge transformations. In particular, the quotient of the group of world-line diffeomorphisms by its connected identity homotopy component is isomorphic to the group $\mathbb{Z}_2$ of two elements. This quotient group is also known as the modular group.

Hence, given these two general requirements, the action of the system should be both a spacetime and a world-line scalar. Form the latter point of view, the spacetime coordinates $x^\mu(\tau)$ are nothing but scalar "field" degrees of freedom on the world-line. One way to construct a world-line scalar action is to couple in an invariant manner these degrees of freedom to an intrinsic world-line metric $g(\tau) = e^2(\tau)$, $e(\tau)$ being the intrinsic world-line einbein. Consequently, the action reads as

$$S[x^\mu, e] = \int_{\tau_i}^{\tau_f} d\tau \; L(\dot{x}^\mu, e) \;, \tag{256}$$

with

$$L(\dot{x}^\mu, e) = \sqrt{g}\left[\frac{1}{2}g^{-1}\dot{x}^\mu\dot{x}^\nu\eta_{\mu\nu} - \frac{1}{2}m^2\right] = \frac{1}{2}|e|^{-1}\dot{x}^2 - \frac{1}{2}|e|m^2 . \tag{257}$$

Here, $m > 0$ stands for a parameter with the dimension of mass, which will indeed turn out to correspond to the particle's mass. At this stage however, it appears to play the role of a one-dimensional cosmological constant for this metric theory, *i.e.*, a theory of gravity on the one-dimensional world-line. Note also that the requirement of Poincaré invariance, in particular under spacetime translations, forbids any term dependent on the coordinates x^μ rather than its τ-derivatives. Finally, a dot above a quantity denotes a τ-derivative, since the parameter τ is to be viewed as the time evolution parameter of the system's dynamics. Nevertheless, the actual physical time is the measurement of the time component x^0 of the particle's spacetime trajectory. It is the requirement of a manifestly Poincaré covariant formulation which necessitates the gauge symmetry in world-line reparametrizations, but this latter symmetry also allows to obtain a description which, physically, is independent of the world-line parametrization. Only the gauge invariant relations between the components of the particle's trajectory $x^\mu(\tau)$ are physically relevant.

The above action provides the linear formulation of the system, since the ensuing equations of motion are linear, as is easily established. From the world-line point of view, Poincaré invariance defines an internal global symmetry whose Noether charges are given by

$$P_\mu = \frac{\partial L}{\partial \dot{x}^\mu} \quad , \quad M_{\mu\nu} = P_\mu x_\nu - P_\nu x_\mu , \tag{258}$$

for the particle's energy- and orbital angular-momentum, respectively. In particular, the Euler-Lagrange equations of motion are nothing but the statement of the conservation of the Noether energy-momentum of the particle, $dP_\mu/d\tau = 0$, for which we leave it as a straightforward exercise to construct the solutions given the above choice of boundary conditions.

On the other hand, under world-line diffeomorphisms, in addition to the transformations (255), the einbein variation is

$$\tilde{e}(\tilde{\tau}) = \frac{d\tau}{d\tilde{\tau}}e(\tau) . \tag{259}$$

In infinitesimal form, one then finds

$$\tilde{\tau} = \tau - \eta(\tau) \quad , \quad \delta_\eta x^\mu = \tilde{x}^\mu(\tau) - x^\mu(\tau) = \eta\dot{x}^\mu(\tau) \quad , \quad \delta_\eta e(\tau) = \frac{d}{d\tau}\left(\eta(\tau)e(\tau)\right) , \tag{260}$$

so that the algebra of small Lagrangian gauge symmetries is given by

$$[\delta_{\eta_1}, \delta_{\eta_2}] = \delta_{\eta_1 \dot{\eta}_2 - \eta_2 \dot{\eta}_1} \ . \tag{261}$$

Note that this algebra of Lagrangian diffeomorphisms is nonabelian.

The Hamiltonian formulation

Turning to the Hamiltonian formulation, phase space is equipped with the canonical brackets

$$\{x^\mu(\tau), P_\nu(\tau)\} = \delta^\mu_\nu \quad , \quad \{e(\tau), \pi_e(\tau)\} = 1 \ , \tag{262}$$

with the conjugate momenta

$$P_\mu(\tau) = \frac{\partial L}{\partial \dot{x}^\mu(\tau)} \quad , \quad \pi_e(\tau) = \frac{\partial L}{\partial \dot{e}(\tau)} = 0 \ , \tag{263}$$

hence leading to the primary constraint $\pi_e = 0$. Note that this constraint actually follows from the fact that $e(\tau)$ is a Lagrange multiplier for a constraint, as the forthcoming analysis will confirm. The canonical Hamiltonian is

$$H_0 = \dot{x}^\mu P_\mu + \dot{e}\pi_e - L = \frac{1}{2}|e| \left[P^2 + m^2\right] \ . \tag{264}$$

The primary Hamiltonian is thus

$$H_* = \frac{1}{2}|e| \left[P^2 + m^2\right] + u\pi_e \ , \tag{265}$$

u being some *a priori* unknown function. Consistent time evolution of the primary constraint $\pi_e = 0$ then requires

$$\dot{\pi}_e = \{\pi_e, H_*\} = -\left(\mathrm{sign}\, e\right) \frac{1}{2} \left[P^2 + m^2\right] \ , \tag{266}$$

thus leading to the secondary constraint

$$\phi = \frac{1}{2} \left[P^2 + m^2\right] = 0 \ . \tag{267}$$

However, requiring the consistent time evolution of this second constraint does to lead to any further condition, nor any restriction on the associated functions defining the contribution of the linear combination of the constraints to the total Hamiltonian. Hence, the complete set of constraints is given by $\pi_e = 0$ and $\phi = 0$, which are first-class since

$$\{\pi_e, \pi_e\} = 0 \ , \ \{\pi_e, \phi\} = 0 \ , \ \{\phi, \phi\} = 0 \ . \tag{268}$$

The Hamiltonian dynamics is thus generated by the total Hamiltonian

$$H_T = \frac{1}{2} \left[|e| + \tilde{\lambda}\right] \left[P^2 + m^2\right] + u\pi_e \ , \tag{269}$$

where $u(\tau)$ and $\tilde{\lambda}(\tau)$ are the associated Lagrange multipliers. Correspondingly, the first-order action is

$$S[x^\mu, P_\mu; e, \pi_e; \tilde{\lambda}, u] = \int_{\tau_i}^{\tau_f} d\tau \left[\dot{x}^\mu P_\mu + \dot{e}\pi_e - \frac{1}{2}\left[|e| + \tilde{\lambda}\right]\left[P^2 + m^2\right] - u\pi_e\right] . \tag{270}$$

However, this form makes it obvious that the (e, π_e) sector may be decoupled altogether, since in fact the einbein $|e|$ indeed may be absorbed into the definition of the Lagrange multiplier for the first-class constraint $\phi = 0$, exactly as was anticipated above. Setting then $u = \dot{e}$ and $\lambda = |e| + \tilde{\lambda}$, one finally reaches the basic Hamiltonian formulation of the relativistic massive scalar particle in the form the first-order action

$$S[x^\mu, p_\mu; \lambda] = \int_{\tau_i}^{\tau_f} d\tau \left[\dot{x}^\mu P_\mu - \lambda\phi\right] = \int_{\tau_i}^{\tau_f} d\tau \left[\dot{x}^\mu P_\mu - \frac{1}{2}\lambda\left[P^2 + m^2\right]\right] . \tag{271}$$

The (e, π_e) sector has indeed been decoupled, leaving over only one first-class constraint $\phi = (P^2 + m^2)/2$ whose Lagrange multiplier λ should thus play the same role as that of the world-line einbein, as shall be confirmed hereafter. Furthermore, phase space consists solely of the sector of spacetime degrees of freedom (x^μ, P_μ) with the previous canonical Poisson brackets, while the first-class Hamiltonian H vanishes identically, $H = 0$, as befits any reparametrization invariant theory since H ought to be the generator of time reparametrizations. Hence, since the total Hamiltonian is only given by the first-class constraint, $H_T = \lambda\phi$, the latter should also correspond to the generator of world-line reparametrizations in the Hamiltonian formulation, an expectation confirmed below.

The Hamiltonian equations of motion are simply

$$\dot{x}^\mu = \lambda P^\mu \quad , \quad \dot{P}_\mu = 0 \quad , \quad P^2 + m^2 = 0 \; . \tag{272}$$

Performing the Hamiltonian reduction of the momenta $P_\mu = \dot{x}_\mu/\lambda$, a substitution in (271) then finds

$$S[x^\mu; \lambda] = \int_{\tau_i}^{\tau_f} d\tau \left[\frac{1}{2}\lambda^{-1}\dot{x}^2 - \frac{1}{2}\lambda m^2\right] , \tag{273}$$

namely precisely the original Lagrangian action in the linear formulation, with the Lagrange multiplier λ playing now the role of the einbein degree of freedom, except for the fact it is no longer the absolute value of the einbein that appears in the action. Consequently, the Hamiltonian formulation of the system is not invariant under the large gauge symmetries of the original

Lagrangian formulation. This gauge symmetry will have to be enforced separately at the end of the analysis, whether at the classical level or after its canonical quantization.

Note that when also solving for the constraint $P^2 + m^2 = 0$ after the Hamiltonian reduction, namely with $\lambda = \sqrt{-\dot{x}^2}/m$, finally the action reduces to

$$S[x^\mu] = -m \int_{\tau_i}^{\tau_f} d\tau \sqrt{-\dot{x}^2} \,. \tag{274}$$

In fact, this action provides the nonlinear formulation of the same system, in which the total world-line length between its boundary points is measured this time in terms of the metric induced on the world-line by the spacetime Minkowski metric in which the world-line is embedded through the functions $x^\mu(\tau)$, rather than the intrinsic metric defined through the choice of einbein $e(\tau)$ or $\lambda(\tau)$. The reader is invited to consider the analysis of constraints starting from this nonlinear formulation. Among other results, it follows that the constraint $\phi = [P^2 + m^2]/2$ then appears immediately as a primary constraint, with a vanishing canonical Hamiltonian, and that no further constraints arise. Hence, the same basic Hamiltonian formulation as the one above is recovered, providing an explicit illustration of the fact that the primary and secondary character of constraints depends on the Lagrangian formulation used, and that gauge invariant systems possess different though physically equivalent Lagrangian and Hamiltonian formulations, but only a single basic Hamiltonian one. Note however that the nonlinear formulation applies only to massive particles, whereas the linear one remains valid even for massless particles.

The sole first-class constraint of the system, $\phi = [P^2 + m^2]/2 = 0$, is the generator of a local Hamiltonian gauge symmetry of the system, which can but only correspond to small world-line diffeomorphisms. To establish the exact correspondence, let us consider the infinitesimal Hamiltonian gauge transformations,

$$\delta_\epsilon x^\mu = \{x^\mu, \epsilon\phi\} = \epsilon P^\mu \quad , \quad \delta_\epsilon P_\mu = 0 \quad , \quad \delta_\epsilon \lambda = \dot{\epsilon} \,, \tag{275}$$

where $\epsilon(\tau)$ is an arbitrary infinitesimal function, which must vanish at the end points, $\epsilon(\tau_{i,f}) = 0$, when enforcing the above choice of boundary conditions for the particle's trajectory. Given these expressions, the associated finite transformations are readily found to be

$$x'^\mu(\tau) = x^\mu(\tau) + h(\tau) P^\mu(\tau) \quad , \quad P'^\mu(\tau) = P^\mu(\tau) \quad , \quad \lambda'(\tau) = \lambda(\tau) + \frac{dh(\tau)}{d\tau} \,, \tag{276}$$

$h(\tau)$ being an arbitrary function such that $h(\tau_{i,f}) = 0$ when enforcing the above boundary conditions, as may easily be confirmed by checking the invariance of the action (271). Given these results, we may now consider their relation to the reparametrization gauge symmetry in the Lagrangian formulation.

First, notice that contrary to the Lagrangian diffeomorphism algebra which is nonabelian, the Hamiltonian one is abelian, $\{\phi, \phi\} = 0$, on account of the antisymmetry property of Poisson brackets. Consequently, these two algebraic structures are not identical, even though there is a unique correspondence, but no identity, between the relevant transformations of the degrees of freedom. Indeed, given the above finite Hamiltonian reparametrization with parameter $h(\tau)$, the corresponding finite Lagrangian reparametrization such that $\tau = f(\tilde{\tau})$ is constructed from the relation[10]

$$h(\tau) = \int_{\tau}^{f(\tau)} d\tau' \, \lambda(\tau') \ , \tag{277}$$

with in particular $f(\tau_{i,f}) = \tau_{i,f}$ and $h(\tau_{i,f}) = 0$ when enforcing the boundary conditions $x^\mu(\tau_{i,f}) = x^\mu_{i,f}$. In the case of infinitesimal transformations such that $f(\tau) = \tau + \eta(\tau)$, this correspondence reduces to

$$\epsilon(\tau) = \lambda(\tau)\, \eta(\tau) \ , \tag{278}$$

possibly with the conditions $\epsilon(\tau_{i,f}) = 0 = \eta(\tau_{i,f})$. Hence, the advocated one-to-one correspondence but not necessarily identity between Lagrangian and Hamiltonian small gauge symmetries is established for this particular system. As a matter of fact, such a correspondence remains valid for reparametrization invariant theories in whatever dimension, thus including general relativity. In contradistinction in the case of nonabelian Yang-Mills theories, this correspondence between the two classes of gauge symmetries becomes in fact an identity, since the relevant gauge symmetries are then internal ones, independent of the spacetime evolution of the system.

The total proper-time, or proper-length, of the particle's trajectory for the specified boundary conditions,

$$\gamma = \int_{\tau_i}^{\tau_f} d\tau \, \lambda(\tau) \ , \tag{279}$$

is indeed also a gauge invariant quantity. In fact, it is the sole gauge invariant degree of freedom in the Lagrange multiplier sector, and characterizes the different metric structures that may be defined on the world-line. Let us thus refer to it as the Teichmüller parameter of the system.[10] As such, it also appears explicitly in the solutions to the equations of motion, given by

$$x^\mu(\tau) = x_i^\mu + \frac{\Delta x^\mu}{\gamma}\int_{\tau_i}^{\tau} d\tau'\,\lambda(\tau')\ ,\quad P^\mu(\tau) = \frac{\Delta x^\mu}{\gamma}\ ,\quad \Delta x^\mu = x_f^\mu - x_i^\mu\ , \tag{280}$$

while the constraint $P^2 + m^2 = 0$ requires that

$$\sqrt{-\left(\Delta x\right)^2} = m|\gamma|\ . \tag{281}$$

Note how the choice of Lagrange multiplier indeed parametrizes the freedom in the choice of world-line parametrization, hence the gauge freedom of the system. Given this solution, the gauge invariant content of the spacetime sector of the formulation may also be identified. Thus the spacetime trajectory is given by

$$\vec{x}(x^0) = \vec{x}_i + \Delta\vec{x}\,\frac{x^0 - x_i^0}{\Delta x^0}\ , \tag{282}$$

which is indeed a gauge invariant relation, independent of the choice of world-line parametrization in τ. Only the specific relation between the physical time x^0 and the world-line evolution parameter τ is gauge dependent and thus dependent on the choice of Lagrange multiplier or world-line einbein, namely

$$x^0(\tau) = x_i^0 + \frac{\Delta x^0}{\gamma}\int_{\tau_i}^{\tau} d\tau'\,\lambda(\tau')\ . \tag{283}$$

Note also that the above solution for the energy-momentum of the particle shows that positive (respectively, negative) values for the quantity γ correspond to the propagation forward (respectively, backward) in time of the particle, namely in a quantum parlance, to a particle (respectively, antiparticle) as opposed to its antiparticle (respectively, particle). This brings us back to the issue of large gauge symmetries, which are not realized in the Hamiltonian formulation. Clearly, large world-line diffeomorphisms induce a change of orientation in the world-line, hence a change of sign of the einbein $e(\tau)$ or Lagrange multiplier $\lambda(\tau)$, and thus also of the Teichmüller parameter γ. Consequently, the $\mathbb{Z}_2$ modular group of orientation reversing diffeomorphisms modulo orientation preserving ones, namely the class of large gauge transformations, acts on Teichmüller space simply as $\gamma \to -\gamma$, thereby distinghuishing a particle description as opposed to its antiparticle.[10] Invariance under these modular transformations of the Teichmüller parameter thus needs to be enforced when considering specific configurations of the unoriented scalar particle. Positive energy solutions then propagate forward in time with the modular invariant restriction $\gamma > 0$.

5.3 Gauge fixing, reduced phase space and Gribov problems

Faddeev's reduced phase space

The redundant degrees of freedom inherent to the gauge symmetries of a constrained dynamics are certainly a challenge to the proper gauge invariant quantization of such systems. *A priori*, one possible approach would be first to solve for the gauge constraints ϕ_α, and only then quantize the reduced phase space degrees of freedom, which are certainly then physical since no gauge symmetry freedom remains. Let us then introduce a set of gauge fixing conditions $\Omega_\alpha = 0$ whose number is equal to that of the first-class constraints ϕ_α (note that this requires the set of first-class constraints to be irreducible, namely locally linearly independent), and which are such that the matrix of brackets of this whole set of constraints is regular,

$$\det \{\Omega_\alpha, \phi_\beta\} \not\approx 0 . \tag{284}$$

In other words, by introducing the additional conditions, the whole set of constraints has been turned into second-class ones.

That such restrictions "freeze" or fix the gauge symmetries of the system may be seen from complementary points of view. First, consider arbitrary infinitesimal gauge transformations of the gauge fixing conditions

$$\delta_\zeta \Omega_\alpha = \{\Omega_\alpha, \zeta^\beta \phi_\beta\} . \tag{285}$$

Thus, if the condition (284) is met, there do not exist infinitesimal gauge transformations leaving the gauge fixing conditions invariant, namely the only solution to the equations $\delta_\zeta \Omega_\alpha = 0$ is trivial, $\zeta^\alpha = 0$. In other words, the conditions $\Omega_\alpha = 0$ do indeed fix the gauge freedom, albeit for small infinitesimal gauge transformations only. From an alternative point of view, consider now the time evolution of these conditions,

$$\frac{d\Omega_\alpha}{dt} \approx \frac{\partial \Omega_\alpha}{\partial t} + \{\Omega_\alpha, H\} + \{\Omega_\alpha, \phi_\beta\}\lambda^\beta . \tag{286}$$

Thus once again, if the condition (284) is met, the requirement that the gauge fixing conditions $\Omega_\alpha = 0$ remain valid at all times implies the unique determination of the Lagrange multipliers $\lambda^\alpha(t)$. Since these functions are known to parametrize the gauge freedom of the system throughout its time history, it thus follows that the conditions $\Omega_\alpha = 0$ imply a specific choice for these functions, namely a specific gauge fixing of the system.

Consequently, it appears that conditions $\Omega_\alpha = 0$ such that (284) is obeyed imply a gauge fixed formulation of the system, in which all the redundant features inherent to such symmetries are explicitly resolved. The latter is simply achieved by working out the Dirac brackets associated to the condition (284) and to the second-class constraints $\Omega_\alpha = 0 = \phi_\alpha$. The ensuing reduced phase space description based on these Dirac brackets is known as Faddeev's reduced phase space formulation of gauge invariant theories.[24] Only gauge invariant physical degrees of freedom remain dynamical within such a formulation.

As an explicit illustration, let us consider again the relativistic scalar particle, with the gauge fixing condition

$$\Omega = x^0(\tau) - \left[x_i^0 + \frac{\Delta x^0}{\Delta\tau}\left(g(\tau) - \tau_i\right) + h_0(\tau)P^0(\tau)\right] , \tag{287}$$

with of course $\Delta\tau = \tau_f - \tau_i$. Here, $g(\tau)$ (respectively, $h(\tau)$) is some arbitrary function such that $g(\tau_{i,f}) = \tau_{i,f}$ (respectively, $h(\tau_{i,f}) = 0$). From previous results for this system, it is clear that $g(\tau)$ parametrizes the gauge freedom in the choice of world-line parametrization, whereas $h_0(\tau)$ parametrizes an arbitrary finite small gauge transformation of this gauge fixing condition. Consequently, the final reduced phase space description associated to this choice should be independent of these two functions.

Since for a massive particle,

$$\{\Omega, \phi\} = P^0 \neq 0 \ , \ \ \phi = \frac{1}{2}\left[P^2 + m^2\right] = 0 \Rightarrow P^0 = \eta\sqrt{\vec{P}^2 + m^2} \ , \ \ \eta = \pm 1 , \tag{288}$$

the two conditions $\Omega = 0 = \phi$ together indeed define a set of second-class constraints. Solving then for the time component degrees of freedom $x^0(\tau)$ and $P^0(\tau)$ from these two constraints, the Dirac brackets for the space components are readily determined to coincide with the original canonical brackets,

$$\{x^i(\tau), P_j(\tau)\}_D = \delta^i_j . \tag{289}$$

The evolution of the system in terms of the physical time x^0 rather than the world-line parameter τ is then generated by the reduced Hamiltonian

$$H_{\text{reduced}} = \eta\sqrt{\vec{P}^2 + m^2} , \tag{290}$$

so that for any observable F on this reduced phase space,

$$\frac{dF}{dx^0} = \frac{\partial F}{\partial x^0} + \{F, H_{\text{reduced}}\}_D . \tag{291}$$

All the gauge dependent features are indeed no longer involved in this gauge fixed description of the system. Rather, they only appear in the gauge dependent relations, namely the τ parametrization of the physical time,

$$x^0(\tau) = x_i^0 + \frac{\Delta x^0}{\Delta\tau}\left[g(\tau) - \tau_i\right] + h_0(\tau)P^0(\tau) , \tag{292}$$

in which the energy degree of freedom is given by

$$P^0(\tau) = \eta\sqrt{\vec{P}^2(\tau) + m^2} \;\; , \;\; \eta = \pm 1 \;, \tag{293}$$

as well as the choice of world-line einbein and Teichmüller parameter,

$$\lambda(\tau) = \eta\frac{\Delta x^0}{\Delta\tau}\frac{\dot{g}(\tau)}{\sqrt{\vec{P}^2(\tau) + m^2}} \; + \; \dot{h}_0(\tau) \;,\; \gamma = \eta\frac{\Delta x^0}{\Delta\tau}\int_{\tau_i}^{\tau_f} d\tau\,\frac{\dot{g}(\tau)}{\sqrt{\vec{P}^2(\tau) + m^2}} \;, \tag{294}$$

in ways that are totally in agreement with the role played by the functions $g(\tau)$ and $h_0(\tau)$ and the gauge transformation properties of these quantities (the quantity $P^0(\tau)$ is gauge invariant on its own already). Note that for configurations solving the equations of motion, we always have $\gamma = \Delta x^0/P^0$, which is indeed independent of the functions $g(\tau)$ and $h_0(\tau)$, as it should.

Admissible gauge fixing and Gribov problems

In order to properly assess[10] what any gauge fixing procedure actually achieves, it is necessary to better understand the redundancy features inherent to the gauge symmetry properties related to the first-class constraints. For this purpose, let us consider the whole of phase space $\{z_A(t)\}$ together with the set of Lagrange multipliers $\lambda^\alpha(t)$, thus defining a large space of functions. Small gauge transformations generated by the first-class constraints directly act on that space, thereby organizing it into a whole set of disjoint gauge orbits without any intersections. The space of all such gauge orbits is nothing but the quotient of the whole space $\{z_A, \lambda^\alpha\}$ by the action of the small local gauge symmetry, and thus represents the ensemble of all physically distinct gauge invariant configurations possibly accessible to the system throughout its time evolution history. What one would in fact hope to be feasible should be the dynamical description of the system, as well as its quantization, on that quotient space of gauge orbits, rather than the space $\{z_A, \lambda^\alpha\}$ with all its inherent redundancy features related to the gauge transformations acting on it. In practice however, the topology and analytical properties of the space of gauge orbits are just too intricate to contemplate such an approach towards gauge invariant dynamics.

Hence, the basic idea of gauge fixing is to identify within the original space $\{z_A, \lambda^\alpha\}$ a subset chosen in such a way that each of its elements is just one and only one representative for each of all the possible gauge orbits of the system. Namely, any gauge fixing should in effect implicitly define some gauge slicing of the space $\{z_A, \lambda^\alpha\}$ which would intersect each of its gauge orbits once and only once. Provision can be made for those cases in which the gauge slice intersects some gauge orbits more than once, but then in such a manner that the gauge slice and its intersections are counted with an orientation leading *in fine* to an effective count of intersections which still adds up to a single one. Clearly, when such a gauge fixing is achieved, it is an admissible one, meaning that the dynamics of the system reduced onto the gauge slice is totally equivalent to the original dynamics formulated either on the space of gauge orbits or equivalently within the space $\{z_A, \lambda^\alpha\}$ with proper account for the gauge symmetries. Clearly, in order to assess whether a given gauge fixing procedure is admissible, it is imperative first to properly identify the space of gauge orbits of the system and its characterization in terms of gauge invariant quantities constructed from the variables $\{z_A, \lambda^\alpha\}$ and which may then serve as coordinates parametrizing the space of gauge orbits.

In practice however, given some gauge fixing procedure, namely a restriction on the variables $\{z_a, \lambda^\alpha\}$ which in effect "freezes" the gauge freedom of the system and leads thereby to an effective reduced phase space formulation involving then physical degrees of freedom only (Faddeev's reduced phase space being the archetype example), there is no guarantee whatsoever that an admissible gauge fixing is achieved. Indeed, any such gauge fixing procedure in effect singles out through some gauge slicing a certain subset of the space $\{z_A, \lambda^\alpha\}$ for which no gauge freedom is left. This gauge slice intersects the gauge orbits in a certain manner specific to the gauge fixing procedure, which may be characterized in terms of a specific covering of the space of gauge orbits.[10] By covering is meant a certain domain or subset of that space, as well as some measure over that domain which represents the possibly multiple degeneracy in the count of intersected orbits. Due to the gauge invariance properties of the original formulation of the system, this covering of the gauge orbits induced by some gauge fixing procedure is all that is relevant for the characterization of that gauge fixing. Two gauge fixing procedures leading to an identical covering of the space of gauge orbits are thus said to be gauge equivalent. The dynamical descriptions achieved through two gauge equivalent gauge fixings are physically equivalent.

However, gauge fixing procedures leading to different coverings of the space of gauge orbits imply dynamical descriptions of the system which are not physically equivalent, even though they both are gauge invariant. In particular, it is only for an admissible covering, namely a gauge fixing which in effect singles out each of the gauge orbits once and only once, that the associated gauge invariant dynamics is physically equivalent to that of the original system.

Whenever a gauge fixing procedure is not admissible in this very specific sense, it is said to suffer from a Gribov problem.[25] In fact, one ought to distinguish two types of Gribov problems.[10] The Gribov problem of type I is of a local character, and occurs whenever among the selected gauge orbits some are selected more than once, thereby leading to an overcount of the corresponding physical configurations accessible to the dynamics. Likewise, the Gribov problem of type II is of a global character, and occurs whenever some of the gauge orbits are not selected by the gauge fixing procedure, thereby forbidding the system to dynamically access some of its physical configurations. In other words, a Gribov problem of type I occurs whenever some gauge orbits are counted more than once (relative to the others), while a Gribov problem of type II occurs when some orbits are not counted at all.

By definition, an admissible gauge fixing is one which does not suffer a Gribov problem of either type. However, an arbitrarily chosen gauge fixing procedure may suffer a Gribov problem of type I or of type II, or even of both types, in which case it is not admissible. Even though any gauge fixing procedure leads to a gauge invariant formulation of the system, when a Gribov problem arises the gauge invariant dynamics which is being described is no longer that of the original system, since it no longer includes the same set of physically distinct configurations accessible to the dynamics. In that sense, one may say that it is only an admissible gauge fixing which is a physically correct gauge fixing, even though any gauge fixing procedure with a Gribov problem leads nonetheless to a gauge invariant description. Gauge invariance is not all there is to gauge invariant systems!

Consequently, whenever considering a given gauge fixing procedure, one must also determine whether it is admissible or not, namely whether it suffers Gribov problems of type I and of type II. This issue may be addressed only on a case by case basis.

As an illustration, let us consider Faddeev's reduced phase space gauge fixing. It is often said in the literature that the condition (284) is a sufficient condition for an admissible gauge fixing. However, this is not correct, and in fact (284) defines only a necessary condition for admissibility, but not necessarily a sufficient one. Indeed, as was discussed previously, (284) is necessary in order that a gauge fixing be achieved for which small infinitesimal gauge transformations are fixed. Nevertheless, this still allows the possibility of nontrivial small finite gauge transformations leaving invariant the gauge fixing conditions, as indeed established by Gribov,[25] as well as large finite gauge transformations. Such a possibility amounts to a gauge slicing in which some gauge orbits are intersected more than once, even though when accounting for an oriented slicing the effective count of intersections may still be acceptable.[26]

Furthermore, (284) does not guarantee either that all gauge orbits are included at least once. Hence, Faddeev's gauge fixing procedure is far from being protected from Gribov problems of either type, and in general does not lead to an admissible gauge fixing. The generic situation is indeed that Faddeev's reduced phase space approach is plagued by Gribov problems.[10,27,28] Even for a system as simple as the relativistic scalar particle, it is shown below that this gauge fixing procedure is not admissible, casting doubt on all other instances where it is being applied for reparametrization invariant theories, namely theories of the gravitational interaction. In fact, this lack of admissibility also applies to the quantization of nonabelian Yang-Mills theories.[25]

Other gauge fixing procedures have been developed over the years, the main reason being that Faddeev's approach usually breaks manifest Poincaré invariance in field theory. This has led to the so-called BRST-BFV Hamiltonian formulation of gauge theories,[10,29] in which gauge fixing is achieved in a different manner. Nonetheless, the issue of Gribov problems arises within that framework as well,[10,27,28] and needs to be assessed on a case by case basis, a problem which requires a comprehensive understanding of the structure of the space of gauge orbits.

Hence, Gribov problems are subtle, difficult but essential features which must be addressed in order to establish the admissibility of a chosen gauge fixing of any gauge invariant dynamics, whatever the gauge fixing procedure being envisaged. Within the context of nonabelian Yang-Mills theories, these issues do not affect any perturbative analysis of quantum properties, since perturbation theory around the ground state amounts to a perturbation within the neighbourhood of vanishing fields, so that the fixing of the small infinitesimal gauge symmetries should suffice. However, it is most likely that nonperturbative phenomena should be highly dependent on Gribov problems, which must thus be avoided in order to gain a genuine and physically correct understanding of such phenomena.

To conclude, let us reconsider the relativistic massive scalar particle. Given the small finite gauge transformations (276) of the variables $x^\mu(\tau)$, $P^\mu(\tau)$ and $\lambda(\tau)$ as well as the choice of boundary conditions on the spacetime coordinates, $x^\mu(\tau_{i,f}) = x^\mu_{i,f}$, it is clear that the Teichmüller parameter γ defines the coordinate labelling the space of gauge orbits within the space of Lagrange multipliers $\lambda(\tau)$. Furthermore, it is readily established[10] that any gauge fixing procedure which induces an admissible gauge fixing in the latter space also induces an admissible gauge fixing of the whole of the variables $\{x^\mu(\tau), P^\mu(\tau), \lambda(\tau)\}$ of the system. This remark thus provides a tool to assess the admissibility of gauge fixing procedures: simply consider the inferred set of values for γ. Given Faddeev's gauge fixing associated to the choice (287), we found

$$\gamma = \eta \frac{\Delta x^0}{\delta \tau} \int_{\tau_i}^{\tau_f} d\tau \, \frac{\dot{g}(\tau)}{\sqrt{\vec{P}^2(\tau) + m^2}} \, . \tag{295}$$

Since this expression implies the upper bound

$$|\gamma| \leq \frac{|\Delta x^0|}{m} \, , \tag{296}$$

it is clear that this gauge fixing procedure suffers a Gribov problem of type II. Furthermore, it also suffers a Gribov problem of type I, on account of the degeneracy in γ values as the system probes all those physical configurations for which the function $\vec{P}^2(\tau)$ remains identical.[10,28] Note that by construction of the gauge fixing procedure, these Gribov problems do not affect the actual classical physical solution to the equations of motion, since the gauge slice always selects the gauge orbit to which that solution belongs, given the chosen boundary conditions. Nevertheless, at the quantum level, these Gribov problems imply that the physically correct quantum amplitudes are not obtained for Faddeev's reduced phase space gauge fixing of this system,[10,28] as may easily be anticipated from the point of view of the path integral representation of quantum amplitudes.

5.4 Dirac's quantization

The canonical quantization of constrained dynamics, known as Dirac's quantization,[8] amounts to the canonical quantization of the basic Hamiltonian formulation of such systems as it has been developed in the above presentation.[10] The space of quantum states $|\psi>$ is a representation space for the commutation relations of the basic phase space degrees of freedom operators $\hat{z}_A$ in the Schrödinger picture,

$$[\hat{z}_a, \hat{z}_B] = i\hbar \hat{C}_{AB}(\hat{z_A}) \, , \tag{297}$$

with all the ensuing operator ordering issues whenever these brackets are noncanonical. Time evolution of quantum states $|\psi, t>$ is generated through the Schrödinger equation,

$$i\hbar \frac{d}{dt} |\psi, t> = \hat{H}_T(t) \, |\psi, t> \quad , \quad \hat{H}_T(t) = \hat{H} + \lambda^\alpha(t) \hat{\phi}_\alpha \, , \tag{298}$$

$\hat{H}$ and $\hat{\phi}_\alpha$ being the first-class Hamiltonian and constraint operators, while $\lambda^\alpha(t)$ are the associated Lagrange multipliers which still play their role as a parametrization of the freedom in applying any gauge transformation as the system evolves in time through the space of quantum states. If the chosen operator ordering of composite quantities is anomaly free, namely if these operators retain their gauge symmetry properties at the quantum level,

$$[\hat{H}, \hat{\phi}_\alpha] = i\hbar \hat{C}_\alpha{}^\beta \hat{\phi}_\beta \ , \quad [\hat{\phi}_\alpha, \hat{\phi}_\beta] = i\hbar \hat{C}_{\alpha\beta}{}^\gamma \hat{\phi}_\gamma \ , \tag{299}$$

with in particular the quantities $\hat{C}_\alpha{}^\beta$ and $\hat{C}_{\alpha\beta}{}^\gamma$, which could indeed be operators themselves in the general case of an open algebra, standing to the left of the constraints in the r.h.s. of these commutation relations, then the quantization of the system is consistent and compatible with its classical gauge invariance properties. However, it could happen that this is not possible, in which case the ensuing gauge anomaly terms appearing as additional contributions of order at least $\hbar^2$ in the r.h.s. of these commutation relations render the physical interpretation of the quantum theory at least problematic, if not inconsistent altogether. In the following discussion, it is assumed that a gauge covariant quantization has been achieved.

The same issue arises for gauge invariant physical observables, namely for the operators to be associated to first-class quantities. Here again, this gauge invariant status remains valid provided only operator ordering makes it possible that the commutation relations are still given by the correspondence principle,

$$[\hat{f}, \hat{\phi}_\alpha] = i\hbar \hat{f}_\alpha{}^\beta \, \hat{\phi}_\beta \ , \tag{300}$$

given the classical bracket $\{f, \phi_\alpha\} = f_\alpha{}^\beta \phi_\beta$. Examples of quantum physical observables are thus the first-class Hamiltonian $\hat{H}$ and constraint $\hat{\phi}_\alpha$ operators. Consequently, in the same way as at the classical level, physical observables are defined as being the gauge equivalence classes of first-class operators under gauge transformations generated by the first-class constraints. The constraints themselves belong to the trivial class.

So far, these issues are analogous to those that arise for the canonical quantization of an ordinary system. For gauge invariant systems, the additional feature is that of gauge transformations generated by the first-class constraints, under which physical configurations should remain invariant. Hence in the quantized system, gauge invariant physical states are those quantum states which are left invariant by small finite gauge transformations, namely which are annihilated by the first-class contraint operators,

$$\hat{\phi}_\alpha \, |\psi, t> \, = \, 0 \, . \tag{301}$$

Note that in some cases, this requirement proves to be too restrictive by not leaving over any state. A weaker condition, which in fact is sufficient for a consistent physical interpretation, is that the matrix elements of the constraint operators for physical states vanish identically,

$$< \psi, t|\hat{\phi}_\alpha|\chi, t >= 0 \ . \tag{302}$$

Provided the constraint algebra is anomaly free, it is clear that these definitions of physical states are consistent with the dynamics, namely the physical character of a state is preserved under time evolution induced through the Schrödinger equation and the first-class total Hamiltonian operator $\hat{H}_T$. In effect, $\hat{H}$ and $\hat{\phi}_\alpha$ are then commuting operators on the subspace of physical states. In particular, the constraints themselves define gauge equivalence classes of physical observables of vanishing value for physical states.

When the above programme is completed, one says that Dirac's quantization of a constrained system has been achieved. However, this still leaves open the issue of the quantum dynamics of such systems, namely the description of the time dependency of the system which amounts to the understanding of the physical properties of the evolution operator associated to its Schrödinger equation. Clearly, the potential difficulty is that when considering the time propagation of quantum states, a proper representation of the actual physical content of the system should include as the sole contributing intermediate states only one quantum physical state for each of the possible gauge orbits. However, the evolution operator based on the total Hamiltonian $\hat{H}_T$ does also propagate all gauge noninvariant states both as intermediate as well as external states. To put it within the framework of the path integral representation of the evolution operator in which a summation over all possible configurations is effected, the actual physical content should follow from an effective integration only over the space of gauge orbits with an equal weight given to each orbit. Otherwise, quantum states other than physical ones contribute to the propagator as intermediate states. This is the specific issue which seems to require some gauge fixing procedure, with its potential Gribov problems as the generic difficulty to be addressed on a case by case basis.

It is often claimed that the gauge fixed path integral representation, hence the gauge fixed quantization, is independent of the gauge fixing procedure, whether for Faddeev's reduced phase space approach (the Faddeev theorem[10,24]) or the BRST-BFV approach (the Fradkin-Vilkovisky theorem[10,29]). However, such a statement is misleading,[10,27,28] since what these theorems in fact establish, and nothing more, is that the resulting path integral representations, and thus also the associated quantized formulations of the system, are gauge invariant. Indeed, what a gauge fixing procedure implies is a specific covering of the space of gauge orbits, so that gauge equivalent gauge fixing procedures (whose induced coverings of gauge orbits are thus identical) do indeed lead to identical quantum formulations and path integral representations. However for gauge nonequivalent gauge fixings, thus inducing different coverings of gauge orbits, necessarily the corresponding quantized formulations of the system, even though each is gauge invariant, are themselves different and thus not physically nor gauge equivalent, thereby leading necessarily to different path integral representations, since throughout its time history the system then explores a different set of physical configurations which is left accessible to it through the gauge fixing procedure. It is only for the class of admissible gauge fixings that the correct quantized and path integral formulation of the system is achieved. Any other nonadmissible gauge fixing leads to a different quantized system, albeit always a gauge invariant one.

As an explicit example of Dirac's quantization, let us consider the relativistic scalar particle. Its canonical quantization is defined by the Heisenberg algebra commutation relations, in the Schrödinger picture,

$$[\hat{x}^\mu, \hat{P}_\nu] = i\hbar\delta^\mu_\nu \quad , \quad \hat{x}^\dagger_\mu = \hat{x}_\mu \quad , \quad \hat{P}^\dagger_\mu = \hat{P}_\mu \ , \tag{303}$$

while the first-class Hamiltonian and constraint are

$$\hat{H} = 0 \quad , \quad \hat{\phi} = \frac{1}{2}\left[\hat{P}^2 + m^2\right] \ . \tag{304}$$

In the configuration space representation of the Heisenberg algebra, quantum states are thus described by their wave function $\psi(x^\mu;\tau)$, which solves the Schrödinger equation

$$i\hbar\frac{\partial\psi(x^\mu;\tau)}{\partial\tau} = \frac{1}{2}\lambda(\tau)\left[-\hbar^2\partial_x^2 + m^2\right]\psi(x^\mu;\tau) \ , \tag{305}$$

$\lambda(\tau)$ being the einbein Lagrange multiplier. Hence, physical states, defined to be annihilated by the reparametrization constraint $\hat{\phi}|\psi,\tau >= 0$, or

$$\left[-\hbar^2\partial_x^2 + m^2\right]\psi(x^\mu;\tau) = 0 \ , \tag{306}$$

are independent of the world-line parameter τ, as they should indeed. Hence, once recovers the fact that the single quantum relativistic scalar particle's dynamics is governed by the Klein-Gordon equation for its wave function in configuration space.

Turning then to the quantum dynamics issue, it should be such that for the physical gauge invariant states, their causal unitary quantum evolution operator be given by Feynman's propagator for a scalar field. However, unless an admissible gauge fixing of the above formulation is effected, this is certainly not the result which one obtains from the above Dirac quantization of the relativistic particle. For instance, following Faddeev's reduced phase space approach based on the gauge fixing condition (287), one may show[10,27,28] that the correct Feynman propagator indeed does not follow, but rather that it suffers precisely the Gribov problems of type I and II which were already described previously in relation to that choice of gauge fixing, namely a bounded integration over the space of gauge orbits characterized by the finite range of Teichmüller parameter values γ as well as a nonuniform integration measure over those gauge orbits that are accounted for because of the degeneracy in the obtained values for γ. In contradistinction, when an admissible gauge fixing is possible (as is the case for this system within the BRST-BFV approach[10,27,28]), then the Feynman propagator is indeed readily recovered, provided the role of large gauge transformations is also properly accounted for, as shall be discuss hereafter. Hence, this simple example confirms the fact that the issue of admissibility and Gribov problems is by no means a trivial and irrelevant one, since otherwise the correct gauge invariant physical content of the system is not recovered.

5.5 Klauder's physical projector: gauge invariant quantum dynamics without gauge fixing

Given all the difficulties surrounding gauge fixing and Gribov problems, it is legitimate to ask whether any gauge fixing is at all a necessity. In fact, it is possible, entirely within Dirac's quantization scheme and nothing more, to circumvent the problem simply by not addressing it, which is most welcome given its most than intricate subtleties! Recall that at the classical level, the analysis of constraints proceeded from the idea[9] that when starting from initial data lying within the constraint hypersurface, namely starting from an initial physical state, time evolution must be such that at each time increment one is projected back onto the constraint hypersurface, even though *a priori* it could be that the complete dynamics generated by the total Hamiltonian could pull the physical trajectory away from the physical subspace. Hence, the identification of the set of constraints makes sure that physical trajectories stay within the physical subspace.

As commented above, this is precisely the issue which faces canonical quantization and quantum dynamics induced by the quantum evolution operator associated to the Schrödinger equation, which is thus given by the time-ordered exponential

$$U(t_2,t_1) = T\,e^{-\frac{i}{\hbar}\int_{t_1}^{t_2}dt\,\hat{H}_T(t)}\quad,\quad \hat{H}_T(t)=\hat{H}+\lambda^\alpha(t)\hat{\phi}_\alpha\ . \tag{307}$$

A priori, this operator could be such that given some initial physical state, its time evolved product would no longer lie within the physical subspace.

However, in order to make sure that physical states stay within that subspace, it would suffice to project them back onto that subspace using an appropriate physical projection operator after each time increment.[9]

In fact, such a physical projector may readily be constructed.[9] Since the constraints $\hat{\phi}_\alpha$ are the generators of small gauge transformations, small finite global symmetry transformations on the space of quantum states are obtained from the operators

$$G(\theta^\alpha) = e^{-\frac{i}{\hbar}\theta^\alpha \hat{\phi}_\alpha} \ , \tag{308}$$

θ^α being the associated symmetry group parameters. Such an operator does indeed appear for each time step in the above total evolution operator (307), the parameters then being the values $\lambda^\alpha(t)$ at that time step, which effect an arbitrary symmetry transformation as the system evolves in time through the first-class Hamiltonian $\hat{H}$.

Note that on account of the definition of physical states, the value (namely either the eigenvalue or the expectation value) of the operators $G(\theta^\alpha)$ acting on physical states is always unity, expressing the gauge invariance of these states. In particular, the same property shows that when considered for such states, in effect the complete evolution operator (307), which otherwise also propagates nonphysical states, reduces to the unitary operator

$$e^{-\frac{i}{\hbar}(t_2-t_1)\hat{H}} \ , \tag{309}$$

irrespective of the choice for the Lagrange multipliers $\lambda^\alpha(t)$.

In order to define now the physical projector, clearly it suffices to consider all the small finite symmetry transformations $G(\theta^\alpha)$ summed over the space of all such transformations, namely

$$\mathbb{E} = \int [dU(\theta^\alpha)] \ e^{-\frac{i}{\hbar}\theta^\alpha \hat{\phi}_\alpha} \ , \tag{310}$$

where $[dU(\theta^\alpha)]$ stands for the normalized group invariant measure over the manifold of small finite symmetry transformations (in the case of a compact Lie group, this is the Haar measure). Consequently, one has

$$\mathbb{E}^2 = \mathbb{E} \quad , \quad \mathbb{E}^\dagger = \mathbb{E} \ , \tag{311}$$

which are indeed the properties characteristic of a projection operator. Given this construction, it should be clear that acting with $\mathbb{E}$ on any quantum state, all its gauge noninvariant components are averaged out through the group integration, leaving over only its gauge invariant physical component. Hence, $\mathbb{E}$ is indeed the physical projector of the system.[9] Furthermore, whenever of application, large gauge transformations may also be included in its construction, so that a truly gauge invariant projector onto all physical states invariant under small as well as large gauge transformations is achieved.

This operator may now be used to construct the physical propagator or evolution operator of the quantized system, which thus only propagates physical gauge invariant states, namely

$$U_{\rm phys}(t_2,t_1) = U(t_2,t_1)\mathbb{E} = \mathbb{E}U(t_2,t_1)\mathbb{E} = \mathbb{E}e^{-\frac{i}{\hbar}(t_2-t_1)\hat{H}}\,\mathbb{E}$$
$$= \mathbb{E}e^{-\frac{i}{\hbar}(t_2-t_1)\mathbb{E}\hat{H}\mathbb{E}}\,\mathbb{E}\,, \tag{312}$$

where the different expressions follow from the properties of $\mathbb{E}$ as well as the fact that the operators $\hat{H}$ and $\hat{\phi}_\alpha$ commute on the physical subspace. Clearly, this physical evolution operator obeys the usual and necessary unitary and involution properties characteristic of such an operator generating time evolution. Furthermore, given its very last representation, it obviously propagates physical states only, both as external as well as intermediate states. The physical projector together with this physical propagator thus provide the complete answer to the issue of the genuine gauge invariant physical quantum dynamics of a gauge invariant system. Nonetheless, the construction is entirely set only within Dirac's quantization framework, without any need for any gauge fixing procedure of any sort, thereby avoiding from the outset the difficult issue of Gribov problems. In fact, the quantum dynamical formulation is gauge invariant by construction, and in effect amounts precisely to an dynamics over the space of gauge orbits with an admissible covering. The physical projector approach to the gauge invariant dynamics is necessarily void of any Gribov problem,[30] maintains manifest Poincaré covariance when present, and avoids all the technical difficulties of functional determinants, ghost contributions and the like following from any gauge fixing procedure.

In addition, the method also extends[9] to second-class constraints, avoiding the difficulties often raised by the quantization of Dirac brackets, as well as to reducible or nonregular constraints. Furthermore, following the usual time slicing route, it is also possible to set up path integral representations of matrix elements of the physical evolution operator and physical observables, leading to convenient calculational methods complementary to the quantum operator ones.

As with all the general discussion of constrained systems, the above programme for the construction of the physical projector must be considered and developed on a case by case basis. Its explicit definition often requires specifications. For example,[9] if the spectrum of a constraint, say $\hat{\phi}$, is continuous in the neighbourhood of its zero eigenvalue, the projector is not normalizable and corresponds to a projector density that could be defined as follows

$$\mathbb{E}_0 = \lim_{\delta \to 0} \frac{1}{2\delta} \mathbb{E}[-\delta < \hat{\phi} < \delta] \ , \tag{313}$$

where the operator $\mathbb{E}[-\delta < \hat{\phi} < \delta]$ stands for the projector onto the subspace spanned by those states whose $\hat{\phi}$ eigenvalue lies within the shown interval. For instance, if $\hat{\phi} = \hat{q}$, $\hat{q}$ being the position operator of the Heisenberg algebra, one has

$$\mathbb{E}_0 = \lim_{\delta \to 0} \frac{1}{2\delta} \int_{-\delta}^{\delta} dq \, |q><q| = \lim_{\delta \to 0} \frac{1}{2\delta} \int_{-\infty}^{\infty} d\xi \, e^{i\xi\hat{q}} \, \frac{\sin(\delta\xi)}{\pi\xi} = \int_{-\infty}^{\infty} \frac{d\xi}{2\pi} \, e^{i\xi\hat{q}} \ , \tag{314}$$

where in the third expression the integral representation of the step function is introduced, leading to the last expression which is indeed nothing but the $\delta(q)$ function in operator form. Note that the exponentiated operator $e^{i\xi\hat{q}}$ is the generator of translations in the position, the associated group parameter ξ being thus integrated over with a normalization of the integration measure such that $\mathbb{E}_0$ is a nonnormalizable operator density.

As a particular example, the constraint $\hat{P}^2 + m^2 = 0$ of the parametrized relativistic scalar particle does possess a continuous spectrum including the zero eigenvalue, so that the physical projector for that system follows such a construction. Furthermore, since the first-class Hamiltonian $\hat{H} = 0$ for that reparametrization invariant system vanishes, the physical evolution operator $U_{\rm physical}(\tau_f, \tau_i)$ in fact coincides with the physical projector, hence

$$U_{\rm physical}(\tau_f, \tau_i) = \mathbb{E}_0 = \int_{-\infty}^{\infty} \frac{d\gamma}{2\pi} e^{-\frac{1}{2} i\gamma(\hat{P}^2 + m^2)} \ , \tag{315}$$

where the symmetry parameter γ is nothing else but the Teichmüller coordinate of the space of gauge orbits. Note that this physical evolution operator is independent of the world-line τ coordinate, as it should on account of the gauge invariance of the formulation. Furthermore, the integration over the space of gauge orbits is indeed admissible, since each of the gauge orbits are accounted for with an equal relative weight, hence once and only once. No gauge fixing has been effected, but nevertheless the correct gauge invariant quantum dynamics is readily obtained through the physical projector.

In particular, let us consider the configuration space matrix elements of this operator, which should thus be in direct correspondence with the Feynman propagator for a scalar field theory. Thus,

$$< x_f^\mu | U_{\rm physical}(\tau_f, \tau_i) | x_i^\mu > = \int_{(\infty)} \frac{d^D p^\mu}{(2\pi)^D} \, e^{i\Delta x \cdot p} \int_{-\infty}^{\infty} \frac{d\gamma}{2\pi} \, e^{-\frac{1}{2} i\gamma(p^2 + m^2)} \ . \tag{316}$$

So far however, we have only accounted for the small gauge symmetries, namely those that preserve the world-line orientation and are generated by the first-class constraint $\hat{\phi}$. However, one should still enforce gauge invariance under large gauge transformations reversing the world-line orientation, and use the corresponding physical projector. This projector is obtained by restricting the γ range of integration to the real positive axis, for the reasons having been discussed previously in that respect. Hence finally, the full gauge invariant spacetime propagator of the unoriented relativistic scalar particle is given by

$$\int_{(\infty)} \frac{d^D p^\mu}{(2\pi)^D}\, e^{i\Delta x \cdot p} \int_0^\infty \frac{d\gamma}{2\pi}\, e^{-\frac{1}{2}i\gamma(p^2+m^2)} = \frac{1}{\pi} \int_{(\infty)} \frac{d^D p^\mu}{(2\pi)^D}\, e^{i\Delta x \cdot p}\, \frac{i}{p^2+m^2-i\epsilon}\,, \tag{317}$$

a result which indeed, up to the normalization factor $1/\pi$, coincides exactly with the Feynman propagator for a scalar field with its canonical normalization. The proper projection onto the physical subspace of quantum states has been achieved in a most straightforward manner, without any gauge fixing nor ghost system whatsoever.[30]

So far, the physical projector approach has been applied to some well-known integrable systems, as well as to some of the issues surrounding the problems of quantum gravity.[31] For example, it has been applied[23,32] to the gauge invariant mechanical models in 0+1 dimensions described previously, with Lagrangian

$$L = \frac{1}{2g^2}\left[\dot{q}_i^a + f^{abc}\lambda^B q_i^c\right]^2 - \frac{1}{2}\omega^2\left(q_i^a\right)^2 \,, \tag{318}$$

in the cases of the gauge groups SO(2) and SO(3). It could prove to be of interest to extend the analysis to all compact Lie algebras, as well as quartic coupling interactions as they arise from the dimensional reduction of the matrix formulation of M-theory. Coherent state techniques for nonabelian groups would certainly be of relevance, as well as the general methods of dynamical integrable systems. The nonperturbative solution of the Schwinger model (a U(1) gauge invariant field theory coupled to a massless fermion in 1+1 dimensions) has also been recovered through the physical projector without the necessaity of any gauge fixing.[33] Likewise, the physical projector has been applied[34] to the quantization of the U(1) invariant Chern-Simons theory, one of the simplest topological quantum field theories in which only a finite number of gauge invariant states, dependent only on the differential topology of the spacetime manifold but not its geometry, is to be projected out from an infinite set of quantum states.

These successes thus bode well for the relevance of this recent approach towards the quantization of constrained dynamics. It would certainly be a worthwhile project to extend its use to quantum field theories in a perturbative quantization and determine what this implies for a modification in the Feynman rules as usually derived within some gauge fixed framework. However and most undoubtedly, it is in the nonperturbative realm of quantum phenomena hithertoo not fully comprehended that this new method offers the widest prospects for original new results and insights into the dynamics of gauge invariant theories.

6 Chern-Simons Quantum Field Theory

Even though general relativity has not been addressed to any extent so far, let us briefly consider one of the issues surrounding a formulation of quantum gravity using the Einstein-Hilbert action as a reference. Presumably, some consistent definition of pure quantum gravity would provide a specific meaning to the otherwise formal path integral representation which effects a summation over all configurations of the system, namely over all the possible geometrical structures associated to a spacetime manifold of given topology and differential structure,

$$\int [\mathcal{D}g_{\mu\nu}]\, e^{i\frac{1}{\kappa}\int d^nx\,\sqrt{g}R} , \qquad (319)$$

where $g_{\mu\nu}$ stands for the metric tensor, κ for a normalization proportional to Newton's constant, and R for the Riemann scalar curvature of the considered metric. The definition of such a path integral requires specification, if only to avoid double counting of configurations that are diffeomorphic equivalent under spacetime reparametrizations, the local gauge symmetry of general relativity.

In these terms, it thus appears that pure quantum gravity would be a theory whose physical properties are not only independent of the spacetime coordinate system, but more importantly, independent of any spacetime geometrical data.[5,6] Thus, pure quantum gravity would be a system whose physical observables are only dependent on the topological and differentiable spacetime structures, namely genuine diffeomorphic topological invariants. It is by following ideas such as these that Witten has uncovered the existence of so-called Topological Quantum Field Theories (TQFT), whose quantum observables consist only of (diffeomorphic) topological invariants.[5,6,7] Even though in their field theory formulation such systems possess an infinite number of degrees of freedom, their actual gauge invariant physical content is that of a finite number of quantum states in direct relation to the diffeomorphic topological data of the manifold on which the fields are considered. Two large classes of such models have been identified,[7] namely TQFT's whose formulation requires metric data but whose gauge invariance is so large that their physics is independent of the geometry nevertheless, and TQFT's whose formulation does not require a metric structure on the base manifold. Such TQFT's are also in direct relation to theories of the general relativistic or of the Yang-Mills type through the character of the gauge symmetries that they possess. These classes of quantum field theories have grown into a topic of great interest in mathematical physics, with applications within fundamental physics of potential great relevance.

One such example is that of pure Chern-Simons theories on a manifold of 2+1 dimensions.[6] In fact, pure gravity in that spacetime dimension may be brought within such a framework, the Yang-Mills symmetry being then based on a noncompact Lie group.[35] The case of compact Lie groups is directly relevant to quite a number of fields in pure mathematic and theoretical physics.[6,36] By lack of time and space, here the full account of the application of the physical projector quantization of the U(1) Chern-Simons theory as it was presented at the Workshop is not reproduced, referring the interested reader to the original publication for details.[13,34] Only a few general comments will be provided.

In terms of the same notations as introduced previously for any Yang-Mills theory, the 2+1 dimensional pure Chern-Simons action reads

$$S[A^a_\mu] = N_k \int_{\mathbb{R}\times\Sigma} dx^0 dx^1 dx^2\, \epsilon^{\mu\nu\rho} \left[A^a_\mu F^a_{\nu\rho} + \frac{1}{3} f^{abc} A^a_\mu A^b_\nu A^c_\rho \right] , \qquad (320)$$

where the gauge coupling constant has been absorbed into the normalization of the gauge fields A^a_μ, while N_k is some normalization factor for the action, the index k anticipating the fact that this quantity needs to take a quantized value. Finally, $\epsilon^{\mu\nu\rho}$ stands for the totally antisymmetric tensor such that $\epsilon^{012} = +1$. Furthermore, the topology of the three-dimensional manifold on which the fields are considered is that of the direct product of the real line $\mathbb{R}$, the coordinate x^0 being considered as the time evolution parameter, with a compact Riemann surface Σ of given topology, the case of a two-torus being the simplest and also of relevance. The reason for this specific choice of topology is that the dynamics will be considered from the Hamiltonian point of view.

Note that the definition of the action is independent of any metric structure. In particular, the index $\mu = 0, 1, 2$ is not to be raised nor lowered in different expressions. In addition, the action is locally gauge invariant by construction. Under small gauge transformations, the action only changes by a local surface term without consequence. For large gauge transformations however, it changes by a constant shift which is proportional to the winding number of the corresponding gauge transformation. This is one reason why the normalization factor N_k needs to be quantized in the quantum theory, since otherwise the quantity e^{iS} being summed over in the path integral is not single-valued under large gauge transformations,

As is generic for TQFT's, the above action is in fact dependent only on the diffeomorphic topological invariant data of the underlying three-dimensional manifold. This is best demonstrated from the Euler-Lagrange equations of motion which read

$$F^a_{\mu\nu} = 0 \ . \tag{321}$$

The classical solutions are thus indeed nothing but the gauge equivalence classes of flat gauge connections on the considered three-dimensional manifold. It is well known that this modular space is characterized in purely topological terms, namely the G-holonomies of all noncontractible cycles within the manifold. Hence for the chosen topology $\mathbb{R} \times \Sigma$, this set of solutions is a finite dimensional space of continuous variables. Pure quantum Chern-Simons theory is thus nothing but quantum mechanics of the modular space of flat gauge connections, indeed a genuine purely diffeomorphic topological system.[6] Among the infinite set of degrees of freedom A^a_μ, there remains only a finite number of gauge invariant physical degrees of freedom. From this fact alone, one may *a priori* only conclude that the physical states of the quantized system span a discrete infinite vector space. However, as we shall see, as a consequence of the compactness of phase space because of large gauge transformations, the actual quantum space of gauge invariant states itself is also finite dimensional. Quantum pure Chern-Simons theory possesses only a finite number of quantum physical states.

Note that the above action is already in Hamiltonian form. Indeed, all time derivatives of the fields appear linearily in the Lagrangian density. Consequently, the configurations A^a_μ are in fact already the phase space degrees of freedom, whose brackets are directly identified from the action to be

$$\{A^a_1(x^0,\vec{x}), A^b_2(x^0,\vec{y})\} = \frac{1}{2N_k}\delta^{ab}\,\delta^{(2)}(\vec{x}-\vec{y}) \ , \tag{322}$$

while the first-class Hamiltonian vanishes identically, $H = 0$, as befits the Hamiltonian of any reparametrization invariant theory, and the first-class constraints, generators of the small Yang-Mills gauge transformations, are

$$\phi^a = -2N_k \left[\partial_1 A^a_2 - \partial_2 A^a_1 - f^{abc} A^b_1 A^c_2\right] \ , \tag{323}$$

whose Lagrange multipliers are A_0^a. A direct calculation then confirms the gauge algebra generated by these quantities

$$\{\phi^a(x^0,\vec{x}),\phi^b(x^0,\vec{y})\} = f^{abc}\phi^c(x^0,\vec{x})\,\delta^{(2)}\,(\vec{x}-\vec{y})\ . \tag{324}$$

To make this brief discussion as simple as possible, let us now particularize to the U(1) gauge symmetry and the two-torus topology, $\Sigma = T_2$. In such case, given a local trivialization of the torus based on choice of basis of the homology group of 1-cycles, a Fourier mode analysis of the periodic fields $A_{1,2}$ is natural. One then finds that under small gauge transformations only the nonzero Fourier modes are varied, leaving invariant the Fourier zero modes $A_{1,2}(x^0)$. In contradistinction, large U(1) gauge transformations, falling themselves into U(1) homotopy classes because of the torus topology, leave invariant the nonzero modes while the zero modes are then shifted by integer multiplies of 2π, depending on the homotopy class of the gauge transformation. Consequently, it follows that the actual gauge invariant phase space is nothing but the zero mode sector taking its values on a two-torus itself, namely a compact phase space. This is not a type of phase space encountered in usual Hamiltonian dynamics, whose quantization requires specific methods which have been designed for such a purpose within the framework of so-called geometric quantization, discussed in Prof. S.T. Ali's lectures in these Proceedings. However, through the use of the physical projector, it remains possible to use the straightforward methods of canonical quantization, and then enforce invariance under large gauge transformations through the physical projector, and thereby in effect quantize a compact phase space.

Physical phase space being compact, it is to be expected that the number of gauge invariant physical states itself be finite, since each quantum state occupies a specific quantum of volume of phase space. Because of the normalization of the above brackets in terms of the factor N_k, this fact translates in the quantization of that factor as

$$N_k = \frac{\hbar}{4\pi}\,k\ , \tag{325}$$

k being an arbitrary integer. However, the same requirement follows from the construction of the quantum operators generating large gauge transformations of quantum states. Through the physical projector which sums over both all small and all large gauge transformations, a finite set of quantum physical states and their explicit wave function representations is readily identified. One then recovers exactly the states that have been constructed through an admissible gauge fixing procedure of the same theories. Futher details may be found in the original publication and references therein.[13,34]

In conclusion, the physical projector is capable of identifying among an infinite number of quantum states the finite subset of physical states in the case of a TQFT, and this within a framework which is simply that of Dirac's canonical quantization of constrained dynamics, without the necessity of any gauge fixing procedure whatsoever, thereby avoiding from the outset the potential difficulties of Gribov problems.

This rather simple system provides a nontrivial example of the relevance of purely topological features to the physics of gauge invariant quantum field theories. Extended to theories of gravity along the ideas mentioned in the introduction to this section, such a situation gives some credence to the suggestion[5,6,35] that quantum topology and its generic finite number of quantum states is at the basis of actual pure quantum gravity, whereas geometry, and thus in particular quantum gravity in which geometrical concepts acquire their physical meaning through some mechanism akin to that of spontaneous symmetry breaking, enters the picture through the introduction of matter interactions, thereby leading also to an infinite number of quantum states. It could be that the path towards a theory of quantum gravity coupled to all other fundamental interactions goes first through the restricted framework of purely topological quantum gauge field theories followed by their coupling to the dynamics of interacting ones.

7 The Closed Bosonic String

As another illustration of the plausible great relevance of topology to the quest for a quantum geometric framework for the unification of all quantum interactions and particles, in the remainder of these notes we shall briefly discuss the quantization of bosonic strings on a Minkowski spacetime.[2,3,37] Most of the considerations to be presented extend to fermionic and superstrings as well. More specifically, among many such indications, here we shall only address the features of so-called T-duality that arise whenever a closed string theory is propagating within a spacetime of which some of its spatial dimensions have been compactified into a torus geometry. As shall be discussed, the topology of these spaces is such that geometries whose radii are inversely related to one another through some fundamental length scale become physically indistinguishable, suggesting that in the realm of a theory of quantum gravity coupled to matter, the pointwise concepts of differentiable continuous manifolds have to be extended at short-distance by some new concepts at the basis of quantum geometry. Presumably, quantum geometry is where quantum interactions, quantum particles and quantum topology meet in fundamental physics.

As in our discussion of the relativistic scalar particle, the choice for the signature of the Minkowski metric is $\eta_{\mu\nu} = \mathrm{diag}\,(-+...+)$, $\mu, \nu = 0, 1, 2\cdots, D-1$, D being the spacetime dimension. The spacetime coordinates of the string world-sheet embedded into spacetime are denoted $x^\mu(\tau, \sigma) = x^\mu(\xi)$, $\xi^\alpha = (\tau, \sigma)$, $\alpha = 0, 1$, being the dimensionless world-sheet coordinates with τ considered as the time evolution parameter and σ restricted to the interval $0 \le \sigma \le \pi$. Thus in the world-sheet, τ (respectively, σ) is time-like (respectively, space-like). In these notes, we shall only consider the closed bosonic string, all quantities then being periodic in σ with periodicity π.

For the same reasons as in the case of the relativistic scalar particle, the action of the system must be a spacetime scalar for the Poincaré group, as well as a world-sheet scalar for world-sheet reparametrizations. In particular, the redundancy in the physical time x^0 and the world-sheet time evolution parameter τ is to be resolved through the latter gauge invariance properties of the system. Consequently, the most natural choice for the action principle is to measure the total area of the string world-sheet swept out between two initial and final configurations. This area may be measured either in terms of the metric induced on the world-sheet by the ambiant spacetime Minkowski metric, or else by some additional intrinsic world-sheet metric. The first choice leads to the Nambu-Goto action with its nonlinear equations of motion, and the second choice to the Polyakov action and its linear equations of motion. Only the Nambu-Goto action will be discussed presently, leaving it as an exercise to develop the same analysis for the Polyakov action. In either case, one is in fact dealing, from the world-sheet point of view, with a two-dimensional theory of quantum gravity coupled to a collection of scalar fields.

7.1 The nonlinear Nambu-Goto action

Given the Minkowski spacetime line element $ds^2 = \eta_{\mu\nu} dx^\mu dx^\nu$, the induced world-sheet line element reads

$$ds^2 = \gamma_{\alpha\beta}\, d\xi^\alpha\, d\xi^\beta \quad , \quad \gamma_{\alpha\beta} = \partial_\alpha x^\mu\, \partial_\beta x^\nu\, \eta_{\mu\nu} \; . \tag{326}$$

The signature of this induced metric $\gamma_{\alpha\beta}$ being $(-+)$, the local reparametrization invariant area element is $d^2\xi\,\sqrt{-\det\gamma_{\alpha\beta}}$. Hence, causal string propagation is defined through the Nambu-Goto action

$$S[x^\mu] = \frac{-1}{2\pi\alpha'}\int_{\tau_1}^{\tau_2} d\tau \int_0^\pi d\sigma \sqrt{-\det\gamma_{\alpha\beta}} = \int_{\tau_1}^{\tau_2} d\tau \int_0^\pi d\sigma \mathcal{L}\,(\dot{x}, x') \; , \tag{327}$$

with

$$\mathcal{L}(\dot{x}, x') = \frac{-1}{2\pi\alpha'}\sqrt{(\dot{x}\cdot x')^2 - \dot{x}^2 x'^2} \; . \tag{328}$$

In these expressions, a dot above a quantity stands for a derivative with respect to τ, as usual, while a prime stands for a derivative with respect to σ. Furthermore, the coefficient α' has the dimensions of a length squared, measures the string tension and is related to the so-called Regge slope. This is the parameter which sets the physical scale of the system, for instance the mass scale of string oscillating modes. Since, as we shall see, the closed bosonic string always includes a massless spin 2 mode, it would be natural to associate the string tension α' to the Planck scale.

Remark

Introducing an intrinsic world-sheet metric $g_{\alpha\beta}$ of signature $(-+)$, the Polyakov action reads

$$S[x^\mu, g_{\alpha\beta}] = \frac{-1}{4\pi\alpha'} \int_{\tau_1}^{\tau_2} d\tau \int_0^\pi d\sigma \sqrt{-\det g_{\alpha\beta}}\, g^{\alpha\beta}\, \partial_\alpha x^\mu \partial_\beta x^\nu \eta_{\mu\nu} \,. \tag{329}$$

Note that the usual Einstein-Hilbert action, proportional to

$$\frac{1}{4\pi} \int d^2\xi \sqrt{-g}\, R^{(2)} \,, \tag{330}$$

is not included, since this latter contribution measures the Euler characteristic of the world-sheet, hence is a surface term that does not affect the equations of motion. Furthermore, a world-sheet cosmological term is not included either,

$$\frac{-1}{4\pi\alpha'} \int d^2\xi \sqrt{-g}\, \mu^2 \,, \tag{331}$$

since at the classical level there exists a solution to the associated equations of motion provided only $\mu^2 = 0$. Hence, this two-dimensional theory of gravity is in fact Weyl invariant in the absence of such a term, namely invariant under local changes in the scale factor of the metric, $g_{\alpha\beta}(\xi) \to e^{\chi(\xi)}\, g_{\alpha\beta}(\xi)$. This is the symmetry which at the quantum level restricts the spacetime dimension to the critical value $D = 26$, when the Polyakov action is quantized in a manifestly reparametrization invariant manner.[2,3,37,38]

Finally, note how describing the propagation of the bosonic string in a curved spacetime of background metric $G_{\mu\nu}(x^\mu)$ is readily achieved, by direct substitution of $\eta_{\mu\nu}$ in the above expressions. The Polyakov action is then the natural starting point for a study of the low energy effective field theory description of strings coupled to background fields.[2,3]

Let us restrict to the Nambu-Goto action. Since from the world-sheet point of view, one is dealing with a local field theory possessing as internal symmetry the Poincaré group of Minkowski spacetime, it follows from

Noether's theorem that there exist currents and charges associated to spacetime translations and rotations which are conserved for the classical configurations solving the equations of motion. The energy-momentum Noether current and charge are

$$P^\alpha_\mu = \frac{\partial \mathcal{L}}{\partial(\partial_\alpha x^\mu)} \quad , \quad P_\mu = \int_0^\pi d\sigma \, P^{\alpha=0}_\mu \ , \tag{332}$$

with

$$\begin{aligned} P^0_\mu &= \tfrac{-1}{2\pi\alpha'} \tfrac{1}{\sqrt{(\dot{x}\cdot x')^2 - \dot{x}^2 x'^2}} \left[(\dot{x}\cdot x')x'_\mu - x'^2 \dot{x}_\mu\right] \ , \\ P^1_\mu &= \tfrac{-1}{2\pi\alpha'} \tfrac{1}{\sqrt{(\dot{x}\cdot x')^2 - \dot{x}^2 x'^2}} \left[(\dot{x}\cdot x')\dot{x}_\mu - \dot{x}^2 x'_\mu\right] \ . \end{aligned} \tag{333}$$

The angular-momentum current and charge are

$$M^\alpha_{\mu\nu} = P^\alpha_\mu x_\nu - P^\alpha_\nu x_\mu \quad , \quad M_{\mu\nu} = \int_0^\pi d\sigma M^{\alpha=0}_{\mu\nu} \ . \tag{334}$$

Any solution to the equations of motion is thus such that

$$\partial_\alpha P^\alpha_\mu = 0 \quad , \quad \partial_\alpha M^\alpha_{\mu\nu} = 0 \quad , \quad \frac{dP_\mu}{d\tau} = 0 \quad , \quad \frac{dM_{\mu\nu}}{d\tau} = 0 \ . \tag{335}$$

Given the fact that the Lagrangian density is dependent only on the ξ^α derivatives of the coordinates $x^\mu(\xi)$ (because of the required invariance under spacetime translations), the Euler-Lagrange equations of motion are nothing but the conservation equations for the energy-momentum Noether currents,

$$\partial_\alpha P^\alpha_\mu = 0 \ . \tag{336}$$

These equations are to be accompanied by boundary conditions. One set of boundary conditions specifies the initial and final string configurations, while the boundary conditions in σ (which are required since the two-dimensional world-sheet is compact, in contradistinction with ordinary field theories which are required to vanish at infinity) are nothing but the periodicity requirement in $\sigma \to \sigma + \pi$. Hence,

$$x^\mu(\tau_{1,2}, \sigma) = x^\mu_{1,2}(\sigma) \quad , \quad x^\mu(\tau, \sigma + \pi) = x^\mu(\tau, \sigma) \ . \tag{337}$$

It is at this point that other choices of boundary conditions in σ are possible, such as for open strings, or more generally for strings ending of fixed branes leading to so-called Dp-branes.[2] None of these shall be addressed here.

It thus appears that this set of equations of motion is highly nonlinear, and thus difficult to solve explicitly. In addition, all the above quantities are not independent, but in fact obey the following constraints,

$$\left[P^0_\mu \pm \frac{\partial_\sigma x_\mu}{2\pi\alpha'}\right]^2 = 0 \ , \quad \left[P^1_\mu \pm \frac{\partial_\tau x_\mu}{2\pi\alpha'}\right]^2 = 0 \ , \tag{338}$$

as may be checked by direct calculation. In fact, since P^0_μ is nothing else than the momentum conjugate to x^μ, the first pair of constraints are primary constraints for the Hamiltonian formulation, and are thus expected to be the generators for small world-sheet reparametrizations as is indeed confirmed by the explicit Hamiltonian analysis. On the other hand, the second pair of constraints is related to the first through large world-sheet reparametrizations which exchange the τ and σ coordinates while preserving the world-sheet orientation, such as

$$\tilde{\tau} = \tau_1 + \frac{\tau_2 - \tau_1}{\pi}\sigma \ , \quad \tilde{\sigma} = \frac{\pi}{\tau_2 - \tau_1}(\tau_2 - \tau) \ . \tag{339}$$

In the same way as for the relativistic scalar particle, it is thus possible to consider oriented strings theories which are required to be invariant under both small and large orientation preserving world-sheet reparametrizations, and unoriented strings invariant under both orientation preserving and reversing world-sheet diffeomorphisms.

Since by construction the system is invariant under small world-sheet reparametrizations, namely a small local gauge symmetry, it follows that the general solution to the equations of motion involves arbitrary functions of ξ^α, in fact two such functions related to the two independent coordinates $\xi^\alpha = (\tau, \sigma)$. In order to construct explicit solutions, it is thus necessary to first fix this large gauge freedom, and then afterwards eventually restore it if necessary.

7.2 Conformal gauge fixing

In order to specify some gauge fixing of the system, let us consider the quantities $\partial_\tau x^\mu$ and $\partial_\sigma x^\mu$. Clearly, they define the vectors tangent to the world-sheet associated to the (τ, σ) coordinate system set-up on that manifold. Hence, it is always possible, by an appropriate local change of coordinates, to bring these two tangent vectors to be locally perpendicular with respect to the spacetime Minkowski metric, namely $\gamma_{01} = 0$, and then by a local rescaling of each of the coordinates τ and σ to set the Lorentz invariant length of each of these vectors to be identical up to their sign since one is time-like and the other space-like, namely $\gamma_{00} + \gamma_{11} = 0$. In terms of the coordinates, we thus have

$$\dot{x} \cdot x' = 0 \ , \quad \dot{x}^2 + x'^2 = 0 \iff \left(\dot{x} \pm x'\right)^2 = 0 \ . \tag{340}$$

However, given this geometrical description of this choice of coordinates, there still remains quite some large gauge freedom, namely choices of coordinates ξ^α which locally amount to a local change of scale for the induced metric $\gamma_{\alpha\beta}$, while the orthogonality conditions are preserved. In the case of an euclidean signature metric, such transformations in two dimensions correspond to conformal or analytic transformations. Hence in the present content, the above choice of gauge fixing is also called conformal gauge fixing, even though the term pseudo-conformal would be more fitting given the Minkowski signature $(-+)$ on the world-sheet. Note however that the conformal gauge is yet not a complete gauge fixing, since conformal transformations inducing a local change of scale in the induced metric $\gamma_{\alpha\beta}$ are still possible. This gauge redundancy will be resolved in the light-cone gauge to be discussed in the next section.

Once the conformal gauge fixing (340) effected, the equations of motion become linear, since one finds

$$P^0_\mu = \frac{\dot{x}_\mu}{2\pi\alpha'} \quad , \quad P^1_\mu = -\frac{x'_\mu}{2\pi\alpha'} \ , \tag{341}$$

thus leading to the Klein-Gordon equations of D free massless scalar fields in two dimensions,

$$\left[\partial^2_\tau - \partial^2_\sigma\right] x^\mu(\tau, \sigma) = 0 \ . \tag{342}$$

However, these equations still have to be supplemented with the gauge fixing conditions $(\dot{x} \pm x')^2 = 0$. Consequently, the general solution is always of the form

$$x^\mu(\tau, \sigma) = x^\mu_L(\tau + \sigma) + x^\mu_R(\tau - \sigma) \ , \tag{343}$$

namely a linear superposition of left- and right-moving modes propagating along the two disjoint branches of the two-dimensional light-cone and noninteracting with one another. This decoupling of massless chiral modes in two dimensions is generic to all closed string theories, and put to great advantage in the construction of string theories in different spacetime dimensions.[2,3]

A simple Fourier mode analysis of the system then readily leads to the following general solution in the conformal gauge,

$$\begin{aligned} x^\mu(\tau, \sigma) = \sqrt{2\alpha'}\,\Big[\ & q^\mu + \alpha^\mu_0(\tau - \sigma) + \bar{\alpha}^\mu_0(\tau + \sigma) \\ & + \tfrac{1}{2} i {\sum_n}' \tfrac{1}{n} \left(\alpha^\mu_n e^{-2in(\tau - \sigma)} + \bar{\alpha}^\mu_n e^{-2in(\tau + \sigma)}\right)\Big] \ , \end{aligned} \tag{344}$$

where the summation runs over all positive and negative integers n except for $n=0$ as indicated by the prime on the summation symbol, while the different constant factors are integration constants such that

$$\alpha_n^{\mu *} = \alpha_{-n}^{\mu} \ , \quad \bar{\alpha}_n^{\mu *} = \bar{\alpha}_{-n}^{\mu} \ , \alpha_0^{\mu} = \frac{1}{2}\sqrt{2\alpha'}P^{\mu} = \bar{\alpha}_0^{\mu} \ . \tag{345}$$

Furthermore, the conformal gauge fixing conditions translate into the constraints

$$L_n = 0 \ , \quad \bar{L}_n = 0 \ , \tag{346}$$

where

$$L_n = \frac{1}{2}\sum_m \alpha_{n-m}^{\mu}\alpha_{m\mu} \ , \quad \bar{L}_n = \frac{1}{2}\sum_m \bar{\alpha}_{n-m}^{\mu}\bar{\alpha}_{m\mu} \ . \tag{347}$$

In particular, the zero mode contraints $L_0 = 0 = \bar{L}_0$ are equivalent to

$$\frac{1}{2}\alpha' M^2 = N + \bar{N} \ , \quad N = \bar{N} \ , \tag{348}$$

with the excitation level number quantities

$$N = \sum_{n=1}^{\infty} \alpha_{-n}^{\mu}\alpha_{n\mu} \ , \quad \bar{N} = \sum_{n=1}^{\infty} \bar{\alpha}_{-n}^{\mu}\bar{\alpha}_{n\mu} \ . \tag{349}$$

As we shall see later on, the quantities L_n and $\bar{L}_n$ are nothing but the generators of the remaining reparametrization invariance, namely conformal symmetry, in the conformal gauge, known as the Virasoro generators. In particular, the sum of the Virasoro zero modes $L_0 + \bar{L}_0 = 0$ leads to the mass spectrum of the solutions, while their difference, $L_0 - \bar{L}_0 = 0$, expresses the invariance of the system under constant shifts in the σ coordinate, $\sigma \to \sigma + \sigma_0$, since the string is closed in that direction of the world-sheet.

Note that from the expressions for N and $\bar{N}$, it is not obvious that all oscillating solutions to the equations of motion have a positive definite mass, given the semi-definite signature of the Minkowski metric. This issue may be resolved only by solving all these Virasoro constraints, or equivalently, by completely fixing the conformal gauge freedom remaining in the conformal gauge. This is the purpose of the next section.

The fact that the Virasoro constraints are related to conformal transformations is readily established. As indicated previously, the constraints $[P_\mu^0 \pm \dot{x}_\mu/2\pi\alpha']^2 = 0$ are the generators of world-sheet reparametrizations. In the conformal gauge, their expression is proportional to $(\dot{x} \pm x')^2 = 0$, namely the gauge fixing conditions themselves, which in terms of their Fourier modes in σ coincide with the Virasoro quantities $L_n = 0 = \bar{L}_n$. Hence, the reason why we still have the Virasoro constraints to enforce in the conformal gauge is that this gauge fixing is not yet complete. Indeed, (pseudo)conformal reparametrizations $\tilde{\xi} = \tilde{\xi}(\xi)$ such that,

$$\frac{\partial\tilde{\tau}}{\partial\tau} = \frac{\partial\tilde{\sigma}}{\partial\sigma} \ , \ \frac{\partial\tilde{\tau}}{\partial\sigma} = \frac{\partial\tilde{\sigma}}{\partial\tau} \ , \tag{350}$$

leave the conformal gauge fixing conditions invariant, by inducing only a local rescaling of the induced metric, as may easily be checked. Note that these relations also imply that

$$\left[\partial_\tau^2 - \partial_\sigma^2\right]\tilde{\tau} = 0 \ , \ \left[\partial_\tau^2 - \partial_\sigma^2\right]\tilde{\sigma} = 0 \ , \tag{351}$$

namely once again the massless Klein-Gordon equations.

7.3 Light-cone gauge fixing

A complete gauge fixing of the system requires some further condition in addition to the conformal ones, $(\dot{x} \pm x')^2 = 0$. Since conformal transformations also obey the free massless Klein-Gordon equation on the world-sheet, as has just been established, a complete gauge fixing would be achieved by setting some linear combination of the spacetime coordinates x^μ equal to some combination of the world-sheet coordinates (τ, σ). However, since σ is free to be shifted by an arbitrary amount while x^μ is then invariant, only a combination involving τ may be envisaged.

In order to give a geometrical interpretation to such a gauge fixing, let us consider a specific constant spacetime vector n^μ. In terms of this vector, the condition that the combination $n_\mu x^\mu$ takes a constant value determines a specific hyperplane of dimension $(D-1)$ in spacetime, perpendicular to the direction of n^μ. This hyperplane should intersect the string world-sheet, an occurrence that may be associated to a specific value of τ as a function of the value for the constant $n_\mu x^\mu$. Hence, let us consider the additional gauge fixing condition

$$n_\mu x^\mu(\tau, \sigma) = 2\alpha' n_\mu P^\mu \, \tau \ , \tag{352}$$

where the coefficient in the r.h.s. multiplying τ is identified from the expression for the total energy-momentum P^μ. A further constant term $\sqrt{2\alpha'} n_\mu q^\mu$ could also be added to the r.h.s. of this condition, but may always be reabsorbed into a redefinition of the value $\tau = 0$ by a constant shift, which is a local world-sheet symmetry. That this additional condition indeed leads to a complete gauge fixing in combination with the conformal gauge fixing conditions is readily established as follows.

For a given value of τ, the condition (352) identifies a specific curve lying within the world-sheet as being the line of intersection of the world-sheet with the hyperplane defined by (352). Since this curve is associated to a constant value for τ, the curve is parametrized in σ in a certain manner. However, as the value for τ changes, this line of intersection also changes accordingly, specifying the parametrization in τ of the world-sheet. Given now the conformal gauge fixing conditions $(\dot{x}+x')^2=0$, the parametrization in σ for each of the lines of intersection is then also uniquely specified, thereby singling out from among the whole set of conformal reparametrizations $(\tilde{\tau},\tilde{\sigma})$ the unique world-sheet parametrization for which all three gauge fixing conditions are met.

Clearly, such a gauge fixing is no longer manifestly spacetime Poincaré invariant, given the role of the constant vector n^μ. Hence, only the little group, namely the subgroup of the Lorentz group leaving this constant vector invariant, is still a manifest symmetry of the formulation of the system. Among all possible choices for the vector n^μ, it proves convenient to work with a light-like one which, by an appropriate space rotation may always be taken to be

$$n^\mu=\frac{1}{\sqrt{2}}(1,0,\cdots,-1)\quad,\quad n^2=0\,. \tag{353}$$

In such a case, the little group is isomorphic to the euclidean group $\mathrm{E}(D-2)_n$, with as subgroup the set $\mathrm{SO}(D-2)_n$ of all rotations in the space directions perpendicular to the light-like vector n^μ. Properties of physical observables under this latter symmetry group are readily identified from the space indices

$$i=1,2,\cdots,D-2\,, \tag{354}$$

carried by diverse quantities. This manifest symmetry will suffice for our purposes.

Hence within this gauge fixing known as the light-cone gauge fixing of string theory, it proves useful to introduce the following notations for any two spacetime vectors u^μ and v^μ,

$$u^\pm=\frac{1}{\sqrt{2}}\left(u^0\pm u^{D-1}\right)\,,\ u^i\,,\ i=1,2,\cdots,D-2\,,\ u\cdot v=-u^+v^--u^-v^++u^iv^i\,, \tag{355}$$

so that the light-cone gauge fixing conditions now read

$$(\dot{x}\pm x')^2=0\quad,\quad x^+=2\alpha' P^+\,\tau\,. \tag{356}$$

When written out, these conditions imply that the actual physical degrees of freedom are the transverse string coordinates $x^i(\tau,\sigma)$ as well as the zero modes q^- and P^+, while all other degrees of freedom, namely $x^\pm(\tau,\sigma)$ (except for the previous two zero modes) are gauge degrees of freedom expressed in terms of the physical ones.

When considered in terms of the explicit solutions to the equations of motion, the set of physical modes is thus

$$q^i \ , \ P^i \ ; \ \alpha^i_{n\neq 0} \ , \ \bar{\alpha}^i_{n\neq 0} \ ; \ q^- \ , \ P^+ \ , \tag{357}$$

with the mode expansion

$$\begin{aligned} x^i(\tau,\sigma) = \sqrt{2\alpha'}\,[\ q^i + \left(\alpha^i_0 + \bar{\alpha}^i_0\right)\tau \\ + \tfrac{1}{2} i {\sum_n}' \tfrac{1}{n}\left(\alpha^i_n e^{-2in(\tau-\sigma)} + \bar{\alpha}^i_n e^{-2in(\tau+\sigma)}\right)\big] \ , \end{aligned} \tag{358}$$

and the relations

$$\alpha^i_0 = \frac{1}{2}\sqrt{2\alpha'}P^i = \bar{\alpha}^i_0 \ . \tag{359}$$

As to the other two longitudinal components $x^\pm(\tau,\sigma)$, when expressed in similar mode expansions, one has the relations

$$q^+ = 0\ , \ \alpha^+_n = \frac{1}{2}\sqrt{2\alpha'}P^+\delta_{n,0} = \bar{\alpha}^+_n\ ;\alpha^-_n = \frac{2}{\sqrt{2\alpha'}P^+}\,L^\perp_n\ , \ \bar{\alpha}^-_n = \frac{2}{\sqrt{2\alpha'}P^+}\,\bar{L}^\perp_n\ , \tag{360}$$

with the transverse Virasoro generators

$$L^\perp_n = \frac{1}{2}\sum_m \alpha^i_{n-m}\alpha^i_m \quad , \quad \bar{L}^\perp_n = \frac{1}{2}\sum_m \bar{\alpha}^i_{n-m}\bar{\alpha}^i_m \ , \tag{361}$$

q^- and P^+ being the only two other independent and physical degrees of freedom.

Note that the above expressions do indeed solve the Virasoro constraints $L_n = 0 = \bar{L}_n$ of the conformal gauge, thus demonstrating that gauge fixing has been completed. Furthermore in this form, it appears now obvious that the mass spectrum is indeed positive definite, since one readily determines, from the relation $M^2 = -P^2 = 2P^+P^- - P^iP^i$,

$$\frac{1}{2}\alpha' M^2 = N^\perp + \bar{N}^\perp \quad , \quad N^\perp = \bar{N}^\perp \ , \tag{362}$$

with of course

$$N^\perp = \sum_{n=1}^{\infty} \alpha^i_{-n}\alpha^i_n \quad , \quad \bar{N}^\perp = \sum_{n=1}^{\infty} \bar{\alpha}^i_{-n}\bar{\alpha}^i_n \ , \tag{363}$$

and in which the level matching condition is again the expression of the invariance of the closed string dynamics and spectrum under constant shifts in the σ coordinate, $\sigma \to \sigma + \sigma_0$.

It would now be possible to work out the Hamiltonian formulation of the system, first in the general setting in which all constraints are identified and classified in terms of their first- or second-class character, and then following either the conformal or the light-cone gauge fixings. The details of such an analysis are left as a useful exercise for the interested reader.[37,38] Let it suffice to say here that conjugate momenta are nothing but the $\alpha = 0$ components of the Noether energy-momentum current,

$$x^\mu(\tau,\sigma) \quad , \quad \pi_\mu(\tau,\sigma) = \frac{\partial \mathcal{L}}{\partial \dot{x}^\mu(\tau,\sigma)} = P^0_\mu(\tau,\sigma) \; , \tag{364}$$

while their canonical brackets

$$\{x^\mu(\tau,\sigma), \pi_\nu(\tau,\sigma')\} = \delta^\mu_\nu \, \delta(\sigma - \sigma') \; , \tag{365}$$

translate into the following brackets for the modes defining the solutions to the equations of motion in the conformal gauge

$$\{\sqrt{2\alpha'} q^\mu, P^\nu\} = \eta^{\mu\nu} \quad , \quad \{\alpha^\mu_n, \alpha^\nu_m\} = -in\eta^{\mu\nu}\delta_{n+m,0} = \{\bar{\alpha}^\mu_n, \bar{\alpha}^\nu_m\} \; . \tag{366}$$

Furthermore, the system of constraints reduces to the two first-class primary constraints

$$\phi_\pm = \frac{1}{2}\pi\alpha' \left[\pi^\mu \pm \frac{\partial_\sigma x^\mu}{2\pi\alpha'}\right]^2 \; , \tag{367}$$

which are the generators of small world-sheet reparemetrizations, while the first-class Hamiltonian density vanishes identically, $\mathcal{H} = 0$, thus implying that the total Hamiltonian of the system is given by

$$\int_0^\pi d\sigma \left[\lambda^+\phi_+ + \lambda^-\phi_-\right] \; , \tag{368}$$

$\lambda^\pm(\tau,\sigma)$ being the associated Lagrange multipliers. The conformal gauge corresponds to the choice $\lambda^+ = 1 = \lambda^-$, in which the Hamiltonian equations of motion become equivalent to those of free massless scalar fields $x^\mu(\tau,\sigma)$ on the world-sheet, hence leading back to the solutions given previously, as well as the mode brackets listed above. In terms of these quantities, in then also follows that the Virasoro generators obey the algebra,

$$\{L_n, L_m\} = -i(n-m)L_{n+m} \quad , \quad \{\bar{L}_n, \bar{L}_m\} = -i(n-m)\bar{L}_{n-m} \; , \tag{369}$$

which is indeed that of the conformal algebra in two dimensions. In particular, time translations are generated by the total Hamiltonian in the conformal gauge,

$$H = 2[L_0 + \bar{L}_0] \ , \tag{370}$$

while the generator for constant translations in σ is $L_0 - \bar{L}_0$, each of these quantities thus being in direct correspondence with the expressions for the mass spectrum $\alpha' M^2/2$ and the level matching conditions $N = \bar{N}$.

Finally, the light-gauge is obtained through Faddeev's reduced phase space approach by introducing two further gauge fixing conditions, namely $x^+ = 2\alpha' P^+ \tau$ and $\pi^+ = P^+/\pi$, and determining then the corresponding Dirac brackets. The above representation of the system, with in particular its physical mode degrees of freedom, is then readily recovered, as the interested reader may easily verify as a useful exercise of his understanding of constrained dynamics.

7.4 Dirac's conformal quantization

Fundamental operator algebra

Given the Hamiltonian formulation of the system within the conformal gauge, its Dirac quantization is defined by the set of basic commutation relations for its mode degrees of freedom,

$$[\sqrt{2\alpha'} q^\mu, P^\nu] = i\eta^{\mu\nu} \ , \ q^{\mu\dagger} = q^\mu \ , \ P^{\mu\dagger} = P^\mu \ , \tag{371}$$

$$[\alpha^\mu_n, \alpha^\nu_m] = n\eta^{\mu\nu}\delta_{n+m,0} = [\bar{\alpha}^\mu_n, \bar{\alpha}^\nu_m] \ , \ \alpha^{\mu\dagger}_n = \alpha^\mu_{-n} \ , \ \bar{\alpha}^{\mu\dagger}_n = \bar{\alpha}^\mu_{-n} \ . \tag{372}$$

The zero mode algebra is nothing but the Heisenberg algebra, for which, given a particle interpretation to be associated to the string spectrum, one chooses a momentum eigenstate basis representation, hence

$$P^\mu \, |p^\mu >= p^\mu \, |p^\mu > \ , \ < p|p' >= \delta^{(D)}(p - p') \ . \tag{373}$$

The nonzero mode algebra is the tensor product over all positive integers $n \geq 1$ of Fock space algebras, for which we shall use the Fock space representation with vacuum $|\Omega>$ annihilated by all operators α^μ_n and $\bar{\alpha}^\mu_n$,

$$\alpha^\mu_n \, |\Omega >= 0 \ , \ \bar{\alpha}^\mu_n \, |\Omega >= 0 \ , \tag{374}$$

the negative moded operators α^μ_{-n} and $\bar{\alpha}^\mu_{-n}$, $n \geq 1$, being the creation operators. Hence, the basis of all quantum states is spanned by the Fock vacua $|\Omega; p>$ as well as all their Fock excitations. In particular, note that because of the appearance of the Minkowski metric in the nonzero mode Fock space algebras, the space of quantum states includes negative norm states, such as $\alpha^{\mu}_{-n}|\Omega;p>$, $n \geq 1$, whose norm is proportional to $\eta^{\mu\nu}$ up to the function $\delta^{(D)}(p-p')$. A consistent causal and quantum unitary theory requires however that no negative norm state contributes to physical amplitudes. One may only hope that this issue of negative norm states and physical consistency is to be resolved through the existence of the gauge symmetries in world-sheet reparametrizations.

Physical states

At the classical level, physical configurations in the conformal gauge are identified through the Virasoro constraints. At the quantum level, these quantities being composite, their actual definition requires a choice of operator ordering for which one chooses of course the usual normal ordering for Fock space creation and annihilation operators, inclusive now of the zero mode operators with the q^{μ} operators always brought to the left of all P^{μ} operators. Given this choice of normal ordering, the Virasoro operators read

$$L_n = \frac{1}{2}\sum_m :\alpha^{\mu}_{n-m}\alpha_{m\mu}: \quad , \quad \bar{L}_n = \frac{1}{2}\sum_m :\bar{\alpha}^{\mu}_{n-m}\bar{\alpha}_{m\mu}: \quad . \tag{375}$$

Clearly, normal ordering only affects the zero mode Virasoro operators L_0 and $\bar{L}_0$, hence the physical spectrum of the theory.

In order to define physical states, it turns out that requiring all the Virasoro operators to vanish is too strong a restriction, whereas it suffices to only require that all positive moded operators actually vanish. Indeed, such a condition is tantamount to requiring the Virasoro constraints in a weak sense, namely that the matrix elements of all Virasoro operators vanish for physical external states, which is also sufficient from the quantum Virasoro algebra point of view to be discussed hereafter. Hence, gauge invariant physical states of the closed bosonic string are defined by the following set of operator constraints

$$L_n\,|\psi_{\rm phys}>=0 \quad , \quad \bar{L}_n\,|\psi_{\rm phys}>=0 \quad , \quad n \geq 1 \; , \tag{376}$$

as well as

$$\left[L_0+\bar{L}_0-2a\right]\,|\psi_{\rm phys}>=0 \quad , \quad \left[L_0-\bar{L}_0\right]\,|\psi_{\rm phys}>=0 \; , \tag{377}$$

where the constant a stands for the unknown normal ordering constant that arises for the Virasoro zero modes, and which must be identical for the left- and right-moving sectors of the theory.. Note that these two zero mode conditions are equivalently expressed as

$$\frac{1}{2}\alpha' M^2 = N + \bar{N} - 2a \ , \quad N = \bar{N} \ , \tag{378}$$

with the excitation level number operators

$$N = \sum_{n=1}^{\infty} \alpha^{\mu}_{-n}\alpha_{n\mu} \ , \quad \bar{N} = \sum_{n=1}^{\infty} \bar{\alpha}^{\mu}_{-n}\bar{\alpha}_{n\mu} \ . \tag{379}$$

Thus in particular, if the normal ordering subtraction constant a happens to be strictly positive, the physical spectrum will include tachyonic states, beginning with the physical Fock vacuum.

Poincaré and conformal algebras

In order to allow a consistent interpretation of quantum string excitations in terms of relativistic quantum particle states, it is necessary that the Poincaré algebra be realized on the space of states, and in particular on the subspace of physical states. Thus, one needs to explicitly check whether the Poincaré algebra is recovered for the quantum Noether charges P^μ and $M^{\mu\nu}$, a fact which is readily established. Furthermore, since it is straightforward to verify that the Poincaré generators commute with all Virasoro operators, the Poincaré algebra is also obtained for the subspace of gauge invariant physical states. Hence, at each excitation level $N = \bar{N}$, all physical quantum states span specific irreducible Poincaré representations of definite mass M^2 and "spin" values, the latter being characterized in terms of the spatial rotation subgroup of the corresponding little group which is SO$(D-2)$ for a massless particle and SO$(D-1)$ for a massive particle.

As far as the conformal algebra is concerned, it follows from normal ordering in the Virasoro zero modes that the conformal algebra acquires a conformal anomaly or central extension, hence leading to the Virasoro algebra with central charge $c = D$,

$$\begin{aligned} [L_n, L_m] &= (n-m)L_{n+m} + \tfrac{1}{12}D(n^3 - n)\delta_{n+m,0} \ , \\ [\bar{L}_n, \bar{L}_m] &= (n-m)\bar{L}_{n+m} + \tfrac{1}{12}D(n^3 - n)\delta_{n+m,0} \ . \end{aligned} \tag{380}$$

Note that because of these algebraic relations, when solving for the physical constraints $L_{n\geq 1} = 0 = \bar{L}_{n\geq 1}$, it suffices to solve only for the modes $n = 1, 2$, since all other modes may then be recovered through these commutation relations.

The no-ghost theorem

Finally, we have to address the issue of the possibility of negative norm physical states, namely the fact that among all quantum states which obey the physical Virasoro constraints, there may remain some states of strictly negative norm, spelling disaster for the physical consistency of these theories. The no-ghost theorem[2,3,37] establishes that, at tree level, the absence of any physical state of negative norm requires that

- $a \leq 1$;
- if $a = 1$: $D \leq 26$;
- if $a < 1$: $D < 26$.

Even though establishing this general result is not straightforward, one may explicitly check that such conditions are indeed necessary by solving the physical state conditions for the first few excitations levels, which in itself is also a worthwhile exercise.

Furthermore, when one then considers one-loop corrections to quantum string amplitudes, one quickly comes to realize that quantum unitarity also requires[2,3] the exact value $D = 26$, hence also $a = 1$ given the above statement of the no-ghost theorem. More specifically, when these two conditions $D = 26$ and $a = 1$ are met, any physical state is given by the sum of a strictly positive norm physical states, as well as a zero norm physical state, which itself then decouples from any physical amplitude either as an external or as an intermediate state. Consequently, normal ordering of operators and quantum unitarity of the manifestly Poincaré convariant conformal gauge quantization of bosonic strings requires the critical spacetime dimension $D = 26$ for a physically consistent interpretation of string excitations as being relativistic quantum particle states of definite mass and spin.

Note that the charaterization of physical states in terms of components of strictly positive and vanishing norm is also that which arises within the Gupta-Bleuler quantization of quantum electrodynamics. In that case, Gauss' law (the first-class constraint generating the local internal U(1) gauge symmetry) is imposed for positive moded components of the gauge field, with the consequence that physical quantum photon states are given by the superposition of a strictly positive norm component corresponding to a transverse photon polarization state, and a zero norm component corresponding to a longitudinal photon polarization. The above characterization in the string case is thus an extension of this result to higher spin massless as well as massive states.

When both $D = 26$ and $a = 1$, one finds that the physical ground state is at zero excitation level, $N = 0 = \bar{N}$, and corresponds to the Fock vacuum $|\Omega; p>$ such that $\alpha' m^2/2 = -2$, hence a tachyonic scalar particle. At the first excitation level $N = 1 = \bar{N}$, one has a collection of strictly positive norm massless physical states, one such state corresponding to a massless graviton with 299 independent physical polarization components, another to a massless antisymmetric tensor state of 276 components, and finally a scalar known as the dilaton with a single polarization state, thus leading to a total of 576 positive norm physical states. Likewise, it is possible to identify all such positive norm physical states at higher excitation levels. Note that all physical states lie along so-called Regge trajectories, namely linear trajectories relating the $\alpha' m^2$ and spin values of these states, the string tension α' indeed playing the role of the Regge slope, and the subtraction constant $a = 1$ that of the intercept for the lowest lying Regge trajectory.

It is quite remarkable that the spacetime physical spectrum of the quantized closed bosonic string includes a massless spin 2 state, the quanta of a metric field usually associated to gravitational interactions in a field theory setting. Indeed, when considering the low energy effective interactions (in comparison to the energy scale set by the string tension α') of these states, their effective action is precisely that of the low energy graviton modes of general relativity expanded around Minkowski spacetime.[2,3] It is thus perfectly consistent to identify these closed string states with the gravitons of the gravitational interaction. Had we quantized the open bosonic string, in a likewise manner we would have uncovered massless spin 1 states whose low energy effective interactions are those of massless Yang-Mills gauge bosons! In other words, and this is indeed a generic feature of all string theories, it appears that the world-sheet symmetries, in the present instance those under world-sheet reparametrizations, translate at the level of the spacetime spectrum into the usual local gauge symmetries and their bosonic carriers of interactions which have proved to provide the basic physics principle for a quantum field theory description of all fundamental interactions and particles. Undoubtedly, there is some profound lesson to be gathered from such a nontrivial result. Many more such fascinating convergences of basic facts have been uncovered within string theories, thus suggesting that this framework may well have brought us to the brink of the long sought-for formalism for a fundamental unification of all quantum interactions and matter. Only time will tell, through the work of the quantum geometers of the XXIst century.

7.5 Light-cone quantization

In the light-cone gauge, the quantized system is defined by the commutation relations

$$\left[\sqrt{2\alpha'}q^-, P^+\right] = -i \ , \ \left[\sqrt{2\alpha'}q^i, P^j\right] = i\delta^{ij} \ ,$$
$$\left[\alpha^i_n, \alpha^j_m\right] = n\delta^{ij}\ \delta_{n+m,0} = \left[\bar{\alpha}^i_n, \bar{\alpha}^j_m\right] \ , \tag{381}$$

including the by now usual hermiticity properties of these operators. Consequently, the Fock space representation of this algebra is based on Fock vacua $|\Omega; p^i, p^+ >$ which are normalized eigenstates of the momentum operators P^+ and P^i and which are annihilated by the positive moded operators $\alpha^i_{n\geq 1}$ and $\bar{\alpha}^i_{n\geq 1}$, the action of the creation operators α^i_{-n} and $\bar{\alpha}^i_{-n}$, $n \geq 1$, spanning the remainder of the Fock space basis.

Given this algebra, it follows that all these quantum states are physical, are of strictly positive norm, and that they correspond solely to transverse string excitation modes. Finally, the mass spectrum of these states is given by

$$\frac{1}{2}\alpha' M^2 = N^\perp + \bar{N}^\perp - 2a \quad , \quad N^\perp = \bar{N}^\perp \ , \tag{382}$$

with the excitation level number operators $N^\perp$ and $\bar{N}^\perp$ defined as previously in terms of the transverse creation and annihilation operators α^i_n and $\bar{\alpha}^i_n$ only. Here, a stands again for the required normal ordering subtraction constant that arises for the transverse Virasoro zero modes $L^\perp_0$ and $\bar{L}^\perp_0$. In the same manner as before, normal ordering is defined by bringing all position and creation operators, q^-, q^i, α^i_{-n} and $\bar{\alpha}^i_{-n}$, to the left of all momentum and annihilation ones, P^+, P^i and α^i_n, $\bar{\alpha}^i_n$, $n \geq 1$.

Clearly, this characterization of physical states coincides with that reached within the above conformal gauge quantization for those physical states of strictly positive norm. However, in contradistinction with the latter quantization, the present one no longer possesses a manifest Poincaré covariant formulation, and one is forced to check whether the Poincaré algebra is realized nonetheless on the space of quantum states, albeit in a nonlinear fashion. Since only the little group $E(D-2)_n$ of the light-like vector n^μ used to define the light-cone gauge is still a manifest spacetime symmetry of this quantization, one needs to check whether the commutation relations which involve the operators M^{-i} and $M^{0(D-1)}$ take the values required by the Poincaré algebra. In turns out that it is only for the commutators $[M^{-i}, M^{-j}] = 0$ that this requirement is not necessarily met, leading to the critical values[2,3,37]

$$D-2=24 \quad , \quad a=1 \; , \tag{383}$$

again in agreement with the critical values required in the conformal gauge for a consistent manifestly Poincaré covariant quantization of the bosonic string.

Even though this result is by no means trivial to establish, once again it is possible to show that it is necessary by working out the first few excitation levels of the system. Consider thus the states at $N^\perp = 1 = \bar{N}^\perp$, namely

$$\alpha^i_{-1}\,\bar{\alpha}^j_{-1}\,|\Omega; p^+, p^i > \; , \tag{384}$$

whose mass is such that $\alpha' m^2/2 = 2(1-a)$. There are thus $(D-2)^2$ such states. However, spacetime covariant properties of this collection of states are obtained only if they are massless, since otherwise they should belong to some spin representation of SO$(D-1)$ rather than SO$(D-2)$ which clearly is impossible. Consequently, one must have $a=1$, implying that the physical ground state is tachyonic. Furthermore, an heuristic ζ-function evaluation of the infinite series defining the normal ordering constant a then also leads to the value $D-2=24$.

Given these critical values, it then follows that the above states at level $N^\perp = 1 = \bar{N}^\perp$ correspond to a massless spin 2 graviton with 299 polarization states, an antisymmetric tensor with 276 states and a scalar with a single polarization state, hence a total of $576 = 24^2$ physical states, in complete agreement with the count in the conformal gauge.

As a matter of fact, it is possible to introduce the partition function of the system which counts the number of positive norm physical states at each excitation level, and study further fascinating properties of this simple string theory. However, we shall refrain from presenting these considerations here.

As a conclusion concerning the quantization of the closed bosonic string, let us point out that these conformal and light-cone gauge fixing procedures are affected by Gribov problems,[37,38] which, however, may be circumvented in the actual construction of physical amplitudes. Furthermore, when including then the proper Faddeev-Popov or BFV Hamiltonian ghost systems following from these gauge fixings, one may check that the total conformal algebra is recovered at the quantum level provided once again the critical conditions $D=26$ and $a=$ are imposed.[2,3,37] Indeed, it is reparametrization and conformal invariance which guarantees the quantum consistency of the system, so that the conformal algebra should better not be affected by an anomalous central extension contribution when properly including all relevant degrees of freedom. In that respect, it would certainly be quite interesting to apply the physical projector approach free of any gauge fixing procedure to, first, the bosonic string, and then to all its supersymmetric cousins in all dimensions ranging from the critical $D = 10$ one down to $D = 4$ or even $D = 2$.

8 Toroidal Compactification of the Closed Bosonic String

As the above discussion has established, the closed bosonic spectrum includes a massless spin 2 graviton. In fact, if one ignores the problem raised by the presence of a tachyonic state which may be projected out for theories of physical relevance, it has been established[2,3] that the perturbative expansion of string theory, when properly renormalized, is indeed a finite one. In other words, string theory defines a perturbative finite quantum theory for quantum gravity coupled to other interactions and matter degrees of freedom, including dynamical geometric degrees of freedom. This rather remarkable conclusion has been one of the strongest motivations to pursue this framework as a possible formulation for the problem of the ultimate unification.

Given this fact and the suggestion made previously that topology is also called to play a fundamental role in a physically relevant and consistent formulation of quantum geometry, to conclude these notes let us discuss the simplest example which shows that our usual geometrical concepts have to be extended when considered from the string theory point of view. Some embodiement of the sought-for principles of quantum geometry must already be lying hidden behind the properties of these theories. For this purpose, this section briefly considers how the previous discussion of the closed bosonic string is modified even when only one of the spatial dimensions is compactified into a circle geometry.

8.1 Toroidal compactification in field theory

To begin with, let us consider a simple free massless scalar field $\phi(x^\mu, y)$ evolving over a spacetime manifold of which one of the spatial dimensions is compactified into a circle of radius R,

$$y \equiv y + 2\pi R \ , \tag{385}$$

while the remaining spacetime dimensions define a Minkowski spacetime of some given dimension. Since the massless Klein-Gordon equation reads

$$\left[\partial_x^2 + \partial_y^2\right] \phi(x^\mu, y) = 0 \ , \tag{386}$$

a Fourier mode expansion of the field over the compactified direction,

$$\phi(x^\mu, y) = \sum_n \phi_n(x^\mu)\, e^{i\frac{n}{R}y} \ , \tag{387}$$

implies that from the lower dimensional point of view the modes $\phi_n(x^\mu)$obey the massive Klein-Gordon equation

$$\left[\partial_x^2 - m_n^2\right] \phi_n(x^\mu) = 0 \quad , \quad m_n = \frac{|n|}{R} \ . \tag{388}$$

Consequently, upon compactification of a field theory onto a compactified space, in the lower dimensional description there appear infinite so-called Kaluza-Klein (KK) towers of massive states whose mass values are determined by the momentum values of the field along the compactified directions. In the decompactification limit $R \to \infty$, these massive KK towers coalesce back into the massless modes of the initial massless field in the complete spacetime which recovers its Minkowski geometry. *A contrario*, in the compactification limit $R \to 0$, the KK towers become infinitely massive, and thus decouple altogether from the lower dimension dynamics, since the extra dimension then collapses to a point in that limit.

These features of dimensional compactification are generic to any field theory. For instance given some vector field $A_M = (A_\mu, \phi)$ in the higher dimensional theory, the compactified theory will include vector, A_μ, and scalar, ϕ, components each of which possesses a similar KK mode expansion in terms of massless and massive states associated to the momentum values of the original fields along the compactified directions. The same clearly applies to a two-index tensor A_{MN}, leading to KK towers of tensor, vector and scalar states in the lower dimensional theory, $A_{\mu\nu}$, $A_\mu^{(1)}$, $A_\mu^{(2)}$ and ϕ.

Furthermore, within the context of a pure general relativity theory in the higher dimensional spacetime, which is thus invariant under arbitrary coordinate transformations in that space, whenever the compactified space possesses some continuous symmetry, as for example the U(1) symmetry of a circle or the SO(n) symmetry of a sphere, the KK reduction implies the appearance of massless gauge bosons associated to the corresponding local internal gauge symmetries. Indeed, one is then free to perform at each lower dimensional spacetime point a different coordinate redefinition within the compactified space, without changing the physics of the system. In other words, there does appear an internal local Yang-Mills gauge symmetry, the internal space being nothing but the compactified directions of the initial theory. This is the basis of the original Kaluza-Klein programme for the fundamental unification of gravity with all other Yang-Mills interactions within a purely geometric and field theory framework. Given that consistent string theories need to be formulated in higher dimensional spacetimes, this programme recovers suddenly new relevance and urgency within this new formalism which supersedes that of ordinary field theory, inclusive of quantum dynamics.

8.2 Toroidal compactification in string theory

Given the large critical spacetime dimensions for consistent quantum string theories, spatial compactification is a natural approach towards the four-dimensional physical world. Let us thus reconsider our previous discussion of the closed bosonic string, but this time with one the space components, say $\mu = 25$, compactified into a circle of radius R.[2,3] Correspondingly, in the conformal or light-cone gauge, one has the mode expansion

$$x^{25}(\tau,\sigma) = \sqrt{2\alpha'}\,[\; q^{25} + \alpha_0^{25}(\tau-\sigma) + \bar{\alpha}_0^{25}(\tau+\sigma) \\ +\tfrac{1}{2}i{\sum_n}'\tfrac{1}{n}\left(\alpha_n^{25}e^{-2in(\tau-\sigma)} + \bar{\alpha}_n^{25}e^{-2in(\tau+\sigma)}\right)] \;. \tag{389}$$

However, given the possibility of winding configurations

$$x^{25}(\tau,\sigma+\pi) = x^{25}(\tau,\sigma) + 2\pi R m \;, \tag{390}$$

of integer winding number m around the compactified circle, the zero modes α_0^{25} and $\bar{\alpha}_0^{25}$ need no longer be equal, but they actually differ by a multiple of the winding number. Given that the quantum momentum operator P^{25} is then also quantized as n/R, n being the momentum quantum, the zero modes take the values

$$\alpha_0^{25} = \frac{1}{2}\sqrt{2\alpha'}\left[\frac{n}{R} - \frac{R}{\alpha'}m\right] \;,\; \bar{\alpha}_0^{25} = \frac{1}{2}\sqrt{2\alpha'}\left[\frac{n}{R} + \frac{R}{\alpha'}m\right] \;. \tag{391}$$

The mass spectrum and level matching conditions are modified accordingly,

$$\frac{1}{2}\alpha' M^2 = N+\bar{N}+\frac{1}{2}\left[\left(\frac{\sqrt{\alpha'}}{R}n\right)^2 + \left(\frac{R}{\sqrt{\alpha'}}m\right)^2\right] - 2 \;,\; N-\bar{N} = nm \;, \tag{392}$$

while the other Virasoro conditions defining quantum physical states remain unaffected since only the zero modes are modified by the winding state contributions.

In the $(n=0, m=0)$ sector, the lowest lying physical state is the tachyonic ground state $|\Omega; p>$ at level $N = 0 = \bar{N}$. At level $N = 1 = \bar{N}$, one finds again the massless symmetric graviton, antisymmetric tensor and scalar for the strictly positive norm physical states obtained from

$$\alpha^\mu_{-1}\bar{\alpha}^\nu_{-1}\,|\Omega; p> \;, \tag{393}$$

but new massless vector and scalar states also appear, namely those related to

$$\alpha^\mu_{-1}\bar{\alpha}^{25}_{-1}\,|\Omega;p> \;,\; \alpha^{25}_{-1}\bar{\alpha}^{\mu}_{-1}\,|\Omega;p> \;,\; \alpha^{25}_{-1}\bar{\alpha}^{25}_{-1}\,|\Omega;p> \;, \tag{394}$$

which are consequence of the circle compactification (as is most straightforwardly established in the light-cone gauge). These are the massless KK states that follow the compactification of the original symmetric, antisymmetric and dilaton degrees of freedom, in exactly the same manner as discussed above in the case of the compactification of such fields from higher dimensions. The fact that these states have to be massless is consequence of the U(1) symmetry of the circle compactification, which translates into the U(1)×U(1) local Yang-Mills gauge invariance for the compactified theory, since the higher dimensional string theory includes a gravitational sector. This conclusion is in full accord with the Kaluza-Klein programme briefly outlined above.

Besides these states, there also exist the towers of KK states with $n \neq 0$, as well as the sector of winding states $m \neq 0$, the latter being a feature totally specific to the compactification of closed string theories, and totally absent from the field theory discussion in which point-like rather than string-like objects are being described in a quantized formulation. In particular, the contribution of winding states to the mass spectrum is a measure of the energy required to stretch and wind a closed string around the compactified dimension.

In fact, this is not the only difference with the field theory case. In the decompactification limit $R \to \infty$, the towers of KK states do indeed coalesce back into the continuous spectrum of momentum eigenstates for the compactified direction. However at the same time the winding states $m \neq 0$ become infinitely massive and decouple from the theory. *A contrario*, in the compactification limit $R \to 0$, the towers of KK states become infinitely massive as they do in the field theory case and thus decouple, but this time the winding states $m \neq 0$ coalesce into a continuum of states with a vanishing contribution to the mass spectrum. In other words, even when the compactified dimension has degenerated into a single point, there is still a trace of that extra dimension in the spectrum of the compactified theory. The notion of the dimension of spacetime appears to be no longer an absolute geometrical nor topological concept within the context of string theory.

This interchange of the decompactification and compactification limits is in fact valid even for whatever finite value of the circle radius R. Indeed, given the above expressions for the mass spectrum, it is clear that this spectrum, and in fact the whole of the string dynamics, remains totally invariant under the transformation

$$n \leftrightarrow m \quad , \quad \frac{\sqrt{\alpha'}}{R} \leftrightarrow \frac{R}{\sqrt{\alpha'}} \, , \tag{395}$$

in which momenta and winding states are exchange as well as the values of the radius with the value α'/R. This symmetry is known as T-duality and again is totally specific to string theories.[39] In fact, this is quite a fascinating symmetry, since it exchanges small distance with large distance physics, and calls into question our usual concepts of local geometry and topology. Note that this a symmetry of the quantized theory, and should thus be one of the expected manifestations of a quantum geometry of spacetime, with its accompanying quantum gravitational interactions as fluctuations in the quantum geometry. If not the topology, the compactified geometry is modified under T-duality without any consequence whatsoever for the quantum dynamics of string theory.

The existence of T-duality also suggests that within quantum geometry there ought to exist some smallest distance scale beyond which physical process may no longer be probed, being equivalent then to physical processes at the inverse distance scale. This smallest distance scale is thus set by the self-dual point under T-duality, associated to the compactification radius

$$R = \sqrt{\alpha'} \ . \tag{396}$$

In fact at the self-dual point, the U(1)×U(1) gauge symmetry of the compactified theory for a generic value of R is enhanced to a SU(2)×SU(2) Yang-Mills symmetry, whose rank is still that of the generic U(1)×U(1) Yang-Mills symmetry. For instance at the massless level, the self-dual value implies the following extra states

n	m	N	$\bar{N}$
1	1	1	0
1	−1	0	1
−1	1	0	1
−1	−1	1	0
2	0	0	0
−2	0	0	0
0	2	0	0
0	−2	0	0

(397)

The first four collections of states include four massless vector ones, which together with the two massless vectors of the generic case, combine into the six massless gauge bosons of the SU(2)×SU(2) Yang-Mills symmetry in 25-dimensional Minkowski spacetime. Likewise, the first four collections of states include also four massless scalars as do the last four collections, leading to a total of nine massless scalars when the generic massless scalar is also accounted for. These scalar states fit into the $(3,3)$ Higgs representations under SU(2)×SU(2). In other words, at $R = \sqrt{\alpha'}$, the system possesses the

SU(2)×SU(2) Yang-Mills symmetry which gets hidden by spontaneous Higgs symmetry breaking whenever the compactification radius R takes a different value.[2,3]

Hence, this simplest example of spatial compactification of string theory already points to quite fascinating quantum geometric properties realized within the realm of quantized string theory, which presumably are nothing but some facets of what a formulation and understanding of quantum geometry has to offer with regards to the hidden secrets for the physics of quantum gravity unified with all other quantum interactions and particles.

9 Conclusions

The principle aim of these notes has been to provide a brief outline, restricted to bosonic degrees of freedom only, of the relativistic and quantum concepts that are at the basis of our present understanding of all fundamental quantum interactions and elementary particles. The general considerations that have led during the XX$^{\rm th}$ century to the identification of relativistic quantum Yang-Mills gauge field theories as the appropriate framework for a consistent causal and quantum unitary description of relativistic quantum point-particles and their interactions have been recalled. The same convergence of ideas centered onto the fundamental concept of the local gauge symmetry principle applies to the gravitational interaction, which, when described within general relativity and its extensions all based on the dynamics of the geometry of spacetime, has been successful so far only at the classical level, while a full-fledged theory for quantum gravity is still eluding us. It appears that the physicist of the XXI$^{\rm st}$ century has arrived at the cross-roads of the three fundamental paths that have guided him during the previous one, and which may be characterized in terms of the three fundamental constants c, $\hbar$ and G_N. It seems that in spite of the amazing successes of the marriage of c with $\hbar$, it is close to impossible to force these sets of ideas to happily live within a *ménage à trois*. Some new paradigm of geometrical and topological concepts is most probably called for within the realm of the quantum gravitational interaction coupled to all other quantum interactions and particles.

As another but complementary aim of these notes, the general issues surrounding the quantization of constrained systems, which include all possible gauge invariant theories based on a field theory formulation, have been described, providing the basic tools necessary for such a study in general. In particular, having shown that the potential difficulties which follow from gauge fixing procedures for such theories are often unavoidable, an alternative and recent approach based on a physical projector[9] onto the gauge invariant quantum configurations of such systems and free of the necessity of gauge fixing, has been advocated as a powerful new tool with which to address these difficult issues, especially with regards to nonperturbative aspects of strongly interacting Yang-Mills theories.

Yang-Mills, and more generally local gauge invariant theories have also shown that topological features, either of spacetime or of the field configuration space, do play a fundamental role in the proper understanding of such interactions. With the discovery of topological quantum field theories,[5,6,35] void of any genuine dynamics but not of any quantum physics nonetheless, it is conceivable that pure quantum gravity could be the physics of quantum topology rather than of spacetime geometry, and that it is by coupling quantum topology to matter and interactions that the quantum geometric properties of spacetime should arise, local relativistic quantum field theories with gauge invariances being their appropriate low energy effective description.

As one illustration among possibly many others that have not been discussed at the Workshop, some of these issues have briefly been touched on within the context of bosonic string theory. Specific fascinating new features having to do with the gravitational sector of such systems and its interplay with the geometry and topology of spacetime have been described in the simplest terms available. Many more such issues have arising within that context, such as for example the possible noncommutative character of spacetime itself within string theory.[40]

It is equipped with this understanding of the world of the fundamental quantum interactions and particles, and the role played by topology within the relativistic gauge invariant quantum field theoretic framework describing this world today, that the physicist of the XXIst century in quest of the ultimate unification is to set out into the unchartered territory towards a truly genuine formulation and understanding of what quantum geometry will turn out to be, the final unification of the relativistic quantum and the relativistic continuum, the completed symphony of the three constants c, $\hbar$ and G_N which have guided us already through the three fundamental conceptual revolutions of XXth century physics.

Acknowledgements

Prof. John R. Klauder is gratefully acknowledged for his many relevant comments concerning the contents of these lectures, and for his supportive interest over the years in applications of the physical projector. Thanks also go out to Prof. M.N. Hounkonnou for his insistence to give these lectures, and to all participants to the Workshop for their active participation, questions and discussions throughout the presentation of this material.

It is hoped that this write-up will entice them to raise ever more inquisitive issues, embark onto their own adventures into the quantum geometer's world of XXIst century physics, and contribute with the mathematics and physics world community to the completion of this unfinished symphony by uncovering with a definite african beat some of its music scores, the full colours of its harmonies being Nature's own.

References

1. S. Weinberg, *What is Quantum Field Theory, and What Did We Think It Is?*, Talk given at the Conference on the Historical Examination and Philosophical Reflections on the Foundations of Quantum Field Theory, Boston (Massachusetts, USA) March 1-3, 1996, `hep-th/9702027`.
2. For a review and references to the original literature, see, J. Polchinski, *String Theory* (Cambridge University Press, Cambridge, 2000), 2 Volumes.
3. For a review and references to the original literature, see, M.B. Green, J.H. Schwarz and E. Witten, *Superstring Theory* (Cambridge University Press, Cambridge, 1987), 2 Volumes.
4. For a review and references to the original literature, see for example, C. Rovelli, *Living Rev. Rel.*, 1 (1998), `gr-qc/9710008`.
5. E. Witten, *Comm. Math. Phys.* **117** , 353 (1988);
E. Witten, *Phys. Lett.* **B206**, 601 (1988);
E. Witten, *Comm. Math. Phys.* **118**, 411 (1988);
J.M.F. Labastida, M. Pernici and E. Witten, *Nucl. Phys.* **B310**, 611 (1988).
6. E. Witten, *Comm. Math. Phys.* **121**, 351 (1989).
7. For a review and references to the original literature, see for example, D. Birmingham, M. Blau, M. Rakowski and G. Thompson, *Physics Reports* **209**, 129 (1991).
8. P.A.M. Dirac, *Lectures on Quantum Mechanics* (Belfer Graduate School of Science, Yeshiva University, New York, 1964).
9. J.R. Klauder, *Ann. Phys.* **254**, 419 (1997);
J.R. Klauder, *Nucl. Phys.* **B547**, 397 (1999);
J.R. Klauder, *Quantization of Constrained Systems, Lect. Notes Phys.* **572**, 143 (2001), `hep-th/0003297`.
10. For a detailed discussion and references to the original literature, see, J. Govaerts, *Hamiltonian Quantisation and Constrained Dynamics* (Leuven University Press, Leuven, 1991).

11. J.R. Klauder, *Beyond Conventional Quantization* (Cambridge University Press, Cambridge, 2000).
12. J. Govaerts and V.M. Villanueva, *Int. J. Mod. Phys.* **A15**, 4903 (2000).
13. J. Govaerts, Proc. First International Workshop on Contemporary Problems in Mathematical Physics, 31 October-5 November 1999, eds. J. Govaerts, M.N. Hounkonnou and W.A. Lester, Jr. (World Scientific, Singapore, 2000), pp. 244-259.
14. J.R. Klauder and B.-S. Skagerstam, *Coherent States: Applications in Physics and Mathematical Physics* (World Scientific, Singapore, 1985).
15. R.P. Feynman and A.R. Hibbs, *Quantum Mechanics and Path Integrals* (McGraw-Hill Book Company, New York, 1965).
16. For a discussion and references, see for example,
M. Florig and S.J. Summers, *Proc. London Math. Soc.* **80**, 451 (2000), `math-ph/0006011`;
G. Sardanashvily, *Nonequivalent Representations of Nuclear Algebra of Canonical Commutation Relations. Quantum Fields*, `hep-th/0202038`;
A. Corichi, J. Cortez and H. Quevedo, *On the Relation between Fock and Schrödinger Representations for a Scalar Field*, `hep-th/0202070`;
A. Iorio, G. Lambiase and G. Vitiello, *Hopf Algebra, Thermodynamics and Entanglement in Quantum Field Theory*, `quant-ph/0207040`.
17. S. Weinberg, *The Quantum Theory of Fields* (Cambridge University Press, Cambridge, 1995), 3 Volumes.
18. M.E. Peskin and D.V. Schroeder, *An Introduction to Quantum Field Theory* (Perseus Books Publishing, Cambridge, Massachusetts, 1995).
19. C. Itzykson and J.-B. Zuber, *Quantum Field Theory* (McGraw-Hill Book Company, New York, 1980).
20. P. Ramond, *Field Theory: a Modern Primer* (Benjamin-Cummings Publishing, Reading, Massachusetts, 1981).
21. M.J.G. Veltman, *Nucl. Phys.* **B7**, 637 (1968);
G. 't Hooft, *Nucl. Phys.* **B35**, 167 (1971);
G. 't Hooft and M.J.G. Veltman, *Nucl. Phys.* **B44**, 189 (1972);
G. 't Hooft and M.J.G. Veltman, *Nucl. Phys.* **B50**, 318 (1972).
22. L. Faddeev and R. Jackiw, *Phys. Rev. Lett.* **60**, 1692 (1988);
J. Govaerts, *Int. J. Mod. Phys.* **A5**, 3625 (1990).
23. J. Govaerts and J.R. Klauder, *Ann. Phys.* **274**, 251 (1999).
24. L.D. Faddeev, *Theor. Math. Phys.* **1**, 1 (1970).
25. V.N. Gribov, *Nucl. Phys.* **B139**, 1 (1978);
I.M. Singer, *Comm. Math. Phys.* **60**, 7 (1978).
26. K. Fujikawa, *Prog. Theor. Phys.* **61**, 627 (1979);
P. Hirschfeld, *Nucl. Phys.* **B157**, 37 (1979);

M.B. Halpern and J. Koplok, *Nucl. Phys.* **B132**, 239 (1978).
27. J. Govaerts, *Int. J. Mod. Phys.* **A4**, 173 (1989);
J. Govaerts, *Int. J. Mod. Phys.* **A4**, 4487 (1989);
J. Govaerts and W. Troost, *Class. Quantum Grav.* **8**, 1723 (1991).
28. J. Govaerts, *The Cosmological Constant of One-Dimensional Matter Coupled Quantum Gravity is Quantized*, preprint STIAS-02-002, `hep-th/0202134`.
29. E.S. Fradkin and G.A. Vilkovisky, *Phys. Lett.* **B55**, 224 (1975);
I.A. Batalin and G.A. Vilkovisky, *Phys. Lett.* **B69**, 309 (1977);
E.S. Fradkin and T.E. Fradkina, *Phys. Lett.* **B72**, 343 (1978) 343;
I.A. Batalin and E.S. Fradkin, *Rivista Nuovo Cimento* **9**, 1 (1986).
30. J. Govaerts, *J. Phys.* **A30**, 603 (1997).
31. J.R. Klauder, *J. Math. Phys.* **40**, 5860 (1999);
G. Watson and J.R. Klauder, *J. Math. Phys.* **41**, 8072 (2000);
J.R. Klauder, *J. Math. Phys.* **42**, 4440 (2001);
J.R. Klauder, *Class. Quant. Grav.* **19**, 817 (2002);
G. Watson and J.R. Klauder, *Class. Quant. Grav.* **19**, 3617 (2002).
32. V.M. Villanueva, J. Govaerts and J.-L. Lucio-Martinez, *J. Phys.* **A33**, 4183 (2000).
33. G.Y.H. Avossevou and J. Govaerts,*The Schwinger Model and the Physical Projector: a Nonperturbative Quantization without Gauge Fixing*, contribution to these Proceedings.
34. J. Govaerts and B. Deschepper, *J. Phys.* **A33**, 1031 (2000).
35. E. Witten, *Nucl. Phys.* **B311**, 46 (1988);
J.H. Horne and E. Witten, *Phys. Rev. Lett.* **62**, 501 (1989);
E. Witten, *Nucl. Phys.* **B323**, 113 (1989).
36. See for example, and references therein,
G.A. Goldin and D.H. Sharp, *Phys. Rev. Lett.* **76**, 1183 (1996).
37. J. Govaerts, Proc. 2nd Mexican School of Particles and Fields, Cuernavaca - Morelos (Mexico), 4-12 December 1986, eds. J.-L. Lucio and A. Zepeda (World Scientific, Singapore, 1987), pp. 247-442;
J. Govaerts, *Nuclear Physics B (Proc. Suppl.)* **11**, 186 (1989);
J. Govaerts, Proc. Cargèse Summer Institute 1989 "Particle Physics", Cargèse (Corsica, France), 18 July-4 August 1989, eds. M. Levy, D. Speiser, J.-L. Basdevant, R. Gastmans, M. Jacob and J. Weyers (Plenum, New York, 1990), pp. 141-216.
38. J. Govaerts, *Int. J. Mod. Phys.* **A4**, 173 (1989).
39. For a review and references to the original literature, see for example,
A. Giveon, M. Porrati and E. Rabinovici, *Physics Reports* **244**, 77 (1994).
40. N. Seiberg and E. Witten, *JHEP* **9909**, 032 (1999).

Parallel Sessions: Part I

Theoretical Methods of Modern Classical and Quantum Physics

DO CROSS-SECTIONS DETERMINE PHASE SHIFTS UNIQUELY?

DAVID ATKINSON
Groningen University, The Netherlands
E-mail: atkinson@phys.rug.nl

It is shown that scattering phase shifts are not uniquely determined by the differential cross-section. It can occur that two inequivalent sets of phase shifts correspond to exactly the same cross-section. The ambiguity is not limited to a finite number of nonvanishing phase shifts; a method is sketched for the generation of amplitudes with an exponentially decreasing tail of high partial-wave amplitudes.

1 Query

A nuclear physicist measures the scattering of spinless particles and finds on analyzing his data that the following phase shifts (in radians) describe his results well: $\delta_0 = 0.88779$, $\delta_1 = -0.56640$, $\delta_2 = 0.23447$, and $\delta_\ell = 0$ for $\ell \geq 3$. A research student of the nuclear physicist analyses the same experimental results, but she obtains $\delta_0 = 0.10930$, $\delta_1 = -0.76993$, $\delta_2 = 0.23447$, again with all the higher phase shifts vanishing. Has the research student, or the nuclear physicist made a mistake, or are they both correct?

This problem was posed in the textbook 'Quantum Mechanics – A Self-Contained Course', by David Atkinson and Mahouton Norbert Hounkonnou, Rinton Press, Princeton, 2001 (see Exercise 8.10). The solution, part of the calculations given here, will appear in Volume 3 of this work, 'Exercises in Quantum Mechanics and Quantum Field Theory – A Self-Contained Book of Questions and Answers', David Atkinson, Mahouton Norbert Hounkonnou and Porter Wear Johnson, which is scheduled to appear in the Spring of 2003.

2 Crichton's Ambiguity

An amplitude containing only S, P and D waves can be written

$$f_k(\theta) = f_0(k) + 3z f_1(k) + \tfrac{5}{2}(3z^2 - 1) f_2(k)\,,$$

where $z = \cos\theta$ and the partial wave amplitudes satisfy

$$f_\ell(k) = \frac{1}{k}\, \mathrm{e}^{i\delta_\ell(k)} \sin\delta_\ell(k)\,.$$

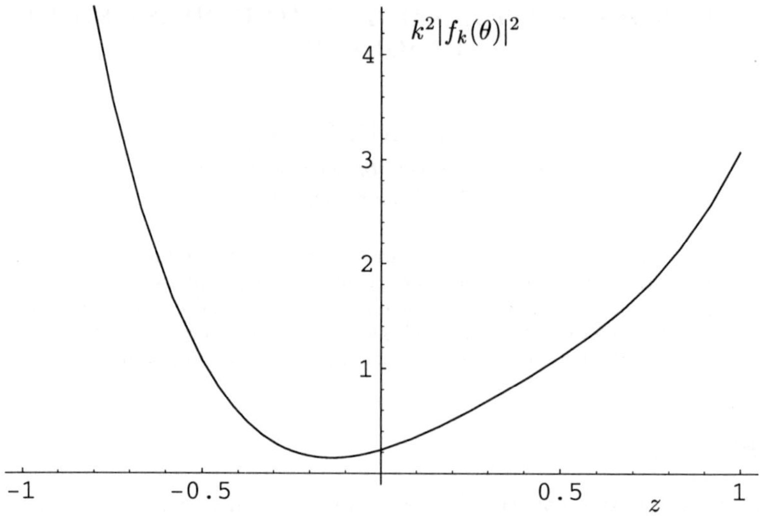

Figure 1. The Crichton cross-section.

The scattering differential cross-section is given by

$$\frac{d\sigma}{d\Omega} = |f_k(\theta)|^2 ,$$

and in Figure 1 we plot

$$\begin{aligned} k^2|f_k(\theta)|^2 &= \left|f_0(k) + 3zf_1(k) + \tfrac{5}{2}(3z^2-1)f_2(k)\right|^2 \\ &= \sin^2\delta_0 + 9z^2\sin^2\delta_1 + \tfrac{25}{4}(3z^2-1)^2\sin^2\delta_2 \\ &\quad +6z\sin\delta_0\sin\delta_1\cos(\delta_0-\delta_1) + 5(3z^2-1)\sin\delta_0\sin\delta_2\cos(\delta_0-\delta_2) \\ &\quad +15z(3z^2-1)\sin\delta_1\sin\delta_2\cos(\delta_1-\delta_2) , \end{aligned} \tag{1}$$

with the phase shifts obtained by the nuclear physicist. Exactly the same curve is obtained with the alternative set of phase shifts obtained by the student. This example, first discovered by Crichton,[1] illustrates the fact that the differential scattering cross-section does not in general uniquely determine the phase shifts.

With the aim of examining the general situation when only S, P and D waves are nonzero, we set[2]

$$k^2|f_k(\theta)|^2 = k^2|f_k'(\theta)|^2 , \tag{2}$$

where $f'_k(\theta)$ has a partial-wave expansion like that of $f_k(\theta)$, excepting only that the phase shifts, δ_ℓ, are replaced by δ'_ℓ. For Eq.(2) to be true, it must be so that the fourth-order polynomial equation holds that is formed by setting the right side of Eq.(1) equal to the corresponding expression with δ'_ℓ in place of δ_ℓ. Since the fourth-order polynomials on either side of the equation must be equal for all values of z, we may equate the coefficients of the various powers of z. To ensure equality of the power z^4, we require

$$\sin^2 \delta_2 = \sin^2 \delta'_2 \,,$$

and so $\delta_2 = \pm\delta'_2$ (the phase shifts are in general only defined modulo π, and the only solutions of the above equation in the interval $-\pi/2 < \delta_2 \le \pi/2$ are as given, i.e., only the sign is undetermined). However, if one changes the sign of every phase shift, the cross-section does not change, and this is referred to as the *trivial* ambiguity in the determination of the phase shifts by the data. Accordingly, we can say that $\delta_2 = \delta'_2$, aside from the trivial ambiguity, which we do not count as an interesting alternative solution, since it is always possible.

To determine the four phase shifts, δ_0, δ'_0, δ_1 and δ'_1, in terms of δ_2, we have four nonlinear equations, namely the equality of the coefficients of the powers z^3, z^2, z and 1. Equating coefficients of z^3, we find

$$\sin \delta_1 \cos(\delta_1 - \delta_2) = \sin \delta'_1 \cos(\delta'_1 - \delta_2) \,,$$

where a common factor $\sin \delta_2$ has already been canceled. From this we find

$$\sin 2\delta_1 \cos \delta_2 - \cos 2\delta_1 \sin \delta_2 = \sin 2\delta'_1 \cos \delta_2 - \cos 2\delta'_1 \sin \delta_2 \,.$$

Set now $\alpha_1 = \delta_1 + \delta'_1$ and $\epsilon_1 = \delta_1 - \delta'_1$. Straightforward manipulations yield

$$\cos \alpha_1 \sin \epsilon_1 \cos \delta_2 + \sin \alpha_1 \sin \epsilon_1 \sin \delta_2 = 0 \,.$$

A common factor $\sin \epsilon_1$ can be canceled, on condition that it is not zero, and we find $\tan \alpha_1 = -\cot \delta_2 = \tan(\delta_2 - \frac{\pi}{2})$, with the solution

$$\alpha_1 = \delta_2 - \frac{\pi}{2} \,. \tag{3}$$

Next, on equating coefficients of z^0, we have

$$\sin^2 \delta_0 - 5 \sin \delta_0 \sin \delta_2 \cos(\delta_0 - \delta_2) = \sin^2 \delta'_0 - 5 \sin \delta'_0 \sin \delta_2 \cos(\delta'_0 - \delta_2) \,, \tag{4}$$

and, on equating coefficients of z^2,

$$3 \sin^2 \delta_1 + 5 \sin \delta_0 \sin \delta_2 \cos(\delta_0 - \delta_2) = 3 \sin^2 \delta'_1 + 5 \sin \delta'_0 \sin \delta_2 \cos(\delta'_0 - \delta_2) \,.$$

The sum of the last two equations is

$$\sin^2 \delta_0 + 3 \sin^2 \delta_1 = \sin^2 \delta'_0 + 3 \sin^2 \delta'_1 \,, \tag{5}$$

which in fact could also have been deduced from the optical theorem, which can be expressed in the form

$$\sum_{\ell=0}^{\infty}(2\ell+1)\sin^2\delta_\ell = \sum_{\ell=0}^{\infty}(2\ell+1)\sin^2\delta'_\ell\,.$$

After a little work, Eq.(4) can be transformed into

$$\sin 2\delta_2\{\sin 2\delta_0 - \sin 2\delta'_0\} + \{\cos 2\delta_2 - \tfrac{3}{5}\}\{\cos 2\delta_0 - \cos 2\delta'_0\} = 0\,,$$

which, in terms of $\alpha_0 = \delta_0 + \delta'_0$ and $\epsilon_0 = \delta_0 - \delta'_0$, reads

$$\sin 2\delta_2 \cos\alpha_0 \sin\epsilon_0 + \{\cos 2\delta_2 - \tfrac{3}{5}\}\sin\alpha_0 \sin\epsilon_0 = 0\,.$$

A common factor $\sin\epsilon_0$ can be canceled, on condition that it is not zero, so

$$\tan\alpha_0 = \frac{\sin 2\delta_2}{\cos 2\delta_2 - \frac{3}{5}}\,. \tag{6}$$

Finally, on equating coefficients of z, we have

$$\sin\delta_0 \sin\delta_1 \cos(\delta_0 - \delta_1) = \sin\delta'_0 \sin\delta'_1 \cos(\delta'_0 - \delta'_1)$$

$$\text{or} \qquad \sin(\alpha_0 - \alpha_1)\sin(\epsilon_0 - \epsilon_1) = \sin\alpha_0 \sin\epsilon_0 + \sin\alpha_1 \sin\epsilon_1\,. \tag{7}$$

It is relatively easy to cast Eq.(5) in the form

$$\sin\alpha_0 \sin\epsilon_0 + 3\sin\alpha_1 \sin\epsilon_1 = 0\,, \tag{8}$$

and we have to solve Eqs.(7)–(8) for ϵ_0 and ϵ_1. They have the form

$$\sin\epsilon_0 = \kappa\,\sin\epsilon_1 \qquad \sin(\epsilon_0 - \epsilon_1) = \Delta\,\sin\epsilon_1\,,$$

respectively, with

$$\kappa = -3\,\frac{\sin\alpha_1}{\sin\alpha_0} \qquad \Delta = -2\,\frac{\sin\alpha_1}{\sin(\alpha_0 - \alpha_1)}\,.$$

Clearly these equations together imply

$$\kappa \sin\epsilon_1 \cos\epsilon_1 - \cos\epsilon_0 \sin\epsilon_1 = \Delta \sin\epsilon_1\,,$$

from which a factor $\sin\epsilon_1$ may be canceled. Hence

$$\cos\epsilon_0 = \kappa\cos\epsilon_1 - \Delta$$

$$1 = \cos^2\epsilon_0 + \sin^2\epsilon_0 = \kappa^2[\sin^2\epsilon_1 + \cos^2\epsilon_1] - 2\Delta\kappa\cos\epsilon_1 + \delta^2\,.$$

This leads directly to

$$\cos\epsilon_0 = \frac{\kappa^2 - \Delta^2 - 1}{2\Delta} \qquad \cos\epsilon_1 = \frac{\kappa^2 + \Delta^2 - 1}{2\kappa\Delta}\,.$$

With $\delta_2 = 0.23447$, the above solutions yield $\alpha_0 = 0.99709$, $\alpha_1 = 1.80527$ for the sums and $\epsilon_0 = \pm 0.77850$, $\epsilon_1 = \pm 2.93807$ for the differences. The signs of ϵ_0 and ϵ_1 are not determined by the inverse cosines, so we need to check the four possibilities against Eq.(8). With plus signs, we find $\sin\alpha_0 \sin\epsilon_0 = 0.589782$ and also $3\sin\alpha_1 \sin\epsilon_1 = 0.589782$, so it is clear from Eq.(8) that we need to choose the signs of ϵ_0 and ϵ_1 to be opposite. With $\epsilon_0 = 0.77850$ and $\epsilon_1 = -2.93807$, we find $\delta_0 = 0.88779$ and $\delta_1 = -0.56640$, in agreement with the phase shift analysis of the nuclear physicist. With the same value of δ_2, we find the alternative S and P wave phase shifts to be $\delta_0' = 0.10930$ and $\delta_1' = 2.37167$, in agreement (modulo π) with the analysis of the research student, since $2.37167 - \pi = -0.76993$.

3 Infinitely Many Partial Waves

Is the Crichton ambiguity limited to the situation in which only a finite number of phase shifts is nonzero? If that were so, the phenomenon would be merely a curiosity, of no real consequence to phase shift analysis and model building. To investigate a special case in which there are infinitely many nonvanishing phase shifts,[3] consider an amplitude that has the expansion

$$f_k(\theta) = \frac{1}{2i}\frac{z - z_1}{1 - z_1}\left\{\gamma - 1 - 3z - \sum_{\ell=2}^{\infty}(2\ell+1)\epsilon_\ell P_\ell(z)\right\}, \qquad (9)$$

where z_1 and γ are complex but the ϵ_ℓ are real. An alternative amplitude, $f_k'(\theta)$, is defined by the same expression, except that z_1 is replaced by z_1^*, its complex conjugate. Since

$$\left|\frac{z - z_1^*}{1 - z_1^*}\right| = \left|\frac{z - z_1}{1 - z_1}\right|$$

for real z, it is clear that $|f_k(\theta)|^2 = |f_k'(\theta)|^2$ so both amplitudes correspond to the same cross-section. The S and P partial-wave amplitudes are

$$f_0 = -\frac{1}{2i}\frac{1 + z_1(\gamma - 1)}{1 - z_1}$$

$$f_1 = \frac{1}{6i}\frac{\gamma - 1 + 3z_1 - 2\epsilon_2}{1 - z_1},$$

and this leads to

$$S_0 = 1 + 2if_0 = -\frac{z_1\gamma}{1 - z_1}$$

$$S_1 = 1 + 2if_1 = \frac{2 + \gamma - 2\epsilon_2}{3(1 - z_1)}.$$

The unitarity requirements $|S_0|^2 = 1 = |S_1|^2$ imply

$$|\gamma|^2 = \left|\frac{1-z_1}{z_1}\right|^2$$

$$\mathrm{Re}\,\gamma = \frac{9|1-z_1|^2 - |\gamma|^2 - 4(1-\epsilon_2)^2}{4(1-\epsilon_2)}.$$

For $\ell \geq 2$, we find

$$S_\ell = 1 - \frac{\eta_\ell - z_1\epsilon_\ell}{1-z_1},$$

where

$$\eta_\ell = \frac{(\ell+1)\epsilon_{\ell+1} + \ell\epsilon_{\ell-1}}{2\ell+1}, \tag{10}$$

with the understanding that $\epsilon_1 = 1$. The unitarity condition $|S_\ell|^2 = 1$ yields

$$\eta_\ell^2 - 2\eta_\ell + 2\mathrm{Re}\,z_1\left(\eta_\ell + \epsilon_\ell - \eta_\ell\epsilon_\ell\right) + |z_1|^2\left(\epsilon_\ell^2 - 2\epsilon_\ell\right) = 0\,. \tag{11}$$

In the case $\epsilon_3 << 1$, we have from Eq.(10) that

$$\eta_2 = \frac{3\epsilon_3 + 2\epsilon_1}{5} \approx \frac{2}{5},$$

and, on inserting this into Eq.(11), we find

$$\epsilon_2 \approx \frac{2(4 - 5\mathrm{Re}\,z_1)}{5(3\mathrm{Re}\,z_1 - 5|z_1|^2)},$$

on condition that $\epsilon_2 << 1$. Evidently this condition is satisfied if $\mathrm{Re}\,z_1$ is close to $\frac{4}{5}$. With $\mathrm{Re}\,z_1 = \frac{4}{5}$ exactly, $\epsilon_2 = 0$ exactly, and all the higher partial wave amplitudes vanish too. The polynomial that remains is precisely the *SPD* case that we examined above (i.e. Crichton's ambiguity). With $\mathrm{Re}\,z_1 = \frac{4}{5} + \zeta$, where $|\zeta| << 1$, we have an amplitude with an infinite number of partial waves, which corresponds however to a cross-section that differs very little from the pure *SPD* case, and for which there exists an ambiguous set of scattering phase shifts.

A further restriction should be imposed, however, namely that the infinite partial-wave series be convergent. A necessary requirement for this is that $\epsilon_\ell \to 0$ as $\ell \to \infty$, and so also $\eta_\ell \to 0$ in this limit. The quadratic equation (11) reduces then to the linear relation $\eta_\ell = u\,\epsilon_\ell$, where

$$u = \mathrm{Re}\,z_1 + \frac{[\mathrm{Im}\,z_1]^2}{\mathrm{Re}\,z_1 - 1},$$

and so Eq.(10) becomes, in this limit,

$$(\ell+1)\epsilon_{\ell+1} - (2\ell+1)\, u\epsilon_\ell + \ell\epsilon_{\ell-1} = 0\,.$$

The general solution of this second-order recurrence relation is

$$\epsilon_\ell = aP_\ell(u) + bQ_\ell(u)\,,$$

where $Q_\ell(u)$ is the Legendre function of the second kind. When $|u| > 1$ this vanishes asymptotically only if $a = 0$, i.e. only if the contribution of the Legendre function of the first kind is absent. It has been shown that it is possible to tune z_1 and ϵ_2 in such a way that this happens[3]. Hence even when there is an infinite number of partial waves, a measurement of the differential cross-section does not necessarily determine the phase shifts uniquely. It should be stressed that only a very small part of the ambiguity in question has been investigated to date, and much more work remains to be done.

Acknowledgments

I would like to thank the Institute of Mathematics and Physical Sciences of the National University of Bénin for its hospitality, and the Faculty of Natural Sciences and the Office for International Relations, both of the University of Groningen in the Netherlands, for encouragement and for financial support.

References

1. J.H. Crichton, *Nuovo Cimento* **A45**, 256 (1966).
2. D. Atkinson, P.W. Johnson, N. Mehta and M. de Roo, *Nucl. Phys.* B **55**, 125 (1973).
3. D. Atkinson, L.P. Kok and M. de Roo, *Phys. Rev.* D **17**, 2492 (1978).

A MATHEMATICAL SOLUTION TO THE BAND GAP PROBLEM

D. BAGAYOKO, G.L. ZHAO AND S. HASAN

Department of Physics, Southern University and A&M College
P. O. Box 11776, Baton Rouge, LA 70813, USA
E-mail: Bagayoko@aol.com

Using the Rayleigh Theorem, we have resolved the long-standing problem consisting of 30%-50% theoretical underestimates of the band gaps of non-metallic materials. We describe the Bagayoko, Zhao and Williams (BZW) method that rigorously circumvents the basis set and variational effect presumed to be a cause of these underestimates. We present ab-initio computational results that are in agreement with experiment for diamond (C), silicon (Si), silicon carbides (3C-SiC and 4H-SiC), and other semiconductors (GaN, $BaTiO_3$, AlN, ZnSe, ZnO). We illustrate the predictive capability of the BZW method in the case of the newly discovered cubic phase of silicon nitride (c-Si_3N_4) and of selected carbon nanotubes ((10,0) and (8,4)). Our conclusion underscores the inescapable need for the BZW method in ab-initio calculations that employ a basis set in a variational approach. Current nanoscale trends amplify this need. We estimate that the potential impact of applications of the BZW method in *advancing our understanding* of non-metallic materials, in *informing experiment*, and particularly in *guiding device design and fabrication is simply priceless.*

1 Introduction

The fundamental understanding and device applications of carbon-based and other non-metals have suffered from a pervasive inability of ab-initio theoretical calculations to produce accurate *unoccupied energy levels or bands*. The 30%-50% underestimates of the band gaps of these materials[1,2] are just a symptom of presumed, theoretical errors in the unoccupied energies. The inability of these calculations to reproduce measured optical properties of semiconductors is another symptom. Using the Rayleigh theorem[3,4] and the physical content of the Hamiltonian, we describe a basis set and variational effect[5,6,7,8,9,10] that is a source of errors in calculated unoccupied energies of atoms, molecules, clusters, semiconductors and insulators. We present the Bagayoko, Zhao and Williams (BZW) method[5,6,7] that rigorously avoids the above effect and provides a predictive capability for variational calculations that utilize a basis set.

In the remainder of this contribution, we first discuss the persistence of the noted underestimation problem from the advent of quantum calculations to 1998, along with various attempts to solve it. This background is followed by a detailed description of the basis set and variational effect, a presentation of our general method and of the Bagayoko, Zhao and Williams method that resolved the problem. We then present electronic properties for diamond (C), silicon (Si), silicon carbide (3C- and 4H-SiC), barium titanate ($BaTiO_3$), gallium nitride (GaN), zinc oxide (ZnO), and aluminum nitride (AlN). Unlike most previous attempts, particularly the ones utilizing local density functional potentials, our BZW results agree with experiment not only for the unoccupied energy levels and the band gaps, but also for the optical transitions, the dielectric function, and the effective mass. We present our predictions for the electronic and related properties of a newly discovered cubic phase of silicon nitride (c-Si_3N_4) and of selected carbon nanotubes. Our conclusion, in light of the growing trend toward the nanoscale, underscores the need for the BZW method in calculations that employ a basis set in a variational approach of the Rayleigh-Ritz type. It also notes the unavoidable nature of the BZW method for descriptive or predictive computations of properties of nanostructures. This conclusion finally intimates the huge time and money savings that could result from this work – in terms of guiding the design and fabrication of molecular, nanoscale, and other devices.

2 Theoretical Underestimates of Unoccupied Energies

2.1 The Problem: The Unoccupied Energy Problem

"*Calculated Electronic Properties of Metals*"[11] has been available since 1978. In this book, Moruzzi, Janak, and Williams provided the calculated electronic energy bands, charge densities, and related properties of metals. Their results, obtained with density functional potentials, generally agree with experimental measurements and with other theoretical findings. We are not aware of a similar book on semiconductors and insulators. One reason for the absence of such a reference manual stems from the fact that from the dawn of quantum calculations to 1998, there existed no easy way to calculate accurately the unoccupied energy levels of bands of semiconductors or insulators. Indeed, most density functional calculations of electronic properties of these materials, including diamond and other carbon-based systems, led to band gaps that were 30% to 50% smaller than the measured values[1,2,3,4,5,6,7]. *This band gap problem* is actually a symptom of a more general problem stemming from the inaccuracy of the calculated unoccupied energy levels or bands. Such a problem, understandably, reverberates throughout other calculated properties of these systems, including optical transition energies and the dielectric function. The seriousness of this problem may be best understood by noting that most device-utilization of semiconductors or insulators involves excited states. *The unoccupied band problem* has therefore been a major stumbling block to efficient device design and fabrication. Even though it is not as well-known as the *band gap problem*, the above underestimation of unoccupied energies also affected theoretical results for atoms, molecules, and clusters. Additionally, it is pervasive in calculated properties of nuclei in the framework of the shell model.

In the case of semiconductors and insulators, the above problem has been ascribed to limitations of density functional potentials, with emphasis on the local ones. These potentials,[12,13,14,15,16] from the early 1970's to present, continue to be extensively utilized in theoretical descriptions of many body systems. Other approaches to the representation of the potential, i.e., those based on Hartree Fock formalism, did not resolve the problem and are often accompanied by other difficulties. For instance, the limitations of the Hartree Fock method are described by Callaway;[17] they include inordinately large band widths.

2.2 Attempts to resolve the problem

Several approaches have been utilized in an attempt to remedy the theoretical underestimation of unoccupied energy levels or bands. They included the use of density functional potentials that are non-local, as done in the generalized gradient approximation (GGA).[14] Some approaches applied the "*scissors approximation*" to unoccupied bands obtained with local density functional potentials.[18] This approximation essentially consists of rigidly shifting upward the calculated unoccupied energy levels or bands.[18] The self-interaction correction (SIC) and the self-interaction with relaxation correction are other approaches[2] that have been tried with limited success. Perhaps one of the most widely used approach that clearly goes beyond density function theory is the dressed Green function (G) and screened Coulomb (W) interaction formalism that employs non-local, energy-dependent, and non-Hermitian operators. This GW method[1] is perhaps the most successful of the ones referenced above, even though it also has its limitations. In particular, the GW method tends to give band gaps that are slightly larger than the experimental values. Many GW calculations are not self-consistent; the self-consistent ones are reported[19] to increase the overestimation of the band gaps of semiconductors and insulators.

In 1998, we had obtained calculated energy bands that reproduced experimentally measured properties, including the band gap, optical transition energies, and the dielectric function of barium titanate.[5,6] The extent of this agreement with experiment, for all calculated properties, warranted a search for an explanation; luck and over-complete or incomplete basis sets were not satisfactory explanations – in light of the pervasiveness of the agreement with experiment. Our prior experiences with calculations with several basis sets of different sizes[9] led us to the explanation of the relative positions of the conduction bands as compared to the occupied bands. In particular, Bagayoko had utilized[9] the Rayleigh theorem to explain the rigid shift of band energies, up or down, depending on whether the basis size is reduced or enlarged, respectively. We identified and characterized a basis set and variational effect that explains most of the *unoccupied energy problem*. This effect is described in the next section.

2.3 An explanation of the problem: a basis set and variational effect

It has been explained[7] that in a typical ab-initio, self-consistent calculation that employs the linear combination of atomic orbital method, electronic eigenfunctions are expanded using basis sets derived from atomic calculations[9,17]. Charge densities and potentials are constructed *using only the functions belonging to occupied states.*[17] The Hamiltonian matrix is generated and diagonalized. Key output quantities are discrete energy levels or continuous bands and related wave functions. The resulting output wave functions for the occupied states are employed to generate a new charge density and the computations are repeated. The process of using the output of iteration (n) – *for occupied states* – to construct the input for iteration $(n+1)$ continues until self-consistency is reached. Various measures are utilized to define self-consistency, i.e., when basic quantities, including charge densities, potentials, eigen-energies, etc., are respectively unchanged from one iteration to the next.

The trial basis sets are found by various authors by augmenting the atomic orbitals with polarization and diffuse orbitals[7] whose numbers, until the work of Bagayoko, Zhao and Williams,[5,6,7] had no particular limit. In fact, from a completeness standpoint, the larger these numbers the better, provided no "catastrophic sinking of energy levels" or negative Millican population numbers occur for occupied states.[20]

The above arbitrariness in the selection of the size of the basis set, by virtue of the Rayleigh theorem and the use of the wave functions for the occupied states to generate the Hamiltonian, leads straightforwardly to the basis set <u>and</u> variational effect noted above. Indeed, the Rayleigh theorem states that when an eigenvalue equation is solved using a basis set of dimension N and one of dimension $N+1$, such that all the orbitals in the first basis are included in the second, larger one, then the eigenvalues resulting from the calculation with $(N + 1)$ orbitals will be respectively lower or equal to their corresponding ones from the calculation with N orbitals. Specifically, if the eigenvalues in each calculation are ordered from the lowest to the highest, then $E_1^{N+1} \leq E_1^N$, $E_2^{N+1} \leq E_2^N$, $E_3^{N+1} \leq E_3^N$, etc. Physically, these inequalities indicate that a given variational eigenvalue approaches the corresponding exact eigenvalue from above.

Clearly, the above theorem suggests that calculations should employ basis sets that are as large as possible. This fact and the preoccupation with ensuring the quantum mechanical completeness of the basis set partly explain the reason that the following effect was not quantitatively circumscribed for many decades. (The rather vague expression of "basis set effects" is not to be confused with the basis set and variational effect as defined and circumscribed herein using a mathematical theorem.) A fact that places a limit on the size of basis sets in variational calculations, however, stems from the *utilization of the wave functions of the occupied states to generate the charge density, the potential, and the Hamiltonian matrix*. This matrix contains the physics of the calculations. We define a minimal basis set as one that contains just enough orbitals to account for all the electrons in the system under study. Beginning with a minimal basis set, if one performs several calculations with basis sets that are each obtained by augmenting the previous, smaller one, then one is expected to reach a size such that the charge density, the potential, and the Hamiltonian no longer change. One does. Past this point, the use of larger basis sets, by virtue of the Rayleigh theorem, will still lead to a lowering of some unoccupied energy levels or bands, even though the occupied ones no longer change. *There lies the basis set and variational effect, i.e., the extra lowering of some unoccupied energy levels of bands, by virtue of the Rayleigh theorem,*[3,4,9] *when unnecessarily large basis sets are employed in self-consistent variational calculations of the Rayleigh-Ritz type.*

To facilitate the forthcoming presentation of the Bagayoko, Zhao and Williams (BZW) method for avoiding this effect, we define the *optimal basis set* of dimension N as that of the first of the calculations, with increasing basis sets, such that the next calculation produces the same charge density, potential, Hamiltonian, and occupied energy levels or bands. *This optimal basis set* defines a new form of convergence, i.e., that of the Hamiltonian and of the occupied energy levels or bands vis-à-vis the dimension of the basis set. It was shown[7] that calculations with basis sets larger than the optimal set, if otherwise done correctly, do not change the charge density, the potential, the Hamiltonian, or the occupied energies. They do continue indefinitely to lower some unoccupied energies, given that the fundamental theorem of algebra dictates the appearance of new eigenvalues upon the increase of the size of the basis set or of the dimension of the Hamiltonian matrix.

3 Method

3.1 General method: the potential and the formalism

The general method employed in our calculations has been extensively described in the literature.[5,6,7,8,9,10] Our self-consistent, computational formalism consists of the linear combination of atomic orbitals (LCAO). We employed the electronic structure package of the group of Professor Bruce Harmon at the Department of Energy (DOE) Laboratory at Iowa State University.[21,22] We utilized Gaussian orbitals and therefore refer to our approach as a linear combination of Gaussian orbitals (LCGO). In describing many body system, the choice of the potential is a critical one. In all the calculations discussed below, we used the Ceperley-Alder type[15] of local density potentials as parameterized by Vosko, Wilk and Nusair.[16]

3.2 The Bagayoko, Zhao and Williams (BZW) method

We introduced this method in 1998 for completely circumventing, from first principle, the above identified basis set and variational effect. The BZW method basically requires a minimum of three self-consistent calculations that utilize basis sets of different sizes. It begins with the minimum basis set, i.e., the basis set needed to account for all the electrons of the atomic or ionic species that are present in the molecule, cluster, or solid under study. Completely self-consistent calculations are carried out. For the second calculation, the minimal basis set is augmented with one or more atomic orbitals that belong to the next and lowest lying energy levels in the atomic or ionic species. The self-consistent bands from calculations I and II are compared, graphically and numerically. In general, there will be qualitative (shape and branching) and quantitative (numerical values) differences between the occupied bands from calculations I and II. A third calculation is performed, using the basis set for calculation II as augmented with orbitals representing the next lowest-lying atomic energy levels.

The above process is continued until the comparison of the occupied energy levels leads to no qualitative or quantitative difference. When the results (i.e., occupied energies only) from calculation N and those from calculation $(N+1)$ agree within the computational error, then the optimum basis set is that of calculation N. This selection rests on the fact that a lowering of unoccupied levels, as in calculation $(N+1)$, after the occupied levels converged with respect to the size of the basis set, as in calculation N, is dictated by the Rayleigh theorem and does not necessarily reflect the physics of the system. Such a lowering may not be the manifestation of any fundamental interaction, but rather the expression of the basis set and variational effect identified above. In fact, it was shown[7] that calculation $(N+1)$ and others with larger basis sets do not change the self-consistent charge density or potential from their values as obtained in calculation N that utilized the optimal basis set. In multi-species systems like GaN, two or more orbitals may be added at a time if the affected atomic levels are close in energy. While we utilized the occupied energy bands for the determination of convergence with respect to the size of the basis set, other parameters could be employed. They may include the charge density for occupied states and particularly the potential.

3.3 Computational details

This paper is intended to present the essentials related to the BZW method, as done above, and the related results for a variety of materials. Consequently, we refer the reader to previous publications[6,7,8,10,23] that provide extensive computational details for each material. In particular, lattice structures, specific lattice parameters, the various s, p and d orbitals included in the calculations, and the numbers of iterations are among these details. We discuss both the angular and radial components of these orbitals.

4 Results: Calculated Electronic Properties of Non-Metals

We list in Tables 1 and 2 the calculated band gaps of diamond, 3C- and 4H-SiC, and of several other semiconductors. Space limitation does not allow us to delve into the details of these results and of many others that are available in our previous publications.[6,7,8,9,10] For each material listed below, the reader should refer to these sources[6,7,8,9,10,23,24,25] for details. Other results in these papers include the electronic energy bands, the density of states, effective masses, charge transfers, contour plots of charge densities, and optical transition energies. For $BaTiO_3$ and 3C-SiC, we provided the dielectric function and the total energy curves, respectively.

The verified theoretical predictions[26,27] of Dr. Carter T. White's group at the Naval Research Laboratories (NRL) on properties of carbon nanotubes partly inspired our interest in these materials. We report below the band gaps for carbon nanotubes (10,0) and (8,4). Carbon nanotubes (10,0) belong to a symmorphic group. Its diameter is 7.83 Å and there are 40 atoms in the unit-cell. Carbon nanotube (8,4) belongs to a nonsymmorphic group. The

Table 1. Calculated electronic properties (i.e., band gaps) of selected carbon-based and other semiconductors versus measured values. All the calculations employed the BZW method. The results below are therefore obtained with the optimal basis sets for the affected materials. Data for c-Si_3N_4 and for carbon nanotubes (8,4) and (10,0) are predictions. Further details are available in the identified publications. The symbol Eg stands for the band gap. Effective masses, M^*, have symmetry and directional (parallel and perpendicular) labels as subscripts. The free electron mass is m_0. Energies are in electron volts (eV). The concept of practically measurable band gap has been explained elsewhere,[7] as per tail structures in the density of states. When a range is provided for the theoretical band gap, the lower limit indicates the theoretical minimum gap while the upper limit shows the largest experimental value, depending on the sensitivity of measurement instrument and method and of subsequent analysis. The Table is continued in Table 2.

	Calculation	**Measurement**
$BaTiO_3$ (Ref.6)		
Eg	2.6 eV	2.8, 3.0 eV
$M^*_{p,\parallel}$	7.5 m_0	very anisotropic
$M^*_{p,\perp}$	1.2 m_0	
$M^*_{n,\parallel}$	3.4 m_0	
$M^*_{n,\perp}$	1.2 m_0	1.0 - 1.5 m_0
GaN (Ref.7)		
Eg	3.4 eV	$\sim$ 3.4 eV
M^*_n	0.22 $\pm$ 0.03 m_0	0.2 $\pm$ 0.02 m_0
Si (Ref.7)		
Eg	1.02 eV	1.14, 1.17 eV
Band width (valence)	12.1 eV	12.5 eV
M^*_{nt}	0.20 $\pm$ 0.03 m_0	0.19 m_0
M^*_{nl}	0.93 $\pm$ 0.03 m_0	0.98 m_0
ZnO		
Eg	3.2 eV	3.4 eV
AlN (Ref.24)		
Eg	5.5 - 6.2 eV	3.9 - 6.2 eV
Carbon nanotube (8,4)		
Eg	0.90 eV	Unknown to us

diameter of (8,4) nanotube is 8.29 Å. The chiral angle of (8, 4) relative to the zigzag direction is 19.1°. There are 112 atoms in its unit-cell. While (10,0) has a direct band gap of 0.95 eV at the gamma point, the 0.9 eV gap of (8,4) is off the gamma point. The highest energy state of the valence band and

Table 2. Continuation of Table 1.

	Calculation	**Measurement**
Diamond (C) (Ref.7)		
Eg	5.05 eV	5.3, 5.48 eV
W_v (valence band width)	21.35 eV	21± 1 eV, 24.2 eV
$M^*_{n,\parallel}$	$1.1 \pm 0.2\ m_0$	$1.4\ m_0$
$M^*_{n,\perp}$	$0.30 \pm 0.03\ m_0$	$0.36\ m_0$
3C-SiC (Ref.8)		
Eg	2.24 eV	2.2, 2.4 eV
M^*_x	$0.72 \pm 0.04\ m_0$	$0.677 \pm 0.015\ m_0$
M^*_{xw}	$0.22 \pm 0.02\ m_0$	$0.247 \pm 0.011\ m_0$
4H-SiC (Ref.8)		
Eg	3.11 eV	3.2, 3.3 eV
$M^*_{n,\perp}$	$0.41 \pm 0.02\ m_0$	$0.42\ m_0$
$M^*_{n,\parallel}$	$0.31 \pm 0.02\ m_0$	$0.33, 0.29\ m_0$
$M^*_{M\Gamma\vert}$	$0.62 \pm 0.03\ m_0$	$0.58 \pm 0.01\ m_0$
$M^*_{MK\vert}$	$0.27 \pm 0.02\ m_0$	$0.31 \pm 0.01\ m_0$
c-Si_3N_4 (Ref.23)		
Eg	3.68 eV	(Not yet available)
ZnSe (Ref.25)		
Eg	2.6 - 3.1 eV	2.8 eV
Carbone nanotube (10,0)		
Eg	0.95 eV	Unknown to us

the lowest state of the conduction band are located at about $0.07X$, where $X = (1, 0, 0)\frac{\pi}{a}$. These nanotube band gaps, and that of 3.68 eV for the newly discovered cubic phase of silicon nitride (c-Si_3N_4), are predictions – given that we are unaware of any available experimental values, to date.

5 Conclusion

As per the above results and others, the described BZW method resolved the long standing theoretical underestimation, by 30% -50%, of the band gaps of semiconductors and insulators, included carbon-based materials and carbon nanotubes. As such, the method provides a tool for significant contributions of theoretical calculations in (a) the correct description of non-metallic materials, (b) the predictions of electronic, optical, and related properties of semiconductors, (c) industrially relevant band gap engineering and related design and fabrication of devices, and (d) the search for novel materials. In particular, current trends toward the nanoscale, in light of the prevailing nature of quantum effects at the atomic and molecular levels, seem to dictate the use of this method for computations that utilize basis sets in a variational approach of the Rayleigh-Ritz type.

Acknowledgments

This work was funded in part by the Department of the Navy, Office of Naval Research (ONR), through the Timbuktu Academy (Grant No. N00014-98-1-0748), and by NASA, through PIPELINES (Grant No. NAG5-8552 and Cooperative Agreement No. CC13-00010). The presentation of this work at the Second International Conference on Contemporary Problems in Mathematical Physics was funded by the Center for Theoretical Studies of Physical Systems (CTSPS), directed by Professor Alfred Z. Msezane, at Clark Atlanta University, Atlanta, Georgia, USA.

References

1. A. Rubio, J.L. Corkill, M.L. Cohen, E.L. Shirley and S.G. Louie, *Phys. Rev.* **B48**, 11810 (1993).
2. D. Vogel, P. Kriiger and J. Pollmann, *Phys. Rev.* **B55**, 12836 (1997).
3. S.H. Gould, *Variational Methods for Eigenvalue Problems* (University of Toronto Press, Toronto, Canada, 1957), Chapter 2.
4. S.G. Mikhlin, *The Numerical Performance of Variational Methods* (Wolters-Noordhoff, Groningen, 1971), Chapters 1, 2 and 7.
5. D. Bagayoko, G.L. Zhao and T.D. Williams, *LDA prediction of electronic properties of* $BaTiO_3$, *Bull. APS* **43 (1)**, 846 (1998).
6. D. Bagayoko *et al.*, *Ab-initio calculations of the electronic structure and optical properties in ferroelectric tetragonal* $BaTiO_3$, *Journal of Physics: Condensed Matter* **10**, 5645 (1998).
7. G.L. Zhao, D. Bagayoko and T.D. Williams, *Local density functional prediction of electronic properties of GaN, Si, C, and* RuO_2, *Phys. Rev.* **B60**, 1563 (1999).
8. G.L. Zhao and D. Bagayoko, *Electronic structure and charge transfer in 3C- and 4H-SiC*, *New Journal of Physics* **2**, 16.1-16.12 (2000), `http://www.njp.org`.
9. D. Bagayoko, *Contraction of gaussian basis sets and the total energy*

of FCC copper, *Int. J. of Quantum Chemistry: Quantum Chemistry Symposium* **17**, 527 (1983).
10. D. Bagayoko and G.L. Zhao, *Predictive ab-initio computations of properties of ferroelectric materials*, *Int. Jour. Mod. Phys.* **B13**, 3767 (1999).
11. L. Moruzzi, J.F. Janak and A.R. Williams, *Calculated Electronic Properties of Metals* (Pergamon, New York, 1978).
12. H. Hohenberg and W. Kohn, *Phys. Rev.* **136**, 3864 (1964).
13. W. Kohn and L.J. Sham, *Phys. Rev.* **140**, A1133 (1965).
14. J. Perdew and A. Zunger, *Phys. Rev.* **B23**, 5048 (1981).
15. D.M. Ceperley and B.J. Alder, *Phys. Rev. Lett.* **45**, 566 (1980).
16. S.H. Vosko, L. Wilk and M. Nusair, *Can. J. Phys.* **58**, 1200 (1980).
17. J. Callaway, *Quantum Theory of the Solid State*, Second Edition (Academic Press, Boston, 1991).
18. K.A. Johnson and N.W. Ashcroft, *Phys. Rev.* **B58**, 15548 (1998).
19. W.-D. Schöne and A.G. Eguiluz, *Phys. Rev. Lett.* **81**, 1662 (1998).
20. D. Bagayoko *et al.*, *3d impurities in aluminum*, *Phys. Rev.* **B54**, 12184 (1996).
21. B.N. Harmon, W. Weber and D.R. Hamann, *Phys. Rev.* **B25**, 1109 (1982).
22. G.L. Zhao *et al.*, *Phys. Rev.* **B40**, 7999 (1989).
23. D. Bagayoko and G.L. Zhao, *Predicted electronic properties of cubic silicon nitride (c-Si_3N_4)*, to appear in *Physica C* (2001).
24. Y. Luo, *Ab-Initio Computation of the Electronic Structure of Wurtzite Aluminum Nitride*, Master's Degree Thesis in Physics, Southern University and A&M College, Baton Rouge, Louisiana, 70813, USA (2000).
25. L. Torrence, *The BZW Method and the Electronic Properties of Zinc Selenide (ZnSe)*, Master's Degree Thesis in Physics, Southern University and A&M College, Baton Rouge, Louisiana, 70813, USA (2001).
26. C.T. White and J.W. Mintmire, *Density of states reflects diameter in nanotubes*, *Nature* **394**, 29 (1998).
27. J.W. Mintmire and C.T. White, *Universal density of states for carbon nanotubes*, *Phys. Rev. Lett.* **81**, 2506 (1998).

HILBERT TRANSFORM OR KRAMERS-KRONIG RELATIONS APPLIED TO SOME ASPECTS OF LINEAR AND NON-LINEAR PHYSICS

G. DEBIAIS

Groupe de Physique Fondamentale, Université de Perpignan
52, Avenue de Villeneuve, F-66860 Perpignan, France
E-mail: debiais@in2p3.univ-perp.fr

1 Introduction

The ordinary Hilbert transform[1,2] is defined by the Cauchy principal-value integral,

$$g(x) = \frac{1}{\pi} P \int_{-\infty}^{+\infty} \frac{f(t)}{t-x} dt = \mathcal{H} f(x), \tag{1}$$

where $f(x)$ is a given function defined for every real x, P standing for the principal-value. This transformed $\mathcal{H}$ is explicitly invertible and the inverse transform is its own negative opposite,

$$f(x) = \frac{-1}{\pi} P \int_{-\infty}^{+\infty} \frac{g(t)}{t-x} dt = -\mathcal{H} g(x). \tag{2}$$

When the variable t is a frequency or an energy, these transformations relate to the well-known Kramers-Kronig relations with their wide field of application. After some theoretical considerations to recall the derivation of these relations within the Theory of Distributions, their importance is illustrated through their application to four examples in linear and nonlinear physics.[3,4,5,6,7]

2 Theoretical Framework

The Hilbert and Kramers-Kronig relations have been known for a long time and are well documented. The former may be established based on the classical Fourier transform, or else on a study of analytical signals. The latter have been discovered, practically simultaneously, by H. Kramers and R. Kronig[8,9] around 1927 through work on the dispersion theory of X rays. These results are briefly summarized hereafter, followed by some examples of applications of these relations.

2.1 Hilbert relations

Fourier's simple integral formula writes as

$$f(x) = \int_0^{+\infty} [a(t') \cos xt' + b(t') \sin xt']\, dt', \tag{3}$$

with,

$$a(t') = \frac{1}{\pi} \int_{-\infty}^{+\infty} f(u) cosut' du, \quad b(t') = \frac{1}{\pi} \int_{-\infty}^{+\infty} f(u) sinut' du. \tag{4}$$

The conjugate integral of $f(x)$ is,

$$g(x) = \int_0^{+\infty} [b(t') \cos xt' - a(t') \sin xt']\, dt'. \tag{5}$$

At least formally, one always has,

$$\begin{aligned} g(x) &= \frac{1}{\pi} \int_0^{+\infty} dt' \int_{-\infty}^{+\infty} sin(u-x)t' f(u) du \\ &= \lim_{\lambda \to \infty} \frac{1}{\pi} \int_0^{+\lambda} dt' \int_{-\infty}^{+\infty} sin(u-x)t' f(u) du \\ &= \lim_{\lambda \to \infty} \frac{1}{\pi} \int_0^{+\infty} \frac{1 - cos(\lambda t)}{t} [f(x+t) - f(x-t)]\, dt. \end{aligned}$$

According to the Riemann-Lebesgue theorem,

$$\lim_{\lambda \to \infty} \int_a^b f(t) cos\lambda t dt \to 0$$

hence,

$$g(x) = \frac{1}{\pi} \int_0^{+\infty} \frac{f(x+t) - f(x-t)}{t} dt. \tag{6}$$

Similarly,

$$f(x) = \frac{-1}{\pi} \int_0^{+\infty} \frac{g(x+t) - g(x-t)}{t} dt \tag{7}$$

These last two relations are equivalent to relations (1) and (2), respectively.

2.2 Kramers-Kronig relations for the dielectric constant

The dielectric constant ϵ expresses the relation that exists between the electrical displacement vector $\mathbf{D}$ in a dielectric medium and the applied electrical field $\mathbf{E}$,

$$\mathbf{D} = \epsilon \mathbf{E}.$$

If the medium is isotropic and homogeneous, ϵ is a scalar quantity, dependent on the thermodynamic state. If the medium is anisotropic, or becomes so some through the action of a sufficiently strong external field, ϵ is a tensor, leading to spatial dispersion. In a particular coordinate system, associated to the direction of the wavevector $\mathbf{k}$, ϵ is a second order symmetric tensor,

$$\epsilon = \begin{bmatrix} \epsilon_{11} & 0 & 0 \\ 0 & \epsilon_{22} & 0 \\ 0 & 0 & 0 \end{bmatrix}. \tag{8}$$

This tensor specifies completely the macroscopic properties of the medium, in relation to the propagation along the direction of the wavevector $\mathbf{k}$ for an homogeneous monochromatic transverse wave whose wavelength is large in comparison to microscopic dimensions.

The most general expression of the scalar dielectric constant, or ϵ_{11} and ϵ_{22} both to be referred to simply as ϵ, as a function of the frequency ν or the angular frequency $\omega = 2\pi\nu$, is derived from the generalized theory of electric susceptibility. This theory relates the properties of nonthermodynamic fluctuations to others quantities that are characteristic of the medium's behaviour when submitted to external actions. It writes,

$$\epsilon(\omega) - \epsilon_\infty = \int_{-\infty}^{+\infty} f(\tau) \exp(j\omega\tau) d\tau,$$

where τ has the dimension of time, ϵ_∞ is the limit of $\epsilon(\omega)$ as $\omega \to \infty$, $f(\tau)$ is a function of time as well as of the properties of the medium which remains finite for all values of τ. In the case of dielectrics, $f(\tau)$ goes to zero when $\tau \to \infty$, whereas the time interval during which it is significantly different from zero is on the order of the relaxation time characteristic of the process. Owing to the principle of causality, "no effect can precede its cause", $f(\tau) = 0$ for $\tau < 0$, and therefore,

$$\epsilon(\omega) - \epsilon_\infty = \int_{0}^{+\infty} f(\tau) \exp(j\omega\tau) d\tau. \tag{9}$$

Given this last expression, and extending ω into a complex variable $\Omega = \omega - j\bar{\omega}$, the following properties for $\epsilon(\Omega)$ are established,

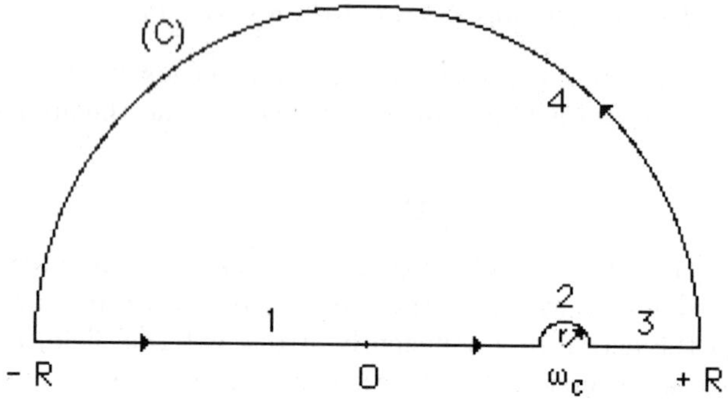

Figure 1. The contour area related to the discussion of the Kramers-Kroning relations.

1. $\epsilon(\Omega)$ is analytic in the upper-half complex plane in Ω as well as on the real axis;

2. $\epsilon(-\Omega) = \epsilon^*(\Omega)$, as well as $\epsilon(-\omega) = \epsilon^*(\omega)$ when Ω is real, where $*$ stands for complex conjugation. In this latter case, by separating the real and imaginary parts of ϵ, one finds $\epsilon'(-\omega) = \epsilon'(\omega)$ and $\epsilon''(-\omega) = -\epsilon''(\omega)$, showing that ϵ' is an even function and ϵ'' an odd function of the frequency;

3. Along the positive imaginary axis in the upper-half complex plane, the function $\epsilon(\Omega)$ reduces to its real part only, while it also decreases monotonously along that axis. In particular, $\epsilon(\Omega)$ has no zeros inside the upper-half complex plane;

4. Following from the thermodynamics law of entropy growth rate in the case of energy losses, the function ϵ'' is positive for any medium and at all frequencies. The sign of ϵ' is also subjected to restrictions.

The Kramers-Kronig relations are readily derived following from these four properties. Consider, in fact, any point ω_c on the positive part of the real axis. Since in the case of a dielectric, $\epsilon(\omega)$ has no pole at the origin, the function

$$\frac{\epsilon(\Omega) - \epsilon_\infty}{\Omega - \omega_c}$$

is analytic in the area represented in Fig.1.

According to Cauchy's theorem, the integral on the closed contour $(C) = (1+2+3+4)$ vanishes. If we let R run to ∞, since $[\epsilon(\Omega) - \epsilon_\infty] \to 0$ when $\Omega \to \infty$, the integral on the semicircle 4 does not contribute. Consequently,

$$\lim_{R\to\infty} \int_C \frac{\epsilon(\Omega) - \epsilon_\infty}{\Omega - \omega_c} d\Omega = \lim_{R\to\infty} \left[\int_1 \frac{\epsilon - \epsilon_\infty}{\omega - \omega_c} d\omega + \int_3 \frac{\epsilon - \epsilon_\infty}{\omega - \omega_c} d\omega \right] + \int_2 \frac{\epsilon - \epsilon_\infty}{\Omega - \omega_c} d\Omega = 0. \tag{10}$$

If the small radius r tends to zero, the term in brackets is the principal part of the integral in the interval $]-\infty, +\infty[$ and the last integration reduces to $-j\pi[\epsilon(\omega_c) - \epsilon_\infty]$. Hence,

$$\epsilon'(\omega_c) - j\epsilon''(\omega_c) - \epsilon_\infty = -\frac{1}{j\pi} P \int_{-\infty}^{+\infty} \frac{\epsilon'(\omega) - j\epsilon''(\omega) - \epsilon_\infty}{\omega - \omega_c} d\omega, \tag{11}$$

while separating imaginary and real parts,

$$\begin{aligned} \epsilon'(\omega_c) - \epsilon_\infty &= \tfrac{1}{\pi} P \int_{-\infty}^{+\infty} \tfrac{\epsilon''(\omega)}{\omega - \omega_c} d\omega, \\ \epsilon''(\omega_c) &= -\tfrac{1}{\pi} P \int_{-\infty}^{+\infty} \tfrac{\epsilon'(\omega) - \epsilon_\infty}{\omega - \omega_c} d\omega. \end{aligned} \tag{12}$$

These relations were derived by H. Kramers and R. Kronig in 1927. Since $\epsilon'(\omega)$ is an even function and $\epsilon''(\omega)$ an odd function of the frequency, these relations may also be given an alternative form,

$$\begin{aligned} \epsilon'(\omega_c) - \epsilon_\infty &= \tfrac{2}{\pi} P \int_0^{+\infty} \tfrac{\omega\epsilon''(\omega)}{\omega^2 - \omega_c^2} d\omega, \\ \epsilon''(\omega_c) &= -\tfrac{2\omega_c}{\pi} P \int_0^{+\infty} \tfrac{\epsilon'(\omega) - \epsilon_\infty}{\omega^2 - \omega_c^2} d\omega. \end{aligned} \tag{13}$$

Note that these expressions remain valid when the variable of integration is no longer the angular frequency ω but, the frequency ν, or the energy $E = h\nu$, or the wavenumber $\sigma = \nu/c$. If the variable is the wavelength $\lambda = c/\nu$, one just has to change the sign of each of the expressions in (13).

2.3 Analytic signal and Hilbert relations

Consider a real signal $x(t)$. This representation includes a dual aspect, namely either a time or a frequency dependency. This means that one may consider the evolution of its amplitude either as a function of time, $x(t)$, or as function of frequency, $X(\nu)$, whose integration can reconstruct the temporal signal. The duality between the two functions is provided by the Fourier transformation, which is invertible, namely in a complex notation,

$$\begin{aligned} X(\nu) &= \int x(t) \exp(-j2\pi\nu t) dt, \\ x(t) &= \int X(\nu) \exp(+j2\pi\nu t) d\nu. \end{aligned} \tag{14}$$

The signal $x(t)$ being real, one has $x(t) = x^*(t)$ or $X(\nu) = X^*(-\nu)$, i.e., $X(\nu)$ possesses a specific hermiticity property. This implies that the knowledge of $X(\nu)$ for positive frequencies $\nu > 0$ only, say, is sufficient to reconstruct the whole signal $x(t)$. Consider the function $Z(\nu) = 2u(\nu)X(\nu)$ where $u(\nu)$ is Heaviside's unit step function. The inverse Fourier transform of $Z(\nu)$ gives the so-called analytic signal $z(t)$. This function $z(t)$ is derived from $x(t)$ through a linear filter which is invariant in time and having a frequency response $2u(\nu)$. The impulsional response of this filter is

$$g(t) = \delta(t) + P\left(j\frac{1}{\pi t}\right), \tag{15}$$

where $\delta(t)$ is the Delta Kronecker signal and P the principal-value of the term between the parentheses. In frequency space, this equation becomes,

$$G(\nu) = 2u(\nu) = G'(\nu) - jG''(\nu), \tag{16}$$

with $G'(\nu) = 1$ and $G''(\nu) = j(\nu/|\nu|)$, $(\nu/|\nu|)$ standing for the sign of ν. One obtains $z(t) = [g \otimes x](t) = x(t) + jy(t)$, where $\otimes$ denotes the convolution product. The real part of $z(t)$ is thus $x(t)$ and its imaginary part $y(t)$ is obtained from $x(t)$ by the filter of response $G''(\nu)$. Consequently,

$$y(t) = P\int \frac{x(\theta)}{\pi(t-\theta)}d\theta, \tag{17}$$

which is nothing but the Hilbert transform of $x(t)$. The inverse filter giving $x(t)$ from $y(t)$ has the frequency response function $[G''(\nu)]^{-1}$, so that,

$$x(t) = -P\int \frac{y(\theta)}{\pi(t-\theta)}d\theta, \tag{18}$$

namely the inverse Hilbert transform.

2.4 Implementation of the method of analysis

As we have just seen, the inverse Hilbert or Kramers-Kronig relations allow the determination of the real part $h'(x)$ of a complex function $h(x)$ of the variable x from its imaginary part $h''(x)$ when known on any large interval, and conversely. Existence conditions for the applicability of these relations are summarized by Bode,[10]

- the function must not have any singularity on the real frequency axis;
- the line integral of $h(x)/x$ along any semi-circle in the upper-half complex plane whose radius goes to infinity must vanish;

- the real part of $h(x)$ is an even function, and its imaginary part an odd function, of the real variable x;

- the Boltzmann principle of superposition can be applied.

To clarify our technique, we briefly recall the example of the complex dielectric constant $\epsilon(\omega)$. It is known to satisfy the conditions above, with in particular ϵ' being an even function of the frequency and ϵ'' an odd function. The modulus ρ of the dielectric constant is defined as

$$\rho = (\epsilon'^2 + \epsilon''^2)^{1/2},$$

or equivalently,

$$\epsilon = \rho \exp(-j\varphi), \tag{19}$$

with $\varphi = \tan^{-1} \epsilon''/\epsilon'$, $\epsilon' = \rho \cos\varphi$ and $\epsilon'' = \rho \sin\varphi$.

If experimental measurements give the dispersion $\epsilon'(\omega)$, the relations (13) allow to deduce the unknown function $\epsilon''(\omega)$, or conversely. However, if experiment gives only the dispersion relation for the modulus $\rho(\omega)$, it is necessary to also calculate the phase angle $\varphi(\omega)$ in order to determine $\epsilon'(\omega)$ and $\epsilon''(\omega)$. To calculate $\varphi(\omega)$, one needs again the inverse Hilbert or Kramers-Kronig relations. Indeed, since $\ln(\epsilon) = \ln(\rho) - j\varphi$, the function $\ln[\epsilon(\omega)]$ satisfies again the necessary conditions for the applicability of these relations as does the function $\epsilon(\omega)$, and, as a result, the imaginary part φ of the real part $\ln(\epsilon)$ is given as

$$\varphi(\omega_c) = -\frac{2\omega_c}{\pi} P \int_0^{+\infty} \frac{\ln \rho(\omega)}{\omega^2 - \omega_c^2} d\omega, \tag{20}$$

where ω_c denotes the considered angular frequency. Through integrating by parts, one obtain an alternative expression which is much better suited for automatic evaluation on a computer,

$$\varphi(\omega_c) = -\frac{1}{\pi} P \int_0^{+\infty} \frac{d \ln \rho(\omega)}{d\omega} \ln\left(\frac{\omega + \omega_c}{\omega - \omega_c}\right) d\omega. \tag{21}$$

Experimentally, the dielectric constant is not known on the entire frequency interval from zero to infinity, but only within a limited range $[\omega_1, \omega_2]$. The calculation of the phase angle may be decomposed in three parts,

$$\varphi = \int_0^{\omega_1} + \int_{\omega_1}^{\omega_2} + \int_{\omega_2}^{\infty} \quad \text{or} \quad \varphi = \varphi_1 + \varphi_2 + \varphi_3\ . \tag{22}$$

In a first step, the integration between ω_1 and ω_2 gives the value φ_2. This calculation may be realized through various methods. The one that we used is as follows. One considers m points on the curve representing the function $\ln[\rho(\omega)]$, *i.e.*, this curved is decomposed into $(m-1)$ segments. If p_i denotes the slope of the ith such segment, one has,

$$\varphi_2(\omega_c) = -\frac{1}{\pi}\left[\sum_{i=1}^{m-1} p_i \int_{\omega_i}^{\omega_{i+1}} \ln\left(\frac{\omega+\omega_c}{\omega-\omega_c}\right) d\omega\right]. \tag{23}$$

This form suggests to introduce the function $f(x_i)$ defined by,

$$f(x_i) = \frac{1}{\pi\omega_i}\int_0^{\omega_i} \ln\left(\frac{\omega+\omega_c}{\omega-\omega_c}\right) d\omega \quad \text{with} \quad x_i = \frac{\omega_c}{\omega_i}. \tag{24}$$

This function may easily be evaluated as follows,

$$f(x_i) = \frac{1}{\pi}\left[(x_i+1)\ln(x_i+1) + (x_i-1)\ln|x_i-1| - 2x_i\ln x_i\right]. \tag{25}$$

The expression for the contribution φ_2 to the phase angle then takes the form,

$$\varphi_2(\omega_c) = \sum_{i=1}^{m-1} p_i\left[\omega_{i-1}f(x_{i-1}) - \omega_i f(x_i)\right], \tag{26}$$

which is readily computed numerically. Note that in such a calculation, the product $(x_i-1)\ln(|x_i-1|)$ goes to zero when x_i goes 1, namely when ω_i goes to ω_c. Obviously, the larger the number m of discretization points, the better the precision of the numerical result.

In a second step, the restriction in the domain of integration must be compensated by the correction terms φ_1 and φ_3 whose contribution is not negligible. These terms may be evaluated based on a mathematical or a semi-empirical law, for which many methods and relations have been proposed. Through experience, we have been led to use expressions determined by Ruiz-Perez[11] in the context of the theory of averages, with the particular advantage to lead to a common representation for all contributions given the restricted interval $[\omega_1, \omega_2]$. These expressions read as follows,

$$\begin{aligned}\varphi_1(\omega_c) &= C[(\omega_1+\omega_c)\ln(\omega_1+\omega_c) + (\omega_c-\omega_1)\ln(\omega_c-\omega_1) - 2\omega_c\ln\omega_c],\\ \varphi_3(\omega_c) &= D\left[\tfrac{\omega_2+\omega_c}{\omega_2\omega_c}\ln\tfrac{\omega_2+\omega_c}{\omega_2\omega_c} + \tfrac{\omega_2-\omega_c}{\omega_2\omega_c}\ln\tfrac{\omega_2-\omega_c}{\omega_2\omega_c} - \tfrac{2}{\omega_c}\ln\tfrac{1}{\omega_c}\right].\end{aligned} \tag{27}$$

where C and D are two constants which must be determined beforehand. For this purpose, it suffices to specify the value, either known *a priori* or to be expected, of the total phase φ at two different frequencies ω' and ω''. This leads to a system of two equations from which to determine the two constants C and D. Generally, and whenever possible, the frequencies ω'

and ω'' are chosen at two extremities of the spectrum, outside the absorption region and thus corresponding to values of φ which are vanishing or in any case negligible. The corrections to the value φ_2 obtained in this manner appear to be reasonably under control, even when the integration conditions are not ideal, which occurs generally when the frequency spectrum being considered is incomplete either in its high or low frequency components.

Hence, this method is applicable to any complex function $h(x)$ of the variable x which meets the Bode conditions. It has largely been used in many areas of physics, providing precious information: in optics (optical functions), electromagnetism (magnetic and electric susceptibilities), nuclear physics (interaction of nucleons and mesons), viscoelasticity, etc. Some of these applications have become classics, such as the determination of the optical constant of a sample through its reflection properties. However, many other areas could likewise benefit from these inverse relations. As far as we are concerned. they have been used successfully in non-linear electrical birefringence, for the first time some ten years ago.

Finally, it should be noted that the impact of the method on the precision with which the phase angle is determined is about 3%. This means that if the modulus is known to 4 or 5% , φ may be evaluated to a precision of 7 to 8%. Such a value compares well with experimental results on the phase angle whose precision typically lies within 12% to 25%.

3 Examples of Application of the Hilbert Relations

The four examples to be discussed hereafter are based on our own calculations and experience. The underlying common theme is the general method of analysis as exposed in Sect.2.4, namely, Bode's conditions are obeyed by a complex function to be determined, allowing the evaluation of its phase angle based on the knowledge of its modulus, thus leading to the real and imaginary parts of the function of interest, and finally its use within a physical context in order to determine specific physical parameters of the corresponding phenomenon. The four examples to be discussed presently are successively introduced as follows.[3,4,5,6,7] In the first example, the optical function of a thin film material is calculated from its reflection spectrum. The second example studies the kinetics of S-haemoglobin polymerization based on the time evolution of its dielectric constants. In the third example, the non-linear response of the Kerr effect or electrical birefringence is considered. Finally, the fourth example is in relation to the exploitation of a LIDAR probe signal of the atmosphere. The first three examples use already published work, while the fourth is based on work performed within an International Collaboration which is soon to be completed.

As already mentioned, it is not in their physical aspects that these examples have a common and shared feature, but rather in the application and interest that lies with the Hilbert relations.

3.1 Study of a spectre of reflection of a thin film

The determination of the optical constant of a semi-infinite material is a classic piece of work in physics. It relies on the expressions for the reflection factor r and the reflection power R, in normal incidence,

$$r = \sqrt{R}\exp(-j\varphi) = \frac{1-n^*}{1+n^*} = \frac{1-n+jk}{1+n-jk}, \tag{28}$$

$$R = \frac{(n-1)^2+k^2}{(1+n)^2+k^2}, \tag{29}$$

n^* being the complex index, n the refraction index, and k the absorption index.

However, on the mathematical side, a difficulty arises because of a property of the starting function. Indeed, $\ln(\omega)$ diverges as ω goes to infinity. However, from properties of ϵ^* and the functions r, $\ln r$ and its derivative $d[\ln(r)]/d\omega$, Stern[12] has shown that in normal incidence one obtains an expression for the phase angle which is analogous to the relation (21). Later, Ruiz-Perez[11] established the same form of expression for arbitrary incidence. Hence, the phase angle φ is also given by,

$$\varphi(\omega_c) = -\frac{1}{\pi}P\int_0^{+\infty} \frac{d\ln\sqrt{R(\omega)}}{d\omega}\ln\left(\frac{\omega+\omega_c}{\omega-\omega_c}\right)d\omega. \tag{30}$$

One may thus establish relations for the dispersive and absorptive component parts of the optical function, namely the indices n and k read as follows,

$$n(\omega) = \frac{1-R(\omega)}{1+R(\omega)-2\sqrt{R(\omega)}cos\varphi(\omega)}, \tag{31}$$

$$k(\omega) = \frac{2\sqrt{R(\omega)}sin\varphi(\omega)}{1+R(\omega)-2\sqrt{R(\omega)}cos\varphi(\omega)}, \tag{32}$$

and likewise for the imaginary and real parts of the dielectric constant,

$$\epsilon' = n^2 - k^2, \tag{33}$$

$$\epsilon'' = 2nk. \tag{34}$$

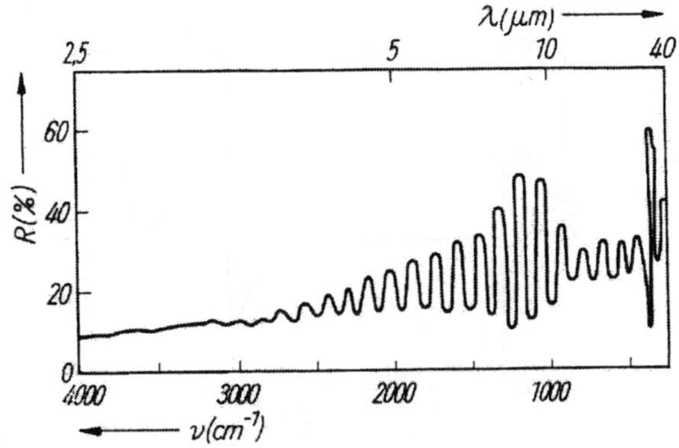

Figure 2. Reflection between 4000 and 250 cm^{-1} of a glass deposited mercury sulphide film of thickness $11,8\mu$m.

The case of a thin film deposited on a substrated is often more difficult to represent given that the expression[13] for its reflecting power **R** has no simple form such as R does, since it not only depends on the film thickness and optical constants but also on the properties of the substrated. An analysis based on the Kramers-Kronig relations remains possible but is quite difficult to apply. Consequently, we have written a computer code for the evaluation of **R** and, by letting k vary, have observed that the transition ($\mathbf{R} \to r$) from ideal transparency, corresponding to $k = 0$, to the semi-infinite behaviour is very sharp, and starts already to arise at $k = 0.5$. This observation then leads to the method to be describe hereafter.

In the spectral region where absorption is negligible, interference fringes appear and their minima are determined by the relation $4nd\sigma = 2m$, d being the thickness of the film, σ the wave-number, m an integer. From this relation the refraction index readily follows, allowing the evaluation of the reflecting power in case the sample were to be semi-infinite, given the relation (29) with $k = 0$. Based on the experimental curve, these values are then extended into the region of strong absorption, leading to a curve for which relations (28) to (32) are applicable.

Fig.2 shows the experimental reflecting power between 4000 and 250 cm^{-1} of a mercury sulphide film of thickness $d = 11.8\mu$m. Fig.3 shows the "reconstructed" reflection power according to the method which has just been described. Figs.4 and 5 display the optical constants n, k, ϵ' and ϵ'' which then follow from our analysis.

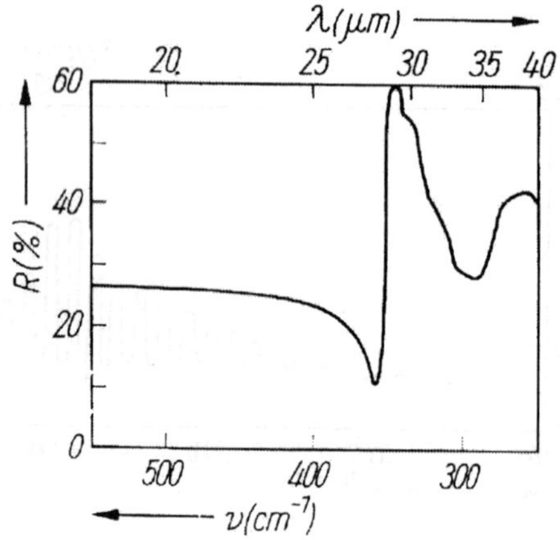

Figure 3. Reflection between 550 and 250 cm^{-1} corresponding to a semi-infinity sample of mercury sulphide after "reconstruction".

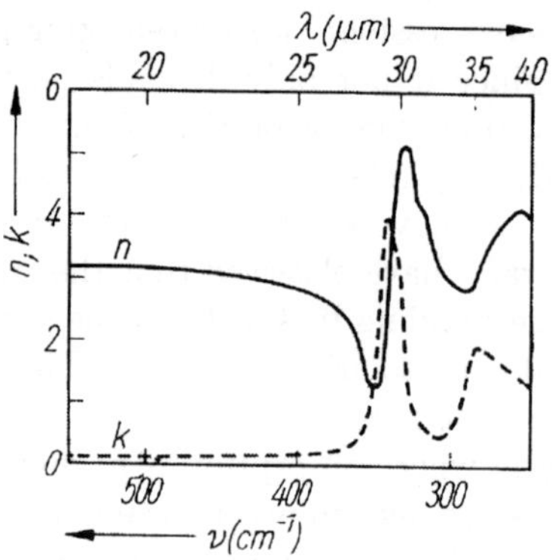

Figure 4. The dispersion optical indices n and k.

3.2 Study of the kinetics of S-haemoglobin polymerization

In was the purpose of one of our publications[14] to identify the inertia and attenuation effects that contribute to the dielectric constant of an ensemble

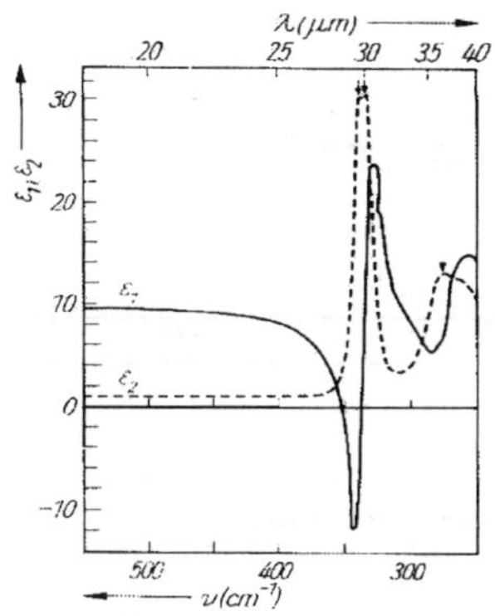

Figure 5. The real and imaginary parts of the dielectric constant ϵ.

of oscillators,[14]

$$\epsilon = \epsilon_\infty + \sum_{i=1}^{n} \frac{S_i \omega_i^2}{(\omega_i^2 - \omega^2) + j\gamma_i\omega_i\omega}, \tag{35}$$

and to compare these to the Debye relation[15] in the absence of inertia,

$$\epsilon = \epsilon_\infty + \sum_{i=1}^{n} \frac{S_i}{1 + j\omega\tau_i}. \tag{36}$$

In these expressions extended to n constituents whose contributions are superposed linearly, we have introduced the quantities,

$$S_i = \left(\frac{N\mu^2}{3kT}\right)_i, \quad \omega_i^2 = \left(\frac{2kT}{I}\right)_i, \quad \tau_i = \left(\frac{\zeta}{2kT}\right)_i, \quad \gamma_i = \omega_i\tau_i = \left(\frac{\zeta}{\sqrt{2IkT}}\right)_i, \tag{37}$$

with, for each species i, N_i the number of dipoles, μ_i the dipole moment, k Boltzman's constant, T the absolute temperature, ω_i the peculiar frequency, I_i the moment of inertia, τ_i the Debye relaxation time, ζ_i the coefficient of rotational friction, and γ_i the attenuation coefficient.

We have then sought in the literature experimental measurements of dielectric constants, and managed to isolate the effect of inertia. This research led to experimental measurements by Delalic *et al.*[16] on deoxygenated S-haemoglobin considered at different stages of aggregation, S-haemoglobin being an abnormal form of haemoglobin which causes sickle-cell anaemia, a plague that affects several African countries.

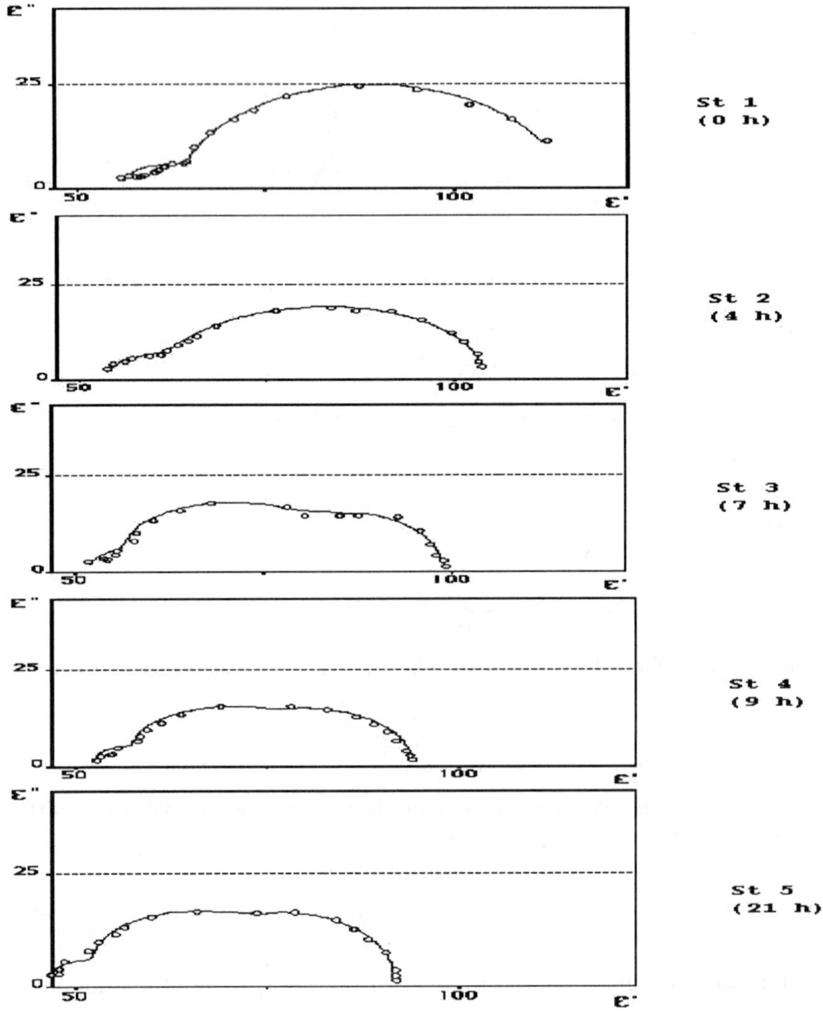

Figure 6. Cole-Cole diagrams, $\epsilon'' = f(\epsilon')$, for five successive aggregation stages at 0 h, 4 h, 7 h, 9 h, 21 h for deoxygenated S-haemoglobin. Delalic's data points are the circular points, while the full lines are the result of our best fit procedure.

Table 1. Time evolution dielectric parameters of deoxygenated S-haemoglobin during its aggregation kinetics.

	E_1(0h)	E_2(4h)	E_3(7h)	E_4(9h)	E_5(21h)
ϵ_0	114	104	99	94	92
ϵ_∞	56	54	52	53	47
$S = \epsilon_0 - \epsilon_\infty$	58	50	47	41	45
C_1	0.85	0.48	0.11	0	0
S_1	49.3	24.0	5.2		
N_1 (MHz)	364	364	364		
γ_1	2000	900	600		
$\tau_1(\mu s)$	0.875	0.393	0.262		
$\zeta_1 \times 10^{20}$ (erg·s)	7.48	3.37	2.25		
C_2	0.15	0.18	0.20	0.20	0.21
S_2	8.7	9.0	9.4	8.2	9.45
N_2 (Mz)	70	70	70	70	70
γ_2	9.0	4.7	3.50	3	3
$\tau_2(\mu s)$	0.0205	0.0107	0.0080	0.0068	0.0068
$\zeta_2 \times 10^{22}$ (erg·s)	17.5	9.15	6.85	5.82	5.82
C_3	0	0.34	0.51	0.55	0.54
S_3		17.0	24.0	22.6	24.3
N_3 (MHz)		5	5	4.9	4.8
γ_3		4.8	2.1	2.3	2.2
$\tau_3(\mu s)$		0.153	0.0668	0.0747	0.0729
$\zeta_3 \times 10^{21}$ (erg·s)		13.1	5.72	6.40	6.25
C_4	0	0	0.18	0.25	0.25
S_4			8.4	10.2	11.25
N_4 (MHz)			0.8	0.85	0.85
γ_4			1.7	1.6	1.6
$\tau_4(\mu s)$			0.338	0.300	0.300
$\zeta_4 \times 10^{20}$ (erg·s)			2.90	2.56	2.56

As discussed in Sect.2.2, the application of the Kramers-Kronig relations is without ambiguity. Hence, given the modulus of the complex dielectric constant, we determine its real $\epsilon'(\omega)$ and imaginary $\epsilon''(\omega)$ parts, and then plot Cole-Cole diagrams of one of these quantities as function of the other, $\epsilon'' = f(\epsilon')$. These curves clearly reveal the presence of several constituents, as a matter of fact four such distinct components. A fit is then applied to these curves using the relations (35) and (36), leading to a systematic adjustment of all the parameters defined previously. The results are shown in Fig.6 for time values of 0 h, 4 h, 7 h, 9 h and 21 h. As a function of time, Table 1 details the evoluton of the dielectric parameters of the deoxygenated S-haemoglobin, as they were derived from our approach.

Note that C_i represents the contribution of the constituent i to the dielectric constant and that $S_i = C_i S$. The moment of inertia of these four constituents is readily determined, as well as their size. In this way, it has been possible to show that constituent 1 was in a monomeric phase which served as starting point for a series of aggregates of increasing size. This monomeric phase is light, and only the Debye relation was applicable. Its characteristics were determined from information available from the literature. It also enabled us to identify the range of validity for the relations (35) and (34) in the context of the physics of the relaxation of the dielectric constant in the dilute phase.

3.3 Electric birefringence in coupled fields

The Kerr effect[17], or electric birefringence, provides another approach towards the investigation of matter in a dilute phase. In order to gain access to information complementary to that obtained from dielectric relaxation, we have also studied the Kerr effect induced by an alternating AC electric field superposed to a continuous DC component,

$$E(t) = E_c + E_0 \cos(\omega t). \tag{38}$$

As a function of time, it is known that birefringence is proportional to the average value of the Legendre polynomial of degree 2. By ignoring the negligible effect of inertia, we have solved the Smolukowski equation expressed in terms of operators and have obtained, through an original approach, expressions for the transitory regime, as well as in the limit of infinite time, expressions for the stationary regime.[18] The latter is comprised of two harmonic components in ω and 2ω, *i.e.*, they represent the non-linear response of the system of the form,

$$< P_2(u) > (\infty) =< P_2(u) >_{st} + \sum_{\alpha=1}^{2} [\Delta n_\alpha(\omega) cos(\alpha\omega t) + \Delta k_\alpha(\omega) sin(\alpha\omega t)], \tag{39}$$

with

$$< P_2(u) >_{st} = \frac{1}{15} < E^2(t) > \left(\frac{\mu}{kT}\right)^2$$
$$\times \left[P + \frac{1}{1+2\left(\frac{E_c}{E_0}\right)^2} \left(\frac{1}{1+\left(\frac{\omega}{2D}\right)^2} + 2\left(\frac{E_c}{E_0}\right)^2 \right) \right], \tag{40}$$

and

$$< E^2(t) > = \frac{E_0^2}{2} + E_c^2,$$

$$\Delta n_1(\omega) = \frac{2}{15} E_0 E_c \left(\frac{\mu}{kT}\right)^2 \left[\frac{P+1/2}{1+\left(\frac{\omega}{6D}\right)^2} + \frac{\frac{1}{2}\left(1-\frac{\omega}{6D}\frac{\omega}{2D}\right)}{\left[1+\left(\frac{\omega}{6D}\right)^2\right]\left[1+\left(\frac{\omega}{2D}\right)^2\right]} \right], \tag{41}$$

$$\Delta k_1(\omega) = \frac{2}{15} E_0 E_c \left(\frac{\mu}{kT}\right)^2 \left[\frac{P+1/2}{1+\left(\frac{\omega}{6D}\right)^2}\left(\frac{\omega}{6D}\right) + \frac{\frac{1}{2}\left(\frac{\omega}{6D}+\frac{\omega}{2D}\right)}{\left[1+\left(\frac{\omega}{6D}\right)^2\right]\left[1+\left(\frac{\omega}{2D}\right)^2\right]} \right],$$

$$\Delta n_2(\omega) = \frac{1}{30} E_0^2 \left(\frac{\mu}{kT}\right)^2 \left[\frac{P}{1+\left(\frac{\omega}{3D}\right)^2} + \frac{1-\frac{\omega}{2D}\frac{\omega}{3D}}{\left[1+\left(\frac{\omega}{2D}\right)^2\right]\left[1+\left(\frac{\omega}{3D}\right)^2\right]} \right],$$
$$\Delta k_2(\omega) = \frac{1}{30} E_0^2 \left(\frac{\mu}{kT}\right)^2 \left[\frac{P}{1+\left(\frac{\omega}{3D}\right)^2}\left(\frac{\omega}{3D}\right) + \frac{\frac{\omega}{2D}+\frac{\omega}{3D}}{\left[1+\left(\frac{\omega}{2D}\right)^2\right]\left[1+\left(\frac{\omega}{3D}\right)^2\right]} \right]. \tag{42}$$

In these expressions, Δn represents the difference between the real parts of the optical indices which correspond to directions which are perpendicular or parallel to the disruptive field $E(t)$, Δk the difference between the imaginary parts of these indices, μ the molecular permanent dipole moment, D the rotational diffusion constant, P a dimensionless parameter defined by $P = kT\Delta\alpha/\mu^2$, $\Delta\alpha$ the difference between the parallel and perpendicular principal electric polarizabilities which is proportional to the induced dipole moment, and k Boltzmann's constant.

We have checked that Bode's conditions are satisfied, and in particular, that the parity conditions are obeyed for each of the two harmonic components taken separately. Consequently, it is justified to consider these two terms on their own in the application of the Kramers-Kronig relations. This work, completly original, is based on experimental measurements of the normalized modulus of birefringence by Bénet[19] on linoleic acid at 20° C and pentachlorodiphenyl at 10° C. As in the preceding study, we obtain the imaginary and real parts Δn_1 and Δk_1, as well as Δn_2 and Δk_2, normalized to unity, and according to a traditional usage, denote them as B'_ω, B''_ω, $B'_{2\omega}$ and $B''_{2\omega}$, respectively. Fig.7 shows the calculation of the phase angles based on

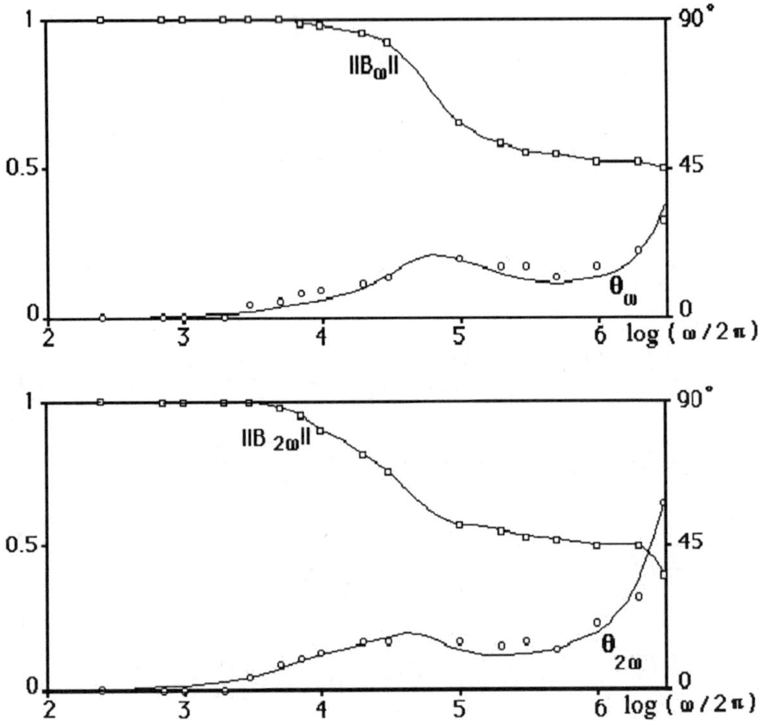

Figure 7. Variation of the phase angles with the frequency (full line) for the two harmonic terms, calculated from the Kramers-Kronig relations and the moduli $||B_\omega||(\omega)$ and $||B_{2\omega}||(\omega)$, for the linoleic acid. Benet's data points are indicate by $\diamond$ and $\circ$.

the birefringence modulus of the linoleic acid for each of the two harmonic terms. Fig.8 displays the corresponding typical Cole-Cole diagrams, namely the imaginary part as a function of the real part. These two figures are limited to the case of the linoleic acid.

In the same manner as in the previous example, a best-fit procedure based on the normalized relations (41) and (42) (namely, divided by $\Delta n_\alpha(0)$, which denotes the value of $\Delta n_\alpha(\omega)$ at $\omega \to 0$, with $\alpha = 1$ or 2) and extended to include several constituents, leads to very important and precise results which may be compared given the double redundancy in the available information. Details may be found in the original publications.[5,6] However, let us note here that the inferred values for the induced dipole moments, through the term

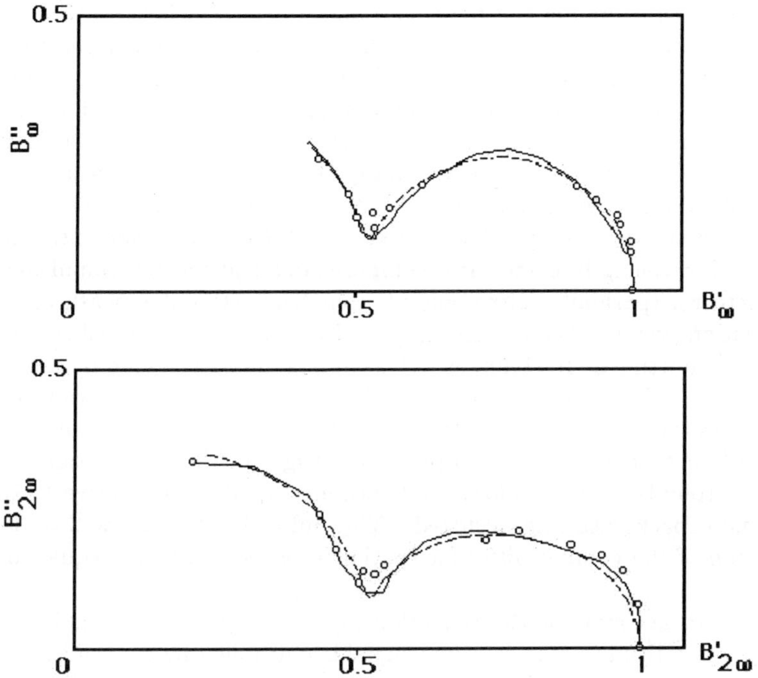

Figure 8. Diagrams for $B'' = f(B')$ for the two harmonic terms. Results following from our method are in full lines; Benet's data points are indicated by ∘; the results of the best-fit procedure are given by the dotted lines.

proportional to P, are in a complementary way totally in agreement with the conclusions based on the relaxation in the dielectric constant. In addition to the originality of the work, we emphasize the precision in the determination of the phase angles, namely 5 to 7%, while these are known experimentally to no better than 20%.

3.4 LIDAR atmospheric signal

Within the framework of our current Collaborations, CELESTE (Cerenkov Low Energy Sampling Timing Experiment)[20] and CAT (Cerenkov AT Thémis),[21] gamma rays originating from some galactic and extragalactic sources in energy ranges of 20 to 300 GeV are being studied. Their detection is made from ground-level to an altitude of 1680 m, close to Perpignan

(France), through computerized telescopes and a field of heliostats that collect the Cerenkov radiation which accompanies the particle shower generated by the interaction of a gamma photon with the atmosphere. In analysis programmes, the atmosphere is assumed to be "standard", *i.e.* , to be characterized by set parameters. A complementary device is currently operational, the so-called LIDAR (Light Detection And Ranging), whose purpose is to probe the atmosphere for a better knowledge of its conditions, and thereby refine the analysis of the detected particle shower. LIDAR uses the radar principle in the optical wavelength regime. It is composed of a transmitter (impulsional laser), a detector (parabolic mirror and photomultipliers) and a measurement and acquisition system. The laser is a typical Nd-Yag laser of 360 mJ running at 10 Hz and emitting at 1064 nm, as well as at 532 nm and 355 nm by using a double and triple frequency multiplier. Its beam divergence is 0,5 mrad. The mirror detects the photons backscattered within its reception cone due to the effect of different phenomena (absorption, Rayleigh or Mie distribution, ...). The photon round-trip time allows to determine the altitude relative to the ground where backscattering occurred. The pulse duration as well as the opening time of the counter allow for spatial resolution, which is equal to 5 meters.

In order to account for the reduction in the reception cone angle as a function of the altitude z, the collected signal is multiplied by z^2, leading to a graph in which the number of photons backscattered within a given solid angle is function of the altitude. This provides quasi an instanteneous snapshot of the atmosphere up to an altitude of some 20 km, leading to a qualitative interesting picture of its actual state. When the sky is clear, the result (see Fig.9) represents quite well the standard atmosphere, *i.e.*, characterized by a quasi-exponentially decreasing evolution. When the weather is overcast, humps and even peaks appear at specific altitudes, corresponding to the presence of fog, aerosols, clouds or even sometimes frozen clouds (see Fig.10). The signature of the latter is very much like that of optical reflection.

First, one needs to filter the signal, since it is accompanied by noise stemming from the electronics and from the atmosphere. The procedure is standard. It is based on the harmonic analysis of the signal from which the higher frequencies are subtracted. Fig.9 shows the result of such a procedure for a clear atmosphere, by keeping only the first 200 harmonics from a total of 4000 following a Fourier series analysis of the data. In the case of an overcast sky, the filtering procedure requires more care since a cut-off in frequencies may alter the actual response of the hazy or cloudy region of the atmosphere. The solution is to filter rather the signal from which the specific response is first subtracted, and then add the latter back to the filtered component. This

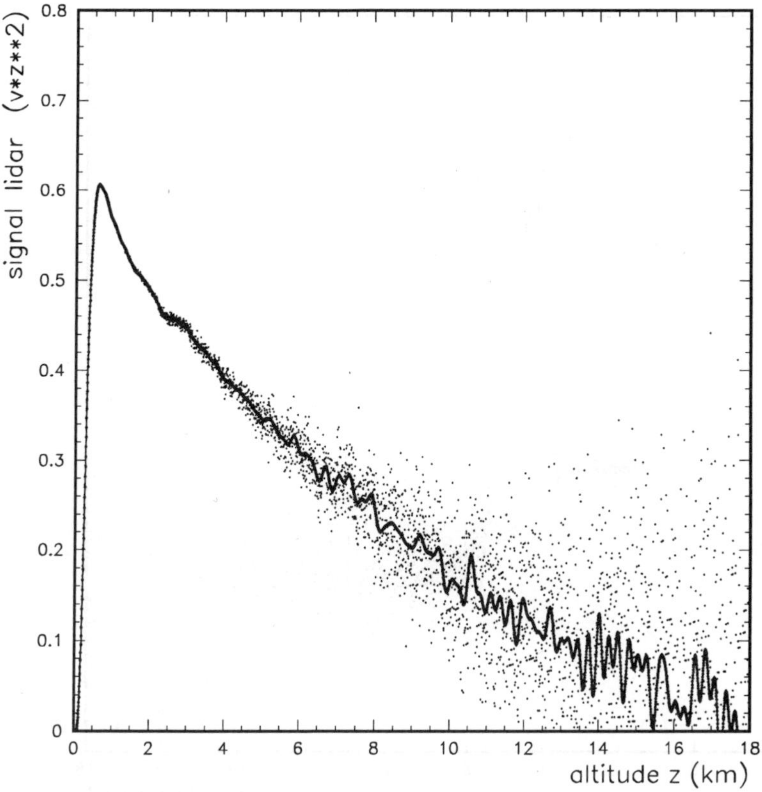

Figure 9. LIDAR signal for a clear atmosphere. Full line: the filtered signal.

remark has its importance, since the collected signal may include the linear superposition of a series of reponses from humid or dry atmosphere, aerosols, haze, clouds or ice, $\cdots$, which would require to be processed separately.

The idea that we followed in our analysis is that photon backscattering is also a reflection process. The backscattered signal, obtained through the laser excitation at a wavelength λ_o, is consided as a complex analytic function written as,

$$s(z, \lambda_0) = \rho(z, \lambda_0) \exp[-j\phi(z, \lambda_0)]. \tag{43}$$

Physically, the atmosphere is a multilayered system. If it is cloudy, one has to include the effects of absorption, leading to a rather complex situation. In the

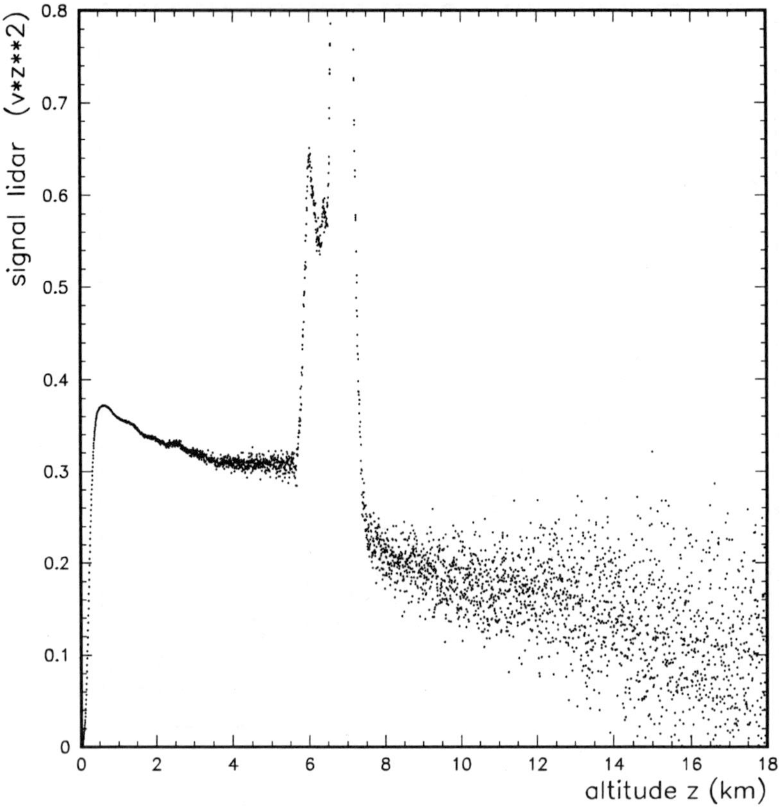

Figure 10. LIDAR signal for a cloudy atmosphere.

case that we analyze, as seen from Fig.9, it is assumed that any absorption is weak and, by recurrence, reflections on successive layers labelled by an integer m are determined according to the rule,

$$R_m = \frac{\rho_m}{\rho_{m-1}} R_{m-1}, \tag{44}$$

where ρ_m is the modulus of the signal detected at the ground, backscattered by layer m at an altitude z_m. Finally, to initialize the calculation, we give to R_1 the following value,

$$R_1 = \left(\frac{n_1 - 1}{n_1 + 1} \right)^2, \tag{45}$$

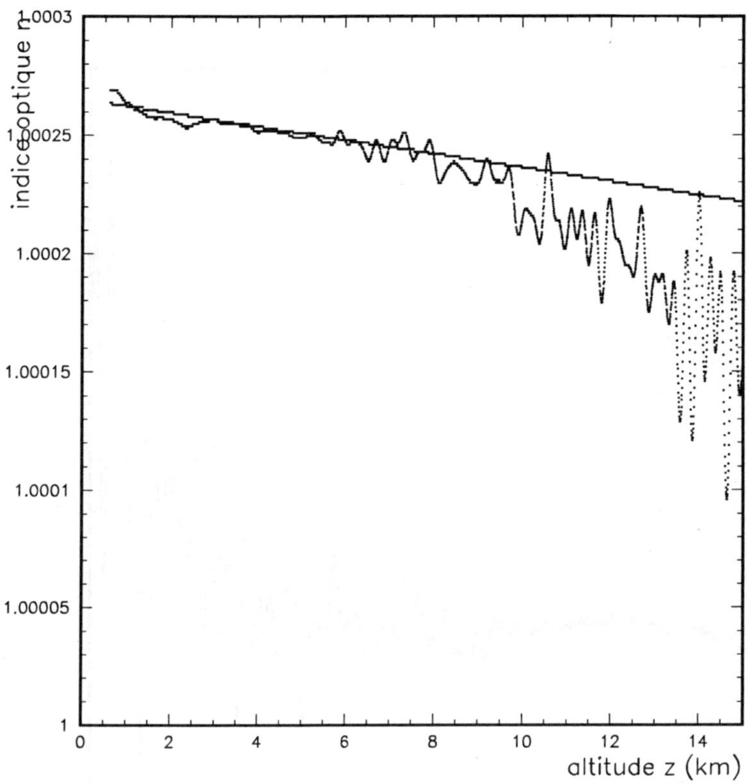

Figure 11. Refraction index for the clear atmosphere. Continuous line: $n(z)$ corresponding to an empirical relation.

in which n_1 may result from a direct interferometric measurement on site or may be given by standard tables.

The Hilbert relations applied to the dispersion $R(z, \lambda_0)$ then lead to the imaginary and real parts of the reflection factor,

$$r(z, \lambda_0) = \sqrt{R(z, \lambda_0)} \exp[-j\phi(z, \lambda_0)] \; , \tag{46}$$

which is an analytic function of the variable z. Finally, by considering that the thickness of each layer is sufficiently large (5 meters), one extracts, to first approximation, the optical indices $n(z, \lambda_0)$ and $k(z, \lambda_0)$ from the relations (31) and (32) valid for given wavelength and altitude. Figs.11 and 12 display the values for the indices $n(z, \lambda_0)$ and $k(z, \lambda_0)$ between 0.5 and 15km. Note the

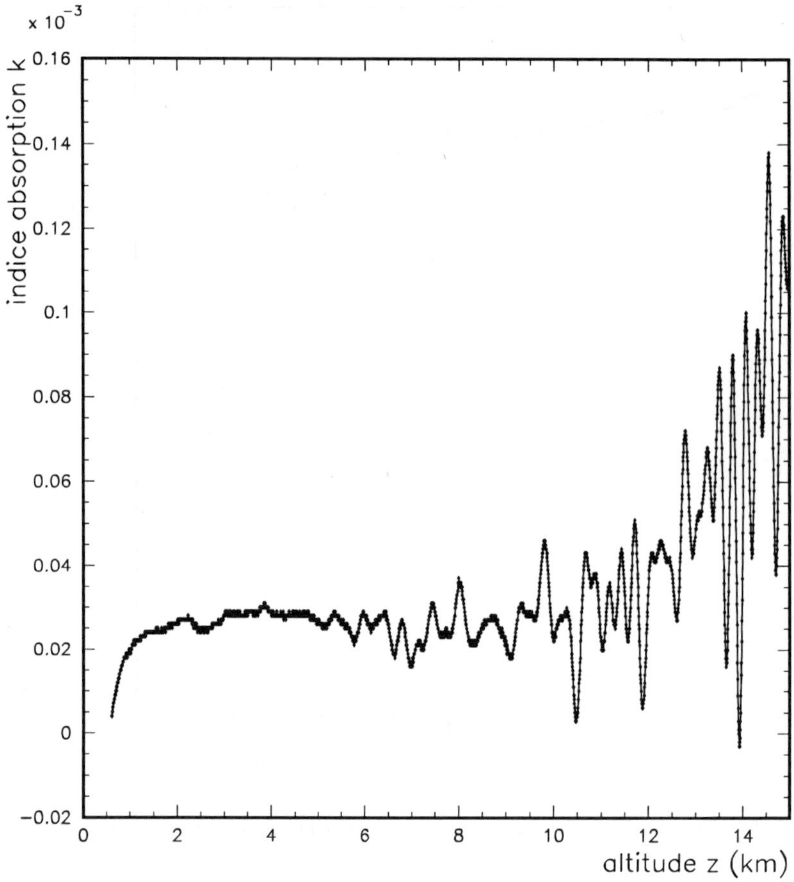

Figure 12. Absorption index for the clear atmosphere

very small values for k, in agreement with our assumptions, while the results at low altitudes follow quite well, the empirical relation,

$$n = \left[(1.00027)^2 - 5.810^{-6}(1.680 + z)\right]^{0.5} \ . \tag{47}$$

In spite of a very simplified model, these results appear to be consistent, suggesting that they could be improved through a series of recursive cycles to better account for the layered character of the experiments involved. Our approach thus proves to be of interest for the CELESTE programme. In the case of an overcast sky, one should consider the separate reponses of the different contributing components and analyze them each on their own. A best-fit should also provide complementary information describing, in addition the quasiexponential contribution of dry atmosphere, the effects of clouds or fog at any given altitude. Work along these lines is now in progress.

4 Conclusion

As these few examples have illustrated, Hilbert relations may very often be used in physical phenomena to lead to important results, based simply on the knowledge of the modulus of a function on some given large interval of some variable. One of their advantages is that they are model independent relations. Their practical implementation is not difficult in principle, but nonetheless requires care and precision. We believe that these relations could be put to great use in a large variety of fields of physics. In the particular case of the LIDAR project, these relations, in combination with simplifying model assumptions, appear to lead to satisfactory results, which we hope to be able to confirm by improving on a microscopic model description of the atmosphere.

Acknowledgements

The author is grateful to Professor B. Fabre, the Group of Fundamental Physics, and members of the CAT and CELESTE Collaborations, for the many fruitful discussions which have helped in the completion of this manuscript.

References

1. A. Erdelyi, ed., *Tables of Integral Transforms* (Mc Graw-Hill, New York, 1954), Volume 2, ch. 15.
2. E.C. Titchmarch, *Introduction to the Theory of Fourier Integrals* (Clarendon Press, Oxford, 1937).
3. G. Debiais, *Phys. Stat. Sol.* **B107**, 729 (1981).
4. G. Debiais and J.L. Dejardin, *Physica* **A175**, 407 (1991).
5. G. Debiais and J.L. Dejardin, *Physica* **A158**, 589 (1989).
6. G. Debiais, Ph. D. Thesis (Université de Perpignan, France, 1992), unpublished.
7. J.L. Dejardin and G. Debiais, *Advances in Chemical Physics*, Volume XCI, I. Prigogine and Stuart A. Rice, eds. (Wiley & Sons, New York, 1995), pp. 241–380.

8. A. Kramers, *Atti del Congresso Internazionale dei Fisici*, Como, 1927, N. Zanichelli, ed. (Bologna 2, 1928), pp. 545.
9. R. Kronig, *J. Opt. Soc. Am.* **12**, 547 (1926).
10. H.W. Bode, *Network Analysis and Feedback Amplifier Design* (van Nostrand, New York, 1945).
11. J. Ruiz-Perez, Ph. D. Thesis (Université de Toulouse, France, 1968).
12. F. Stern, in *Solid State Physics* **15**, Seitz and Turnbull, eds. (Academic Press, 1956), pp. 299–408.
13. G. Debiais, *Phys. Stat. Sol.* **A83**, 269 (1984).
14. G. Debiais and J.L. Dejardin, *J. Mol. Liquids* **49**, 37 (1991).
15. P. Debye, *Polar Molecules*[(Chemical Catalog, New York, 1929).
16. Z. Delalic, S. Takashima, K. Adachi and T. Asakura, *J. Mol. Biol.* **168**, 659 (1983).
17. J. Kerr, *Phil. Mag.* **50**, 337 (1875).
18. J.L. Dejardin and G. Debiais, *Phys. Rev.* **A40**, 1560 (1989).
19. S. Benet, Ph. D. Thesis (Université de Perpignan, France, 1987).
20. CELESTE Collaboration, *Nucl. Phys. (Proc. Suppl.)* **54B**, 362 (1997).
21. CAT Collaboration, *Nucl. Instrum. Meth.* **A416**, 425–437 (1998).

MOLECULAR-BASED MODEL FOR HYSTERESIS WITH RANDOM FORCING

N.G. MEDHIN
Department of Mathematics & Center for Theoretical Studies of Physical Systems
Clark Atlanta University, Atlanta, GA 30314
E-mail: nmedhin@yahoo.com

We consider a molecular-based model for the dynamics of rubber viscoelasticity under white noise perturbation. The dynamic equations are developed based on a stick-slip model which integrates polymer dynamics and continuum mechanics. The molecular model is based on chemically cross-linked (CC)-system of molecules which provides box constraint for entrapped physically constrained molecules (PC-molecules). Since rubber is a system of networked molecular chains, to develop a model we place at each point of the rubber continuum a unit of CC-molecules. The deformation of a unit cell dragging along an entrapped PC-molecule, and the subsequent relaxation of the PC-molecule are used to develop a model for the viscoelastic response of rubber under deformation.

1 Introduction

Various molecular and phenomenological models have been proposed to model small and large deformations in rubber. Johnson and Stacer integrate molecular models and continuum mechanics to develop a continuum model for rubber viscoelasticty.[10] Banks and Medhin used the work of Johnson and Stacer as a starting point to develop a model for the dynamic response of rubber under tensile deformations.[2] Here we incorporate random perturbation in the model presented in Banks and Medhin.[2] The motivation for incorporating random perturbation is the fact that most currently available models of polymer dynamics use Brownian motion in an essential way. Constituent molecules in molecular chains are subjected to random forces from neighbouring molecules. The random forces are assumed to be Gaussian. To make our presentation self contained we repeat some material from Banks and Medhin[2] as needed.

Johnson and Stacer use[10] the reptation model presented in Doi and Edwards.[6] In the reptation model proposed,[6] cross-linked molecules (CC-molecules) deform during a large-step-strain dragging with them entrapped, physically constrained molecules (PC-molecules). The entrapped molecules creep to return to their unstrained lower energy, higher entropy state.

To develop a continuum model Johnson and Stacer consider the CC-molecules as providing constraint unit cells or boxes with entrapped molecular segments, and place the constraint cells at each point of the rubber continuum.

Then, one proceeds to obtain equations for the associated principal stretches of the cells.[2,10]

2 Strain Energy Density and Continuum Dynamic Model

To develop a continuum model a hypothetical constraint cell is placed at each point of the rubber continuum. In addition a PC-molecular segment is assumed entrapped in the cell. We suppose that the CC-constraint tube have length $L(t)$, and the entrapped PC-molecular segment have length $\ell(t)$. Suppose the instantaneous-step-stretch at $0 \leq t_0$ is $\Delta L(t_0)$, and the corresponding stretch of the entrapped molecule is $\Delta\ell(t_0)$. If we relate the stretches in the form $\frac{\Delta\ell(t_0)}{\Delta L(t_0)} = \frac{\ell(t_0)}{L(t_0)}$ we have

$$\Delta\ell(t_0) = \frac{\ell(t_0)}{L(t_0)}\,\Delta L(t_0), \tag{2.1}$$

and in the time interval $t_0 < t < t_1$, we have

$$\ell(t) = \ell(t_0) + \frac{\ell(t_0)}{L(t_0)}\,\Delta L(t_0)e^{-(t-t_0)/\tau},$$

where τ is the relaxation time for the slip motion. Suppose that the CC-box is subjected to further instantaneous step-deformation of magnitude $\Delta L_i = \Delta L(t_i)$ at time t_i, $i = 1, 2, \ldots$. For $t_{n-1} < t < t_n$, we have

$$\ell(t) = \ell_0 + \sum_{i=1}^{n-1} \frac{\ell_i}{L_i}\,\Delta L_i e^{-(t-t_i)/\tau},$$

where $\ell_i = \ell(t_i)$, $L_i = L(t_i)$, $\Delta L_i = \Delta L(t_i)$. As $t_i - t_{i-1} \to 0$ we are led to the formula

$$\ell(t) = \ell_0 + \int_{t_0}^{t} \frac{\ell(s)}{L(s)}\,\frac{dL(s)}{ss}\,e^{-(t-s)/\tau}ds. \tag{2.2}$$

2.1 Strain energy density and stress

Energy is stored in an elastic material in a deformation process. One can compute the stress in the elastic material if the energy density function is known.

Suppose the CC-box has energy density function $W_{CC}(\lambda_{1c}, \lambda_{2c}, \lambda_{3c})$ for the stretches $\lambda_{1c}, \lambda_{2c}, \lambda_{3c}$. Similarly we suppose $W_{PC}(\lambda_{1p}, \lambda_{2p}, \lambda_{3p})$ is the energy density function corresponding to the PC-system. Here, we are thinking of the entrapped molecule as a PC-cell contained within the CC-cell. The total energy density function is $W = W_{CC} + W_{PC}$. The Cauchy stress in the principal direction e_j, where e_j is a unit vector in the x_j-direction, is

$$\tau_j = \lambda_{jc}\frac{\partial W}{\partial \lambda_{jc}} - P, \tag{2.3}$$

where P is the hydrostatic pressure. Since we are considering tensile deformation we assume $\tau_2 = \tau_3 = 0$. Then, we use the fact $\tau_2 = \tau_3 = 0$ to eliminate P from (2.3) and use $\partial\lambda_{jp}/\partial\lambda_{ic} = (\lambda_{jp}/\lambda_{ic})\delta_{ji}$ (see (2.1)) to write

$$\tau_1 = \lambda_{1c}\frac{\partial W_{CC}}{\partial \lambda_{1c}} - \lambda_{2c}\frac{\partial W_{CC}}{\partial \lambda_{2c}} + \lambda_{1p}\frac{\partial W_{PC}}{\partial \lambda_{1p}} - \lambda_{2p}\frac{\partial W_{PC}}{\partial \lambda_{2p}}. \tag{2.4}$$

To obtain a specific model we consider the strain energy density function suggested by Johnson and Stacer[10] based on experimental data of Young and Danik[9] given by

$$W_{CC} = 105.0(I_1 - 3) + 103.0(I_2 - 3), \tag{2.5}$$

$$W_{PC} = 169.0(I_1 - 3) + 0.0138(I_2 - 3)^2 + 7.89(I_2 - 3)^3, \tag{2.6}$$

where

$$I_1 = \lambda_{1c}^2 + \lambda_{2c}^2 + \lambda_{3c}^2 \quad \text{and} \quad I_2 = \lambda_{1c}^2\lambda_{2c}^2 + \lambda_{1c}^2\lambda_{3c}^2 + \lambda_{2c}^2\lambda_{3c}^2.$$

We impose the incompressibility condition $\lambda_1\lambda_2\lambda_3 = 1$ on the parameters $\lambda_{CC} = (\lambda_{1c}, \lambda_{2c}, \lambda_{3c})$, and $\lambda_{PC} = (\lambda_{1p}, \lambda_{2p}, \lambda_{3p})$.

2.2 Dynamic equation

Let u_c represent the deformation of the CC-box and u_p represent the deformation of the PC-box. Letting $\lambda_{1c} = 1 + \partial_x u_c$, and $\lambda_{1p} = 1 + \partial_x u_p$, the engineering stress $\sigma = \tau_1/\lambda_1$, using (2.4)–(2.6), is given as

$$\sigma \approx (1248 - 1014\partial_x u_p)\partial_x u_c + 1014\partial_x u_p. \tag{2.7}$$

From (2.2) we have

$$\frac{d\lambda_{1p}}{dt} = \frac{1}{\tau} - \left[\frac{1}{\tau} - \frac{1}{\lambda_{1c}}\frac{d\lambda_{1c}}{dt}\right]\lambda_{1p}. \tag{2.8}$$

The momentum balance equation is

$$\rho A_c \partial_t^2 u_c - \partial_x\left[A_c(1248 - 1014\,\partial_x u_p)\partial_x u_c\right] - \partial_x(1014\,A_c\partial_x u_p) = q, \tag{2.9}$$

where q is the applied load (in force/unit length), A_c is the cross-sectional area, and ρ the mass density. Using (2.8) to relate u_p and u_c, a linear approximation to the momentum balance equation is obtained:

$$\rho A_c \partial_t^2 u_c - 2262\, A_c \partial_x^2 u_c + \frac{1014\, A_c}{\tau} \int_0^t \partial_x^2 u_c(s,x) e^{-(t-s)/\tau} ds$$
$$= q - 1014\, A_c \partial_x^2 u_c(0,x) e^{-t/\tau}, \tag{2.10}$$

where q represents external load.

3 White Noise Perturbation

Our starting point is (2.10). We replace q in (2.10) by white noise perturbation. Let $(\Omega, \mathcal{F}, P)$ be a complete probability space, and $\{\mathcal{F}_t\}_{t\geq 0}$ be a nondecreasing family of sub-σ-algebra of $\mathcal{F}$, and let $W_k(t)$, $k = 1, 2, 3, \ldots$ be a Wiener process adapted to the filtration $\mathcal{F}_t$, $t \geq 0$.

Let λ_n and φ_n, $n = 1, 2, 3, \ldots$ be eigenvalues and corresponding eigenfunctions of the Sturm-Liouville problem

$$-y'' - \lambda y = 0$$

$$y(a) - h_0 y'(a) = 0, \qquad h_0 \geq 0,$$

$$y(b) + h_1 y'(b) = 0, \qquad h_1 \geq 0,$$

where the φ_n's are orthonormal in $H = L_2(a, b)$.

The white noise perturbation we consider is of the form

$$q(t,x) = \sum_{k=1}^{\infty} \mu_k \frac{d}{dt} W_k(t) \varphi_k(x), \tag{3.1}$$

where

$$\sum_{k=1}^{\infty} \mu_k^2 < \infty. \tag{3.2}$$

Setting

$$v = \frac{2262}{\rho} u_c - \frac{1014}{\rho\tau} \int_0^t u_c(s,x) e^{-(t-s)/\tau} ds, \tag{3.3}$$

we have, from (2.10),

$$v_{tt} + d_1 v_t - d_2 v + w = f, \tag{3.4}$$

where

$$w(t,x) = d_3^2 \int_0^t e^{-d_3(t-s)} \left(v_s + \frac{1}{\tau} v \right) ds, \tag{3.5}$$

and

$$\mathcal{F} = d_4 \sum_{k=1}^{\infty} \mu_k \frac{d}{dt} W_k(t)\varphi_k(x) - \frac{1951}{\rho} d_5^2 e^{-d_5 t} \partial^2 u_c(0,x), \tag{3.6}$$

$$d_1 = \frac{1014}{2262\tau}, \qquad d_2 = \frac{1248}{2262\tau^2}, \qquad d_3 = \frac{1248}{2262\tau},$$

$$d_4 = \frac{2262}{\rho^2 A_c}, \qquad d_5 = \frac{937}{1951\tau}.$$

Consider a solution of (3.4) in the form

$$v(t,x) = \sum_{n=1}^{\infty} v_n(t)\varphi_n(x).$$

Writing

$$f(t,x) = \sum_{n=1}^{\infty} f_n(t)\varphi_n(x),$$

$$w(t,x) = \sum_{n=1}^{\infty} w_n(t)\varphi_n(x),$$

we set

$$\begin{aligned} u_n^1(t) &= v_n(t), \\ u_n^2(t) &= v_n'(t), \\ u_n^3(t) &= w_n(t), \\ \Delta_n &= \frac{2262}{\rho}\lambda_n - \frac{1248}{1951\tau^2}. \end{aligned} \tag{3.7}$$

Then, from (3.7) we have

$$\frac{d}{dt} U_n = A_n U_n + F_n, \tag{3.8}$$

where

$$U_n = [u_n^1, u_n^2, u_n^3]^T,$$
$$F_n = [0, f_n, 0]^T,$$
$$A_n = \begin{pmatrix} 0 & 1 & 0 \\ -\Delta_n & -1014/1951\tau & -1 \\ d_3^2/\tau & d_3^2 & -d_3 \end{pmatrix},$$
$$f_n = \mu_n' \frac{d}{dt} W_n(t)\varphi_n(x) + e^{-t/\tau'}\xi_n(x),$$
$$\xi_n(x) = -\frac{1951}{\rho}(d_5)^2 A_c\langle \partial_x^2 u_c(0,\cdot), \varphi_n\rangle\varphi_n = c_n\varphi_n(x),$$
$$\mu_n' = d_4\mu_n, \qquad \tau' = \left(\frac{1951}{937}\right)\tau,$$

$$U_n = e^{tA_n}U_n(0) + \int_0^t e^{(t-s)A_n}e^{-s/\tau}\begin{pmatrix} 0 \\ c_n \\ 0 \end{pmatrix} ds + \int_0^t e^{(t-s)A_n}\begin{pmatrix} 0 \\ \mu_n' \\ 0 \end{pmatrix} dW_n(s). \tag{3.9}$$

There exists $0 < \tau_0$ such that for $0 < \tau < \tau_0$ the eigenvalues of A_n are all distinct for every n, $n = 1, 2, 3, \ldots$. For any fixed $\tau > 0$ there exists an integer $n_0 > 0$ such that for every integer $n \geq n_0$ all the eigenvalues of A_n are distinct. Without affecting our subsequent conclusion, and for ease of presentation, we will assume henceforth that the eigenvalues of A_n are distinct for $n = 1, 2, 3 \ldots$. That is, if λ_n^1, λ_n^2 and λ_n^3 are the eigenvalues of A_n, then they are distinct. One of the eigenvalues is real, and the other two are complex conjugates, and by the Ruth-Hurwitz Theorem, $\operatorname{Re}\lambda_n^i < 0$.[7]

Let λ_n^1, λ_n^2, and λ_n^3 be the eigenvalues of A_n with corresponding orthonormal eigenvectors V_n^1, V_n^2, and V_n^3. Then,

$$\begin{pmatrix} 0 \\ \mu_n' \\ 0 \end{pmatrix} = \sum_{i=1}^3 \alpha_n^i V_n^i, \tag{3.10}$$

$$\begin{pmatrix} 0 \\ c_n \\ 0 \end{pmatrix} = \sum_{i=1}^3 \beta_n^i V_n^i. \tag{3.11}$$

We note that

$$\int_0^t e^{(t-s)A_n} e^{-s/\tau'} \begin{pmatrix} 0 \\ c_n \\ 0 \end{pmatrix} ds$$

$$= \begin{cases} \displaystyle\sum_{i=1}^{3} \frac{\beta_n^i \tau'}{\tau' \lambda_n^i + 1} \left[e^{t\lambda_n^i} - e^{-t/\tau'} \right] V_n^i, & \text{if } \tau' \lambda_i^n + 1 \neq 0, \\ \displaystyle\sum_{i=1}^{3} \beta_n^i t e^{t\lambda_n^i} V_n^i, & \text{if } \tau' \lambda_i^n + 1 = 0, \end{cases} \tag{3.12}$$

and

$$E\left(\left| \sum_{i=1}^{3} \alpha_n^i \int_0^t e^{(t-s)\lambda_n^i} dW_n(s) V_n^i \right|^2 \right)$$

$$= \sum_{i=1}^{3} |\alpha_n^i|^2 \frac{1}{2 \operatorname{Re} \lambda_n^i} \left[e^{2t \operatorname{Re} \lambda_n^i} - 1 \right]. \tag{3.13}$$

From (3.10) and (3.11) we note that $|\alpha_n^i| \leq |\mu_n'|$, and $|\beta_n^i| \leq |c_n|$.

If we write $\lambda_n^1 = p_n + iq_n$, $\lambda_n^2 = p_n - iq_n$, $\lambda_n^3 = r_n \in \mathbb{R}$, then we can verify that

$$|q_n| \sim \sqrt{\Delta_n} \qquad \text{as } n \to \infty\ , \tag{3.14}$$

$$|p_n| \to \frac{507}{2262\tau} \qquad \text{as } n \to \infty\ , \tag{3.15}$$

$$|r_n| \to \frac{1248}{2262\tau} \qquad \text{as } n \to \infty. \tag{3.16}$$

Let

$$X = L^2(\Omega, H) \times L^2(\Omega, H) \times L^2(\Omega, H).$$

If $\phi = (\phi_1, \phi_2, \phi_3)^T \in X$, then $||\phi||_X^2 = \sum_{i=1}^3 ||\phi_i||_{L^2(\Omega,H)}^2$.

Theorem 3.1 *With U_n as in (3.9), if*

$$\sum_{n=1}^{\infty} (|U_n(0)|^2 + |c_n|^2 + |\mu_n'|^2) < \infty,$$

then, $\sum_{n=1}^{\infty} \varphi_n(x) U_n(t, \omega)$ converges in X to $W(t)$ uniformly on bounded subintervals of $t \geq 0$, and the covariance of $W(t)$ is given by

$$\sum_{n=1}^{\infty} \sum_{i=1}^{3} |\alpha_n^i|^2 \frac{1}{2 \operatorname{Re} \lambda_n^i} \left[e^{2t \operatorname{Re} \lambda_n^i} - 1 \right].$$

Proof The proof follows immediately from (3.10)–(3.16).

In a straightforward manner we can also verify that

Theorem 3.2 *With* $W(t)$ *as in* Theorem 3.1,

$$E(\|W(t)-W(t')\|_X^2) \leq \left(\sum_{n=1}^{\infty}|\mu_n|^2\right)|t-t'|,$$

and

$$\|E(W(t))\|_Y^2 \to 0 \qquad \text{as } t \to \infty,$$

where

$$Y = [L^2(a,b)]^3.$$

Acknowledgement

This research is supported by the NSF under cooperative agreement #HRD-9632844.

References

1. A. Ackleh, H.T. Banks and G.A. Pinter, *Well-Posed Results for Models of Elastomers*, Tech. Rep. CRSC-TR00-21, NCSU, Sept. 2000; *J. Math. Analysis Appl.*, submitted.
2. H.T. Banks and N.G. Medhin, *A Molecular Based Model for Viscoelastic Responses of Rubber in Tensile Deformations*, *Communications on Applied Nonlinear Analysis* **8**, 1–18 (2001).
3. H.T. Banks, G.A. Pinter and L.K. Potter, *Existence of Unique Weak Solutions to a Dynamical System for Nonlinear Elastomers with Hysteresis*, CRSC-TR98-43, NCSU, Nov. 1998; *Differential and Integral Equations* **13**, 1001–1024 (2000).
4. H.T. Banks, G.A. Pinter, L.K. Potter, M.J. Gaitens and L.C. Yanyo, *Modeling of Nonlinear Hysteresis in Elastomers Under Uniaxial Tension*, CRSC-TR99-99, NCSU, Feb. 1999; *J. Intelligent Material Systems and Structures* **10**, 116–134 (1999).
5. E. Coddington and N. Levinson, *Theory of Ordinary Differential Equations* (McGraw-Hill, New York, 1955).
6. M. Doi and S.F. Edwards, *The Theory of Polymer Dynamics* (Oxford University Press, New York, 1986).

7. F.R. Gantmacher, *Matrix Theory* (Chelsea Publishing Company, New York, 1959), Volume 2.
8. A.R. Johnson, *Modeling and Viscoelastic Materials Using Internal Variables, The Shock and Vibration Digest* **31**, 91-100 (1999).
9. A.R. Johnson, C.J. Quigley, D.G. Young and J.A. Danik, *Viscohyperelastic Modeling of Rubber Vulcanization, Tire Sci. Technol.* **21**, 179–199 (1993).
10. A.R. Johnson and R.G. Stacer, *Rubber Viscoelasticity Using the Physically Constrained Systems' Stretches as Internal Variables, Rubber Chemistry and Technology* **66**, 567–577 (1993).
11. R.W. Ogden, *Non-Linear Elastic Deformations* (Ellis Horwood Limited, Chichester, 1984).
12. L.H. Sperling, *Introduction to Physical Polymer Science* (John Wiley & Sons, New York, 1992).
13. L.R.G. Treloar, *The Physics of Rubber Elasticity* (Clarendon Press, Oxford, 1975).
14. K.C. Valanis and R.F. Landel, *The Strain-energy Functions of a Hyperelastic Material in Terms of the Extension Ratios, J. Applied Physics* **38**, 2997–3002 (1967).

A GENERAL FOUR-FERMION EFFECTIVE LAGRANGIAN FOR DIRAC AND MAJORANA NEUTRINO-CHARGED MATTER INTERACTIONS

JEAN EL BACHIR MENDY

Department of Physics, University Cheikh Anta Diop, Dakar, Senegal
E-mail: mendiz19@yahoo.fr

JAN GOVAERTS

Institute of Nuclear Physics, Catholic University of Louvain
2, Chemin du Cyclotron, B-1348 Louvain-la-Neuve, Belgium
E-mail: jan.govaerts@fynu.ucl.ac.be

Given the most general Lorentz invariant four-fermion effective interaction possible for two neutrinos and two charged fermions, whether quarks or leptons, all possible $2 \to 2$ processes involving two neutrinos, whether Dirac or Majorana ones, and two charged fermions are considered. Explicit and convenient expressions are given for the associated differential cross-sections. Such a parametrization should help assess the sensitivity to physics beyond the Standard Model of neutrino beam experiments which are in the design stage at neutrino factories.

1 Introduction

Neutrino physics is called to play an important role in fundamental physics, certainly for the decades to come, ranging from high energy particle physics to astroparticle physics and cosmology. Neutrinos hold the key to some of the present-day mysteries in these fields, with the potential for profound breakthroughs in our search of the unification of all interactions. However in this endeavour, the lead is certainly to come from experiments: they must identify which, if any, of the many theoretical avenues that have been imagined, has actually been chosen by Nature. Thus ambitious projects are in the making, ranging from neutrino telescopes to accelerator experiments and neutrino factories.

Given this context, it should be of interest to have available a general model independent parametrization of large classes of processes within an effective description relevant for a given energy range. For example in the 1950's, such a general four-fermion effective interaction[1] has enabled the identification of the $(V - A)$ structure of the weak interaction in β-decay and is still used in modern precision measurements.[2] A similar parametrization is also of relevance to precision studies in the muon sector looking for physics beyond the Standard Model within purely leptonic processes.[3,4] Likewise, such a parametrization of semi-leptonic muonic interactions is available in the intermediate energy range of nuclear muon capture.[5] Given the eventual advent of neutrino factories, an analogous general parametrization should thus be of interest, in order to assess the potential of any given experiment based on neutrino beams to constrain the parameter space of possible physics beyond the Standard Model (SM) within the neutrino sector.

In this note, we wish to report briefly on such an analysis and some of its results,[6,7] based on the most general four-fermion effective interaction possible of two neutrinos and two charged fermions (whether leptons or quarks) of fixed "flavours", or rather more correctly, of definite mass eigenstates, solely constrained by the requirements of Lorentz invariance and electric charge conservation. For instance, even though this might be realized only in peculiar classes of models beyond the SM, allowance is made for the possibility that both the neutrino fields and their charge conjugates couple in the effective Lagrangian density. Furthermore, the analysis is developed separately whether for Dirac or Majorana neutrinos, with the hope to identify circumstances under which scattering experiments involving neutrinos could help discriminate between these two cases through different angular correlations for differential cross sections, given the high rates to be expected at neutrino factories. As is well known, the "practical Dirac-Majorana confusion theorem" states[8] that within the SM, namely in the limit of massless neutrinos as well as $(V-A)$ interactions only, these two possibilities are physically totally equivalent, and hence cannot be distinguished. On the other hand, relaxing the purely $(V-A)$ structure of the electroweak interaction by including at least another interaction whose chirality structure is different, should suffice to evade this conclusion, even in the limit of massless neutrinos.

The general classes of processes comprise neutrino pair annihilation into charged leptons,[a] the inverse process of neutrino pair production through lepton annihilation, and finally neutrino-lepton scattering. These processes will also be considered whether either one or both pairs of neutrino and lepton flavours, (a, b) and (i, j) respectively, are identical or not. The sole implicit assumption is that the energy available to the reaction is both sufficiently large in order to justify ignoring neutrino and lepton masses, and sufficiently small in order to justify the four-fermion parametrization of the boson exchanges responsible for the interactions. Hence, calculations are performed in the limit of zero mass for all external neutrino and lepton mass eigenstates.

[a] Henceforth, the charged fermions are referred to as leptons, even though exactly the same analysis and results apply to quarks, with due account then for the quark colour degree of freedom and the quark structure of the hadrons involved. Also, charged leptons will simply be called leptons, for short.

Nonetheless, effects that distinguish Majorana from Dirac neutrinos survive in this limit. Note that this massless approximation also justifies our abuse of language in referring to neutrino mass eigenstates as flavour eigenstates.

Sect.2 provides a list of useful relations for Dirac and Majorana spinors. Sect.3 discusses the general four-fermion effective Lagrangian used in our analysis. Sects.4 to 6 then list the results for the three classes of processes mentioned above. Some comments and concluding remarks are made in Sect.7.

2 A Compendium of Properties

2.1 Dirac, Weyl and Majorana spinors

This section present facts relevant to Dirac, Weyl and Majorana spinors. Since all processes are considered in the massless limit, the representation of the Clifford-Dirac algebra $\{\gamma^\mu, \gamma^\nu\} = 2g^{\mu\nu} = 2\,\mathrm{diag}\,(+\,-\,-\,-)$ used throughout is the chiral one ($\mu = 0,1,2,3$; $i = 1,2,3$),

$$\gamma^0 = \begin{pmatrix} 0 & -\mathbb{1} \\ -\mathbb{1} & 0 \end{pmatrix} \ , \ \gamma^i = \begin{pmatrix} 0 & \sigma^i \\ -\sigma^i & 0 \end{pmatrix} \ , \ \gamma_5 = i\gamma^0\gamma^1\gamma^2\gamma^3 = \begin{pmatrix} \mathbb{1} & 0 \\ 0 & -\mathbb{1} \end{pmatrix} \ , \tag{1}$$

σ^i being of course the Pauli matrices. The chiral projectors P_η ($\eta = \pm$) are given by

$$P_\eta = \frac{1}{2}\left[1 + \eta\gamma_5\right] \ , \quad P_\eta^2 = P_\eta \quad , \quad P_\eta P_{-\eta} = 0 \ , \quad \eta = +,- \ . \tag{2}$$

By definition, the charge conjugation matrix C is such that

$$\begin{aligned} C^{-1}\mathbb{1}C &= \mathbb{1}^{\mathrm{T}} \quad , & C^{-1}\gamma_5 C &= \gamma_5^{\mathrm{T}} \ , \\ C^{-1}\gamma^\mu C &= -\gamma^{\mu\mathrm{T}} \quad , & C^{-1}\left(\gamma^\mu\gamma_5\right)C &= \left(\gamma^\mu\gamma_5\right)^{\mathrm{T}} \ , \\ C^{-1}\sigma_{\mu\nu}C &= -\sigma_{\mu\nu}^{\mathrm{T}} \quad , & C^{-1}\left(\sigma_{\mu\nu}\gamma_5\right)C &= -\left(\sigma_{\mu\nu}\gamma_5\right)^{\mathrm{T}} \ , \end{aligned} \tag{3}$$

with $C^{\mathrm{T}} = C^\dagger = -C$ and $CC^\dagger = \mathbb{1} = C^\dagger C$, and is realized in the chiral representation by $C = \mathrm{diag}(-i\sigma^2,\ i\sigma^2)$.

Given a four component Dirac spinor ψ, our definition of the associated charge conjugate spinor is such that

$$\psi_c = \psi^c = \lambda\, C\overline{\psi}^{\mathrm{T}} \ , \tag{4}$$

where λ is some arbitrary unit phase factor, whose value may depend on the spinor field.

Solutions to the free massless Dirac equation may be expanded as follows in the helicity basis in the case of a Dirac spinor $\psi_D(x)$,

$$\psi_D(x) = \int_{(\infty)} \frac{d^3\vec{k}}{(2\pi)^3 2|\vec{k}|} \sum_{\eta=\pm} \left[e^{-ik\cdot x} u(\vec{k},\eta) b(\vec{k},\eta) + e^{ik\cdot x} v(\vec{k},\eta) d^\dagger(\vec{k},\eta) \right] . \tag{5}$$

Here, the fermionic creation and annihilation operators have the Lorentz covariant normalization

$$\left\{ b(\vec{k},\eta), b^\dagger(\vec{k}',\eta') \right\} = (2\pi)^3 2|\vec{k}| \delta_{\eta,\eta'} \delta^{(3)}(\vec{k}-\vec{k}') = \left\{ d(\vec{k},\eta), d^\dagger(\vec{k}',\eta') \right\} , \tag{6}$$

while the plane wave spinors $u(\vec{k},\eta)$ and $v(\vec{k},\eta)$ are given by,

$$u(\vec{k},+) = v(\vec{k},-) = \sqrt{2|\vec{k}|} \begin{pmatrix} \chi_+(\hat{k}) \\ 0 \end{pmatrix} , \ u(\vec{k},-) = v(\vec{k},+) = \sqrt{2|\vec{k}|} \begin{pmatrix} 0 \\ \chi_-(\hat{k}) \end{pmatrix} , \tag{7}$$

with the Pauli bi-spinors

$$\chi_+(\hat{k}) = \begin{pmatrix} e^{-i\varphi/2} \cos\theta/2 \\ e^{i\varphi/2} \sin\theta/2 \end{pmatrix} \quad , \quad \chi_-(\hat{k}) = \begin{pmatrix} -e^{-i\varphi/2} \sin\theta/2 \\ e^{i\varphi/2} \cos\theta/2 \end{pmatrix} , \tag{8}$$

such that $\hat{k}\cdot\vec{\sigma}\,\chi_\eta(\hat{k}) = \eta\,\chi_\eta(\hat{k})$ and $\chi_\eta(\hat{k})\chi_\eta^\dagger(\hat{k}) = (\mathbb{1} + \eta\hat{k}\cdot\vec{\sigma})/2$, φ and θ being the spherical angles for the unit vector $\hat{k} = \vec{k}/|\vec{k}|$ with respect to the axes $i = 1,2,3$, namely $\hat{k} = (\sin\theta\cos\varphi, \sin\theta\sin\varphi, \cos\theta)$.

The value of $\eta = \pm$ coincides with the helicity of the associated massless one-particle states, as well as the chirality of the associated quantum field. Left- or right-handed four component Weyl spinors, with $\eta = -$ and $\eta = +$ respectively, read

$$\psi_\eta(x) = \int_{(\infty)} \frac{d^3\vec{k}}{(2\pi)^3 2|\vec{k}|} \left[e^{-ik\cdot x} u(\vec{k},\eta) b(\vec{k},\eta) + e^{ik\cdot x} v(\vec{k},-\eta) d^\dagger(\vec{k},-\eta) \right] , \tag{9}$$

as implied by the identification

$$\psi_\eta(x) = P_\eta\, \psi_D(x) . \tag{10}$$

Hence, $b^\dagger(\vec{k},\eta)$ and $d^\dagger(\vec{k},\eta)$ are the creation operators of a particle and of an antiparticule, respectively, each of helicity η and momentum $\vec{k}$.

This identification may also be established from the chiral properties of the plane wave spinors,

$$\begin{array}{ll} P_\eta\, u(\vec{k},\eta) = u(\vec{k},\eta) & , \ P_\eta\, u(\vec{k},-\eta) = 0 , \\ P_\eta\, v(\vec{k},\eta) = 0 & , \ P_\eta\, v(\vec{k},-\eta) = v(\vec{k},-\eta) , \\ \overline{u}(\vec{k},\eta)\, P_\eta = 0 & , \ \overline{u}(\vec{k},-\eta)\, P_\eta = \overline{u}(\vec{k},-\eta) , \\ \overline{v}(\vec{k},\eta)\, P_\eta = \overline{v}(\vec{k},\eta) & , \ \overline{v}(\vec{k},-\eta)\, P_\eta = 0 , \end{array} \tag{11}$$

as well as

$$u(\vec{k},\eta)\overline{u}(\vec{k},\eta) = \frac{\mathbb{1}+\eta\gamma_5}{2}\not{k} \quad , \quad v(\vec{k},\eta)\overline{v}(\vec{k},\eta) = \frac{\mathbb{1}-\eta\gamma_5}{2}\not{k} \ . \tag{12}$$

Their properties under charge conjugation are such that $C\overline{u}^{\rm T}(\vec{k},\eta) = v(\vec{k},\eta)$, $C\overline{v}^{\rm T}(\vec{k},\eta) = u(\vec{k},\eta)$, $\overline{v}(\vec{k},\eta) = u^{\rm T}(\vec{k},\eta)C$ and $\overline{u}(\vec{k},\eta) = v^{\rm T}(\vec{k},\eta)C$, these relations being specific to the helicity basis. Charge conjugates of spinors are then given by, say for a Dirac spinor $\psi_D(x)$,

$$\psi_D^c(x) = \int_{(\infty)} \frac{d^3\vec{k}}{(2\pi)^3 2|\vec{k}|} \sum_{\eta=\pm} \left[e^{-ik\cdot x}\lambda u(\vec{k},\eta)d(\vec{k},\eta) + e^{ik\cdot x}\lambda v(\vec{k},\eta)b^\dagger(\vec{k},\eta)\right] \ . \tag{13}$$

As opposed to a Dirac spinor comprised of two independent Weyl spinors of opposite chiralities, namely one of each of the two fundamental representations of the (covering group of the) Lorentz group, $\psi_D(x) = \psi_+(x) + \psi_-(x)$, a Majorana spinor $\psi_M(x)$ is a four component spinor, thus also covariant under Lorentz transformations, but constructed from a single Weyl spinor, say of left-handed chirality[b] $\eta = -$, and which is invariant under charge conjugation[c]

$$\psi_M(x) = \psi_-(x) + \psi_-^c(x) \quad , \quad \psi_M^c(x) = \lambda_M C\overline{\psi}^{\rm T} = \psi_M(x) \ , \tag{14}$$

where the possible spinor dependency of the phase factor λ_M is now emphasized. Consequently, the mode expansion of a Majorana spinor in the helicity basis is,

$$\psi_M(x) = \int_{(\infty)} \frac{d^3\vec{k}}{(2\pi)^3 2|\vec{k}|} \sum_{\eta=\pm} \left[e^{-ik\cdot x}u(\vec{k},\eta)a(\vec{k},\eta) + e^{ik\cdot x}\lambda_M v(\vec{k},\eta)a^\dagger(\vec{k},\eta)\right] \ , \tag{15}$$

---

[b] Since charge conjugation exchanges left- and right-handed chiralities, the chirality of the basic Weyl spinor used in this construction is irrelevant to the definition of a Majorana spinor.

[c] A similar definition starting from a Dirac rather than a Weyl spinor might be contemplated, leading then to two independent Majorana spinors, each of which is obtained in the manner just described from a single distinct Weyl spinor, namely $\psi_M^{(1)} = (\psi_D + \psi_D^c)/\sqrt{2}$ and $\psi_M^{(2)} = -i(\psi_D - \psi_D^c)/\sqrt{2}$, in complete analogy with the real and imaginary parts of a single complex scalar field as well as the physical interpretation of the associated quanta as being particles which are or not their own antiparticles. Specifically, we have $\psi_M^{(1)} = \psi_-^{(1)} + \psi_-^{(1)c}$ and $\psi_M^{(2)} = \psi_-^{(2)} + \psi_-^{(2)c}$ with $\psi_-^{(1)} = (\psi_- + \psi_+^c)/\sqrt{2}$, $\psi_-^{(2)} = -i(\psi_- - \psi_+^c)/\sqrt{2}$, where $\psi_D = \psi_- + \psi_+$. Setting either ψ_- or ψ_+ to zero, the Weyl spinors $\psi_-^{(1)}$, $\psi_-^{(2)}$ hence also the Majorana ones $\psi_M^{(1)}$, $\psi_M^{(2)}$ are then no longer independent, leading back to the construction described in the body of the text.

where the annihilation and creation operators $a(\vec{k},\eta)$ and $a^\dagger(\vec{k},\eta)$ obey the fermionic algebra

$$\left\{a(\vec{k},\eta),a^\dagger(\vec{k}',\eta')\right\} = (2\pi)^3 2|\vec{k}|\delta_{\eta,\eta'}\delta^{(3)}(\vec{k}-\vec{k}') \ . \tag{16}$$

In terms of the quanta of the basic Weyl spinor used in the construction, we have the following correspondence (the complex conjugate of a complex number z is denoted z^* throughout),

$$\begin{array}{lcl} a(\vec{k},-) : b(\vec{k},-) & ; & a^\dagger(\vec{k},-) : b^\dagger(\vec{k},-) \ , \\ a(\vec{k},+) : \lambda_M d(\vec{k},+) & ; & a^\dagger(\vec{k},+) : \lambda^*_M d^\dagger(\vec{k},+) \ , \end{array} \tag{17}$$

showing that $a^\dagger(\vec{k},\eta)$ is the creation operator of a particle of momentum $\vec{k}$ and helicity η which is also its own antiparticle. The charge conjugation phase factor λ_M is seen to corresponds to the so-callsed "creation phase factor".[9]

2.2 Differential cross sections

All $2 \to 2$ processes to be discussed are considered in their center-of-mass (CM) frame, with a kinematics of the form

$$p_1 + p_2 \to q_1 + q_2 \ , \tag{18}$$

the quantities $p_{1,2}$, $q_{1,2}$ standing for the four-momenta of the respective in-coming and out-going massless particles. Given rotational invariance, and the fact that all particles are of spin 1/2 and of zero mass, hence of definite helicity, the sole angle of relevance is the CM scattering angle θ between, say, the momenta $\vec{p}_1$ and $\vec{q}_1$. For all the reactions listed hereafter, the same order is used for the pairs (p_1, p_2) and (q_1, q_2) of the initial and final particles involved, hence leading always to the same interpretation for this angle θ as being the scattering angle between the first particle in each of these two pairs of in-coming and out-going states.

For external particles of definite helicity, the differential CM cross section of all such processes is given by

$$\frac{d\sigma}{d\Omega_{\hat{q}_1}} = \frac{1}{S_f}\frac{1}{64\pi^2 s}|\mathcal{M}|^2 \quad , \quad \frac{d\sigma}{d\cos\theta} = \frac{1}{S_f}\frac{1}{32\pi s}|\mathcal{M}|^2 \ . \tag{19}$$

Here, $\sqrt{s}$ is the reaction total invariant energy, with $s = (p_1+p_2)^2 = (q_1+q_2)^2$, $d\Omega_{\hat{q}_1}$ is the solid angle associated to the outgoing particle of normalized momentum $\hat{q}_1 = \vec{q}_1/|\vec{q}_1|$, $S_f = 2$ or $S_f = 1$ depending on whether the two particles—including their helicity—in the final state are identical or not, respectively, and $\mathcal{M}$ is Feynman's scattering matrix element. Our results are listed in terms of the relevant amplitudes $\mathcal{M}$.

3 The Four-Fermion Effective Lagrangian

Given our assumption concerning the energy and mass scales involved, an effective four-fermion parametrization is warranted, constrained by the sole requirements of Lorentz invariance and electric charge conservation. Since fermion number is not necessarily conserved, one may equally well couple the neutrino fields and their charge conjugates to the charged fermionic fields. For the latter, Dirac fields represent the ordinary charged leptons (or quarks) rather than their antiparticules. It is relative to this choice that the neutrino fields and their charge conjugates are thus specified.

We shall consider all processes involving neutrinos or their antineutrinos of definite flavours a and b, as well as leptons or their antileptons of flavours i and j, all denoted as ν_a, ν_b, ℓ_i^- and ℓ_j^-, respectively. Hence, the total four-fermion effective Lagrangian in the case of Dirac neutrinos is (through a Fierz transformation, its expression may be brought to the charge-retention form),

$$\mathcal{L}_{\text{eff}} = 4\frac{g^2}{8M^2}\left[\mathcal{L}_D + \mathcal{L}_D^\dagger\right] \quad , \quad \mathcal{L}_D = \mathcal{L}_1 + \mathcal{L}_2 + \mathcal{L}_3 + \mathcal{L}_4 \ , \tag{20}$$

each separate contribution being given by

$$\begin{aligned}\mathcal{L}_1 = S_1^{\eta_a,\eta_b}\overline{\nu_a}P_{-\eta_a}\ell_i\,\overline{\ell_j}P_{\eta_b}\nu_b \ &+ V_1^{\eta_a,\eta_b}\overline{\nu_a}\gamma^\mu P_{\eta_a}\ell_i\,\overline{\ell_j}\gamma_\mu P_{\eta_b}\nu_b \\ &+ \tfrac{1}{2}T_1^{\eta_a,\eta_b}\overline{\nu_a}\sigma^{\mu\nu}P_{-\eta_a}\ell_i\,\overline{\ell_j}\sigma_{\mu\nu}P_{\eta_b}\nu_b \ ,\end{aligned} \tag{21}$$

$$\begin{aligned}\mathcal{L}_2 = S_2^{\eta_a,\eta_b}\overline{\nu_a^c}P_{\eta_a}\ell_i\,\overline{\ell_j}P_{\eta_b}\nu_b \ &+ V_2^{\eta_a,\eta_b}\overline{\nu_a^c}\gamma^\mu P_{-\eta_a}\ell_i\,\overline{\ell_j}\gamma_\mu P_{\eta_b}\nu_b \\ &+ \tfrac{1}{2}T_2^{\eta_a,\eta_b}\overline{\nu_a^c}\sigma^{\mu\nu}P_{\eta_a}\ell_i\,\overline{\ell_j}\sigma_{\mu\nu}P_{\eta_b}\nu_b \ ,\end{aligned} \tag{22}$$

$$\begin{aligned}\mathcal{L}_3 = S_3^{\eta_a,\eta_b}\overline{\nu_a}P_{-\eta_a}\ell_i\,\overline{\ell_j}P_{-\eta_b}\nu_b^c \ &+ V_3^{\eta_a,\eta_b}\overline{\nu_a}\gamma^\mu P_{\eta_a}\ell_i\,\overline{\ell_j}\gamma_\mu P_{-\eta_b}\nu_b^c \\ &+ \tfrac{1}{2}T_3^{\eta_a,\eta_b}\overline{\nu_a}\sigma^{\mu\nu}P_{-\eta_a}\ell_i\,\overline{\ell_j}\sigma_{\mu\nu}P_{-\eta_b}\nu_b^c \ ,\end{aligned} \tag{23}$$

$$\begin{aligned}\mathcal{L}_4 = S_4^{\eta_a,\eta_b}\overline{\nu_a^c}P_{\eta_a}\ell_i\,\overline{\ell_j}P_{-\eta_b}\nu_b^c \ &+ V_4^{\eta_a,\eta_b}\overline{\nu_a^c}\gamma^\mu P_{-\eta_a}\ell_i\,\overline{\ell_j}\gamma_\mu P_{-\eta_b}\nu_b^c \\ &+ \tfrac{1}{2}T_4^{\eta_a,\eta_b}\overline{\nu_a^c}\sigma^{\mu\nu}P_{\eta_a}\ell_i\,\overline{\ell_j}\sigma_{\mu\nu}P_{-\eta_b}\nu_b^c \ ,\end{aligned} \tag{24}$$

an implicit summation over the chiralities η_a and η_b being understood of course. It is important to keep in mind that no summation over the flavour indices a and b, nor i and j is implied; all four of these values are fixed from the outset, keeping open still the possibility that a and b might be equal or not, and likewise for i and j.

The overall normalization factor $4g^2/8M^2$ involves a dimensionless coupling constant g as well as a mass scale M, while the factor 4 cancels the two factors 1/2 present in the chiral projection operators $P_{\pm\eta_a}$ and $P_{\pm\eta_b}$. The rationale for this choice of normalization is that in the limit of the SM, g is then the SU(2)$_L$ gauge coupling constant g_L and M the $W^\pm$ mass M_W, with the tree-level relation to Fermi's constant, $G_F/\sqrt{2} = g_L^2/(8M_W^2)$.

A complex value for either of the coupling coefficients $\{S, V, T\}_{1,2,3,4}^{\eta_a,\eta_b}$ leads to CP violation. The indices η_a and η_b correspond to the neutrino helicities η_a or η_b, while the lepton helicities are then identical or opposity depending on the chiral structure of the coupling operator. The tensor couplings $T_1^{\eta_a,\eta_b}$ and $T_4^{\eta_a,\eta_b}$ contribute only if $\eta_a = -\eta_b$, while the couplings $T_2^{\eta_a,\eta_b}$ and $T_3^{\eta_a,\eta_b}$ contribute only if $\eta_a = \eta_b$.

For Majorana neutrinos, the parametrization is

$$\mathcal{L}_{\text{eff}} = 4\frac{g^2}{8M^2}\left[\mathcal{L}_M + \mathcal{L}_M^\dagger\right] , \tag{25}$$

where

$$\begin{aligned}\mathcal{L}_M = S^{\eta_a,\eta_b}\overline{\nu_a}P_{-\eta_a}\ell_i\;\overline{\ell_j}P_{\eta_b}\nu_b \;&+ V^{\eta_a,\eta_b}\overline{\nu_a}\gamma^\mu P_{\eta_a}\ell_i\;\overline{\ell_j}\gamma_\mu P_{\eta_b}\nu_b \\ &+ \tfrac{1}{2}T^{\eta_a,\eta_b}\overline{\nu_a}\sigma^{\mu\nu}P_{-\eta_a}\ell_i\;\overline{\ell_j}\sigma_{\mu\nu}P_{\eta_b}\nu_b\ .\end{aligned} \tag{26}$$

Compared to the definitions above, and given the property $\psi_M^c = \psi_M$, the correspondence between the effective coupling coefficients in the Majorana and the Dirac cases is obvious. Note that in the Dirac case, the total neutrino number is conserved only for couplings of type 1 and 4, $\{S, V, T\}_{1,4}^{\eta_a,\eta_b}$, whereas the couplings of type 2 and 3, $\{S, V, T\}_{2,3}^{\eta_a,\eta_b}$, violate that quantum number by two units.

In the electroweak Standard Model, besides the normalization factor $4g^2/(8M^2) = 4g_L^2/(8M_W^2)$, in order to identify the nonvanishing couplings, different situations must be distinguished depending on whether only $W^\pm$ or only Z_0 exchanges are involved, or both.

Purely $W^\pm$ exchange processes arise when $a = i$, $b = j$, $a \neq b$ and $i \neq j$, in which case the only nonvanishing effectif coupling is the pure $(V - A)$ one, $V_1^{-,-} = -1$. For purely Z_0 neutral current processes which arise when $(a = b) \neq (i = j)$, the only nonvanishing couplings are $S_1^{-,-} = \sin^2\theta_W$ and $V_1^{-,-} = \frac{1}{4}\left(1 - 2\sin^2\theta_W\right)$, θ_W being the electroweak gauge mixing angle. Finally, charged as well as neutral exchanges both contribute only when all four fermion flavours are identical, $(a = b = i = j)$, leading to the only nonvanishing couplings $S_1^{-,-} = \sin^2\theta_W$ and $V_1^{-,-} = \frac{1}{4}\left(-1 - 2\sin^2\theta_W\right)$. In the last two situations, $\mathcal{L}_D$ and $\mathcal{L}_D^\dagger$, or $\mathcal{L}_M$ and $\mathcal{L}_M^\dagger$, are identical. Any extra coupling beyond these ones thus corresponds to some new physics beyond the SM.

The remainder of the calculation proceeds straightforwardly. Given any choice of external states for the in-coming and out-going particles including their helicities, the substitution of the effective Lagrangian operator enables the direct evaluation of the matrix element $\mathcal{M}$ using the Fock algebra of the creation and annihilation operators. Rather than computing $|\mathcal{M}|^2$ through the usual trace techniques, it is far more efficient to substitute for the $u(\vec{k},\eta)$ and $v(\vec{k},\eta)$ spinors. One then readily obtains the value for $\mathcal{M}$ as a function of θ.

4 Neutrino Pair Annihilation

In the Dirac case, neutrino pair annihilations are labelled as,

$(ab)(ij)$ Dirac neutrino annihilations

ab1: $\nu_a + \nu_b \to \ell_i^- + \ell_j^+$,	ab2: $\nu_a + \nu_b \to \ell_i^+ + \ell_j^-$,
ab3: $\nu_a + \bar{\nu}_b \to \ell_i^- + \ell_j^+$,	ab4: $\nu_a + \bar{\nu}_b \to \ell_i^+ + \ell_j^-$,
ab5: $\bar{\nu}_a + \nu_b \to \ell_i^- + \ell_j^+$,	ab6: $\bar{\nu}_a + \nu_b \to \ell_i^+ + \ell_j^-$,
ab7: $\bar{\nu}_a + \bar{\nu}_b \to \ell_i^- + \ell_j^+$,	ab8: $\bar{\nu}_a + \bar{\nu}_b \to \ell_i^+ + \ell_j^-$,

while in the Majorana case,

$(ab)(ij)$ Majorana neutrino annihilations

Mab1: $\nu_a + \nu_b \to \ell_i^- + \ell_j^+$, Mab2: $\nu_a + \nu_b \to \ell_i^+ + \ell_j^-$.

Due to common angular-momentum selection rules, the matrix element $\mathcal{M}$ for all these ten processes is

$$\begin{aligned}
&\mathcal{M}_{(ab)(ij)} = \\
&= -4s\left(\frac{g^2}{8M^2}\right) N_1\,\delta_{ij}\Big\{\delta_{ab}\delta_{\eta_i}^{-\eta_a}\delta_{\eta_j}^{-\eta_b}\left[A_{11}\sin^2\theta/2 + 2\delta_{\eta_a,\eta_b}B_{11}(1+\cos^2\theta/2)\right] \\
&\qquad +\delta_{ab}\delta_{\eta_i}^{\eta_a}\delta_{\eta_j}^{\eta_b}C_{11}\left[(1+\eta_a\eta_b) - (1-\eta_a\eta_b)\cos^2\theta/2\right] \\
&\qquad +\delta_{\eta_i}^{-\eta_b}\delta_{\eta_j}^{-\eta_a}\eta_a\eta_b D_1\left[A_{12}\cos^2\theta/2 + 2\delta_{\eta_a,\eta_b}B_{12}(1+\sin^2\theta/2)\right] \\
&\qquad +\,\delta_{\eta_i}^{\eta_b}\delta_{\eta_j}^{\eta_a}\eta_a\eta_b D_1 C_{12}\left[(1+\eta_a\eta_b) - (1-\eta_a\eta_b)\sin^2\theta/2\right]\Big\} \\
&\quad -4s\left(\frac{g^2}{8M^2}\right) N_2\,\Big\{\delta_{ab}\delta_{\eta_i}^{-\eta_b}\delta_{\eta_j}^{-\eta_a}\left[A_{21}\cos^2\theta/2 + 2\delta_{\eta_a,\eta_b}B_{21}(1+\sin^2\theta/2)\right] \\
&\qquad +\delta_{ab}\delta_{\eta_i}^{\eta_b}\delta_{\eta_j}^{\eta_a}C_{21}\left[(1+\eta_a\eta_b) - (1-\eta_a\eta_b)\sin^2\theta/2\right] \\
&\qquad +\delta_{\eta_i}^{-\eta_a}\delta_{\eta_j}^{-\eta_b}\eta_a\eta_b D_2\left[A_{22}\sin^2\theta/2 + 2\delta_{\eta_a,\eta_b}B_{22}(1+\cos^2\theta/2)\right] \\
&\qquad +\,\delta_{\eta_i}^{\eta_a}\delta_{\eta_j}^{\eta_b}\eta_a\eta_b D_2 C_{22}\left[(1+\eta_a\eta_b) - (1-\eta_a\eta_b)\cos^2\theta/2\right]\Big\}\ ,
\end{aligned} \tag{27}$$

where θ is the scattering angle between the neutrino of flavour a and the charged lepton of flavour i. The particle helicities are η_a, η_b, η_i and η_j, respectively. Tables 1 and 2 list the values for the constant phase factors $N_{1,2}$ and $D_{1,2}$ and the subsets of the scalar, tensor and vector effective couplings constants, in that order, defining the quantities $A_{11,12,21,22}$, $B_{11,12,21,22}$ and $C_{11,12,21,22}$, whether in the case of Dirac or Majorana neutrinos.

5 Neutrino Pair Production

For the sake of completeness, neutrino pair production has also been considered. In the Dirac case, the following list applies,

$$\underline{(ij)(ab)\ \text{Dirac processes}}$$

$$\begin{array}{llll} \text{ij1: } \ell_i^- + \ell_j^+ \to \nu_a + \nu_b & , & \text{ij2: } \ell_i^+ + \ell_j^- \to \nu_a + \nu_b & , \\ \text{ij3: } \ell_i^- + \ell_j^+ \to \nu_a + \bar{\nu}_b & , & \text{ij4: } \ell_i^+ + \ell_j^- \to \nu_a + \bar{\nu}_b & , \\ \text{ij5: } \ell_i^- + \ell_j^+ \to \bar{\nu}_a + \nu_b & , & \text{ij6: } \ell_i^+ + \ell_j^- \to \bar{\nu}_a + \nu_b & , \\ \text{ij7: } \ell_i^- + \ell_j^+ \to \bar{\nu}_a + \bar{\nu}_b & , & \text{ij8: } \ell_i^+ + \ell_j^- \to \bar{\nu}_a + \bar{\nu}_b & , \end{array}$$

while in the Majorana case

$$\underline{(ij)(ab)\ \text{Majorana processes}}$$

$$\text{Mij1: } \ell_i^- + \ell_j^+ \to \nu_a + \nu_b \quad , \quad \text{Mij2: } \ell_i^+ + \ell_j^- \to \nu_a + \nu_b \ .$$

For all these ten processes, the amplitude $\mathcal{M}$ is of the form

$$\begin{aligned} \mathcal{M}_{(ij)(ab)} &= \\ = 4s\left(\frac{g^2}{8M^2}\right) & N_1\,\delta_{ij}\Big\{\delta_{ab}\delta_{\eta_i}^{-\eta_a}\delta_{\eta_j}^{-\eta_b}\left[A_{11}\sin^2\theta/2 + 2\delta_{\eta_a,\eta_b}B_{11}(1+\cos^2\theta/2)\right] \\ & -\delta_{ab}\delta_{\eta_i}^{\eta_a}\delta_{\eta_j}^{\eta_b}C_{11}\left[(1+\eta_a\eta_b) - (1-\eta_a\eta_b)\cos^2\theta/2\right] \\ & +\delta_{\eta_i}^{-\eta_b}\delta_{\eta_j}^{-\eta_a}D_1\left[A_{12}\cos^2\theta/2 + 2\delta_{\eta_a,\eta_b}B_{12}(1+\sin^2\theta/2)\right] \\ & -\ \delta_{\eta_i}^{\eta_b}\delta_{\eta_j}^{\eta_a}D_1C_{12}\left[(1+\eta_a\eta_b) - (1-\eta_a\eta_b)\sin^2\theta/2\right]\Big\} \\ +4s\left(\frac{g^2}{8M^2}\right) & N_2\Big\{\delta_{ab}\delta_{\eta_i}^{-\eta_b}\delta_{\eta_j}^{-\eta_a}\left[A_{21}\cos^2\theta/2 + 2\delta_{\eta_a,\eta_b}B_{21}(1+\sin^2\theta/2)\right] \\ & -\delta_{ab}\delta_{\eta_i}^{\eta_b}\delta_{\eta_j}^{\eta_a}C_{21}\left[(1+\eta_a\eta_b) - (1-\eta_a\eta_b)\sin^2\theta/2\right] \\ & +\delta_{\eta_i}^{-\eta_a}\delta_{\eta_j}^{-\eta_b}D_2\left[A_{22}\sin^2\theta/2 + 2\delta_{\eta_a,\eta_b}B_{22}(1+\cos^2\theta/2)\right] \\ & -\ \delta_{\eta_i}^{\eta_a}\delta_{\eta_j}^{\eta_b}D_2C_{22}\left[(1+\eta_a\eta_b) - (1-\eta_a\eta_b)\cos^2\theta/2\right]\Big\}\ , \end{aligned} \tag{28}$$

θ being the angle between the lepton of flavour i and the neutrino of flavour a. The different factors and coefficients appearing in this expression are detailed in Tables 3 and 4, whether in the case of Dirac or Majorana neutrinos.

6 Neutrino Scattering

In the case of neutrino scattering onto a charged lepton, we list the results only for the classes of processes $(ai)(bj)$ and $(aj)(bi)$, since the other two classes $(bi)(aj)$ and $(bj)(ai)$ may be obtained by appropriate permutations.[6,7]

6.1 (ai)(bj) neutrino scattering processes

In the case of Dirac neutrinos, the list of processes is

$$\underline{(ai)(bj)\ \text{Dirac processes}}$$

$$\begin{array}{llll} \text{ai1: } \nu_a + \ell_i^- \to \nu_b + \ell_j^- & , & \text{ai2: } \nu_a + \ell_i^+ \to \nu_b + \ell_j^+ & , \\ \text{ai3: } \nu_a + \ell_i^- \to \bar{\nu}_b + \ell_j^- & , & \text{ai4: } \nu_a + \ell_i^+ \to \bar{\nu}_b + \ell_j^+ & , \\ \text{ai5: } \bar{\nu}_a + \ell_i^- \to \nu_b + \ell_j^- & , & \text{ai6: } \bar{\nu}_a + \ell_i^+ \to \nu_b + \ell_j^+ & , \\ \text{ai7: } \bar{\nu}_a + \ell_i^- \to \bar{\nu}_b + \ell_j^- & , & \text{ai8: } \bar{\nu}_a + \ell_i^+ \to \bar{\nu}_b + \ell_j^+ & , \end{array}$$

while in the Majorana case

$$\underline{(ai)(bj)\ \text{Majorana processes}}$$

$$\text{Mai1: } \nu_a + \ell_i^- \to \nu_b + \ell_j^- \quad , \quad \text{Mai2: } \nu_a + \ell_i^+ \to \nu_b + \ell_j^+ \ .$$

The general amplitude $\mathcal{M}$ then reads in all ten cases

$$\begin{aligned} &\mathcal{M}_{(ai)(bj)} = \\ &= 4s\left(\frac{g^2}{8M^2}\right)\ N_1\,\delta_{ij}\Big\{\delta_{ab}\delta_{\eta_i}^{\eta_a}\delta_{\eta_j}^{\eta_b}\left[A_{11} - 2\delta_{\eta_a,-\eta_b}B_{11}(\cos^2\theta/2 - \sin^2\theta/2)\right] \\ &\qquad +\delta_{ab}\delta_{\eta_i}^{-\eta_a}\delta_{\eta_j}^{-\eta_b}C_{11}\left[1 + \eta_a\eta_b(\cos^2\theta/2 - \sin^2\theta/2)\right] \\ &\qquad +\delta_{\eta_i}^{-\eta_b}\delta_{\eta_j}^{-\eta_a}\eta_a\eta_b D_1\left[A_{12}\cos^2\theta/2 - 2\delta_{\eta_a,-\eta_b}B_{12}(1+\sin^2\theta/2)\right] \\ &\qquad + \delta_{\eta_i}^{\eta_b}\delta_{\eta_j}^{\eta_a}\eta_a\eta_b D_1 C_{12}\left[(1+\eta_a\eta_b) - (1-\eta_a\eta_b)\sin^2\theta/2)\right]\Big\} \\ &+4s\left(\frac{g^2}{8M^2}\right)\ N_2\,\Big\{\delta_{ab}\delta_{\eta_i}^{-\eta_b}\delta_{\eta_j}^{-\eta_a}\left[A_{21}\cos^2\theta/2 - 2\delta_{\eta_a,-\eta_b}B_{21}(1+\sin^2\theta/2)\right] \\ &\qquad +\delta_{ab}\delta_{\eta_i}^{\eta_b}\delta_{\eta_j}^{\eta_a}C_{21}\left[(1+\eta_a\eta_b) - (1-\eta_a\eta_b)\sin^2\theta/2\right] \\ &\qquad +\delta_{\eta_i}^{\eta_a}\delta_{\eta_j}^{\eta_b}\eta_a\eta_b D_2\left[A_{22} - 2\delta_{\eta_a,-\eta_b}B_{22}(\cos^2\theta/2 - \sin^2\theta/2)\right] \\ &\qquad + \delta_{\eta_i}^{-\eta_a}\delta_{\eta_j}^{-\eta_b}\eta_a\eta_b D_2 C_{22}\left[1 + \eta_a\eta_b(\cos^2\theta/2 - \sin^2\theta/2)\right]\Big\}\ , \end{aligned} \tag{29}$$

θ being the neutrino scattering angle. The list of factors and coefficients appearing in this expression is detailed in Tables 5 and 6, both in the Dirac and in the Majorana case.

6.2 (aj)(bi) neutrino scattering processes

The list of processes in the Dirac case is

$$\underline{(aj)(bi)\ \text{Dirac processes}}$$

$$\begin{array}{llll} \text{aj1: } \nu_a + \ell_j^- \to \nu_b + \ell_i^- & , & \text{aj2: } \nu_a + \ell_j^+ \to \nu_b + \ell_i^+ & , \\ \text{aj3: } \nu_a + \ell_j^- \to \bar{\nu}_b + \ell_i^- & , & \text{aj4: } \nu_a + \ell_j^+ \to \bar{\nu}_b + \ell_i^+ & , \\ \text{aj5: } \bar{\nu}_a + \ell_j^- \to \nu_b + \ell_i^- & , & \text{aj6: } \bar{\nu}_a + \ell_j^+ \to \nu_b + \ell_i^+ & , \\ \text{aj7: } \bar{\nu}_a + \ell_j^- \to \bar{\nu}_b + \ell_i^- & , & \text{aj8: } \bar{\nu}_a + \ell_j^+ \to \bar{\nu}_b + \ell_i^+ & , \end{array}$$

while in the Majorana case

$$\underline{(aj)(bi)\ \text{Majorana processes}}$$

$$\text{Maj1: } \nu_a + \ell_j^- \rightarrow \nu_b + \ell_i^- \quad , \quad \text{Maj2: } \nu_a + \ell_j^+ \rightarrow \nu_b + \ell_i^+ \ .$$

The general scattering amplitude $\mathcal{M}$ is of the form

$$\begin{aligned}
\mathcal{M}_{(aj)(bi)} = & \\
= 4s\left(\frac{g^2}{8M^2}\right) & N_1\,\delta_{ij}\Big\{\delta_{ab}\delta_{\eta_i}^{-\eta_a}\delta_{\eta_j}^{-\eta_b}\left[A_{11}\cos^2\theta/2 - 2\delta_{\eta_a,-\eta_b}B_{11}(1+\sin^2\theta/2)\right] \\
& +\delta_{ab}\delta_{\eta_i}^{\eta_a}\delta_{\eta_j}^{\eta_b}C_{11}\left[(1+\eta_a\eta_b)-(1-\eta_a\eta_b)\sin^2\theta/2\right] \\
& +\delta_{\eta_i}^{\eta_b}\delta_{\eta_j}^{\eta_a}\eta_a\eta_b D_1\left[A_{12} - 2\delta_{\eta_a,-\eta_b}B_{12}(\cos^2\theta/2 - \sin^2\theta/2)\right] \\
& +\ \delta_{\eta_i}^{-\eta_b}\delta_{\eta_j}^{-\eta_a}\eta_a\eta_b D_1 C_{12}\left[1+\eta_a\eta_b(\cos^2\theta/2-\sin^2\theta/2)\right]\Big\} \\
+4s\left(\frac{g^2}{8M^2}\right) & N_2\Big\{\delta_{ab}\delta_{\eta_i}^{\eta_b}\delta_{\eta_j}^{\eta_a}\left[A_{21} - 2\delta_{\eta_a,-\eta_b}B_{21}(\cos^2\theta/2-\sin^2\theta/2)\right] \\
& +\delta_{ab}\delta_{\eta_i}^{-\eta_b}\delta_{\eta_j}^{-\eta_a}C_{21}\left[1+\eta_a\eta_b(\cos^2\theta/2-\sin^2\theta/2)\right] \\
& +\delta_{\eta_i}^{-\eta_a}\delta_{\eta_j}^{-\eta_b}\eta_a\eta_b D_2\left[A_{22}\cos^2\theta/2 - 2\delta_{\eta_a,-\eta_b}B_{22}(1+\sin^2\theta/2)\right] \\
& +\ \delta_{\eta_i}^{\eta_a}\delta_{\eta_j}^{\eta_b}\eta_a\eta_b D_2 C_{22}\left[(1+\eta_a\eta_b)-(1-\eta_a\eta_b)\sin^2\theta/2\right]\Big\}\ ,
\end{aligned} \tag{30}$$

the angle θ being that of the scattered neutrino. Tables 7 and 8 list the relevant factors and coefficients both in the Dirac and in the Majorana case.

7 Concluding Remarks

The above general results[6,7] provide means to assess directly the sensitivity of neutrino beam experiments in the energy range up to a few ten's of GeV's to different fundamental issues of physics in the neutrino sector, whether new interactions beyond the Standard Model, whether the Dirac or Majorana character of neutrinos. By lack of space, only one illustration of the latter instance is presented.

Scalar or tensor couplings being typically less well constrained than vector ones, let us consider an extra scalar interaction, for either of the following two elastic scattering reactions,

$$\nu_\mu + e^- \rightarrow \nu_\mu + e^- \quad , \quad \nu_\mu + \mu^- \rightarrow \nu_\mu + \mu^- \ . \tag{31}$$

These reactions are of the "ai1" type in the $(ai)(bj)$ class, with $a = b \neq i = j$ in the first case, and $a = b = i = j$ in the second. Assuming that beyond the couplings of the SM, $S_1^{+,-}$ is the sole nonvanishing extra interaction, and considering an unpolarized measurement, the sum over all polarization states reads,

$$\sum_{\rm pol.} |\mathcal{M}|^2 = (4s)^2 \left(\frac{g^2}{8M^2}\right)^2 \times \\ \times \left\{ \left[4{\rm Re}\, V_1^{-,-}\right]^2 + \left[{\rm Re}\, S_1^{-,-}\right]^2 (1+\cos\theta)^2 + \tfrac{1}{4}|S_1^{+,-}|^2 (1\pm\cos\theta)^2 \right\} , \tag{32}$$

where in the last term the upper sign corresponds to the Dirac case, and the lower sign to the Majorana case. Hence indeed, any interaction whose chirality structure differs from the SM one leads to processes in which the angular dependency discriminates between Dirac and Majorana neutrinos. Taking as an illustration a value $|S_1^{+,-}| = 0.10$ which is a typical upper-bound on such a coupling in the leptonic $(e\mu)$ sector[4], one finds a 10% sensitivity in the forward-backward asymmetry, certainly a possibility worth to be explored further within the context of realistic foreseen experimental conditions. In other words, neutrino factories may offer an alternative to neutrinoless double β-decay[10] in establishing the Dirac or Majorana character of neutrinos.

The main purpose of this work[6,7] has been to provide the general results for the Feynman amplitudes for all possible $2 \to 2$ processes with two neutrinos. On that basis, it should now be possible to develop a detailed and dedicated analysis, extending similar work within restricted classes of effective couplings,[11] of the potential reach of different such reactions towards the above physics issues, inclusive of the possible discrimination between Dirac and Majorana neutrinos, given a specific design both of neutrino beams and their intensities, and of detector set-ups. Besides the great interest to be found in neutrino scattering experiments, the possibilities offered by intersecting neutrino beams should also not be dismissed offhand without first a dedicated assessment as well, the more so since they could possibly run in parasitic mode in conjunction with other experiments given a proper geometry for the neutrino beams.

Acknowledgements

J.E.B.M. acknowledges the financial support of the "Coopération Universitaire au Développement, Commission Interuniversitaire Francophone" (CUD-CIUF) of the Belgian French Speaking Community which enabled him to pursue his Ph. D. at the "Institut de Mathématiques et de Sciences Physiques (IMSP)" (Benin), and wishes to thank the Institute of Nuclear Physics (Catholic University of Louvain, Belgium) for its hospitality while this work was being pursued. J.G. wishes to thank the C.N. Yang Institute for Theoretical Physics (State University of New York at Stony Brook, USA) for its hospitality during the Summer 2001 while part of this work was completed.

References

1. J.D. Jackson, S.B. Treiman and H.W. Wyld Jr., *Phys. Rev* **106**, 517 (1957);
   J.D. Jackson, S.B. Treiman and H.W. Wyld Jr., *Nucl. Phys.* **4**, 206 (1957).
2. For a review, see for example
   J. Deutsch and P. Quin, in *Precision Tests of the Standard Electroweak Model*, ed. P. Langacker (World Scientific, Singapore, 1995), p. 706.
3. W. Fetscher, H.-J. Gerber and K.F. Johnson, *Phys. Lett.* **B173**, 102 (1986).
4. D.E. Groom *et al.* (Particle Data Group), *Eur. Phys. Jour.* **C15**, 1 (2000).
5. J. Govaerts and J.-L. Lucio-Martinez, *Nucl. Phys.* **A678**, 110 (2000).
6. J. El Bachir Mendy, Ph.D. Thesis (Institut de Mathématiques et de Sciences Physiques, Republic of Benin, December 2001), unpublished.
7. J. El Bachir Mendy and J. Govaerts, *Neutrino-Charged Matter Interactions: a General Four-Fermion Effective Parametrization*, `hep-ph/0201280`.
8. B. Kayser and R.E. Shrock, *Phys. Lett.* **B112**, 137 (1982).
9. B. Kayser, *Phys. Rev.* **D30**, 1023 (1984);
   B. Kayser, F. Gibrat-Debu and F. Perrier, *The Physics of Massive Neutrinos*, Lecture Notes in Physics **25** (World Scientific, Singapore, 1989).
10. See for example,
    H.V. Klapdor-Kleingrothaus and A. Staudt, *Non-Accelerator Particle Physics* (Institute of Physics, Bristol, 1995);
    H.V. Klapdor-Kleingrothaus and S. Stoica, *Double-Beta Decay and Related Topics* (World Scientific, Singapore, 1996).
11. See for example, and references therein,
    M.C. Gonzalez-Garcia *et al.*, *Phys. Rev. Lett.* **82**, 3202 (1999);
    N. Fornengo, M. Maltoni, R. Tomàs Bayo and J.W.F. Valle, *Phys. Rev.* **D65**, 013010 (2001);
    P. Huber, T. Schwetz and J.W.F. Valle, *Confusing Non-Standard Neutrino Interactions with Oscillations at a Neutrino Factory*, `hep-th/0202048`;
    Z. Berezhiani, R.S. Raghavan and A. Rossi, *Probing Non-Standard Couplings of Neutrinos at the Borexino Detector*, `hep-ph/0111138`;
    Z. Berezhiani and A. Rossi, *Phys. Lett.* **B535**, 207 (2002).

Table 1. The constant factors appearing in (27) for the first five $(ab)(ij)$ Dirac neutrino annihilation processes.

	ab1	ab2	ab3	ab4	ab5
N_1	$\eta_a\lambda_a^*$	$\eta_a\lambda_b^*$	η_a	η_a	η_a
A_{11}	$S_2^{\eta_b,\eta_a}$	$S_3^{\eta_b,\eta_a *}$	$S_1^{-\eta_b,\eta_a}$	$S_4^{-\eta_b,\eta_a *}$	$S_4^{\eta_b,-\eta_a}$
B_{11}	$T_2^{\eta_b,\eta_a}$	$T_3^{\eta_b,\eta_a *}$	$T_1^{-\eta_b,\eta_a}$	$T_4^{-\eta_b,\eta_a *}$	$T_4^{\eta_b,-\eta_a}$
C_{11}	$V_2^{\eta_b,\eta_a}$	$V_3^{\eta_b,\eta_a *}$	$V_1^{-\eta_b,\eta_a}$	$V_4^{-\eta_b,\eta_a *}$	$V_4^{\eta_b,-\eta_a}$
D_1	1	1	$\lambda_a^*\lambda_b$	1	1
A_{12}	$S_2^{\eta_a,\eta_b}$	$S_3^{\eta_a,\eta_b *}$	$S_4^{\eta_a,-\eta_b}$	$S_1^{\eta_a,-\eta_b *}$	$S_1^{-\eta_a,\eta_b}$
B_{12}	$T_2^{\eta_a,\eta_b}$	$T_3^{\eta_a,\eta_b *}$	$T_4^{\eta_a,-\eta_b}$	$T_1^{\eta_a,-\eta_b *}$	$T_1^{-\eta_a,\eta_b}$
C_{12}	$V_2^{\eta_a,\eta_b}$	$V_3^{\eta_a,\eta_b *}$	$V_4^{\eta_a,-\eta_b}$	$V_1^{\eta_a,-\eta_b *}$	$V_1^{-\eta_a,\eta_b}$
N_2	$\eta_b\lambda_b^*$	$\eta_b\lambda_a^*$	η_b	η_b	η_b
A_{21}	$S_3^{\eta_b,\eta_a *}$	$S_2^{\eta_b,\eta_a}$	$S_4^{-\eta_b,\eta_a *}$	$S_1^{-\eta_b,\eta_a}$	$S_1^{\eta_b,-\eta_a *}$
B_{21}	$T_3^{\eta_b,\eta_a *}$	$T_2^{\eta_b,\eta_a}$	$T_4^{-\eta_b,\eta_a *}$	$T_1^{-\eta_b,\eta_a}$	$T_1^{\eta_b,-\eta_a *}$
C_{21}	$V_3^{\eta_b,\eta_a *}$	$V_2^{\eta_b,\eta_a}$	$V_4^{-\eta_b,\eta_a *}$	$V_1^{-\eta_b,\eta_a}$	$V_1^{\eta_b,-\eta_a *}$
D_2	1	1	1	$\lambda_a^*\lambda_b$	$\lambda_a\lambda_b^*$
A_{22}	$S_3^{\eta_a,\eta_b *}$	$S_2^{\eta_a,\eta_b}$	$S_1^{\eta_a,-\eta_b *}$	$S_4^{\eta_a,-\eta_b}$	$S_4^{-\eta_a,\eta_b *}$
B_{22}	$T_3^{\eta_a,\eta_b *}$	$T_2^{\eta_a,\eta_b}$	$T_1^{\eta_a,-\eta_b *}$	$T_4^{\eta_a,-\eta_b}$	$T_4^{-\eta_a,\eta_b *}$
C_{22}	$V_3^{\eta_a,\eta_b *}$	$V_2^{\eta_a,\eta_b}$	$V_1^{\eta_a,-\eta_b *}$	$V_4^{\eta_a,-\eta_b}$	$V_4^{-\eta_a,\eta_b *}$

Table 2. The constant factors appearing in (27) for the last three $(ab)(ij)$ Dirac neutrino annihilation processes, and the two Majorana neutrino ones.

	ab6	ab7	ab8	Mab1	Mab2
N_1	η_a	$\eta_a\lambda_b$	$\eta_a\lambda_a$	$\eta_a\lambda_a^*$	$\eta_a\lambda_b^*$
A_{11}	$S_1^{\eta_b,-\eta_a*}$	$S_3^{-\eta_b,-\eta_a}$	$S_2^{-\eta_b,-\eta_a*}$	$S^{-\eta_b,\eta_a}$	$S^{\eta_b,-\eta_a*}$
B_{11}	$T_1^{\eta_b,-\eta_a*}$	$T_3^{-\eta_b,-\eta_a}$	$T_2^{-\eta_b,-\eta_a*}$	$T^{-\eta_b,\eta_a}$	$T^{\eta_b,-\eta_a*}$
C_{11}	$V_1^{\eta_b,-\eta_a*}$	$V_3^{-\eta_b,-\eta_a}$	$V_2^{-\eta_b,-\eta_a*}$	$V^{-\eta_b,\eta_a}$	$V^{\eta_b,-\eta_a*}$
D_1	$\lambda_a\lambda_b^*$	1	1	1	1
A_{12}	$S_4^{-\eta_a,\eta_b*}$	$S_3^{-\eta_a,-\eta_b}$	$S_2^{-\eta_a,-\eta_b*}$	$S^{-\eta_a,\eta_b}$	$S^{\eta_a,-\eta_b*}$
B_{12}	$T_4^{-\eta_a,\eta_b*}$	$T_3^{-\eta_a,-\eta_b}$	$T_2^{-\eta_a,-\eta_b*}$	$T^{-\eta_a,\eta_b}$	$T^{\eta_a,-\eta_b*}$
C_{12}	$V_4^{-\eta_a,\eta_b*}$	$V_3^{-\eta_a,-\eta_b}$	$V_2^{-\eta_a,-\eta_b*}$	$V^{-\eta_a,\eta_b}$	$V^{\eta_a,-\eta_b*}$
N_2	η_b	$\eta_b\lambda_a$	$\eta_b\lambda_b$	$\eta_b\lambda_b^*$	$\eta_b\lambda_a^*$
A_{21}	$S_4^{\eta_b,-\eta_a}$	$S_2^{-\eta_b,-\eta_a*}$	$S_3^{-\eta_b,-\eta_a}$	$S^{\eta_b,-\eta_a*}$	$S^{-\eta_b,\eta_a}$
B_{21}	$T_4^{\eta_b,-\eta_a}$	$T_2^{-\eta_b,-\eta_a*}$	$T_3^{-\eta_b,-\eta_a}$	$T^{\eta_b,-\eta_a*}$	$T^{-\eta_b,\eta_a}$
C_{21}	$V_4^{\eta_b,-\eta_a}$	$V_2^{-\eta_b,-\eta_a*}$	$V_3^{-\eta_b,-\eta_a}$	$V^{\eta_b,-\eta_a*}$	$V^{-\eta_b,\eta_a}$
D_2	1	1	1	1	1
A_{22}	$S_1^{-\eta_a,\eta_b}$	$S_2^{-\eta_a,-\eta_b*}$	$S_3^{-\eta_a,-\eta_b}$	$S^{\eta_a,-\eta_b*}$	$S^{-\eta_a,\eta_b}$
B_{22}	$T_1^{-\eta_a,\eta_b}$	$T_2^{-\eta_a,-\eta_b*}$	$T_3^{-\eta_a,-\eta_b}$	$T^{\eta_a,-\eta_b*}$	$T^{-\eta_a,\eta_b}$
C_{22}	$V_1^{-\eta_a,\eta_b}$	$V_2^{-\eta_a,-\eta_b*}$	$V_3^{-\eta_a,-\eta_b}$	$V^{\eta_a,-\eta_b*}$	$V^{-\eta_a,\eta_b}$

Table 3. The constant factors appearing in (28) for the first five $(ij)(ab)$ Dirac neutrino pair production processes.

	ij1	ij2	ij3	ij4	ij5
N_1	$\eta_a\lambda_a$	$\eta_a\lambda_b$	η_a	η_a	η_a
A_{11}	$S_2^{\eta_b,\eta_a *}$	$S_3^{\eta_b,\eta_a}$	$S_1^{-\eta_b,\eta_a *}$	$S_4^{-\eta_b,\eta_a}$	$S_4^{\eta_b,-\eta_a *}$
B_{11}	$T_2^{\eta_b,\eta_a *}$	$T_3^{\eta_b,\eta_a}$	$T_1^{-\eta_b,\eta_a *}$	$T_4^{-\eta_b,\eta_a}$	$T_4^{\eta_b,-\eta_a *}$
C_{11}	$V_2^{\eta_b,\eta_a *}$	$V_3^{\eta_b,\eta_a}$	$V_1^{-\eta_b,\eta_a *}$	$V_4^{-\eta_b,\eta_a}$	$V_4^{\eta_b,-\eta_a *}$
D_1	1	1	$\lambda_a\lambda_b^*$	1	1
A_{12}	$S_2^{\eta_a,\eta_b *}$	$S_3^{\eta_a,\eta_b}$	$S_4^{\eta_a,-\eta_b *}$	$S_1^{\eta_a,-\eta_b}$	$S_1^{-\eta_a,\eta_b *}$
B_{12}	$T_2^{\eta_a,\eta_b *}$	$T_3^{\eta_a,\eta_b}$	$T_4^{\eta_a,-\eta_b *}$	$T_1^{\eta_a,-\eta_b}$	$T_1^{-\eta_a,\eta_b *}$
C_{12}	$V_2^{\eta_a,\eta_b *}$	$V_3^{\eta_a,\eta_b}$	$V_4^{\eta_a,-\eta_b *}$	$V_1^{\eta_a,-\eta_b}$	$V_1^{-\eta_a,\eta_b *}$
N_2	$\eta_a\lambda_b$	$\eta_a\lambda_a$	η_a	η_a	η_a
A_{21}	$S_3^{\eta_b,\eta_a}$	$S_2^{\eta_b,\eta_a *}$	$S_4^{-\eta_b,\eta_a}$	$S_1^{-\eta_b,\eta_a *}$	$S_1^{\eta_b,-\eta_a}$
B_{21}	$T_3^{\eta_b,\eta_a}$	$T_2^{\eta_b,\eta_a *}$	$T_4^{-\eta_b,\eta_a}$	$T_1^{-\eta_b,\eta_a *}$	$T_1^{\eta_b,-\eta_a}$
C_{21}	$V_3^{\eta_b,\eta_a}$	$V_2^{\eta_b,\eta_a *}$	$V_4^{-\eta_b,\eta_a}$	$V_1^{-\eta_b,\eta_a *}$	$V_1^{\eta_b,-\eta_a}$
D_2	1	1	1	$\lambda_a\lambda_b^*$	$\lambda_a^*\lambda_b$
A_{22}	$S_3^{\eta_a,\eta_b}$	$S_2^{\eta_a,\eta_b *}$	$S_1^{\eta_a,-\eta_b}$	$S_4^{\eta_a,-\eta_b *}$	$S_4^{-\eta_a,\eta_b}$
B_{22}	$T_3^{\eta_a,\eta_b}$	$T_2^{\eta_a,\eta_b *}$	$T_1^{\eta_a,-\eta_b}$	$T_4^{\eta_a,-\eta_b *}$	$T_4^{-\eta_a,\eta_b}$
C_{22}	$V_3^{\eta_a,\eta_b}$	$V_2^{\eta_a,\eta_b *}$	$V_1^{\eta_a,-\eta_b}$	$V_4^{\eta_a,-\eta_b *}$	$V_4^{-\eta_a,\eta_b}$

Table 4. The constant factors appearing in (28) for the last three $(ij)(ab)$ Dirac neutrino pair production processes, and the two Majorana neutrino ones.

	ij6	ij7	ij8	Mij1	Mij2
N_1	η_a	$\eta_a\lambda_b^*$	$\eta_a\lambda_a^*$	$\eta_a\lambda_a$	$\eta_a\lambda_b$
A_{11}	$S_1^{\eta_b,-\eta_a}$	$S_3^{-\eta_b,-\eta_a *}$	$S_2^{-\eta_b,-\eta_a}$	$S^{-\eta_b,\eta_a *}$	$S^{\eta_b,-\eta_a}$
B_{11}	$T_1^{\eta_b,-\eta_a}$	$T_3^{-\eta_b,-\eta_a *}$	$T_2^{-\eta_b,-\eta_a}$	$T^{-\eta_b,\eta_a *}$	$T^{\eta_b,-\eta_a}$
C_{11}	$V_1^{\eta_b,-\eta_a}$	$V_3^{-\eta_b,-\eta_a *}$	$V_2^{-\eta_b,-\eta_a}$	$V^{-\eta_b,\eta_a *}$	$V^{\eta_b,-\eta_a}$
D_1	$\lambda_a^*\lambda_b$	1	1	1	1
A_{12}	$S_4^{-\eta_a,\eta_b}$	$S_3^{-\eta_a,-\eta_b *}$	$S_2^{-\eta_a,-\eta_b}$	$S^{-\eta_a,\eta_b *}$	$S^{\eta_a,-\eta_b}$
B_{12}	$T_4^{-\eta_a,\eta_b}$	$T_3^{-\eta_a,-\eta_b *}$	$T_2^{-\eta_a,-\eta_b}$	$T^{-\eta_a,\eta_b *}$	$T^{\eta_a,-\eta_b}$
C_{12}	$V_4^{-\eta_a,\eta_b}$	$V_3^{-\eta_a,-\eta_b *}$	$V_2^{-\eta_a,-\eta_b}$	$V^{-\eta_a,\eta_b *}$	$V^{\eta_a,-\eta_b}$
N_2	η_a	$\eta_a\lambda_a^*$	$\eta_a\lambda_b^*$	$\eta_a\lambda_b$	$\eta_a\lambda_a$
A_{21}	$S_4^{\eta_b,-\eta_a *}$	$S_2^{-\eta_b,-\eta_a}$	$S_3^{-\eta_b,-\eta_a *}$	$S^{\eta_b,-\eta_a}$	$S^{-\eta_b,\eta_a *}$
B_{21}	$T_4^{\eta_b,-\eta_a *}$	$T_2^{-\eta_b,-\eta_a}$	$T_3^{-\eta_b,-\eta_a *}$	$T^{\eta_b,-\eta_a}$	$T^{-\eta_b,\eta_a *}$
C_{21}	$V_4^{\eta_b,-\eta_a *}$	$V_2^{-\eta_b,-\eta_a}$	$V_3^{-\eta_b,-\eta_a *}$	$V^{\eta_b,-\eta_a}$	$V^{-\eta_b,\eta_a *}$
D_2	1	1	1	1	1
A_{22}	$S_1^{-\eta_a,\eta_b *}$	$S_2^{-\eta_a,-\eta_b}$	$S_3^{-\eta_a,-\eta_b *}$	$S^{\eta_a,-\eta_b}$	$S^{-\eta_a,\eta_b *}$
B_{22}	$T_1^{-\eta_a,\eta_b *}$	$T_2^{-\eta_a,-\eta_b}$	$T_3^{-\eta_a,-\eta_b *}$	$T^{\eta_a,-\eta_b}$	$T^{-\eta_a,\eta_b *}$
C_{22}	$V_1^{-\eta_a,\eta_b *}$	$V_2^{-\eta_a,-\eta_b}$	$V_3^{-\eta_a,-\eta_b *}$	$V^{\eta_a,-\eta_b}$	$V^{-\eta_a,\eta_b *}$

Table 5. The constant factors appearing in (29) for the first five $(ai)(bj)$ Dirac neutrino scattering processes.

	ai1	ai2	ai3	ai4	ai5
N_1	η_a	η_a	$\eta_a\lambda_b^*$	$\eta_a\lambda_a^*$	$\eta_a\lambda_a$
A_{11}	$S_4^{\eta_b,\eta_a*}$	$S_1^{\eta_b,\eta_a}$	$S_3^{-\eta_b,\eta_a*}$	$S_2^{-\eta_b,\eta_a}$	$S_2^{\eta_b,-\eta_a*}$
B_{11}	$T_4^{\eta_b,\eta_a*}$	$T_1^{\eta_b,\eta_a}$	$T_3^{-\eta_b,\eta_a*}$	$T_2^{-\eta_b,\eta_a}$	$T_2^{\eta_b,-\eta_a*}$
C_{11}	$V_4^{\eta_b,\eta_a*}$	$V_1^{\eta_b,\eta_a}$	$V_3^{-\eta_b,\eta_a*}$	$V_2^{-\eta_b,\eta_a}$	$V_2^{\eta_b,-\eta_a*}$
D_1	1	$\lambda_a^*\lambda_b$	1	1	1
A_{12}	$S_1^{\eta_a,\eta_b*}$	$S_4^{\eta_a,\eta_b}$	$S_3^{\eta_a,-\eta_b*}$	$S_2^{\eta_a,-\eta_b}$	$S_2^{-\eta_a,\eta_b*}$
B_{12}	$T_1^{\eta_a,\eta_b*}$	$T_4^{\eta_a,\eta_b}$	$T_3^{\eta_a,-\eta_b*}$	$T_2^{\eta_a,-\eta_b}$	$T_2^{-\eta_a,\eta_b*}$
C_{12}	$V_1^{\eta_a,\eta_b*}$	$V_4^{\eta_a,\eta_b}$	$V_3^{\eta_a,-\eta_b*}$	$V_2^{\eta_a,-\eta_b}$	$V_2^{-\eta_a,\eta_b*}$
N_2	η_b	η_b	$\eta_b\lambda_a^*$	$\eta_b\lambda_b^*$	$\eta_b\lambda_b$
A_{21}	$S_1^{\eta_b,\eta_a}$	$S_4^{\eta_b,\eta_a*}$	$S_2^{-\eta_b,\eta_a}$	$S_3^{-\eta_b,\eta_a*}$	$S_3^{\eta_b,-\eta_a}$
B_{21}	$T_1^{\eta_b,\eta_a}$	$T_4^{\eta_b,\eta_a*}$	$T_2^{-\eta_b,\eta_a}$	$T_3^{-\eta_b,\eta_a*}$	$T_3^{\eta_b,-\eta_a}$
C_{21}	$V_1^{\eta_b,\eta_a}$	$V_4^{\eta_b,\eta_a*}$	$V_2^{-\eta_b,\eta_a}$	$V_3^{-\eta_b,\eta_a*}$	$V_3^{\eta_b,-\eta_a}$
D_2	$\lambda_a^*\lambda_b$	1	1	1	1
A_{22}	$S_4^{\eta_a,\eta_b}$	$S_1^{\eta_a,\eta_b*}$	$S_2^{\eta_a,-\eta_b}$	$S_3^{\eta_a,-\eta_b*}$	$S_3^{-\eta_a,\eta_b}$
B_{22}	$T_4^{\eta_a,\eta_b}$	$T_1^{\eta_a,\eta_b*}$	$T_2^{\eta_a,-\eta_b}$	$T_3^{\eta_a,-\eta_b*}$	$T_3^{-\eta_a,\eta_b}$
C_{22}	$V_4^{\eta_a,\eta_b}$	$V_1^{\eta_a,\eta_b*}$	$V_2^{\eta_a,-\eta_b}$	$V_3^{\eta_a,-\eta_b*}$	$V_3^{-\eta_a,\eta_b}$

Table 6. The constant factors appearing in (29) for the last three $(ai)(bj)$ Dirac neutrino scattering processes, and the two Majorana neutrino ones.

	ai6	ai7	ai8	Mai1	Mai2
N_1	$\eta_a\lambda_b$	η_a	η_a	η_a	η_a
A_{11}	$S_3^{\eta_b,-\eta_a}$	$S_1^{-\eta_b,-\eta_a *}$	$S_4^{-\eta_b,-\eta_a}$	$S^{-\eta_b,-\eta_a *}$	S^{η_b,η_a}
B_{11}	$T_3^{\eta_b,-\eta_a}$	$T_1^{-\eta_b,-\eta_a *}$	$T_4^{-\eta_b,-\eta_a}$	$T^{-\eta_b,-\eta_a *}$	T^{η_b,η_a}
C_{11}	$V_3^{\eta_b,-\eta_a}$	$V_1^{-\eta_b,-\eta_a *}$	$V_4^{-\eta_b,-\eta_a}$	$V^{-\eta_b,-\eta_a *}$	V^{η_b,η_a}
D_1	1	$\lambda_a\lambda_b^*$	1	1	$\lambda_a^*\lambda_b$
A_{12}	$S_3^{-\eta_a,\eta_b}$	$S_4^{-\eta_a,-\eta_b *}$	$S_1^{-\eta_a,-\eta_b}$	$S^{\eta_a,\eta_b *}$	$S^{-\eta_a,-\eta_b}$
B_{12}	$T_3^{-\eta_a,\eta_b}$	$T_4^{-\eta_a,-\eta_b *}$	$T_1^{-\eta_a,-\eta_b}$	$T^{\eta_a,\eta_b *}$	$T^{-\eta_a,-\eta_b}$
C_{12}	$V_3^{-\eta_a,\eta_b}$	$V_4^{-\eta_a,-\eta_b *}$	$V_1^{-\eta_a,-\eta_b}$	$V^{\eta_a,\eta_b *}$	$V^{-\eta_a,-\eta_b}$
N_2	$\eta_b\lambda_a$	η_b	η_b	η_b	η_b
A_{21}	$S_2^{\eta_b,-\eta_a *}$	$S_4^{-\eta_b,-\eta_a}$	$S_1^{-\eta_b,-\eta_a *}$	S^{η_b,η_a}	$S^{-\eta_b,-\eta_a *}$
B_{21}	$T_2^{\eta_b,-\eta_a *}$	$T_4^{-\eta_b,-\eta_a}$	$T_1^{-\eta_b,-\eta_a *}$	T^{η_b,η_a}	$T^{-\eta_b,-\eta_a *}$
C_{21}	$V_2^{\eta_b,-\eta_a *}$	$V_4^{-\eta_b,-\eta_a}$	$V_1^{-\eta_b,-\eta_a *}$	V^{η_b,η_a}	$V^{-\eta_b,-\eta_a *}$
D_2	1	1	$\lambda_a\lambda_b^*$	$\lambda_a^*\lambda_b$	1
A_{22}	$S_2^{-\eta_a,\eta_b *}$	$S_1^{-\eta_a,-\eta_b}$	$S_4^{-\eta_a,-\eta_b *}$	$S^{-\eta_a,-\eta_b}$	S^{η_a,η_b}
B_{22}	$T_2^{-\eta_a,\eta_b *}$	$T_1^{-\eta_a,-\eta_b}$	$T_4^{-\eta_a,-\eta_b *}$	$T^{-\eta_a,-\eta_b}$	T^{η_a,η_b}
C_{22}	$V_2^{-\eta_a,\eta_b *}$	$V_1^{-\eta_a,-\eta_b}$	$V_4^{-\eta_a,-\eta_b *}$	$V^{-\eta_a,-\eta_b}$	V^{η_a,η_b}

Table 7. The constant factors appearing in (30) for the first five $(aj)(bi)$ Dirac neutrino scattering processes.

	aj1	aj2	aj3	aj4	aj5
N_1	η_b	η_b	$\eta_b\lambda_a^*$	$\eta_b\lambda_b^*$	$\eta_b\lambda_b$
A_{11}	$S_1^{\eta_b,\eta_a}$	$S_4^{\eta_b,\eta_a *}$	$S_2^{-\eta_b,\eta_a}$	$S_3^{-\eta_b,\eta_a *}$	$S_3^{\eta_b,-\eta_a}$
B_{11}	$T_1^{\eta_b,\eta_a}$	$T_4^{\eta_b,\eta_a *}$	$T_2^{-\eta_b,\eta_a}$	$T_3^{-\eta_b,\eta_a *}$	$T_3^{\eta_b,-\eta_a}$
C_{11}	$V_1^{\eta_b,\eta_a}$	$V_4^{\eta_b,\eta_a *}$	$V_2^{-\eta_b,\eta_a}$	$V_3^{-\eta_b,\eta_a *}$	$V_3^{\eta_b,-\eta_a}$
D_1	$\lambda_a^*\lambda_b$	1	1	1	1
A_{12}	$S_4^{\eta_a,\eta_b}$	$S_1^{\eta_a,\eta_b *}$	$S_2^{\eta_a,-\eta_b}$	$S_3^{\eta_a,-\eta_b *}$	$S_3^{-\eta_a,\eta_b}$
B_{12}	$T_4^{\eta_a,\eta_b}$	$T_1^{\eta_a,\eta_b *}$	$T_2^{\eta_a,-\eta_b}$	$T_3^{\eta_a,-\eta_b *}$	$T_3^{-\eta_a,\eta_b}$
C_{12}	$V_4^{\eta_a,\eta_b}$	$V_1^{\eta_a,\eta_b *}$	$V_2^{\eta_a,-\eta_b}$	$V_3^{\eta_a,-\eta_b *}$	$V_3^{-\eta_a,\eta_b}$
N_2	η_a	η_a	$\eta_a\lambda_b^*$	$\eta_a\lambda_a^*$	$\eta_a\lambda_a$
A_{21}	$S_4^{\eta_b,\eta_a *}$	$S_1^{\eta_b,\eta_a}$	$S_3^{-\eta_b,\eta_a *}$	$S_2^{-\eta_b,\eta_a}$	$S_2^{\eta_b,-\eta_a *}$
B_{21}	$T_4^{\eta_b,\eta_a *}$	$T_1^{\eta_b,\eta_a}$	$T_3^{-\eta_b,\eta_a *}$	$T_2^{-\eta_b,\eta_a}$	$T_2^{\eta_b,-\eta_a *}$
C_{21}	$V_4^{\eta_b,\eta_a *}$	$V_1^{\eta_b,\eta_a}$	$V_3^{-\eta_b,\eta_a *}$	$V_2^{-\eta_b,\eta_a}$	$V_2^{\eta_b,-\eta_a *}$
D_2	1	$\lambda_a^*\lambda_b$	1	1	1
A_{22}	$S_1^{\eta_a,\eta_b *}$	$S_4^{\eta_a,\eta_b}$	$S_3^{\eta_a,-\eta_b *}$	$S_2^{\eta_a,-\eta_b}$	$S_2^{-\eta_a,\eta_b *}$
B_{22}	$T_1^{\eta_a,\eta_b *}$	$T_4^{\eta_a,\eta_b}$	$T_3^{\eta_a,-\eta_b *}$	$T_2^{\eta_a,-\eta_b}$	$T_2^{-\eta_a,\eta_b *}$
C_{22}	$V_1^{\eta_a,\eta_b *}$	$V_4^{\eta_a,\eta_b}$	$V_3^{\eta_a,-\eta_b *}$	$V_2^{\eta_a,-\eta_b}$	$V_2^{-\eta_a,\eta_b *}$

Table 8. The constant factors appearing in (30) for the last three $(aj)(bi)$ Dirac neutrino scattering processes, and the two Majorana neutrino ones.

	aj6	aj7	aj8	Maj1	Maj2
N_1	$\eta_b\lambda_a$	η_b	η_b	η_b	η_b
A_{11}	$S_2^{\eta_b,-\eta_a *}$	$S_4^{-\eta_b,-\eta_a}$	$S_1^{-\eta_b,-\eta_a *}$	S^{η_b,η_a}	$S^{-\eta_b,-\eta_a *}$
B_{11}	$T_2^{\eta_b,-\eta_a *}$	$T_4^{-\eta_b,-\eta_a}$	$T_1^{-\eta_b,-\eta_a *}$	T^{η_b,η_a}	$T^{-\eta_b,-\eta_a *}$
C_{11}	$V_2^{\eta_b,-\eta_a *}$	$V_4^{-\eta_b,-\eta_a}$	$V_1^{-\eta_b,-\eta_a *}$	V^{η_b,η_a}	$V^{-\eta_b,-\eta_a *}$
D_1	1	1	$\lambda_a\lambda_b^*$	$\lambda_a^*\lambda_b$	1
A_{12}	$S_2^{-\eta_a,\eta_b *}$	$S_1^{-\eta_a,-\eta_b}$	$S_4^{-\eta_a,-\eta_b *}$	$S^{-\eta_a,-\eta_b}$	$S^{\eta_a,\eta_b *}$
B_{12}	$T_2^{-\eta_a,\eta_b *}$	$T_1^{-\eta_a,-\eta_b}$	$T_4^{-\eta_a,-\eta_b *}$	$T^{-\eta_a,-\eta_b}$	$T^{\eta_a,\eta_b *}$
C_{12}	$V_2^{-\eta_a,\eta_b *}$	$V_1^{-\eta_a,-\eta_b}$	$V_4^{-\eta_a,-\eta_b *}$	$V^{-\eta_a,-\eta_b}$	$V^{\eta_a,\eta_b *}$
N_2	$\eta_a\lambda_b$	η_a	η_a	η_a	η_a
A_{21}	$S_3^{\eta_b,-\eta_a}$	$S_1^{-\eta_b,-\eta_a *}$	$S_4^{-\eta_b,-\eta_a}$	$S^{-\eta_b,-\eta_a *}$	S^{η_b,η_a}
B_{21}	$T_3^{\eta_b,-\eta_a}$	$T_1^{-\eta_b,-\eta_a *}$	$T_4^{-\eta_b,-\eta_a}$	$T^{-\eta_b,-\eta_a *}$	T^{η_b,η_a}
C_{21}	$V_3^{\eta_b,-\eta_a}$	$V_1^{-\eta_b,-\eta_a *}$	$V_4^{-\eta_b,-\eta_a}$	$V^{-\eta_b,-\eta_a *}$	V^{η_b,η_a}
D_2	1	$\lambda_a\lambda_b^*$	1	1	$\lambda_a^*\lambda_b$
A_{22}	$S_3^{-\eta_a,\eta_b}$	$S_4^{-\eta_a,-\eta_b *}$	$S_1^{-\eta_a,-\eta_b}$	$S^{\eta_a,\eta_b *}$	$S^{-\eta_a,-\eta_b}$
B_{22}	$T_3^{-\eta_a,\eta_b}$	$T_4^{-\eta_a,-\eta_b *}$	$T_1^{-\eta_a,-\eta_b}$	$T^{\eta_a,\eta_b *}$	$T^{-\eta_a,-\eta_b}$
C_{22}	$V_3^{-\eta_a,\eta_b}$	$V_4^{-\eta_a,-\eta_b *}$	$V_1^{-\eta_a,-\eta_b}$	$V^{\eta_a,\eta_b *}$	$V^{-\eta_a,-\eta_b}$

ON THE MAGNETIC POLARIZATION OF MN MONOLAYER AND MNCO SURFACE ALLOY ON A CO(001) SUBSTRATE

B. M'PASSI MABIALA
Laboratoire des Matériaux et Energies
Département de Physique, B.P. 69, Brazzaville, Congo
E–mail: mpassi_b@yahoo.com

S. MEZA-AGUILAR
Escuela de Ciencias Fisico Matemáticas, Universidad Autónoma de Sinaloa
Bldv. de las Americas y Universitarios, Ciudad Universitaria, Culiacán Sinaloa
C.P. 80010, México

C. DEMANGEAT
Institut de Physique et Chimie des Matériaux de Strasbourg
23, rue du Loess, F-63037 Strasbourg, France

We have investigated the magnetic properties of the Mn monolayer, the $Mn_{0.5}Co_{0.5}$ ordered surface alloy and the $(Mn)_1/(Mn_{0.5}Co_{0.5})$ subsurface alloy on the Co(001)substrate, using the tight-binding linear muffin-tin orbitals method in the generalized gradient approximation of the density functional theory. Ferromagnetic ordering has been found in the surface alloy as well as in the subsurface alloy in agreement with experimental results. The Mn monolayer remains in the $c(2\times2)$ antiferromagnetic configuration as in previous local density calculations.

1 Introduction

Scientific and technical advances have created unprecedented new opportunities for designing artificial materials. Among these are epitaxial grown magnetic ultrathin $3d$ transition metal films stabilized on various non-magnetic and magnetic materials substrates. With the increasing sophistication of our unterstanding of such phenomena as interface anisotropy, giant magnetoresistance and interlayer coupling, the role of the interface magnetic moments is becoming recognized as increasingly important. New impetus in this field has been recently given by the application of scanning tunneling microscopy (STM) studies to the structural characterisation of ultrathin magnetic films. In addition new magnetic microscopy techniques have emerged, notably magnetic force microscopy, spin-sensitive STM techniques, near optical magneto-optical Kerr-effect (MOKE), the application of advanced lithography, and pattern transfer techniques to the fabrication of nanoscale elements based on ultrathin magnetic films.

The $3d$ transition metals films show enhanced magnetic moments to those in bulk materials.[1,2] Most *ab-initio* calculations are centered around $3d$-monolayers on (001)-oriented noble metal substrates as discussed in several articles.[3,4,5] These systems can be considered as experimental realizations of two-dimensional magnets with little influence on the magnetic properties from the substrate. For these monolayer systems it was found, (i) that the magnetic moments of the ultrathin films are very large, approaching the atomic limit, and, (ii) that there are two competing magnetic phases: a $p(1 \times 1)$ ferromagnetic phase and an in-plane $c(2 \times 2)$ antiferromagnetic phase. A general trend was discovered: early transition metals (V, Cr, and Mn) prefer the antiferromagnetic superstructure and late transition metal metals (Fe, Co, and Ni) are stable in the ferromagnetic state. The magnetism is controlled by the hybridization between d electrons on nearest-neighbour sites, represented in terms of a simplified spin model by an in-plane exchange coupling. Mn is a particulary interesting case because according to Hund's rule the magnetic moment of the free atom is as large as 5 μ_B. Mn is also known to form strongly ferromagnetic compounds such as MnSb[6] and to have a magnetic moment of 4.9 μ_B as dilute impurity in Cu.[7] If such a large magnetic (or even close to it) could be experimentally stabilized in ferromagnetic lattice, this could constitute a major advance in atomically engineered materials. More complicated is the study of the interdiffusion between two magnetic elements because the strong *d-d* hybridization perturbs many layers of the magnetic substrate contrary to the magnetic element on noble substrate. A important issue is the coupling of different magnetic moments across a common interface. Investigation of magnetic properties at such interfaces is intimately related to the phenomena of interlayer exchange coupling,[8,9], the giant magneto-resistance,[10,11] and the spin-valve realization[12] for which for example Cr/Fe(001)[13,14] and Mn/Fe(001),[12] respectively, developed to model systems. The outcome on the magnetic coupling is partly contradictory, particulary at the monolayer range because of coverage. At the monolayer range, different growth conditions such as layer-by-layer growth, island growth, and interface alloy formation may occur within small temperature ranges. The Mn/Co/Cu(001) structure provides a model epitaxial system which is well suited to the study of the magnetic interaction between Mn and ferromagnetic Co. Mn ultrathin films have been of interest because Mn forms an ideally two-dimensional surface alloy and the magnetic property is considered to affect geometry strongly. Mn has a complicated magnetic structure, and exhibits a wide variety of structural and magnetic properties that make it is an interesting candidate for film growth. Mn exists in four phases α, β, γ and δ. The α and β phases are complex structures with 29 and 20 atoms, respectively. The γ and δ phases are the simple fcc and bcc structures, respectively.

Up to now, contradictions remain between experimental and theoretical results. O'Brien *et al.*[15] used XMCD and found that a single Mn monolayer is ferromagnetically aligned with respect to the fcc Co(001) substrate. They suspected in an other work[16] the instability of the perfect Mn monolayer on Co(001) and proposed that the formation of a Mn-Co surface alloy, which is consistent with free-energy arguments, may explain this disagreement in the theoretical and experimental exchange coupling between Mn and Co. Choi *et al.*[17] using low–energy electron diffraction recently displayed an ordered Mn-Co alloy at the surface of Co for a 0.3-0.8 equivalent monolayer of Mn on Co(001). From MOKE and XMCD experiments it was concluded that the coupling between Mn and Co in these surface alloys is ferromagnetic in type. Recently, investigations of the magnetism of an ultrathin Mn film on Co(001) and the effect of oxidation by XMCD have confirmed the Mn-Co ferromagnetic coupling without O_2.[18] On the contrary, a competition between an in-plane ordered $c(2\times2)$ configuration and one in which a ferromagnetic Mn monolayer is antiferromagnetically coupled to the Co is predicted by tight-binding model calculations.[19] Ab-initio TB-LMTO-ASA calculation using LSDA[20] was performed and the $c(2\times2)$ antiferromagnetic configuration was confirmed to be the ground state and the difference of the total energy with metastable configurations was not considerable. A magnetic configuration with a Mn in a high moment state $3.2\mu_B$ and ferromagnetically coupled to Co is obtained in agreement with recent MOKE and XMCD experiments. However, this is a metastable solution with total energy slightly higher than the ground state energy displaying an antiferromagnetic coupling between Mn and Co. Stepanyuk *et al.*[21] within a Korringa-Kohn-Rostocker-Green's-function approach, obtained two solutions in the case of a Mn impurity in bulk fcc Co. The solution with antiferromagnetic coupling between the Mn impurity and Co host moments has lower total energy than that with the magnetic moment of the Mn impurity aligned with that of Co. So far, the strong correlation between surface alloying and magnetic polarization at the Mn/Co interface is up to now an almost unsolved problem.

Wüttig *et al.*[22] were the first to identify the magnetism of Mn as the driving force for their buckling and moreover, for the stability of ordered MnCu surface alloy film on Cu(001). Later, Blügel[23] reported on the stabilization of surface alloys through the magnetic component of the generalized gradient approximation (GGA) on exchange and correlation. He found that GGA promotes stability of MnCu surface alloy considerably. Motivated by the recent experimental results of Yonamoto[18] where a ferromagnetic coupling was found at the MnCo interface without O_2, and magnetic measurements of Co monolayers on fcc Cu(001) by Ney *et al.*,[24] the purpose of the present paper is to investigate the magnetic map of the Mn-Co surface alloy on Co(001) using GGA of Langreth-Mehl-Hu[30] and the Perdew-Wang-91.[31] Following Blügel's conclusions about the GGA,[23] we guess that the LDA used by Meza *et al.*[20] and the semi-empirical code used by Noguera *et al.*[19] could be the origin of the disagreement with the experimental results. It is well known that the LDA fails to describe magnetic properties of 3*d* transition metal. It cannot for example give the correct ground state of the bcc ferromagnetic (FM) phase for Fe.[25] TB-LMTO results have been already obtained in GGA for the determination of 3*d* transition-metal monolayer on Cu(111) and Ag(111),[26] and the results obtained appeared suitable.

The aim of this work is to investigate using TB-LMTO-ASA in the GGA approximation the following systems Mn/Co, $Mn_{0.5}Co_{0.5}$/Co and Mn/$Mn_{0.5}Co_{0.5}$/Co in the non-magnetic case as well as in the magnetic one. For the magnetic case all configurations have to be considered and the ground state configuration for each system is obtained from the total energy calculations. The results have to be compared with LDA calculations done by Meza *et al.*[20] and experimental results obtained by O'Brien and coworkers[15,16] and Choi *et al.*[17]

2 Computational Model

We have used a scalar-relativistic version of the *k*-space TB-LMTO method[27] with atomic sphere approximation. This method is based in the density functional theory.[28] In this approach we have first determined the lattice parameter for bulk Co by energy minimization of the total energy. We have tested different functionals, the von Barth-Hedin,[29] Langreth-Mehl-Hu[30] (GGA-LMH) and Perdew-Wang[31] (GGA-PW-91) functionals, and the different values obtained are shown in Table 1. We have retained the GGA-PW-91 functional because it is known to reproduce the magnetic properties of Mn as shown by Asada and Terakuda.[32]

Table 1. Lattice parameter (in *a.u.*) and magnetic moments (μ_B) for the differents approximations for bulk fcc Co.

Approximation	Lattice parameter	magnetic moment
LSDA-vBH	6.54	1.56
GGA-LMH	6.64	1.64
GGA-PW-91	6.81	1.69

The overlayer surface is modeled using repeated slab geometry.[33] We take nine metallic layers separated by five layers of empty spheres for the Mn monolayer on Co(001) and $Mn_{0.5}Co_{0.5}$ alloy on Co(001) (see Fig.1 of Meza-Aguilar and coworkers[20]), and eleven metallic layers for Mn on $Mn_{0.5}Co_{0.5}$/Co(001). This is sufficient to prevent interaction between the slabs which is controlled through vanishing dispersion in the direction perpendicular to the slab and the vanishing charge in the central layer of the empty spheres. The calculations are performed using an increasing number of k points, until final convergence is obtained for at least 338 points in the irreductible Brillouin zone (IBZ). We restrict to two inequivalent atoms per layer.

3 Results and Discussion

We report here the results of calculations done on the Mn monolayer, the 2D $Mn_{0.5}Co_{0.5}$ surface alloy and the $(Mn)_1/(Mn_{0.5}Co_{0.5})$ subsurface on fcc Co(001) substrate in the non-magnetic as well as in the magnetic cases. We have calculated the formation energy (E) following Blügel formula,[23]

$$E = E_{Mn-Co/Co(001)} - 1/2(E_{Co/Co(001)} + E_{Mn/Co(001)}),$$

where $E_{Mn-Co/Co(001)}$, $E_{Co/Co(001)}$ and $E_{Mn/Co(001)}$ are the total energies of the Mn-Co surface alloy on Co(001), the Co(001) slab and the Mn monolayer on Co(001) substrate, respectively.

For the Mn monolayer on Co(001), the following magnetic configurations have been considered as input: $p(1 \times 1) \uparrow$, $p(1 \times 1) \downarrow$ and $c(2 \times 2)$. The converged configurations obtained (see Table 2) show that the antiferromagnetic $c(2 \times 2)$ remains the ground state in agreement with LSDA results[20] and in disagreement with experimental results.[15,16]

Calculations done on non-magnetic Mn monolayer, $Mn_{0.5}Co_{0.5}$ surface alloy on Co(001), have shown that the surface alloy is more stable than the Mn monolayer by -5.86 mRy. Following this trend which is in agreement with the experimental results of Choi *et al.*,[17] we have investigated the magnetic case of the surface alloy. We have considered the case of 2D $Mn_{0.5}Co_{0.5}$ surface ordered alloy (1ML-thick) in the surface plane. This has been done in relation with the results of Choi *et al.*[17] displaying surface alloy for Mn concentrations from 0.3 to 0.8 equivalent ML. Four types of magnetic configurations (Mn↓Co↓, Mn↑Co↓, Mn↑Co↑ and Mn↓Co↑) corresponding to all possible directions of magnetization of Mn and Co surface atoms have been considered. Only two configurations (Mn↓Co↑ and Mn↑Co↑) remain after convergence (see Table 3).

Table 2. Magnetic moments (in μ_B) for Mn/Co(001) and differences of total energies per cell (in mRy) with the Perdew-Wang-91 approximation. The ground state is indicated by 0.

Input	c(2×2)	p(1×1)↑	p(1×1)↓
Energy	0.0	46	159
Atom	Moments	Moments	Moments
Mnb	3.22	3.53	-3.26
Mna	-3.51	3.53	-3.26
Co4b	1.15	1.55	1.17
Co4a	1.14	1.55	1.17
Co3b	1.75	1.77	1.74
Co3a	1.79	1.77	1.74
Co2b	1.71	1.70	1.71
Co2a	1.71	1.70	1.71
Co1b	1.71	1.70	1.70
Co1a	1.71	1.70	1.70

The configuration with ferromagnetic coupling between the Mn surface atoms and the Co atoms is the ground state in agreement with the experimental results.[17]

Table 3. Magnetic moments (in μ_B) for Mn-Co/Co(001) and total energies per cell (in Ry) with GGA-PW-91 approximation.

Input	Mn↓Co↓ Mn↓Co↑	Mn↑Co↓ Mn↑Co↑
Energy	32.7	0.00
Atom	Moments	Moments
Mn	-3.79	3.67
Co5	0.82	1.73
Co4b	1.33	1.59
Co4a	1.33	1.59
Co3b	1.77	1.76
Co3a	1.71	1.73
Co2b	1.71	1.70
Co2a	1.71	1.70
Co1b	1.70	1.70
Co1a	1.71	1.70

The formation energy in the magnetic calculations, reported in Table 4, is in agreement with the trends found in the non-magnetic case, i.e., a stabilization of the $Mn_{0.5}Co_{0.5}$ configuration. The total energies of all the magnetic configurations reported here are lower than the corresponding non-magnetic one. However it is worthy to point out that contrary to the results reported within LDA by Meza *et al.*[20] where the formation energy decreases when going from the non-magnetic case to the magnetic one, here, within GGA the opposite is true.

Table 4. Differences of energies (in mRy) with LSDA (a) and GGA (b) functionals, per cell between the ground state configurations for the $Mn_{0.5}Co_{0.5}$/Co(001) alloy, Mn ML on Co(001), $(Mn_{0.5}Co_{0.5})_2$/Co(001) and Co/Mn/Co(001). The values in parentheses denote those of half-ML coverage. For Mn ML coverage, for Mn on Co(001), Co/Mn/Co(001) and ordered $(Mn_{0.5}Co_{0.5})_2$/Co(001) alloy, it is a simple average with that of Co(001), corresponding to the surface segregation. The origin of energies (noted by 0) for LSDA is the spin polarized Co/Mn/Co(001) system in the Mn ML and half-ML coverage and for GGA it is the spin polarized Mn/Co(001) system in the Mn ML coverage and $Mn_{0.5}Co_{0.5}$/Co(001) system for half-ML coverage.

Approximation	(a)	LSDA
System	non-magnetic	magnetic
Co/Mn/Co(001)	54 (116)	0 (0)
$(Mn_{0.5}Co_{0.5})_2$/Co(001)	129 (154)	15 (8)
$Mn_{0.5}Co_{0.5}$/Co(001)	(173)	(12)
Mn/Co(001)	207 (193)	34 (17)
Approximation	(b)	GGA-PW-91
System	non magnetic	magnetic
$Mn_{0.5}Co_{0.5}$/Co(001)	(422)	(0)
Mn/Co(001)	428 (441)	0 (10)
$(Mn_{0.5}Co_{0.5})_2$/Co(001)	352 (383)	4 (12)
Co/Mn/Co(001)	273 (344)	35 (27)

In order to investigate the very abrupt transition from magnetically ordered to non-magnetic behaviour of the Mn film as function of film thickness observed by O'Brien and coworkers,[15,16] calculations have been done for $(Mn)_n/(Mn_{0.5}Co_{0.5})$ on Co(001) with $n = 1$. All magnetic configurations are to be taken into account. Results obtained are shown in Table 5. The Mn monolayer has the $c(2 \times 2)$ AF but the subsurface alloy remains ferromagnetic, and the mean magnetic moment of Mn surface atom is very close to zero. Calculations for $n = 2, 3$ are presently under investigation.[34]

Table 5. Magnetic moments (in μ_B) for Mn/$Mn_{0.5}Co_{0.5}$/Co(001) and differences of energies per cell (in mRy) with GGA-PW-91 approximation. The ground state configuration is noted 0.00. The first line is the input configuration of Mn surface monolayer. The second line is the input configuration of the Mn-Co subsurface alloy. By lack of convergence, the column with (Input 1: p(1 × 1) ↓) and (Input 2: ↑↓) is not included.

Input 1	c(2 × 2)	c(2 × 2)	p(1 × 1) ↓	p(1 × 1) ↑	p(1 × 1) ↑
Input 2	↑↑	↑↓	↑↑	↑↓	↑↑
Energy	0.00	29	58	49	94
Atom					
Mn2b	-3.37	-3.12	-3.17	3.16	3.38
Mn2a	3.13	3.35	-3.17	3.16	3.38
Co5	2.35	-0.04	0.87	0.95	1.43
Mn	1.02	-2.33	2.57	-2.91	2.44
Co4b	1.70	1.44	1.65	1.54	1.68
Co4a	1.65	1.53	1.65	1.54	1.68
Co3b	1.74	1.74	1.73	1.76	1.75
Co3a	1.71	1.68	1.70	1.68	1.72
Co2b	1.72	1.72	1.70	1.72	1.71
Co2a	1.67	1.71	1.70	1.72	1.71
Co1b	1.70	1.71	1.70	1.73	1.72
Co1a	1.70	1.70	1.68	1.70	1.69

4 Conclusion

We have found, within GGA, that the c(2 × 2) antiferromagnetic coupling between the Mn surface atoms for the Mn monolayer on fcc Co(001) remains in the ground state in agreement with the theoretical results of Meza *et al.*[20] but in disagreement with experiment. Total energy obtained within spin-polarisation calculations as function of the magnetic configurations performed on the Mn monolayer and the 2D ordered $Mn_{0.5}Co_{0.5}$ surface alloy on Co(001) using TB-LMTO-ASA in the GGA-PW-91 approximation have shown that, as well as in previous LSDA calculations,[20] the 2D ordered $Mn_{0.5}Co_{0.5}$ surface ordered alloy is the ground state configuration. A ferromagnetic ground state is, however, found within GGA for the 2D ordered $Mn_{0.5}Co_{0.5}$ surface alloy in agreement with experimental results.[17] This ferromagnetic ground state remains as subsurface alloy in $(Mn)_1/(Mn_{0.5}Co_{0.5})$ on Co(001).

Acknowledgments

B.M. and S.M. wish to acknowledge the support of the Abdus Salam International Centre for Theoretical Physics (ICTP) and Swedish International Development Cooperation Agency (SIDA) for financial support. C.D. wishes to acknowledge Stefan Pick and J.A.C. Bland for fruitful discussions.

References

1. S. Blügel, B. Drittler, R. Zeller and P.H. Dederichs, *Appl. Phys. A: Solids Surf.* **49**, 547 (1989).
2. S. Mirbt, O. Erickson and B. Johannson, *Phys. Rev.* **B52**, 15070 (1995).
3. A.J. Freeman and R.Wu, *J. Magn. Magn. Mat.* **100**, 497 (1991).
4. M. Weinert and S.S. Blügel, in *Magnetic Multilayers*, L.H. Bennett and R.E. Watson, eds. (World Scientific, Singapore, 1993).
5. T. Asada, G. Bihlmayer, S. Handschuh, S. Heinze, Ph. Kurz and S.S. Blügel, *J. Phys.: Condens. Matter* **11**, 9347 (1999).
6. M.C. Cadeville and J.L. Moran-Lopez, *Phys. Reports* **153**, 331 (1987).
7. J.D. Cohen and C.P. Slichter, *Phys. Rev.* **B22**, 45 (1980).
8. P. Grünberg, R. Schreiber, Y. Pang, M.B. Brodsky and H. Sowers, *Phys. Rev. Lett.* **57**, 2442 (1986).
9. S.S.P. Parkin, N. Pore and K.P. Roche, *Phys. Rev. Lett.* **64**, 2304 (1990).
10. M.N. Baibich, J.M. Broto, A. Fert, F. NGuyen Van Dau, F. Petroff, P. Etienne, G. Creuset, A. Friederich and J. Chazelas, *Phys. Rev. Lett.* **61**, 2472 (1988).
11. G.Binasch, P. Grünberg, F. Saurenbach and W. Zinn, *Phys. Rev.* **B39**, 4828 (1989).
12. B. Dieny , V.S. Speriosu, S.S.P. Parkin, B.A. Gurney, D.R. Wilhoit and D. Mauri, *Phys. Rev.* **B43**, 1297 (1991).
13. J. Unguris, R.J. Celotta and D.T. Pierce, *Phys. Rev. Lett.* **67**, 140 (1991).
14. P. Grünberg, S. Demokritov, A. Suss, R. Schreiber, J.A. Wolf and S.T. Purcell, *J. Magn. Magn. Mat.* **104**, 1734 (1992).
15. W.L. O'Brien and B.P. Tonner, *Phys. Rev.* **B50**, 2963 (1994).
16. W.L. O'Brien and B.P. Tonner, *Phys. Rev.* **B58**, 3191 (1998).
17. B.Ch. Choi, P.J. Bode and J.C.A. Bland, *J. Appl. Phys.* **85**, 5063 (1999); *Phys. Rev.* **B59** , 7029 (1999).
18. Yoshiki Yonamoto, Toshihiko Yokoyama, Kenta Amemiya, Daiju Matsumura and Toshiaki Ohta, *Phys. Rev.* **B63**, 214406 (2001).

19. A. Noguera, S. Bouarab, A. Mokrani, C. Demangeat and H. Dreyssé, *J. Magn. Magn. Mat.* **156**, 21 (1996).
20. S. Meza-Aguilar, O. Elmouhssine, H. Dreyssé and C. Demangeat, *Phys. Rev.* **B63**, 064421 (2001).
21. V.S. Stepanyuk, R. Zeller, P.H. Dederichs and I. Mertig, *Phys. Rev.* **B49**, 5157 (1994).
22. M. Wuttig, Y. Gauthier and S. Blügel, *Phys. Rev. Lett.* **70**, 3619 (1993).
23. S. Blügel, *Appl. Phys. Mater. Sci. Process.* **A63**, 595 (1996).
24. A. Ney, P. Poulopoulos, M. Farle and K. Baberschke, *Phys. Rev.* **B62**, 11336 (2000).
25. C.S. Wang, B.M. Klein and H. Krakauer, *Phys. Rev. Lett.* **54**, 1852 (1985).
26. P. Krüger, M. Taguchi, S. Meza-Aguilar, *Phys. Rev.* **B61**, 15277 (2000).
27. O.K. Andersen and O. Jepsen, *Phys. Rev. Lett.* **53**, 2571 (1984);
O.K. Andersen, Z. Pawlowska and O. Jepsen, *Phys. Rev.* **B34**, 5253 (1986).
The standard TB-LMTO-ASA code (version 47) developed by O.K. Andersen *et al.* at M.P.I. Stuttgart, Germany, was used.
28. P. Hohenberg and W. Kohn, *Phys. Rev.* **136**, B864 (1964);
W. Kohn and L.J. Sham, *Phys. Rev.* **140**, A1133 (1965).
29. U. von Barth and L. Hedin, *J. Phys.* **C5**, 1629 (1972).
30. D.C. Langreth and M.J. Mehl, *Phys. Rev. Lett.* **47**, 446 (1981);
C.D. Hu and D.C. Langreth, *Phys. Scr.* **32**, 391 (1985).
31. J.P. Perdew, Y. Wang and E. Engel, *Phys. Rev. Lett.* **66**, 508 (1991).
32. T. Asada and K. Terakuda, *Phys. Rev.* **B47**, 15992 (1993).
33. M.A. Khan, *J. Phys. Soc. Jpn.* **62**, 1682 (1993).
34. B. M'Passi-Mabiala, S. Meza-Aguilar and C. Demangeat, submitted to *Surf. Sc.* (2002).

NOVEL ANALYTICAL REGGE POLES TRAJECTORIES CALCULATION FOR SINGULAR POTENTIALS: THE LENNARD-JONES POTENTIAL

S.N. NABOKO[1], Z. FELFLI[2], N.B. AVDONINA[3] AND A.Z. MSEZANE[2]

[1] *Department of Mathematical Physics, St. Petersburg State University St. Petersburg, Russia*

[2] *Center for Theoretical Studies of Physical Systems, Clark Atlanta University Atlanta, GA 30314, USA*

[3] *Department of Physics and Astronomy, University of Pittsburgh Pittsburgh PA 15260, USA*

A novel, completely analytical method is presented for the accurate calculation of Regge poles trajectories for singular potentials, *viz.* potentials more singular than r^{-2} at the origin. The approach assumes the existence of a close pair of turning points of the effective potential responsible for the Regge poles in the complex angular momentum plane and exploits the Bohr-Sommerfeld quantization condition. Analytical expressions for the Lennard-Jones potentials are obtained and compared with published results.

1 Introduction

The great success of the complex angular momentum description of scattering problems involving heavy particles such as atoms and molecules[1] as well as electrons[2] has inspired the development of various, particularly analytic methods for the calculation of Regge poles trajectories as demonstrated in recent works[3,4,5,6,7,8,9,10,11,12,13,14] and references therein. The main difficulty in calculating the Regge poles trajectories for singular potentials, *viz.* potentials more singular than r^{-2} at the origin, is that one deals with a singular eigenvalue problem for a non-Hermitian Schrödinger operator with currently no general simple working method. Here we continue our development of a methodology to calculate Regge poles trajectories for singular potentials, using the Lennard-Jones potentials as illustration. This may be viewed as a continuation of previous ideas.[14]

2 Theory

2.1 Accurate calculation of turning points positions for the Lennard-Jones potentials

The Lennard-Jones potentials are given by

$$q(r) = 4\varepsilon\left[\left(\frac{\sigma}{r}\right)^{12} - \left(\frac{\sigma}{r}\right)^{6}\right], \quad \sigma > 0, \ \ \varepsilon > 0,$$

where the parameter ε is the well's depth. To investigate separately the influence of the second term in $q(r)$ let us introduce a new parameter x in front of the term $(\sigma/r)^6$. Surely the physical value of x is equal to unity,

$$q(r) = 4\varepsilon\left[\left(\frac{\sigma}{r}\right)^{12} - x\left(\frac{\sigma}{r}\right)^{6}\right]. \tag{1}$$

The exact turning points equation is (use $x = 1$ for a start)

$$\left(\frac{r}{\sigma}\right)^{12} - \frac{a}{\sigma^2}\left(\frac{r}{\sigma}\right)^{10} - \frac{4\varepsilon}{k^2}\left[1 - \left(\frac{r}{\sigma}\right)^{6}\right] = 0 \tag{2}$$

where

$$a := l(l+1)/k^2. \tag{3}$$

Let us introduce the
Hypothesis 1. Suppose that for $|k| \gg 1$ there exists a couple of turning points (TP) $r_\pm$ that are relatively close together,

$$|r_+ - r_-| \ll |r_\pm|,$$

determining a series of Regge poles ($n = 0, 1, 2, \ldots$). We have to check this hypothesis eventually. This hypothesis immediately leads to the condition

$$\frac{d}{dr}\left(\left(\frac{r}{\sigma}\right)^{12} - \frac{a}{\sigma^2}\left(\frac{r}{\sigma}\right)^{10} - \frac{4\varepsilon}{k^2}\left[1 - \left(\frac{r}{\sigma}\right)^{6}\right]\right) \approx 0, \tag{4}$$

which implies that

$$\frac{12}{\sigma}\left(\frac{r}{\sigma}\right)^{11} - \frac{a}{\sigma^2}10\left(\frac{r}{\sigma}\right)^{9}\frac{1}{\sigma} + \frac{4\varepsilon}{k^2}6\left(\frac{r}{\sigma}\right)^{5}\frac{1}{\sigma} \approx 0. \tag{5}$$

Combining both equations (2) and (5) one can easily cancel the term containing the unknown parameter a (*i.e.*, l) to obtain

$$\left(\frac{r}{\sigma}\right)^{12} + \frac{20\varepsilon}{k^2}\left[1 - \left(\frac{r}{\sigma}\right)^{6}\right] + \frac{2\varepsilon}{k^2}6\left(\frac{r}{\sigma}\right)^{6} \approx 0.$$

This is a quadratic equation in the variable $(r/\sigma)^6$ and therefore

$$\left(\frac{r}{\sigma}\right)^6 = \frac{4\varepsilon}{k^2} \pm \sqrt{\left(\frac{4\varepsilon}{k^2}\right)^2 - \frac{20\varepsilon}{k^2}}.$$

Denoting by[10]

$$\delta := \sqrt{4 - \frac{5k^2}{\varepsilon}} \tag{6}$$

we have

$$\left(\frac{r}{\sigma}\right)^6 \cong \frac{4\varepsilon}{k^2}(1 \pm \delta/2).$$

Choose[a]

$$\left(\frac{r}{\sigma}\right)^6 \cong \frac{4\varepsilon}{k^2}(1 - \delta/2),$$

where the branch $\sqrt{4 - 5k^2/\varepsilon}$ has been chosen such that $\sqrt{4 - 5k^2/\varepsilon} \to +i\sqrt{5/\varepsilon}k$ as $k \to \infty$. We obtain eventually

$$\left(\frac{r}{\sigma}\right)^6 \cong \frac{10}{2+\delta} \tag{7}$$

since

$$4\varepsilon/k^2 \equiv 20/(4 - \delta^2)$$

by definition of δ.

Formula (7) gives us the approximate value of the TP (both r_+ and r_-) under our hypothesis of relative closeness. This hypothesis immediately leads to the calculation of the leading term for Regge poles (or rather Regge poles trajectories) asymptotics as $k^2 \to \pm\infty$. Indeed substituting the value of (r/σ) from Eq.(7) into Eq.(2) one gets

$$l(l+1) \equiv k^2 a \cong k^2\sigma^2 \left[\left(\frac{10}{2+\delta}\right)^2 - \frac{20}{4-\delta^2}\left(1 - \frac{10}{2+\delta}\right)\right] \cdot \left(\frac{10}{2+\delta}\right)^{-\frac{10}{6}}$$

$$\equiv k^2\sigma^2 \left(\frac{10}{2+\delta}\right)^{-\frac{2}{3}} 12 \cdot \frac{3-\delta}{4-\delta^2} \tag{8}$$

---

[a]One should check that only this choice of the branch of the two solutions is the correct one. This can be done numerically by the investigation of the Stokes Lines connection of the corresponding turning points.

This expression (modulo $(l+1/2)^2 \cong l(l+1)$) exactly coincides with the leading term for $l(k)$ from formula (23) from previous work,[10]

$$(l+1/2)^2 \cong (3\beta - 1)/4\alpha^4.$$

For more precise calculations of Regge Poles trajectories we must use in addition to our hypothesis the Bohr-Sommerfeld condition.

Similarly to Eq.(7) we naturally introduce the following notation for our couple $r_\pm$ (relatively close TP)

$$\left(\frac{r_\pm}{\sigma}\right)^6 = \frac{10}{2+\delta}(1+\widetilde{\Delta}_\pm) \tag{9}$$

with the hope that the values of $\widetilde{\Delta}_\pm$ are small enough at high energies. Let us substitute Eq.(9) into the exact TP equation (2) including our new parameter x ($\widetilde{\Delta}_\pm$ is replaced by $\widetilde{\Delta}$), to obtain

$$\left(\frac{10}{2+\delta}\right)^2 (1+\widetilde{\Delta})^2 \;-\tfrac{a}{\sigma^2}\left(\tfrac{10}{2+\delta}\right)^{10/6}(1+\widetilde{\Delta})^{10/6}$$
$$-\tfrac{20}{4-\delta^2}\left[1 - x\tfrac{10}{2+\delta}(1+\widetilde{\Delta})\right] = 0$$

or

$$(1+\widetilde{\Delta})^2 - \frac{a}{\sigma^2}\left(\frac{10}{2+\delta}\right)^{-1/3}(1+\widetilde{\Delta})^{10/6} - \frac{1}{5}\left(\frac{2+\delta}{2-\delta}\right)\left[1 - x\frac{10}{2+\delta}(1+\widetilde{\Delta})\right] = 0. \tag{10}$$

Suppose that $|\widetilde{\Delta}| \ll 1$ (we have to check this eventually[b]). One can easily calculate the coefficients in Eq.(10) of the various powers of the small parameter $\widetilde{\Delta}$,

$$\left[\frac{6}{5} - \frac{6}{5}t - \frac{4}{5}\frac{1}{2-\delta} + \frac{2x}{2-\delta}\right] + 2\widetilde{\Delta}\left[1 - t + \frac{x}{2-\delta}\right] + \widetilde{\Delta}^2\left[1 - \frac{2}{3}t\right] + O(\widetilde{\Delta}^3) = 0, \tag{11}$$

where

$$t := \frac{5}{6}\frac{a}{\sigma^2}\left(\frac{2+\delta}{10}\right)^{1/3}. \tag{12}$$

The coefficient in front of $\widetilde{\Delta}^3$ is of order t, but according to our previous calculation of the leading term (see Eq.(8)) we have

$$a \sim \sigma^2 \left(\frac{10}{2+\delta}\right)^{-2/3} 12\frac{3-\delta}{4-\delta^2}, \tag{13}$$

---

[b] Indeed the smallness of $\widetilde{\Delta}$ implies that our Hypothesis of the relative closeness of this couple of TP holds true

or

$$t \sim 5 \cdot 2 \left(\frac{10}{2+\delta} \right)^{-1} \frac{3-\delta}{4-\delta^2} \equiv \frac{3-\delta}{2-\delta} \sim 1 \tag{14}$$

as $k \to \infty$ (remember that $\delta \to \infty$ as $k \to \infty$). Hence $t \to 1$ as $k \to \infty$ and the term $|O(\widetilde{\Delta}^3)|$ in Eq.(11) can be estimated from above as $const \cdot (\widetilde{\Delta})^3$, where $const$ is uniform in k, $|k| \gg 1$. Of course at the end, after obtaining the formula for the RP, we must check all the assumptions, which so far are simply more or less reasonable. For example, clearly $t \to 1$, as $k \to \infty$. Let us introduce the new notations related to our equation for the TP

$$\begin{aligned} u_1 &:= 1 - t + \tfrac{x}{2-\delta}, \\ u_2 &:= 1 - t + \tfrac{\frac{5}{3}x - \frac{2}{3}}{2-\delta}. \end{aligned} \tag{15}$$

By the way, $u_1 = u_2$ provided $x = 1$. Eq.(11) can be rewritten in the form

$$\widetilde{\Delta}^2 \left(1 - \frac{2}{3}t \right) + 2\widetilde{\Delta} u_1 + \frac{6}{5} u_2 + O(\widetilde{\Delta}^3) = 0.$$

Denote by

$$v_{1,2} := u_{1,2}/(1 - 2t/3) \approx u_{1,2} \cdot 3, \qquad \text{as} \qquad k \to \infty, \tag{16}$$

so that Eq.(11) eventually becomes

$$\widetilde{\Delta}^2 + 2\widetilde{\Delta} v_1 + \frac{6}{5} v_2 + O(\widetilde{\Delta}^3) = 0. \tag{17}$$

This is almost a quadratic equation (therefore we have exactly two roots $\widetilde{\Delta}_\pm$ for large values of k) for small $\widetilde{\Delta}$. Suppose $t \approx 1$; we get that v_1 and $v_2 \approx 0$ if $|k| \gg 1$. Hence, the coefficients of $(\widetilde{\Delta})^1$, $(\widetilde{\Delta})^0$ and $O(\widetilde{\Delta}^3)$ are very small. For the truncated equation (17) $\widetilde{\Delta}^2 + 2\widetilde{\Delta} v_1 + \frac{6}{5} v_2 = 0$, one gets

$$\widetilde{\Delta}_\pm \cong -v_1 \pm \sqrt{v_1^2 - \frac{6}{5} v_2} \cong \pm \sqrt{-\frac{6}{5} v_2}, \tag{18}$$

provided v_1, v_2 are small enough and $v_1 \asymp v_2$ (indeed for $x = 1$ we have $v_1 \equiv v_2$). More exactly

$$\widetilde{\Delta}_\pm = -v_1 \pm \sqrt{v_1^2 - \frac{6}{5} v_2 - O(\widetilde{\Delta}^3)}. \tag{19}$$

Since

$$\widetilde{\Delta}_\pm \cong \sqrt{-\frac{6}{5} v_2} \Rightarrow O(\widetilde{\Delta}^3) \cong O\left(v_2^{3/2} \right)$$

is a minor term here. Using the formula $\widetilde{\Delta}_{\pm} \cong \pm\sqrt{-\frac{6}{5}v_2}$, where we have $v_2 \cong (1-t)/(1-\frac{2}{3}t)$, omitting the term $x/(2-\delta) = O(1/k)$, as $k \to \infty$, and taking into account the Bohr-Sommerfeld condition one can obtain just like in previous work[14]

$$l + \frac{1}{2} \cong k^{5/6}\sqrt{\frac{6}{5}} \cdot 5^{1/12}(-4\varepsilon)^{1/12}\sigma + i\left(n + \frac{1}{2}\right)\sqrt{10}.$$

This formula coincides with the first two terms $\left(\frac{3\beta-1}{4\alpha^4} - \frac{2}{\alpha^2}\left(n+\frac{1}{2}\right)\right)$ of Eq.(23) of previous work[10] and, in fact gives RT for the power-like potential $q = 4\varepsilon\left(\frac{r}{\sigma}\right)^{12}$ without the x-term ($x = 0$). Therefore the influence of the x-term in $q(r)$, Eq.(1) probably should be of the smaller order. We will see later that it is of the order $O(1/k^{1/6})$, as $k \to \infty$, in the formula for values of l. Hence, we have to calculate everything more precisely if we wish to investigate the influence of the x-term in Eq.(1) for the RT.

We had for the power-like potential $4\varepsilon\left(1/r^{12}\right)$ that

$$\begin{aligned} v_2 &\cong -\frac{5}{6}\widetilde{\Delta}_1^2 \cong \left(-\frac{5}{6}\right)\cdot\left(n+\frac{1}{2}\right)\frac{8/i}{\sqrt{10}}\left(\frac{5}{6}\right)^{5/12}\left(-\frac{\sigma^{12}}{6\cdot 4\varepsilon}\right)^{1/12}\cdot\frac{1}{k^{5/6}} \\ &\cong const\cdot\frac{1}{k^{5/6}}. \end{aligned} \tag{20}$$

Therefore, it is reasonable that in the case of the Lennard-Jones potentials the value of v_2 stays of the order $O\left(1/k^{5/6}\right)$, as $k \to \infty$. We now introduce new reasonable hypotheses (in addition to the main one, concerning the relative closeness of TP). All of them must be checked at the end of the calculations. We need them to understand better which order of terms is important for calculation concerning the preserving of the x-term in Eq.(19).

We saw already that

$$v_1 \asymp v_2 = O\left(1/k^{5/6}\right), \qquad \text{as} \qquad k \to \infty.$$

The last conclusion has been obtained from the Bohr-Sommerfeld condition. Moreover

$$v_1 = \frac{1-t}{1-2t/3} + \frac{x}{2-\delta}\,\frac{1}{1-2t/3} = \frac{1-t}{1-2t/3} + O(1/k) \tag{21}$$

since $\delta \sim i\sqrt{\frac{5}{\varepsilon}}\,k$ (see Eq.(6)), and $t \to 1$ as $k \to \infty$ (Eq.(14)). Therefore

$$\frac{1-t}{1-2t/3} = O\left(\frac{1}{k^{5/6}}\right) - O(1/k) = O(1/k^{5/6}).$$

Furthermore, we also saw that

$$1 - t = O(1/k^{5/6}).$$

Now, we know how small the value of $(1-t)$ should be as $k \to \infty$. Consider the parameter v_2. From the previous fact we get

$$v_2 \equiv \frac{1-t}{1-2t/3} + \frac{5/3x - 2/3}{(2-\delta)} \cdot \frac{1}{1-2t/3} \equiv O(1/k^{5/6}) + O(1/k)$$

since $(1-t) = O(1/k^{5/6})$ and $\delta \asymp k$ (see Eq.(6)). Hence

$$v_2 = \frac{1-t}{1-2t/3}\left(1 + \left(\frac{5}{3}x - \frac{2}{3}\right)O(1/k^{1/6})\right)$$

because $\delta \asymp k$ and $(1-t) \asymp O(1/k^{5/6})$ (we have both estimates from above and *below*).

Consequently, the expressions for v_1 and v_2 reduce to

$$\begin{aligned} v_1 &\equiv \tfrac{1-t}{1-2t/3}\left(1 + x \cdot O(1/k^{1/6})\right), \\ v_2 &\equiv \tfrac{1-t}{1-2t/3}\left(1 + \left(\tfrac{5}{3}x - \tfrac{2}{3}\right). \cdot O(1/k^{1/6})\right) \end{aligned} \tag{22}$$

Therefore, if we wish *to take into account* the role of the parameter x, we *have to* calculate the term $v_{1,2}/(1-t)$ with accuracy up to the order $O(1/k^{1/6})$. The same holds true for $\widetilde{\Delta}_\pm$ since

$$\widetilde{\Delta}_\pm \cong \pm\sqrt{-\frac{6}{5}v_2} = \pm\sqrt{-\frac{6}{5}}\sqrt{\frac{1-t}{(1-2t/3)}} \cdot \sqrt{\left(1 + \left(\frac{5}{3}x - \frac{2}{3}\right)O(1/k^{1/6})\right)}.$$

Indeed, using Eq.(18) for $\widetilde{\Delta}_\pm$ one gets

$$\begin{aligned} \widetilde{\Delta}_\pm &= -v_1 \pm \sqrt{v_1^2 - 6/5v_2 + O(\widetilde{\Delta}_\pm^3)} \\ &= O(1/k^{5/6}) \pm \sqrt{-6v_2/5} \cdot \sqrt{1 - 6/5\frac{v_1^2}{v_2} + \frac{1}{v_2} \cdot O(\widetilde{\Delta}_\pm^3)} \\ &\equiv O(1/k^{5/6}) \pm \sqrt{-\frac{6}{5}v_2} \cdot \sqrt{1 + O(1-t) + \frac{O(\widetilde{\Delta}_\pm^3)}{(1-t)}} \\ &= O(1/k^{5/6}) \pm \sqrt{-\frac{6}{5}v_2} \cdot \sqrt{1 + O(1/k^{5/6}) + k^{5/6}O\left((1/k^{5/12})^3\right)}. \end{aligned}$$

Here we used

$$1 - t \asymp O(1/k^{5/6}) \quad (\textit{both} \text{ sides estimates})$$

and

$$\widetilde{\Delta}_\pm \asymp \sqrt{-6v_2/5} \asymp \sqrt{1-t} \asymp O\Big(1/k^{5/12}\Big).$$

Eventually we obtain

$$\begin{aligned}\widetilde{\Delta}_\pm &= O\Big(1/k^{5/6}\Big) \pm \sqrt{-\frac{6}{5}v_2}\Big(1+O\Big(1/k^{5/6}\Big)+O\Big(1/k^{5/12}\Big)\Big)^{1/2}\\ &\equiv O\Big(1/k^{5/6}\Big) \pm \sqrt{-\frac{6}{5}v_2}\Big(1+O\Big(1/k^{5/12}\Big)\Big)^{1/2}.\end{aligned}$$

Since $5/6 > 5/12$, hence

$$\widetilde{\Delta}_\pm = \frac{1}{k^{5/6}}\cdot O(1) \pm \sqrt{-\frac{6}{5}v_2}\Big(1+O\Big(1/k^{5/12}\Big)\Big) \qquad \text{as} \qquad (1+y)^{1/2} \cong 1+y/2.$$

Splitting the first term in the following way (due to $v_2 \asymp (1-t) \asymp O(\frac{1}{k^{5/6}}i)$)

$$\frac{1}{k^{5/6}} \equiv \frac{1}{k^{5/12}}\cdot\frac{1}{k^{5/12}} = \sqrt{-\frac{6}{5}v_2}\cdot O(1)\cdot\frac{1}{k^{5/12}},$$

one gets

$$\widetilde{\Delta}_\pm = \sqrt{-\frac{6}{5}v_2}\Big(O\Big(\frac{1}{k^{5/12}}\Big) \pm 1 + O\Big(\frac{1}{k^{5/12}}\Big)\Big) \equiv \pm\sqrt{-\frac{6}{5}v_2}\Big(1+O\Big(\frac{1}{k^{5/12}}\Big)\Big). \tag{23}$$

Remember that the influence of the parameter x on v_2 has the following form

$$v_2 = \frac{1-t}{1-2t/3}\Big(1+\Big(\frac{5}{3}x-\frac{2}{3}\Big)O\Big(1/k^{1/6}\Big)\Big).$$

The same is true for

$$\sqrt{-\frac{6}{5}v_2} = \Big(-\frac{6}{5}\frac{1-t}{1-2t/3}\Big)^{1/2}\Big(1+\Big(\frac{5}{3}x-\frac{2}{3}\Big)O\Big(1/k^{1/6}\Big)\Big)^{1/2}$$

since $(1+g)^{1/2} \cong 1+g/2$. Summing up the considerations, we conclude that if our aim is to investigate the role of the parameter x in the formula for turning points and, therefore for Regge Poles trajectories,

1. We have to calculate v_2 with accuracy (in the *multiplicative form*) up to the $\Big(1+O\Big(1/k^{1/6}\Big)\Big)$. The x-term here is included in the "error" $O\Big(1/k^{1/6}\Big)$;

2. we should not determine $\widetilde{\Delta}_\pm$ with more accuracy than $\left(1+O\left(1/k^{1/6}\right)\right)$ due to Eq.(23). The "error" of $O\left(1/k^{5/12}\right)$ here is too small compared with the term of $O\left(1/k^{1/6}\right)$ and *cannot cancel it*. So the influence of x will be preserved even if we replace $\widetilde{\Delta}_\pm$ simply by $\sqrt{-6v_2/5}$ without calculating the "error" of the order of $O\left(1/k^{5/12}\right)$ in Eq.(23) *at all*. It is very convenient and saves a lot of work. Surely, if we intend to calculate the TP ($\Rightarrow$ RP) *more precisely* we must calculate $\widetilde{\Delta}_\pm$ via v_2 also *more accurately*. It is a tedious calculation, but sufficiently simple. We note that the less calculations are used to obtain the desired result the better.

2.2 Bohr-Sommerfeld condition for the Lennard-Jones potentials

Just as in Eq.(8) of previous work,[14] the following Bohr-Sommerfeld condition holds in the case of Lennard-Jones potentials,

$$-\pi\left(n+\frac{1}{2}\right)+O\left(\frac{1}{n}\right)=k\left(\frac{r_\pm}{\sigma}\right)^6\sqrt{A(r_\pm)}\left(\frac{r_+}{\sigma}-\frac{r_-}{\sigma}\right)^2\cdot i\frac{\pi}{8}\sigma^2,\qquad n\in\mathbb{Z},\tag{24}$$

where n is the semiclassical integer parameter, and

$$2A(r_\pm)\cong\frac{d^2}{dr^2}\left(\left(\frac{r}{\sigma}\right)^{12}-\frac{a}{\sigma^2}\left(\frac{r}{\sigma}\right)^{10}-\frac{4\varepsilon}{k^2}\left[1-\left(\frac{r}{\sigma}\right)^6\right]\right),\tag{25}$$

provided that $r=r_\pm$, where all terms should be calculated up to the accuracy of at least the order $\left(1+O\left(1/k^{1/6}\right)\right)$. We will see later that such a simple approximation of the BS condition leads to an integral value that is still correct and gives an accuracy of order $\left(1+O\left(1/k^{5/12}\right)\right)$, which is much better than the necessary $\left(1+O\left(1/k^{1/6}\right)\right)$. Indeed

$$\begin{aligned}\left(\frac{r_\pm}{\sigma}\right)^6&=\frac{10}{2+\delta}(1+\widetilde{\Delta}_\pm)\cong\frac{10}{2+\delta}\left(1\pm\sqrt{-\frac{6}{5}v_2}\right)\\&=\frac{10}{2+\delta}\left(1+O\left(\sqrt{v_2}\right)\right)=\frac{10}{2+\delta}\left(1+O\left(1/k^{5/12}\right)\right),\end{aligned}$$

so the correction of the term $\left(r_\pm/\sigma\right)^6$ due to the changing r running over the line connecting the two points (r_+,r_-) is of the order $\left(1+O\left(1/k^{5/12}\right)\right)$ and $5/12>1/6$. The same argument is valid for $A(r_\pm)$. Hence, the consideration allows us to avoid again some tedious calculations and use a very simple version of the integral calculation in the right-hand side of the BS condition.

2.3 Calculation of the factor $A(r_\pm)$

From Eqs.(25), (9) and (18), we get

$$\begin{aligned}
2A(r_\pm) &\cong 12\cdot 11\left(\frac{r_\pm}{\sigma}\right)^{10}\frac{1}{\sigma^2} - 10\cdot 9\frac{a}{\sigma^2}\left(\frac{r_\pm}{\sigma}\right)^{8}\frac{1}{\sigma^2} + \frac{4\epsilon}{k^2}6\cdot 5\left(\frac{r_\pm}{\sigma}\right)^{4}\frac{1}{\sigma^2} \\
&= \frac{1}{\sigma^2}\Bigg[12\cdot 11\cdot\left(\frac{10}{2+\delta}\right)^{10/6}\cdot\left(1+O\left(1/k^{5/12}\right)\right) \\
&\quad -\frac{a}{\sigma^2}10\cdot 9\left(\frac{10}{2+\delta}\right)^{8/6}\left(1+O\left(1/k^{5/12}\right)\right) \\
&\quad +x\frac{4\epsilon}{k^2}6\cdot 5\left(\frac{10}{2+\delta}\right)^{4/6}\left(1+O\left(1/k^{5/12}\right)\right)\Bigg] \\
&= \frac{1}{\sigma^2}\left[\frac{10}{2+\delta}\right]^{10/6}\Bigg[\left(12\cdot 11 - 10\cdot 9\cdot\frac{6}{5}t\right)\cdot\left(1+O\left(1/k^{5/12}\right)\right) \\
&\quad +xO\left(\frac{1}{k^{2/6+2/3}}\right)\Bigg].
\end{aligned}$$

Since $(12\cdot 11 - 10\cdot 9\frac{6}{5}\cdot 1) \neq 0$ and $2/6+2/3 > 5/12$ we get that

$$2A(r_\pm) \simeq \frac{1}{\sigma^2}\left[\frac{10}{2+\delta}\right]^{10/6}[12(11-9t)]\cdot\left(1+O\left(1/k^{5/12}\right)\right), \tag{26}$$

and therefore we can omit the x-term here. Since $t-1 = O\left(1/k^{5/6}\right)$, then

$$2A(r_\pm) \simeq \frac{1}{\sigma^2}\left(\frac{10}{2+\delta}\right)^{10/6} 24\left(1+O\left(1/k^{5/12}\right)\right) \tag{27}$$

as $5/6 > 5/12$, but it is still accurate enough as $5/12 > 1/6$.

2.4 Calculation of the term $\left(\frac{r_+}{\sigma} - \frac{r_-}{\sigma}\right)^2$

Using Eq.(9) we calculate this term as follows

$$\begin{aligned}
\left(\frac{r_+}{\sigma} - \frac{r_-}{\sigma}\right) &\equiv \left(\frac{r_+}{\sigma}\right)^{6/6} - \left(\frac{r_-}{\sigma}\right)^{6/6} \\
&= \left(\frac{10}{2+\delta}\right)^{1/6}\left((1+\widetilde{\Delta}_+)^{1/6} - (1+\widetilde{\Delta}_-)^{1/6}\right) \\
&= \left(\frac{10}{2+\delta}\right)^{1/6}\left(\frac{1}{6}(\widetilde{\Delta}_+ - \widetilde{\Delta}_-) + O(\widetilde{\Delta}_\pm^2)\right)
\end{aligned}$$

$$= \left(\frac{10}{2+\delta}\right)^{1/6} \frac{1}{6} \left(\pm\sqrt{-\frac{6}{5}v_2}\left(1+O\left(1/k^{5/12}\right)\right) + O\left(1/k^{5/6}\right)\right)$$

$$= \left(\frac{10}{2+\delta}\right)^{1/6} \frac{1}{6} \left(\pm\sqrt{-\frac{6}{5}v_2}\right) \left(1+O\left(1/k^{5/12}\right)\right).$$

Therefore

$$\left(\frac{r_+}{\sigma} - \frac{r_-}{\sigma}\right)^2 = \left(\frac{10}{2+\delta}\right)^{1/3} \frac{1}{36} \left(-\frac{6}{5}v_2\right) \left(1+O\left(1/k^{5/12}\right)\right). \tag{28}$$

2.5 Formula for the Regge trajectories

Let us substitute all the equations into the BS condition (Eq.(24)) and get

$$-\pi\left(n+\frac{1}{2}\right) + O\left(\frac{1}{n}\right) =$$

$$= \frac{k}{\left(\frac{10}{2+\delta}\right)\left(1+O(1/k^{5/12})\right)}$$

$$\times\sqrt{\frac{1}{\sigma^2}\left(\frac{10}{2+\delta}\right)^{10/6} \cdot 24\left(1+O(1/k^{5/12})\right)}$$

$$\times \left(\frac{10}{2+\delta}\right)^{1/3} \frac{1}{36}\left(-\frac{6}{5}v_2\right)\left(1+O\left(1/k^{5/12}\right)\right) i\frac{\pi}{8}\sigma^2$$

$$= k\cdot\left(\frac{10}{2+\delta}\right)^{-1+5/6+1/3} \cdot \frac{1}{\sigma}\sqrt{24}\left(-\frac{1}{30}v_2\right)$$

$$\times i\frac{\pi}{8}\sigma^2\left(1+O\left(1/k^{5/12}\right)\right)$$

or

$$-\left(n+\frac{1}{2}\right) + O\left(\frac{1}{n}\right) = k\cdot\left(\frac{10}{2+\delta}\right)^{1/6} \frac{\sqrt{6}}{15}\frac{1}{8}\sigma(-iv_2)\cdot\left(1+O\left(1/k^{5/12}\right)\right).$$

We then get the formula for the parameter v_2

$$v_2 = \frac{-\left(n+\frac{1}{2}\right)+O\left(\frac{1}{n}\right)}{k\cdot\left(\frac{10}{2+\delta}\right)^{1/6}\frac{\sqrt{6}}{15}\frac{1}{8}(-i\sigma)}\left(1+O\left(1/k^{5/12}\right)\right), \tag{29}$$

and, therefore

$$\frac{1-t+\left(\frac{5}{3}x-\frac{2}{3}\right)/(2-\delta)}{1-2t/3} = \frac{-i\left(n+\frac{1}{2}\right)+O\left(\frac{1}{n}\right)}{k\cdot\left(\frac{10}{2+\delta}\right)^{1/6}\frac{\sqrt{6}}{15}\frac{1}{8}\sigma}\left(1+O\left(1/k^{5/12}\right)\right).$$

Since $(t-1)=O\big(1/k^{5/6}\big)$ we have

$$3(1-t)+\frac{3\big(\frac{5}{3}x-\frac{2}{3}\big)}{2-\delta}=\frac{-i\big(n+\frac{1}{2}\big)+O\big(\frac{1}{n}\big)}{k\cdot\big(\frac{10}{2+\delta}\big)^{1/6}}\,\frac{15\cdot 8}{\sigma\sqrt{6}}\Big(1+O\Big(1/k^{5/12}\Big)\Big),$$

or

$$1-\frac{a}{\sigma^2}\left(\frac{2+\delta}{10}\right)^{1/3}\frac{5}{6}+\frac{\frac{5}{3}x-\frac{2}{3}}{2-\delta}=\frac{-i\big(n+\frac{1}{2}\big)+O\big(\frac{1}{n}\big)}{k\cdot\big(\frac{10}{2+\delta}\big)^{1/6}}\cdot\frac{5\cdot 8}{\sigma\sqrt{6}}\Big(1+O\Big(1/k^{5/12}\Big)\Big). \tag{30}$$

Calculating the value of a from Eq.(3) one has that

$$l(l+1)\equiv k^2\sigma^2\left(\frac{a}{\sigma^2}\right)=\frac{6}{5}\left(\frac{10}{2+\delta}\right)^{1/3}k^2\sigma^2+k^2\sigma^2\frac{6}{5}\left(\frac{10}{2+\delta}\right)^{1/3}\frac{\big(\frac{5}{3}x-2/3\big)}{2-\delta}$$
$$+\frac{i\big(n+\frac{1}{2}\big)+O\big(\frac{1}{n}\big)}{k\cdot\big(\frac{10}{2+\delta}\big)^{1/6}}\cdot\sqrt{6}\cdot 8\sigma k^2\Big(1+O\Big(1/k^{5/12}\Big)\Big),$$

or

$$l(l+1)=\frac{6}{5}\left(\frac{10}{2+\delta}\right)^{1/3}k^2\sigma^2\Big[1+\frac{\frac{5}{3}x-\frac{2}{3}}{2-\sigma}$$
$$+\frac{i\big(n+\frac{1}{2}\big)+O\big(\frac{1}{n}\big)}{k\cdot\big(\frac{10}{2+\delta}\big)^{1/6}}\cdot\frac{5\cdot 8}{\sigma\sqrt{6}}\Big(1+O\Big(1/k^{5/12}\Big)\Big)\Big],$$

$$\sqrt{l(l+1)}=\sqrt{\frac{6}{5}}\left(\frac{10}{2+\delta}\right)^{1/6}k\sigma\Big[1+\frac{\frac{5}{3}x-\frac{2}{3}}{2(2-\delta)}$$
$$+\frac{i\big(n+\frac{1}{2}\big)+O\big(\frac{1}{n}\big)}{k\cdot\big(\frac{10}{2+\delta}\big)^{1/6}}\frac{5\cdot 4}{\sigma\sqrt{6}}+O\left(\frac{1}{k^{15/12}}\right)\Big].$$

Eventually we get the formula for Regge trajectories

$$\left(l+\frac{1}{2}\right)\approx\sqrt{l(l+1)}=\sqrt{\tfrac{6}{5}}\left(\tfrac{10}{2+\delta}\right)^{1/6}k\sigma+\big[i(n+1/2)+O(1/n)\big]\cdot\tfrac{20}{\sqrt{5}}$$
$$+\frac{\big(\frac{5}{3}x-\frac{2}{3}\big)}{2(2-\delta)}\cdot\sqrt{\tfrac{6}{5}}\left(\tfrac{10}{2+\delta}\right)^{1/6}k\sigma+O\left(\tfrac{1}{k^{5/12}}\right), \tag{31}$$

where the third term is due to the second part of the potential.

The last expression can be decomposed into a few terms of the following orders

$$O\Big(k^{5/6}\Big)+O\Big(k^{0}\Big)+O\Big(k^{-1/6}\Big)+O\Big(k^{-5/12}\Big).$$

Note that $x=2/5$ is a singular point where the third term in Eq.(31) disappears.

3 Concluding Remarks

In conclusion, some remarks regarding our assumptions are appropriate. Now let us check that our assumptions are satisfied provided that the Eq.(31) holds true.

i) Part of our Hypothesis 1 concerning the relative closeness of the turning points holds true because in Eq.(9) the value $\widetilde{\Delta}_{\pm}$ is small enough for $|k| \gg 1$. The second part was not proved although we used it for the calculation of the integral in the Bohr-Sommerfeld condition. More or less it can be proved rigorously numerically by the analysis of the Stokes Lines topology. Note that this analysis is rather complicated "technically" without using a computer.

ii) The estimate of the term $O(\widetilde{\Delta}_{\pm}^3)$ in Eq.(9) follows immediately from the smallness of $\widetilde{\Delta}_{\pm}$ discussed in i).

iii) From the closeness of the parameter t to 1, one can get readily from Eq.(31) the value of a. This also gives an estimate for $(t-1)$.

iv) All the assumptions used in the approximate calculation of $A(r_{\pm})$ and $\left(\frac{r_+}{\sigma} - \frac{r_-}{\sigma}\right)^2$ are satisfied due to i)-iii).

So all additional assumptions which have been used in obtaining Eq.(31) do not contradict the answer for Regge poles and the latter is correct provided the second part of the Hypothesis 1 holds true.

Acknowledgment

The research was supported by the National Science Foundation and U.S. DoE, Division of Chemical Sciences, Office of Energy Research (AZM).

References

1. J.N.L. Connor, *J. Chem. Soc. Faraday Trans.* **86**, 1627 (1990).
2. Z. Felfli, A.Z. Msezane and D. Bessis, *Phys. Rev. Lett.* **81**, 963 (1998).
3. A. Amaha and K.E. Thylwe, *Phys. Rev.* **A50**, 1420 (1994).
4. N. Fröman and P.O. Fröman, *Phys. Rev.* **A43**, 3563 (1991).
5. N. Andersson, *J. Phys.* **A26**, 5085 (1993).
6. T.C. Germann and S. Kais, *J. Chem. Phys.* **106**, 599 (1997).
7. S. Kais and G. Beltrame, *J. Chem. Phys.* **97**, 2453 (1993).
8. S.A. Sofianos, S.A. Rakityansky and S.E. Massen, *Phys. Rev.* **A60**, 337 (1999).
9. D. Sokolovski, C. Tully and D.S.F. Crothers, *J. Phys.* **A31**, 1 (1998).

10. D. Vrinceanu, A.Z. Msezane and D. Bessis, *Phys. Rev.* **A62**, 022719 (2000) .
11. D. Vrinceanu, A.Z. Msezane and D. Bessis, *Chem. Phys. Lett.* **311**, 395 (1999).
12. D. Vrinceanu, A.Z. Msezane, D. Bessis, J.N.L. Connor and D. Sokolovski, *Chem. Phys. Lett.* **324**, 311 (2000).
13. C.R. Handy and A.Z. Msezane, *J. Phys.* **A34**, 531 (2001).
14. N.B. Avdonina, S. Belov, Z. Felfli, A.Z. Msezane and S. Naboko, *Phys. Rev. A*, submitted (2001).

APPLICATION OF THE GIBBS SAMPLER TO THE CONDITIONAL SIMULATION OF RAIN FIELDS

HUBERT ONIBON, THIERRY LEBEL
LTHE - 1025, B.P. 53 Domaine Universitaire
F-38041 Grenoble Cedex 9, France
E-mail: hubert.onibon@inpg.fr

ABEL AFOUDA
Université d'Abomey-Calavi
B.P. 526 Cotonou, Republic of Benin

GILLES GUILLOT
INRA, Unité de Biométrie, Domaine Saint Paul
F-84914 Avignon Cedex 9, France

The low density of raingauge networks in West-Africa and the availability of climatological variable at large scale only (as outputs of climatological models or remote sensing estimation algorithms) can be bridged by the so-called disagreggation algorithms which are designed to provide rainfall data at any scale given large scale data, and a small set of ground (point scale) measurements. The proposed algorithms is based on a geostatistical model, namely a stochastic model taking explictly into account the spatial structure of the variable. The disagreggation problem is then viewed as a Monte-Carlo simulation problem at the required scale conditionned by all the available information. As the problem of sampling from the multivariate density derived from our model does not have direct solution, the simulation is performed with an iteration of a Markov Chain whose limiting distribution is the desired one. Under the general Metropolis-Hastings scheme we use the well known Gibbs sampler, which allows here to deal with grids with arbitrary large number of points at the cost of small approximations. The behaviour of the model is demonstrated by disagreggation of a few rainfall events observed during the EPSAT-Niger experiment.

1 Objective of this Study

The possible impacts of climate change on water resources and agricultural yields is now well recognized. Available tools for realistic assessment of these impacts are General Circulation Models (GCM) and Regional Climate Models (RCM). Nevertheless and despite considerable progress, rainfall still remains one of the most difficult variables to predict in these models. This limits the accuracy with which other hydrological variables such as runoff can be modeled and/or predicted. Moreover, the results of physically based GCM are available only at large grid scales much coarser than those needed for hydrological assessment and application at basin scales. Theoretical schemes are therefore required to bridge the gap between the coarse resolution of atmospheric scales determined by the atmosphere[1] and the hydrological resolution conditioned by the landscape heterogeneity.[2] The aim of this study is to disagregate the areal storm rain depth produced by GCMs into a high resolution input dataset usable for hydrological models, respecting *a priori* knowledge about the statistical properties of the data (under study).

In this paper, the original rain data are considered to be a random vector Y of the form

$$Y = \Phi(X, \epsilon) \tag{1}$$

where Φ is a non-linear transformation to be defined, ϵ is an independent noise field and X is a gaussian vector.

Several well known methods are available for the simulation of 2D or 3D gausian processes.[3,4,5,6] Due to the various convergence theorems to gaussian distribution, and to their stability through many operations, they are relatively easy to simulate.

In the hydrometeorology context, the storm rainfall probability density function is strongly asymetric.[7,8] This problem is overcome by exploring the gaussian transformed model defined as follows,

Definition 1 *A random function Y is gaussian transformed if there exists a monotonic function Φ referred to hereafter as anamorphosis and a gaussian random function X such that,*

$$Y = \Phi(X) \tag{2}$$

Under these conditions, the random function Y is said to be gaussian transformed and its distribution is characterized by the anamorphosis function Φ, with the expectation m_X and the covariance C_X of the gaussian vector X.

Suppose that G is the gaussian probability function of X and F the probability function of Y then, $G(X) = F(Y) = F[\Phi(X)]$ (see Fig.1). If F is continuous, then the anamorphosis function is invertible and $\Phi^{-1} = G^{-1}oF$. A detailed mathematical description of the gausian anamorphosis model is presented by Freulon[4] and Guillot.[6]

Using the Turning Band Method[3] for the numerical implementation of the gaussian anamorphosis random non-conditional simulation, Guillot and Lebel[2] showed that this model is efficient for the generation of rain fields. The weakness of this numerical approach resides in the requirement of a large set of simulations scenarios for reproducing of the distribution of observed average rainfalls in a conditional context. To overcome this problem, the Gibbs sampler is proposed here for the numerical implementation of the gaussian transformed model.

This paper is organized as follows. Sec.2 is devoted to the mathematical description of the Gibbs algorithm. Subsequently, it is shown in Sec.3 how the stochastic spatial disaggregation model obtained by replacing the Turning Band Method through the association of Gibbs sampler with an acceptation-rejection algorithm provides a convenient answer to the problem of conditional simulation of rain fields. In Sec.4, the model is implemented for the simulation of Sahelian rain fields. Finally, Sec.5 offers general conclusions.

2 The Gibbs Sampler

The Gibbs sampler belongs to the category of Markov Chain Monte Carlo (MCMC) algorithms, which is widely used in statistics for sampling complicated distributions and consequently for simulation purposes. This algorithm was formally developed by Geman and Geman[9] in the context of degraded images restoration. Gelfand and Smith[10] demonstrated its applicability to general bayesian computations. Freulon[4] and Allard[11] developed the first geostatistical applications while Perrault *et al.* [12] applied it to bayesian change-point analysis in hydrometeorological time-series.

2.1 Presentation of the Gibbs algorithm

Let $x = [x_s,\ s \in S]$ which denotes any family of random variable indexed by s. Let Ω be the set of all possible configurations,

$$\Omega = \{\omega = (x_{s1}, ..., x_{sn}).$$

The event $\{X_{s1} = x_{s1}, ..., X_{sn} = x_{sn}\}$ is abbreviated as $\{X = \omega\}$. X is a Markov process if

. $P(X = \omega) > 0$ for all $\omega \in \Omega$ and,

. $P(X_s = x_s | X_r = x_r,\ r \neq s) = P(X_s = X_s | X_r = x_r)$ for $s \in S$ and $(x_{s1}, ..., x_{sN}) \in \Omega$.

The probability density function (pdf) of X is denoted hereafter by $f(x)$. Simplifying again the notations, x is denoted as $(x_1, ..., x_n)$. The vector x without its component x_i is denoted

$$x_{-i} = (x_1, x_2, ..., x_{i-1}, x_{i+1}, ..., x_n).$$

The problem is to sample from this multivariate density function, an n-dimensional random vector x.

Since the system evolves by discrete changes, the starting configuration can be labeled as $x^0 = (x_1^0, x_2^0, ..., x_n^0)$. Assume that the conditional distributions $f(x_i|x_1, x_2, ..., x_{i-1}, x_{i+1}, ..., x_n)$ is known. Then the components of the vector x can be updated successively using the conditional density of those components given the other. In the gaussian case,[8,13] the simple kriging estimate is equal to the conditional expectation and so the updated value of the ith component at iteration j is just the simple kriging x_i^{SK} estimate based on the values of all the other components $(x_1^j, x_2^j, ..., x_{i-1}^j, x_{i+1}^{j-1}, ..., x_n^{j-1})$ plus a gaussian residual u times the kriging standard deviation σ_i^{SK} . In this context, the ith component is generated at the jth iteration by the following equation,

$$x_i^j = x_i^{SK} + \sigma_i^{SK} u \tag{3}$$

with the gaussian conditional density function

$$g(x_i|x_{-i}) = \frac{1}{\sigma_i^{SK}} g(\frac{x_i - x_i^{SK}}{\sigma_i^{SK}}). \tag{4}$$

The key element of a MCMC algorithm is the transition kernel that describes the probability P_i consecutive to the modification of the component x_i at each iteration. For the Gibbs sampler and in the gaussian case, this probability is obtained from[9]

$$dP_i(x, w) = \delta_{x_i}(w_i) \frac{1}{\sigma_i^{SK}} g(\frac{x_i - x_i^{SK}}{\sigma_i^{SK}}) dw_i. \tag{5}$$

More details on the mathematical properties (stationarity and ergodicity in particular) of the Geman and Geman transition probability can be found in Freulon[4] and Gelfand and Smith.[10]

2.2 Simple kriging estimate of the component x_i

For the implementation of the Gibbs sampler in the gaussian case, the distribution of x_i conditional on x_{-i} has the following construction,

$$x_i = x_i^{SK} + \sigma_i^{SK} u, \tag{6}$$

where

$$x_i^{SK} = \sum_{j=1}^{n} \lambda_{ij} x_j, \tag{7}$$

and where λ_{ij} denotes the simple kriging weights determining the conditional mean of the component x_i to be simulated.

We denote C the symmetric covariance matrix formed by all the components of x, $E = (e_{ij})_{\leq i,j \leq n}$ its inverse, $\Lambda = (\lambda_{ij})_{\leq i,i \leq n}$ the kriging weights matrix, V the diagonal matrix of kriging variances : $v_{ii} = (\sigma_i^{SK})^2$ and D the integration domain. From the conventional theory of the conditional multivariate normal distribution and the expression of the inverse of a symmetric matrix, λ_{ij} and σ_i^{SK} are obtained as follows,

$$i \in D \Rightarrow \begin{cases} (\sigma_i^{SK})^2 = \frac{1}{e_{ij}}, \\ \lambda_{ij} = \delta_{ij} - \frac{e_{ij}}{e_{ii}}, \quad \forall j \in D. \end{cases} \tag{8}$$

To express this equation in matrix form, we write

$$\Lambda = I - VC^{-1}. \tag{9}$$

The construction of the covariance matrix must take into account the fact that in general the vector x represents a serially correlated process. When the process to be simulated is stationary, we need to express the inverse of the covariance matrix before the sampling procedure begins. The Gauss-Seidel method, described in detail in Froberg[14] and Gastinel[15] is currently used for solving linear systems. In a recent work, Gali and Gao[13] showed that the Gibbs sampler for a gaussian vector corresponds to the stochastic version of the Gauss-Seidel iteration on the inverse of the covariance matrix. For these reasons, the Gauss-Sidel algorithm will be used to compute the matrix E. Some numerical problems related to the covariance matrix inversion appear when this matrix is too large. In the context of simple kriging estimate, it is possible to solve these problems by working on moving neighbourhood.

3 Application of the Model to the Spatial Disagregation of Rain Fields

Let z denotes the spatial average of the event rainfall and

$$Y = (Y_i)_{i \in N = \{1,2,\ldots,n\}}.$$

the random vector representing the rain depths at the nodes of a regular grid. The purpose of this section is to implement the gaussian anamorphosis model briefly presented in Sec.2.1, for the generation of the vector Y conditioned by the average rainfall z. Therefore, each component of the vector $Y = \Phi(X)$ will be generated according to the following equation,

$$z = \sum_{i=1}^{n} \lambda_i \Phi(x_i) + k\epsilon, \tag{10}$$

where $\lambda = (\lambda_i)_{i \in N}$ is a weight system associated to Y, ϵ a gaussian random variable characterized by the density $g_\epsilon = \frac{1}{k\sqrt{2\pi}} exp(-\frac{1}{2}(\frac{\epsilon}{k})^2)$.

According to the Gibbs sampler which basic concept was presented in Sec.2.1, and to Eq.(10), the ith component of the vector x must be simulated with the conditional density $f(x|x_{-i}, z)$. In term of probability density function, Eq.(10) may be expressed as,

$$f(z|x) = f(\epsilon|x) = g_\epsilon(\frac{1}{k}(z - \sum_{i=1}^{n} \lambda_i \Phi(x_i))) \tag{11}$$

After some transformations based on bayesian properties, Onibon[8] showed that Eq.(11) can be replaced by the proportionality relation

$$f(x_i, x_{-i}|z) = f(x_i|x_{-i}, z) f(x_{-i}|z) \propto g_\epsilon(\frac{1}{k}(z - \sum_{i=1}^{n} \lambda_i \Phi(x_i))) g(x_i|x_{-i}) g(x_{-i}). \tag{12}$$

At the modification of the ith component, the gaussian density $g(x_{-i})$ and the conditional density $f(x_{-i}|z)$ are constants. Then Eq.(12) becomes,

$$f(x_i|x_{-i}, z) \propto g_\epsilon(\frac{1}{k}(z - \sum_{i=1}^{n} \lambda_i \Phi(x_i))) g(x_i|x_{-i}). \tag{13}$$

We also know that

$$g_\epsilon(\frac{1}{k}(z - \sum_{i=1}^{n} \lambda_i \Phi(x_i))) \propto exp(-\frac{1}{2}(\frac{1}{k}(z - \sum_{i=1}^{n} \lambda_i \Phi(x_i)))^2). \tag{14}$$

Finally,

$$f(x_i|x_{-i}, z) \propto exp(-\frac{1}{2}(\frac{1}{k}(z - \sum_{i=1}^{n} \lambda_i \Phi(x_i)))^2) g(x_i|x_{-i}). \tag{15}$$

Eq.(15) shows that, while an analytical expression of the conditional of x_i cannot be obtained, this density is proportional to another density which is easy to calculate. In order to generate each component of the vector x by respecting the conditional density $f(x_i|x_{-i}, z)$, the random gaussian residual noise u will be generated with the acceptation-rejection algorithm.

3.1 Acceptation-rejection algorithm

Introduced in statistic by Von Neuman[16], the acceptation-rejection algorithm is used to simulate random variables characterized by a density function f totally defined up to a multiplicative constant. This algorithm requires the envelopment of f by another density function q. We will denote the dimension of the variable to be simulated by n, (x, y) an element of $\mathbb{R}^n \times \mathbb{R}$ and A_f the intersection of the sub-graph of f and of the semi-space $\mathbb{R}^n \times \mathbb{R}^+$. According to Fig.2, $A_f = \{(x, y) \in \mathbb{R}^n \times \mathbb{R} \;\; 0 \leq y \leq f(x)\}$.

The simulation of a random variable x which probability density function is f, with the acceptation-rejection method, consists in majoring f by a probability density function q which can be easily simulated.[4,17] A constant C can be defined such that,

$$f(x) \leq Cq(x) \;\; x \in \mathbb{R}^n. \tag{16}$$

It has been shown[4,8] that it is necessary to simulate only a point belonging to A_f. Hence, a point belonging to A_{Cq} (Fig.2) must be simulated as follows: two random variables X and W which respective probability density function are q and $I_{]0,1[}$ are simulated. If the simulated point falls within A_{Cq}, then it also falls within A_f. Because $Cq(x) \geq f(x)$, the graph of f lies under the graph of Cq (Fig.2).

The acceptation-rejection algorithm is

a) Sample a variable x with a density q and a random variable w with a density $I_{]0,1[}$;

b) If $wCq(x) \leq f(x)$ accept x. Else, go to the previous step.

The speed of this algorithm depends of the number of candidates to simulate and to test before accepting one value. Ycart[17] shows that the event by which the algorithm is conditioned follows a geometric law which expectation equals C. More generally, q is chosen quite closed to f such that $C = max\{\frac{f(x)}{q(x)}\}$ is as small as possible. This algorithm is therefore of interest if on the one hand q is easy to simulate, and if on the other hand f is very close to q (C close to 1).

3.2 Disagregation algorithm

The rain depths at the nodes of a regular grid must be simulated with the conditional density

$$f(x_i|x_{-i}, z) \propto h_{x_{-i},z}(x_i) = exp(-\frac{1}{2}(\frac{1}{k}(z - \sum_{i=1}^{n} \lambda_i \Phi(x_i)))^2) g(x_i|x_{-i}). \tag{17}$$

If $q(x) = g(x_i|x_{-i})$, then

$$\frac{h_{x_{-i},z}(x_i)}{q(x)} = exp(-\frac{1}{2}(\frac{1}{k}(z - \sum_{i=1}^{n} \lambda_i \Phi(x_i)))^2) \leq 1. \tag{18}$$

Thus, $q(x) = g(x_i|x_{-i})$ is an envelope function of $h_{x_{-i},z}(x_i)$.

When the mean of the initial gaussian anamorphosis vector Y^0 is too different from the mean value z which conditioned the simulation, the percentage of rejection can be high. As a consequence, in order to reduce the computational time, an initial vector Y^0 with mean close to z, must be built by using the Gibbs algorithm in a non-conditional context.[8] More specifically, the disagregation algorithm proposed here is based upon the coupling of the Gibbs sampler and the acceptation-rejection algorithm. The algorithm is described as follows.

Build an initial gaussian vector x^0;

At the jth iteration, and for the ith node,

1) calculate the simple kriging x_i^{SK} and σ_i^{SK} ;
2) generate a standard zero-mean normal random variable u;
3) replace the component x_i^{j-1} by $x_i^j = x_i^{SK} + \sigma_i^{SK} u$;
4) calculate $\frac{h_{x_{-i},z}(x_i)}{q(x)} = exp(-\frac{1}{2}(\frac{1}{k}(z - \sum_{i=1}^{n} \lambda_i \Phi(x_i)))^2)$;
5) generate a uniform random variable w;
6) if $w \leq \frac{h_{x_{-i},z}(x_i)}{q(x)}$, then accept x_i^j. Else go to 2.

Stop the iterations.

3.3 Convergence of the disagregation algorithm

A critical issue when Markov Chain Monte Carlo (MCMC) methods are used is to determine when it is safe to stop the sampling algorithm and to be sure that the distribution law of the model is well reproduced. To solve the key problem of convergence of the Gibbs algorithm, some statistical methods have been developed.[13,17,18,19] The weakness of the outlined theoretical convergence methods lies in the fact that it is difficult to develop an appropriate numerical algorithm for their implementation. For this reason, the automatic convergence monitoring should be avoided. The Markov nature of the Gibbs algorithm means that at convergence, the components of the simulated vector will generally be correlated to each other. For this reason, we suggest at first to run the model over a number of k_0 iterations which are sufficient to reach equilibrium: it is the start-up phase of the iteration process. Subsequently, we perform k_1 iterations to obtain convergence. While several methods have been proposed, few rigorous results allows a clear determination of k_0 and

k_1 . According to the computational algorithm of the disagregation model presented in Sec.3.2, k_0 is nothing else than the number of iterations which are necessary to build an adequate initial vector.

The value of k_1 corresponds to the number of iterations necessary to reach the probability law underlying the model. Freulon[4] showed on an experimental basis that i) the convergence towards the variogram is concomitant with the convergence towards the variance of the field to be simulated, and ii) when this latest condition is satisfied, the histogram is well reproduced. For this reason, the knowledge of the experimental variance of the rain field to be simulated allows one to easily control the number k_1 of iterations following the start-up phase (it goes down to comparing the experimental variance to the variance of the simulated field after each iteration).

4 Implementation of the Model for Sahelian Rain Fields Disagregation

4.1 Cumulative distribution function of point rainfall depths

The use of the gaussian anamorphosis model requires the determination of the point rainfall distribution. During the EPSAT-Niger experiment,[20] the cumulative points event rain depths have been recorded by a network made of 30 raingauges, almost regularly spaced over the Niamey Degree Square area (16000 km^2). Guillot[6] showed that a mixture of a gamma distribution and an atom at zero represents correctly the experimental distribution of the EPSAT-Niger event rain depths. If F_0 denotes the frequency of zero and y_{ij} the cumulative rainfall recorded at the raingauge i during the event j, then the cumulative frequency distribution is,

$$F(y_{ij}) = \begin{cases} F_0 & if\ y_{ij} = 0, \\ F_0 + (1 - F_0)F_\Gamma(y_{ij}) & if\ y_{ij} > 0, \end{cases} \tag{19}$$

where F_Γ represents the probability function of the gamma law whose density function is,

$$f_{\alpha,\lambda}(y_{ij}) = \begin{cases} 0 & if\ y_{ij} = 0, \\ \frac{\lambda^\alpha}{\Gamma(\alpha)} y_{ij}^{\alpha-1} e^{-\lambda y_{ij}} & if\ y_{ij} > 0, \end{cases} \tag{20}$$

Γ being the classical gamma function. The parameters α and λ are obtained as follows,

$$\begin{cases} \alpha = \frac{E^2(Y|Y>0)}{\sigma^2(Y|Y>0)}, \\ \\ \lambda = \frac{E(Y|Y>0)}{\sigma^2(Y|Y>0)}. \end{cases} \tag{21}$$

The parameters of the distribution (F_0, α and λ) are averages computed over the sample of EPSAT-Nige network. Given the homogenous climate and the absence of relief in the region, it is sound to look for a single stationary regional distribution function.

4.2 Spatial structure of Sahelian rain fields

The protocol for the estimation of the weighting coefficients of the simple kringing estimate presented in Sec.2.2 is based on the use of the covariance matrix. For this reason, the spatial structure of the rain fields will be characterized by a covariance function. This choice is also justified by the fact that Guillot[6] has already adjusted a covariance model to the rain fields for the study area. The analysis of the empirical covariance leads to identify two spatial structures characterized by an anisotropy pointing in the E-N-E direction. The corresponding theoretical covariance model is the following,

$$C(h) = \sigma_1^2 e^{-\frac{|h|_1}{s_1}} + \sigma_2^2 e^{-\frac{|h|_2}{s_2}}, \tag{22}$$

where

$$|h|_i = \sqrt{h_E^2 + \frac{h_N^2}{a_i^2}}, \tag{23}$$

(h_E, h_N) being the coordinates of h expressed in the basis of the axis of anisotropy (h_E along the E-N-E axis, h_N along the N-W-N axis), a_i, $i = 1, 2$ the anisotropy ratio of the structure i, s_i is the scale distance parameter and σ_i the scale variance parameter.

4.3 Experimental results

In conformity with the main objective of the stochastic disagregation model implementation, the Niamey Degree Square is decomposed into a grid of 400 regularly spaced points. To provide some indications on the model's efficiency, we applied it to the conditional simulation of observed events over the study area. Its behaviour is tested by comparing the statistical properties of observed and simulated rain fields sampled at the same space frequency. For each event, 50 rain fields were independently simulated.

<u>Comparison of the cumulative distribution functions</u>

As a preliminary step, the realism of the model in terms of cumulative distribution functions is studied. The two examples presented concern the disagregation of the events recorded on the 12/07/90 and on the 08/08/90 over the study region. In Fig.3, we compare the cumulative distribution curves of the first 10 conditional simulations. According to this figure, the internal variability of the model is noticeable and increases with the point rain depth values. This result clearly illustrates an intrinsic property of the disagregation model which is its capacity to generate a large number of distributions for a given spatial average over one entire domain. One can also see in Fig.3 that the observed distribution is in general enveloped by the distributions of the different simulations. The comparison of the observed and the simulated cumulative distribution functions leads to the two following conclusions: i) despite the internal variability of the disagregation model, there is a good agreement between the distributions obtained for some individual simulations and the observed distribution (see Fig.4); ii) the extreme values are in general overestimated by the model.

Spatial structure of the simulated rain fields

The example presented here concerns the event observed on the 12/07/90 over the study area. Fig.5 provides more indications about the spatial organization of this event. The isohyetal curves are in general centered on the network and the high values are located at the center. As it is shown in Fig.6, some simulated rain fields exhibit a very similar pattern. Since the spatial organization of rainfall events are characterized by a high degree of randomness, two events with equal spatial average and equal variance can have very different patterns. We also know that in the model, the point values are randomly simulated and only conditioned by the event average rainfall. When comparing the images shown in Fig.6, we observe a very high variability of the spatial distribution of points rainfall values. This result shows the skill of the model in reproducing this large range of patterns associated to the events with equal magnitude (spatial average) and equal dispersion (variance).

5 Conclusion

In this paper, a disagregation model of rainfall events at small space scale was presented. The basic concept underlying the model is the simulation of an gaussian transformed random vector $Y = \Phi(X)$ conditioned by a spatial average rainfall z. After describing theoretically how the Gibbs sampler works, we established the relationship between the Gibbs algorithm and the proposed spatial disagregation model. To overcome the difficulty of simulating Y by respecting the conditional density $f(Y|z)$, the model is numerically achieved by coupling of the Gibbs sampler and the accepting-rejection algorithm.

The proposed iterative algorithm is interesting from a theoretical point of view and straightforward to implement. In order to validate the model, two examples of simulated rain fields were presented and briefly analyzed. Although the model is not designed to reproduce the features of any *a priori* rain field, and despite its strong internal variability, it has proved to be available tool for the modeling of rainfall fields.[2,8] The present work provides a comprehensive framework for the use of the gaussian transformed model in the hydrological problem of conditional simulation.

References

1. H. Onibon, T. Lebel and A. Afouda, *Space-time rainfall variability in West Africa derived from observations and GCMs*, *IAAHS Publication*, **274**, 483–490 (2002).
2. G. Guillot and T. Lebel, *Approximation of Sahelian rainfall fields with meta-Gaussian random functions. Part 2: parameter estimation and comparison to data*, *Stochastic Environmental Research and Risk Assessment* **13**, 113–130 (1999).
3. G. Matheron, *The intrinsic random functions and their applications*, *Adv. Appl. Prob.* **5**, 211–222 (1973).
4. X. Freulon, *Conditionnement du modèle gaussien par des contraintes ou des randomisées*, Ph. D. Thesis (Ecole Nationale Supérieure des Mines de Paris, France, 1992), unpublished.
5. C. Lantuejul, *Iterative algorithms for conditional simulation*, in *Geostatistical Wollongong'96*, E.Y. Bafi and N.A. Schofield, eds. (Kluwer, Dordrecht, 1997), Vol. 1, pp. 27–40.
6. G. Guillot, *Modélisation statistique des champs de pluie sahéliens. Application à leur désagrégation spatiale et temporelle*, Ph. D. Thesis (Universite Joseph Fourier, Grenoble, France, 1998), unpublished.
7. T. Lebel, I. Braud and J.D. Creutin, *A space-time disaggregation model adapted to sahelian mesoscale convective complexes*, *Water Resour. Res.* **34**, 1711–1726 (1998).
8. H. Onibon, *Simulation conditionnée des champs de pluie au Sahel: application de l'algorithme de Gibbs*, Ph. D. Thesis (Institut National Polytechnique de Grenoble, France, 2001), unpublished.
9. S. Geman and D. Geman, *Stochastic relaxation, Gibbs distribution and the bayesian restoration of images*, *IEEE transactions: Pattern Analysis and Machine Intelligence* **6**, 721–741 (1984).
10. A.E. Gelfand and E.E.M. Smith, *Sampling base approaches to calculating*

marginal densities, *J. Am. Stat. Assoc.* **85**, 398–409 (1990).
11. D. Allard, *Connexité des ensembles aléatoires: application à la simulation de réservoirs pétroliers hétérog'enes*, Ph. D. Thesis (Ecole Nationale Supérieure des Mines de Paris, France, 1993), unpublished.
12. L. Perreault, J. Bernier, B. Bobbe and E. Parent, *Bayesian change-point analysis in hydrometeorological time series. Part 1: The normal model revisited*, *Journal of Hydrology* **235**, 221–241 (2000).
13. A. Galli and H. Gao, *Rate of convergence of the Gibbs sample in the gaussian case*, *Mathematical Geology* **33**, 653–677 (2001).
14. C.E. Froberg, *Introduction to Numerical Analysis* (Addition-Wesley Reading, Massachusetts, 1964).
15. N. Gastinel, *Analyse numérique linéaire* (Herman, Paris, 1966).
16. J. Von Neuman, *Various techniques used in connection with random digits*, *National Bureau of Standards, Applied Mathematics Series* **2**, 6–38 (1951).
17. B. Ycart, *Méthodes de Monte Carlo*, Cours de DEA *Statistiques et modèles aléatoires en économie et en finance* (Université Denis Diderot, France, 2000).
18. C. Ritter and M.A. Tanner, *Facilitating the Gibbs sampler: the Gibbs stopper and the griddy Gibbs samplers*, *Journal of the American Statistical Association* **87**, 861–868 (1992).
19. G.O. Roberts, *Convergence diagnostic of the Gibbs sampler*, in *Bayesian Statistics 4*, J.M. Bernado, J. Berger, A.P. David and A.F.M. Smith, eds. (Oxford University Press, Oxford, 1992), pp. 775–782.
20. N. d'Amato and T. Lebel, *On the characteristics of the rainfall events in the Sahel with a view to the analysis of climatic variability*, *Int. J. Climatol.* **18**, 955–974 (1998).

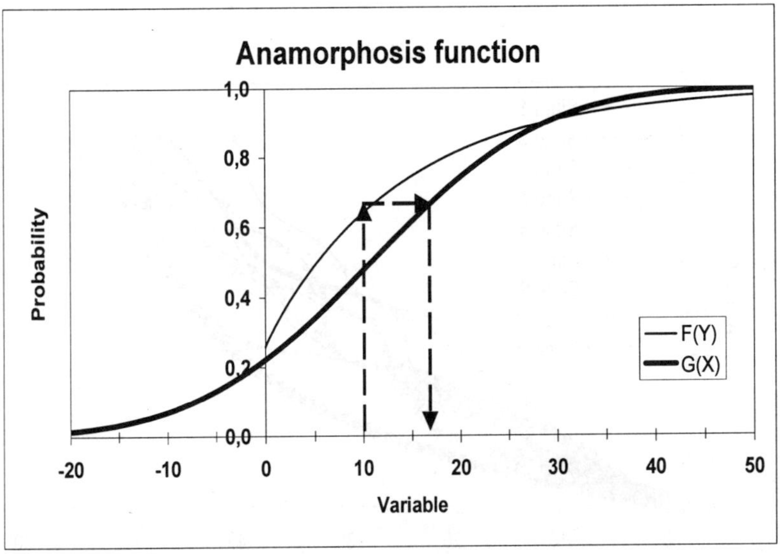

Figure 1. Example of anamorphosis function.

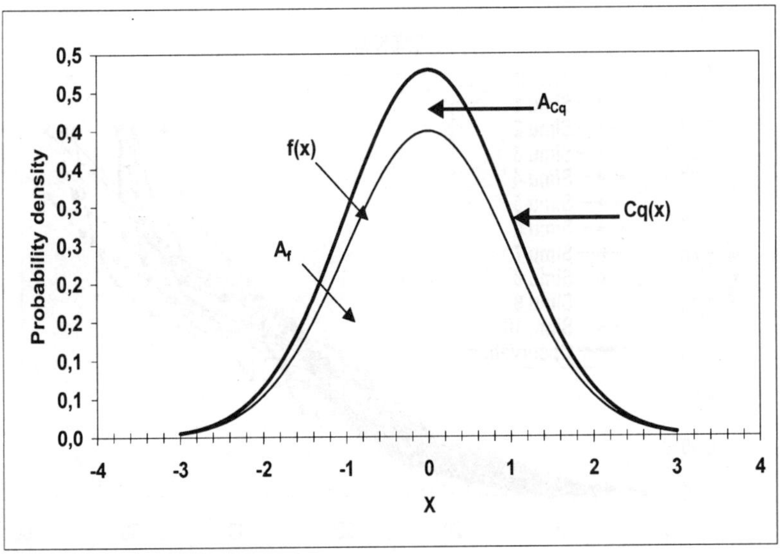

Figure 2. Acceptation-rejection method.

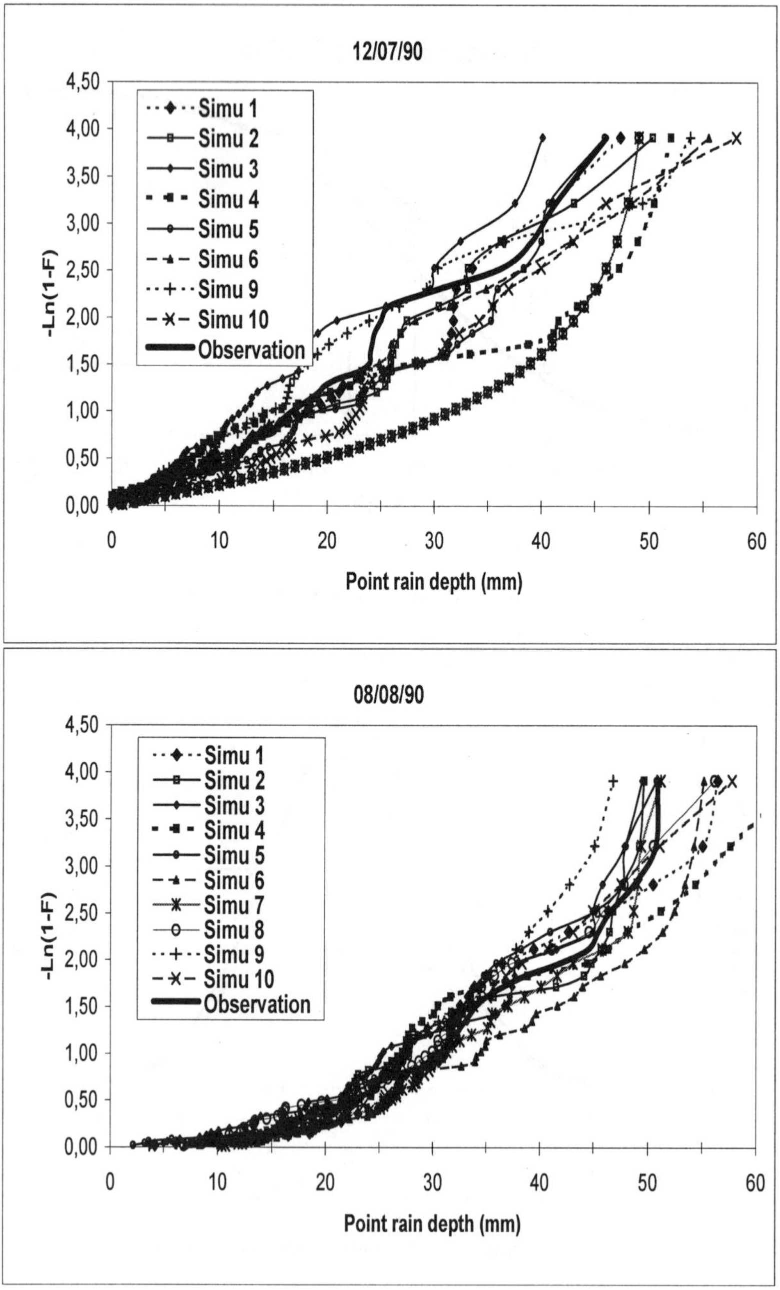

Figure 3. Cumulative distribution functions (internal variability of the model).

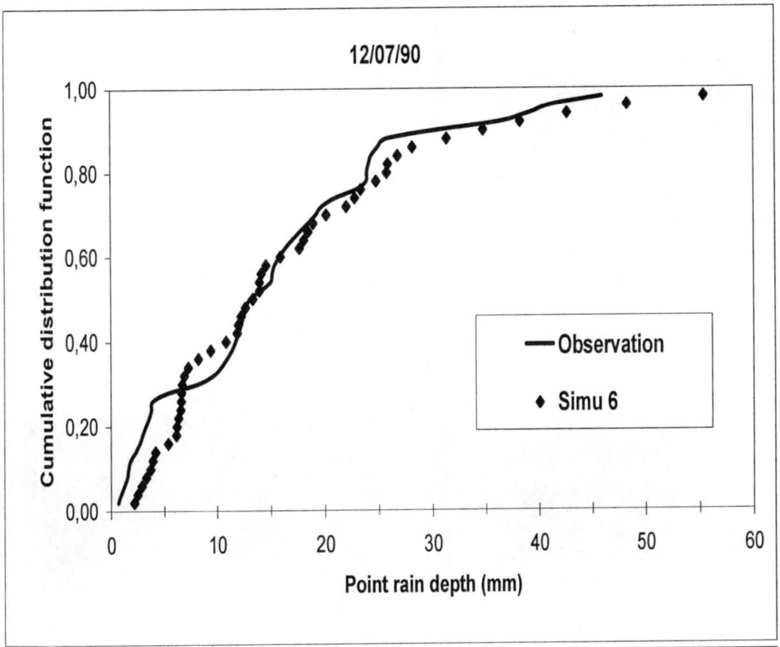

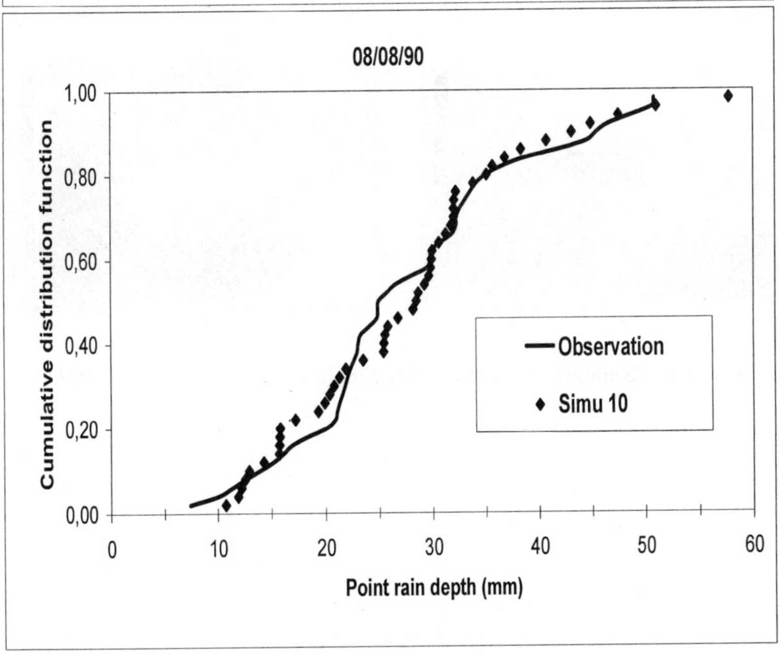

Figure 4. Cumulative distribution functions (observations compared with simulations)

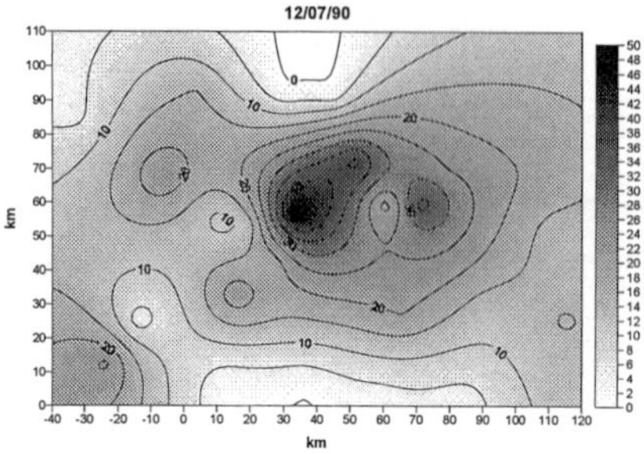

Figure 5. Spatial distribution of the cumulated values observed during the event of 12/07/90.

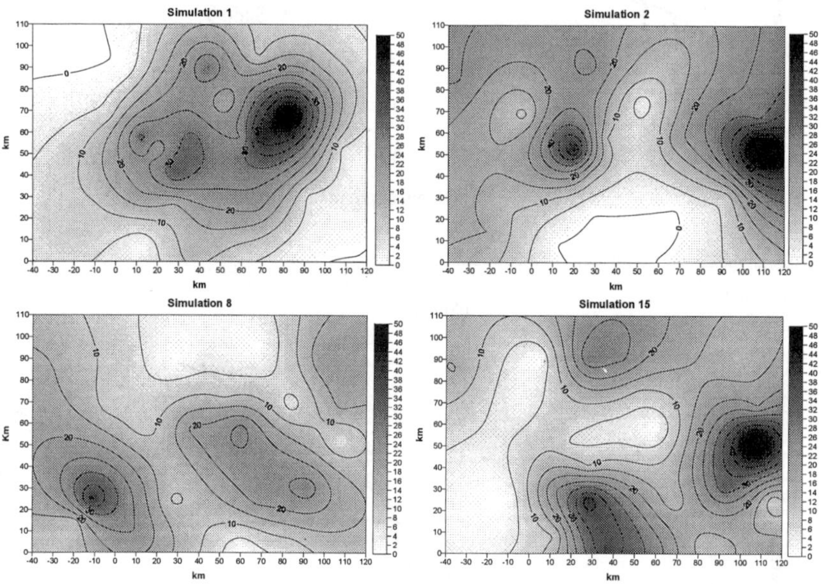

Figure 6. Examples of the constraint simulation of the event of 12/07/90.

SPINODAL DECOMPOSITION UNDER SHEAR FLOW: A SOLUBLE MODEL

N.P. RAPAPA AND N.B. MALIEHE

National University of Lesotho, P.O. Roma 180, Lesotho
E-mail: np.rapapa@nul.ls, nb.maliehe@nul.ls

The dynamics of phase separation for a binary fluid subjected to a uniform shear is studied within the soluble model where the number of components of the order parameter field, n, tends to infinity. Three characteristic length scales are obtained which increase as $[t^\alpha / \ln t]^{1/4}$, where $\alpha = 1$ in directions perpendicular to the shear flow and $\alpha = 5$ in directions parallel to the flow. These length scales are extracted from the the structure factor $S(\mathbf{k}, t)$ which has four maxima. In the shear flow plane there are two long parallel ridges as observed in scattering experiments.

1 Introduction

The dynamics of phase separation when a system (for example binary fluid) is quenched from the high temperature phase into the the low temperature region of two or more ordered phases have been of great interest.[1,2,3] In the absence of shear, the dynamics of phase separation is now well established.[1] Domains of the two equilibrium phases are formed and coarsen with time in a manner well-described by a dynamical scaling phenomenology with a single growing length scale $L(t)$ which generally grows as a power law in time, $L(t) \sim t^a$. The characteristic length scale $L(t)$ is normally associated with the wave vector k_m at which the spherically symmetric structure $S(\mathbf{k}, t)$ is maximum (i.e., $k_m \sim L^{-1}$). Both the equal time pair correlation $C(\mathbf{r}, t)$ and its Fourier transform, the structure factor $S(\mathbf{k}, t)$, obey standard scaling. By standard scaling it is meant that $C(\mathbf{r}, t)$ and $S(\mathbf{k}, t)$ can be written as

$$\begin{aligned} C(\mathbf{r}, t) &= f(r/L)\,, \\ S(\mathbf{k}, t) &= L^d g(kL)\,, \end{aligned} \tag{1}$$

where d is the dimensionality of the system.

For binary fluids, the exponent α takes different values depending on the dynamical regime under study. In order of increasing time, there are "diffusive" ($\alpha = 1/3$), "viscous hydrodynamic" ($\alpha = 1$) and the "inertial hydrodynamic" ($\alpha = 2/3$) regimes.[1,4,5] The crossover between these regimes is determined by the fluid properties such as viscosity and density. Here we will focus on the diffusive regime, in which hydrodynamic effects can be neglected. In the absence of shear, phase separation is described by the Cahn-Hilliard equation[1] for the order-parameter field $\phi(\mathbf{r},t)$, namely[1] $\partial_t\phi = -\nabla^2(\nabla^2\phi + \phi - \phi^3)$.

When the phase-separating system is subjected to a uniform shear,[3] the isotropy in the structure factor is broken as the domains grow faster in the flow direction than in directions perpendicular to the shear flow. This anisotropy is confirmed by simulation[6] and analytical,[7] numerical[8] and experimental results.[9]

If a uniform shear flow is imposed in the x-direction, with shear direction y, the flow velocity is $v_x = \gamma y$, where γ is the shear rate. For an incompressible fluid, the Cahn-Hilliard equation acquires an advective term $\mathbf{v}\cdot\nabla\phi = \gamma y\partial_x\phi$ on the left-hand side. Generalizing to an n-component vector order parameter, this gives

$$\partial_t\vec{\phi} + \gamma y \partial_x\vec{\phi} = -\nabla^2\left(\nabla^2\vec{\phi} + \vec{\phi} - \frac{1}{n}(\vec{\phi})^2\vec{\phi}\right). \tag{2}$$

For a critical mixture quenched into the two-phase region from the homogeneous phase, an appropriate initial condition is a Gaussian random field with zero mean and short-range correlations: $\langle\phi_i(\mathbf{r},0)\phi_j(\mathbf{r}',0)\rangle = \Delta\delta_{ij}\delta(\mathbf{r}-\mathbf{r}')$.

Here we will show that within the large-n limit the structure factor exhibits multiscaling with characteristic length scales such that $L_x(\sim k_{mx}^{-1}) = L_y(\sim k_{my}^{-1}) \sim (t/\ln t)^{1/4}$ and $L_z \sim k_{mz}^{-1} \sim (t/\ln t)^{1/4}$ in directions perpendicular to the flow, and parallel to the flow, respectively. These length scales are extracted from the structure factor $S(\mathbf{k},t)$ which has four maxima at $\mathbf{k} = \pm(k_{mx}, k_{my}, \pm k_{mz}))$, showing that the isotropy is completely broken. In the k_z plane, two parallel ridges are found as observed in experiment.[9] The multiscaling found here is due to large-n approximation as it is the case without shear.[12] For the scalar case without shear hydrodynamic effects, renormalization-group arguments[12] predict $k_{mx}^{-1} = t^{4/3}$ and $k_{my}^{-1} \sim t^{1/3}$, for the viscous hydrodynamic regime Läuger *et al.*[9] found experimentally $k_{mx}^{-1} = t^2$ and $k_{my}^{-1} \sim t$. The ratio $k_{mx}/k_{my} \sim 1/t$ is consistent with analytical,[7] numerical,[8] simulational[6] and experimental results.[9]

Before we proceed we first present the solution to Eq.(2) without shear solved initially by Zenatti.[11]

2 The Large-n Limit without Shear

The simplest way to take the limit $n \to \infty$ is to recognise that, for $n \to \infty$, $\vec{\phi}^2/n$ can be replaced by its average,[1] to give

$$\frac{\partial\phi}{\partial t} = -\nabla^2\left[\nabla^2\phi + a(t)\,\phi\right], \tag{3}$$

where ϕ is any (one) component of $\vec{\phi}$ and

$$a(t) = 1 - \langle \phi^2 \rangle = 1 - \sum_{\mathbf{k}} S(\mathbf{k}, t) \;, \tag{4}$$

with the structure factor $S(\mathbf{k}, t) = \langle \phi_k(t) \phi_{-k}(t) \rangle$. There is an extensive discussion on the limitations of large-n theory in the literature.[13] The solution to Eq.(3) can be easily obtained in $\mathbf{k}$-space. Fourier transforming (3) and solving for ϕ_k gives

$$\phi_k(t) = \phi_k(0) \exp\left[-k^4 t + k^2 b(t)\right] \tag{5}$$

where $a(t) = db/dt$ and the structure factor $S(\mathbf{k}, t)$ becomes

$$S(\mathbf{k}, t) = \Delta \exp\left[-2k^4 t + 2k^2 b(t)\right] \,. \tag{6}$$

Changing the sum to an integral and using (6), the last term on the left hand side of Eq.(4) gives

$$\sum_k S(\mathbf{k}, t) = \Delta S_d \left(\frac{b(t)}{t}\right)^{\frac{d}{2}} \int_0^\infty x^{d-1} \exp\left[-2\frac{b(t)^2}{t}\left(x^4 - x^2\right)\right] dx \,, \tag{7}$$

where $x = k(t/b(t))^{1/2}$ and S_d comes from angular integration. Provided $b^2/t \to \infty$ for $t \to \infty$ (which can be verified *a posteriori*), the integral in (7) can be evaluated by the method of steepest descents[14] in the limit $t \to \infty$ to give

$$\sum_k S(\mathbf{k}, t) = \Delta S_d \left(\frac{b(t)}{2t}\right)^{\frac{d}{2}} \left(\frac{\pi t}{2b(t)^2}\right)^{\frac{1}{2}} \exp\left[\frac{b(t)^2}{2t}\right] . \tag{8}$$

Since we shall find *a posteriori* that $a(t) << 1$ (supported also by dimensional analysis) at late times, the left hand side of (4) is very small for $t \to \infty$, which implies that at this stage $\sum_k S(\mathbf{k}, t) = 1$. Using this result and Eq.(8) to leading order in t,

$$\frac{b(t)^2}{2t} \sim \frac{d}{4} \ln t \,. \tag{9}$$

Using Eqs.(6), (9) and the condition $\sum_k S(\mathbf{k}, t) = 1$, it is straightforward to show that the structure factor can be written in the form

$$S(\mathbf{k}, t) = \text{Const.}\, (\ln L)^{1/2}\, L^{d\psi(q)} \,, \tag{10}$$

where $\mathbf{q} = \mathbf{k}L$, $L = k_m^{-1} = (8t/d \ln t)^{1/4}$ and $\psi(q) = 2q^2 - q^4$. Eq.(10) shows that exponent of the scaled volume $V = L^d$ depends continuously on the scaling variable q, this behaviour is different from the conventional (i.e., standard) scaling where $S(\mathbf{k},t) = Vg(\mathbf{q})$ and is normally referred to as "multiscaling" in the literature.[1,11] Attempts to find multiscaling for finite n were not successful.[15] Bray *et al.*[12] showed, within an approximate calculation that multiscaling is strictly found for $n = \infty$ and that for any finite n the standard scaling is recovered.

3 Dynamics of Phase Separation under Shear

As in the previous section $(\vec{\phi})^2/n$ in (2) is replaced by its mean in the usual way,[1] leading to a self-consistent linear equation. After Fourier transformation this reads

$$\frac{\partial \phi_{\mathbf{k}}}{\partial t} - \gamma k_x \frac{\partial \phi_{\mathbf{k}}}{\partial k_y} = -\mathbf{k^2}[\mathbf{k^2} - a(t)]\phi_{\mathbf{k}}, \tag{11}$$

where ϕ is (any) one component of $\vec{\phi}$, and $a(t) = 1 - \langle \phi^2 \rangle$. The same equation was studied numerically in two dimensions by Corberi *et al.*[8], where it was regarded as a "self-consistent one-loop" approximation to the scalar version of (2).

Eq.(11) can be solved via the change of variables $(k_x, k_y, k_z, t) \to (k_x, \sigma, k_z, \tau)$, where $\tau = t$ and $\sigma = k_y + \gamma k_x t$. The left-hand side of (11) then becomes $\partial \phi_{\mathbf{k}}/\partial \tau$, and the equation can be integrated directly to give, after transforming back to the original variables, $\phi_{\mathbf{k}}(t) = \phi_{\mathbf{k}}(0)\exp[f(\mathbf{k},t)]$, where

$$\begin{aligned} f(\mathbf{k},t) = &-(k_x^2 + k_z^2)^2 t - \frac{2(k_x^2 + k_z^2)}{3\gamma k_x}\left((k_y + \gamma k_x t)^3 - k_y^3\right) \\ &- \frac{1}{5\gamma k_x}\left((k_y + \gamma k_x t)^5 - k_y^5\right) \\ &+ \left(k_x^2 + k_z^2 + (k_y + \gamma k_x t)^2\right) b(t) \\ &- 2\gamma k_x (k_y + \gamma k_x t) c(t) + \gamma^2 k_x^2 e(t), \end{aligned} \tag{12}$$

with $b(t) = \int_0^t dt'\, a(t')$, $c(t) = \int_0^t dt'\, t' a(t')$, and $e(t) = \int_0^t dt'\, t'^2 a(t')$.

The quantity $a(t)$ can be found self-consistently by using its definition: $a(t) = 1 - \langle \phi^2 \rangle = 1 - \sum_{\mathbf{k}} S(\mathbf{k},t)$ as before. The momentum sum can be evaluated for large t using the method of steepest descents. To simplify the analysis we make the following ansatz, which will be justified *a posteriori*. Naive power counting applied to (11) suggests $a(t) \sim t^{-1/2}$, and characteristic length scales $L_y \sim L_z \sim t^{1/4}$, and $L_x \sim \gamma t^{5/4}$ (but we will see that these power laws are modified by logarithms). Guided by the $\gamma = 0$ result from Sect.2, we make the ansatz $a(t) \sim (\ln t/t)^{1/2}$ for $t \to \infty$. Then, to leading logarithmic accuracy, $b(t) \sim (t \ln t)^{1/2}$, $c(t) \to tb(t)/3$, and $e(t) \to t^2 b(t)/5$. Inserting these results in (12), and making the change of variable

$$\gamma k_x = \sqrt{\frac{b}{t^3}}\, u, \quad k_y = \sqrt{\frac{b}{t}}\, v, \quad k_z = \sqrt{\frac{b}{t}}\, w \tag{13}$$

gives

$$S(\mathbf{k}, t) = \Delta \exp\left(2\frac{b^2}{t}\, F(u, v, w)\right) \tag{14}$$

$$\begin{aligned} F(u,v,w) = &-\frac{1}{5u}\left((u+v)^5 - v^5\right) + \frac{8}{15}u^2 + \frac{4}{3}uv + v^2 \\ &-\frac{2}{3}w^2(u^2 + 3uv + 3v^2) + w^2 - w^4 . \end{aligned} \tag{15}$$

Note that contributions to F in equation (15) which vanish as $t \to \infty$ (at fixed u, v, w) have been dropped. The self-consistency equation for $a(t)$ reads

$$\begin{aligned} 1 - a(t) &= \int \frac{d^3k}{(2\pi)^3}\, S(\mathbf{k}, t) \\ &= \frac{\Delta b^{3/2}}{(2\pi)^3 \gamma t^{5/2}} \\ &\quad \times \int du\, dv\, dw \exp\left(2\frac{b^2}{t}\, F(u, v, w)\right) . \end{aligned} \tag{16}$$

Since b^2/t grows like $\ln t$ by assumption, the triple integral can be evaluated by steepest descents for large t. The complete set of stationary points (u, v, w) is listed in Table 1, together with their type (maximum, minimum, saddle point) and the corresponding values of F.

The maximum value of F, $F_m = 23/60$, occurs at four points in uvw-space (see Table 1), corresponding, via (13), to four points in momentum space. The integral in (16) can now be evaluated by steepest descents.[14] Since $a(t)$ in (16) vanishes (like $(\ln t/t)^{1/2}$) for $t \to \infty$ it can be dropped to give

$$1 = \text{const.}\, \frac{\Delta}{\gamma t b^{3/2}} \exp\left(\frac{2F_m b^2}{t}\right), \tag{17}$$

hence to leading logarithmic accuracy,

$$b(t) = \left(\frac{7}{8F_m}\, t \ln t\right)^{1/2} . \tag{18}$$

Eq.(18) is consistent with our original ansatz.

Table 1. Stationary points of $F(u,v,w)$: Max = maximum, Min = minimum, S = saddle point, Sn = saddle point of type n (the matrix of second derivatives has n positive eigenvalues), IS = "inflection saddle point" (one positive, one zero, one negative eigenvalue), IMax = "inflection maximum" (one zero, one negative eigenvalue).

Stationary Points for $F(u,v,w)$ in (u,v,w) space			
Position(u,v,w)	Number	F	Type
$(0,0,0)$	1	0	Min
$\pm(2/\sqrt{3},0,0)$	2	$16/45$ ≈ 0.35556	IS
$\pm(\sqrt{2}-1/\sqrt{3},-1/\sqrt{2},0)$	2	$(37-12\sqrt{6})/180$ ≈ 0.04225	S2
$\pm(\sqrt{2}+1/\sqrt{3},-1/\sqrt{2},0)$	2	$(37+12\sqrt{6})/180$ ≈ 0.36885	S1
$\pm(0,0,1/\sqrt{2})$	2	$1/4$ $= 0.25$	S1
$\pm(\sqrt{3},-1/\sqrt{3},\pm1/\sqrt{6})$	4	$23/60$ ≈ 0.38333	Max
Stationary Points for $F(u,v,0)$ in (u,v) plane			
Position(u,v)	Number	F	Type
$(0,0,0)$	1	0	Min
$\pm(2/\sqrt{3},0,0)$	2	$16/45$ ≈ 0.35556	IMax
$\pm(\sqrt{2}-1/\sqrt{3},-1/\sqrt{2},0)$	2	$(37-12\sqrt{6})/180$ ≈ 0.04225	S
$\pm(\sqrt{2}+1/\sqrt{3},-1/\sqrt{2},0)$	2	$(37+12\sqrt{6})/180$ ≈ 0.36885	Max
Stationary Points for $F(u,0,w)$ in (u,w) plane			
Position(u,w)	Number	F	Type
$(0,0,0)$	1	0	Min
$\pm(2/\sqrt{3},0)$	2	$16/45$ ≈ 0.35556	S
$\pm(0,1/\sqrt{2})$	2	$1/4$ $= 0.25$	S
$(\pm3,\pm1)/2\sqrt{2}$	4	$29/80$ $= 0.3625$	Max

From Eq.(13) we can define characteristic length scales for the three directions: $L_x = \gamma(t^3/b)^{1/2} \sim \gamma(t^5/\ln t)^{1/4}$, and $L_y = L_z = (t/b)^{1/2} \sim (t/\ln t)^{1/4}$, by setting $u = k_x L_x$, $v = k_y L_y$, and $w = k_z L_z$. Finally the structure factor is given by $S(\mathbf{k},t) = \Delta \exp[(2b^2/t)F(u,v,w)]$. Using (17) and (18), the structure factor $S(\mathbf{k},t)$ becomes

$$S(\mathbf{k},t) = \text{const.}\,(\ln V_s)^{3/2} V_s^{F(\mathbf{q})/F_m}, \tag{19}$$

where $V_s(t) = L_x L_y L_z \sim \gamma t^{7/4}/(\ln t)^{3/4}$ is the "scale volume" at time t, and $\mathbf{q} = (k_x L_x, k_y L_y, k_z L_z)$ is the scaled momentum. The corresponding result for zero shear (in space dimension d) is $S(\mathbf{k},t) = \text{const.}\,(\ln L)^{1/2} L^{d\phi(q)}$ given by Eq.(10). Eq.(19) does not have a form consistent with conventional scaling, $S(\mathbf{k},t) = V_s(t) f(\mathbf{q})$, but rather a form of "multiscaling", as in the zero-shear case, where the *power* of the scaling volume V_s depends continuously on the scaling variables. As in the zero-shear case, however, we anticipate that simple scaling will be recovered asymptotically for any finite n, and that the $\ln t$ terms will disappear from the length scales.[12]

3.1 Two-dimensional theory

In measurements of the structure factor by small-angle light scattering, the scattering vector $\mathbf{k}$ is perpendicular to the beam direction. Two common experimental arrangements[9] are with the beam in the z-direction, i.e., perpendicular to both the shear and the flow, or in the y (i.e., shear) direction. The former corresponds to $k_z = 0$, the latter to $k_y = 0$. For $k_z = 0$, $S(k_x, k_y, 0, t)$ is determined by $F(u,v,0)$ (see Fig.1) while for $k_y = 0$, $S(k_x, 0, k_z, t)$ is determined by $F(u,0,w)$.

Here we first consider $F(u,v,w)$ in the uv-plane (i.e., $w = 0$). Repeating the same analysis as in three dimension, a two-dimensional steepest descent integration yields

$$1 = \text{const.}\,\frac{\Delta}{\gamma t b} \exp\left(\frac{2F_m^{2D} b^2}{t}\right), \tag{20}$$

instead of (17), where (from Table 1) $F_m^{2D} = (37 + 12\sqrt{6})/180$, giving $b(t) = (3t \ln t / 4F_m^{2D})^{1/2}$ (to leading logarithmic accuracy) instead of (18). The structure factor is given by $S(\mathbf{k},t) = \text{const.}\,(\ln A_s)\, A_s^{F^{2D}(\mathbf{q})/F_m^{2D}}$, where $A_s = L_x L_y$ is the "scale area", $L_x = \gamma (t^3/b)^{1/2} \sim \gamma (t^5/\ln t)^{1/4}$, $L_y = (t/b)^{1/2} \sim (t/\ln t)^{1/4}$, $\mathbf{q} = (k_x L_x, k_y L_y)$ is the scaled momentum, and $F_m^{2D}(\mathbf{q})$ means $F(u,v,0)$, with $u = k_x L_x$ and $v = k_y L_y$. $F(u,v,0)$ has two parallel ridges as observed in experiment.[9] The ridges are terminated by inflexion maxima at $F = 0.35556$ and the maxima at $F = 0.36885$. Note that the relations $\ln S(\mathbf{k},t) \propto F \ln t$ (plus k-independent terms) and $k_x = u/L_x$ etc., mean that the ridges in S become higher, narrower, and closer together as a function of $\mathbf{k}$ with increasing time. The angle between the ridges and the v-axis is found to be $\tan\phi = 2(1 - 1/\sqrt{6})$. Then from (13), the corresponding angle, θ, in

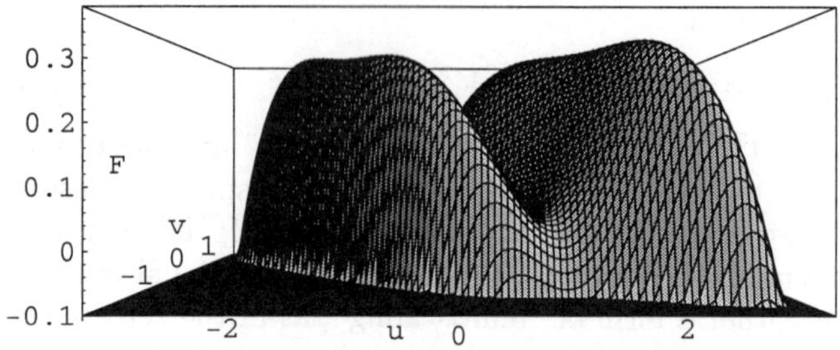

Figure 1. The projection of the function $F(u, v, 0)$ in the uv plane showing the two parallel ridges and the stationary points. Regions for $F < -0.1$ are not shown.

the (k_x, k_y) plane is given by $\tan\theta = (\tan\phi)/\gamma t$, i.e., the ridges tend to align closer to the shear direction as t increases, consistent with experiments[9] and simulations.[6]

For uw-plane, the stationary points are shown in Table 1 also and the function $F(u, 0, w)$ is shown in Fig.2. $F(u, 0, w)$ has four maxima, one minimum and four saddle points. Currently available experimental data cannot resolve this structure. Instead a broadly elliptical scattering pattern is seen in the (k_x, k_z) plane,[9] with the major axis along the k_z direction, and the eccentricity increasing in time, as expected from different growth rates (by a factor t) of L_x and L_z.

Finally, if the beam is parallel to the flow ($k_x = 0$), the shear term drops out of (11). The scattering intensity then has full circular symmetry in the (k_y, k_z) plane.

4 Discussion

The numerical calculations of Corberi *et al.*[8] also reveal the two ridge-like structures, but the ridges are terminated by two peaks whose relative heights *oscillate* in time, such that first one peak, then the other is the higher. The explanation for such oscillations was that they "characterize the steady state". Since no oscillatory behaviour is found in the asymptotics of our analytical solution, we believe that the observed oscillations are slowly-decaying preasymptotic transients.

As far as the shape of the structure factor is concerned, we expect the scaling function $g(\mathbf{q})$ to be described, in broad terms, by $\ln g(\mathbf{q}) \sim F(q_x, q_y, q_z)$, where $F(u, v, w)$ is given by (15), i.e., rather like the large-n theory but without the $\ln V_s \sim \ln t$ factor multiplying F. In particular, the graph of

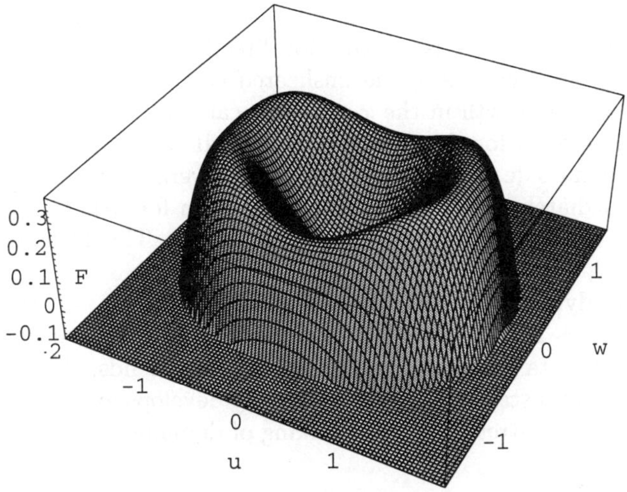

Figure 2. The projection of the function $F(u, 0, w)$ in the uw plane showing the two parallel ridges and the stationary points. Regions for $F < -0.1$ are not shown.

$\ln g(q_x, q_y, 0)$ and $\ln g(q_x, q_y, 0)$ should look very similar to Fig.1 and Fig.2, respectively. These are also confirmed by numerical,[8] simulational[6] and experimental results.[9] We have found that length scales parallel (i.e., L_x) and perpendicular (i.e., L_y, L_y) to the flow are given by $\gamma(t^5 / \ln t)^{1/4}$ and $(t/\ln t)^{1/4}$ respectively. The standard scaling is not found but "multi-scaling", however, this is believed to be an artifact of the large-n theory as in the case of zero shear. We also note that for large-n, there are no domains, therefore the steady state[10] at late times due to stretching and breaking up of domains does not apply here. In order to observe the steady state analytically, one must consider the scalar case. So far extending the problem to conserved scalar case is not trivial (even for zero shear, the conservation laws complicate the calculations[1]). Another still open problem is to consider the effects of the hydrodynamics in the presence of shear.

We end the discussion with some conjectures about the physically realistic case (for binary fluids in the diffusive regime) of a scalar order parameter. These are informed by our exact solution for $n = \infty$, and by the way the $n = \infty$ solution is known to be modified for scalar fields in the *unsheared* case.[12] First we expect that, for any finite n, the structure factor will exhibit asymptotic scaling of the form $S(\mathbf{k}, t) = V_s g(\mathbf{q})$, with $V_s = \Pi_{i=1} L_i$ $(i = x, y, z)$

and $q_i = k_i L_i$, instead of the multiscaling form (19). As in the $n = \infty$ case, we expect the growth of the characteristic scales for directions normal to the flow to obey the same power laws as in the unsheared case, i.e., for scalar fields, $L_y \sim L_z \sim t^{1/3}$. The growth in the x-direction can then be deduced from the assumed scaling form for the structure factor. If we multiply the two terms on the left-hand side of (11) by $\phi_{-\mathbf{k}}(t)$, and average, the result $L_x \sim \gamma t L_y$ follows immediately if we insert the scaling form for $S(\mathbf{k}, t)$ and assume both terms are of the same order in the scaling limit. This leads to the result $L_x \sim \gamma t^{4/3}$.

In the viscous hydrodynamic regime, where without shear $L(t) \sim t$, the same heuristics suggest $L_y \sim L_z \sim t$, $L_x \sim \gamma t^2$. The predictions for the length scales are consistent with data by Läuger *et al.*[9] on polymer blends, though it has been suggested[10] that a stationary state eventually develops for $\gamma t \gg 1$ due to a competition between stretching and breaking of domains.

Acknowledgments

N.P.R. and N.B.M. thank both RCC (Research and Conferences Committee of the National University of Lesotho) and COPROMAPH2 (Second International Workshop on Contemporay Problems in Mathematical Physics, Benin) for financial support.

References

1. A.J. Bray, *Adv. Phys.* **43**, 357 (1994), and references therein.
2. J.D. Gunton, M. San Miguel and P. S. Sahni, *Phase Transitions and Critical Phenomena*, C. Domb and J.L. Lebowitz, eds. **8**, 267 (Academic Press, New York, 1983).
3. A. Onuki, *J. Phys.: Condens. Matter* **9**, 6119 (1997), and references therein.
4. E.D. Siggia, *Phys. Rev.* **A20**, 595 (1979).
5. H. Furukawa, *Phys. Rev.* **A31**, 1103 (1985).
6. D.H. Rothman, *Phys. Rev. Lett.* **65**, 3305 (1990);
   P. Padilla and S. Toxvaerd, *J. Chem. Phys.* **106**, 2342 (1997);
   A.J. Wagner and J.M. Yeomans, *Phys. Rev.* **E59**, 4366 (1999);
   R. Yamamoto and X.C. Zeng, *Phys. Rev.* **E59**, 3223 (1999).
7. N.P. Rapapa and A.J. Bray, *Phys. Rev. Lett.* **83**, 3856 (1999);
   N.P. Rapapa, *Phys. Rev.* **E61**, 247 (2000).

8. F.Corberi, G. Gonnella and A. Lamura, *Phys. Rev. Lett.* **81**, 3852 (1998);
   L. Berthier, *Phys. Rev.* **E63**, 051503 (2001).
9. C.K. Chan, F. Perrot and D. Beysens, *Phys. Rev.* **A43**, 1826 (1991);
   J.Läuger, C. Laubner and W. Gronski, *Phys. Rev. Lett.* **75**, 3576 (1995);
   A.H. Krall, J.V. Sengers and K. Hamano, *Phys. Rev. Lett.* **69**, 1963 (1992).
10. T. Hashimoto, K. Matsuzaka, E. Moses and A. Onuki, *Phys. Rev. Lett.* **74**, 126 (1995);
    K. Matsuzaka, T. Koga and T. Hashimoto, *Phys. Rev. Lett.* **80**, 5441 (1998).
11. A. Coniglio and M. Zannetti, *EuroPhys. Lett.* **10**, 575 (1989).
12. A.J. Bray and K. Humayun, *Phys. Rev. Lett.* **68**, 1559 (1992).
13. A. Coniglio, P. Ruggiero and M. Zannetti, *Phys. Rev. Lett.* **50**, 1046 (1994);
    F. Corberi, A. Coniglio and M. Zannetti, *Phys. Rev.* **E51**, 5469 (1995);
    C. Castellano and M. Zannetti, *Phys. Rev. Lett.* **77**, 2742 (1996); *Phys. Rev.* **E53**, 1430 (1996); *Phys. Rev.* **E58**, 5410 (1998).
14. G. Arfken, *Mathematical Methods for Physicists*, Third Edition (Academic Press, 1985).
15. A. Coniglo, Y. Oono, A. Shinozaki and M. Zannetti, *EuroPhys. Lett.* **18**, 59 (1992);
    M. Mondello and N. Goldenfeld, *Phys. Rev.* **E47**, 2383 (1993);
    M. Siegert and M. Rao, *Phys. Rev. Lett.* **70**, 1956 (1993);
    M. Rao and A. Chakrabarki, *Phys.Rev.* **E49**, 3727 (1994).

THE MATHEMATICS OF AN ALGEBRAIC APPROACH TO THE PHYSICS OF HADRONS

MILTON DEAN SLAUGHTER

Department of Physics, University of New Orleans
New Orleans, LA 70148, USA
E-mail: mslaught@uno.edu

We outline the mathematics and physics associated with a unique algebraic methodology which incorporates Quantum Chromodynamic equal time commutation relations, infinite momentum frame $SU_F(N)$ broken flavor symmetry, and asymptotic level realization of chiral current algebras, which when taken together generate algebraic sum rules consisting of covariant matrix elements (also gauge invariant for matrix elements involving the electromagnetic current) which can be used to study a wide variety of hadronic processes important for the advancement of our knowledge of nuclear and elementary particle physics.

1 Introduction: Outline of Basic Theoretical Concepts

We explore the consequences of using a unique algebraic methodology which incorporates Quantum Chromodynamic (QCD) equal time commutation relations (ETCRs), asymptotic (infinite momentum frame) $SU_F(N)$ *broken flavor symmetry*, and asymptotic level realization of chiral current algebras, which when taken together generate algebraic sum rules consisting of covariant matrix elements (and also *gauge invariant* for matrix elements involving the electromagnetic current) which can be used to study – nonperturbatively – a wide variety of strong and electroweak processes important for the advancement of our knowledge of nuclear and elementary particle physics. Contrary to popular belief in some physics circles, the use of current-algebra techniques to solve difficult nonperturbative problems in nuclear and elementary particle physics continues to be timely, worthwhile, and extremely effective. The significance of our nonperturbative algebraic approach to hadronic physics is especially relevant when one considers that while the Standard Model (SM) of particle physics is the most viable theoretical framework available today for the description of the Weak, Electromagnetic, and Strong Interactions, it is also very clear that the SM is incomplete. In the hadronic world, the SM incorporates Quantum Chromodynamics (QCD), a theory which although very beautiful and elegant and in many ways a direct extrapolation of QED to the hadronic sector, also brings into existence gluons, which unlike photons, directly interact with each other in *highly non-linear, non-abelian* fashion. Indeed, in QCD, gluons may form particles called glueballs, which (non-exotic) can mimic ordinary mesons. Quantum mechanically, it is therefore quite difficult to distinguish physical glueball candidates from such ordinary mesons. Additionally, present theory does not indicate *clearly* what properties that *even a bare* glueball would possess. Since QCD coupling "constants" can be of order one or more, the successful use of field-theoretic, perturbative techniques using Feynman diagrams is extremely difficult. This situation has thus led to the ever-increasing use of technologies such as lattice gauge theory, chiral perturbation theory, supersymmetry theory, and algebraic models which hold out the promise of performing QCD calculations in a nonperturbative fashion. The algebraic methods to be employed in carrying out the proposed research are based on the equal-time commutation relations (ETCRs) [*i.e.* $[V_i, V_j] = [A_i, A_j] = if_{ijk}V_k$, $[V_i, A_j] = if_{ijk}A_k$, $[V_i, V_j^\mu] = [A_i, A_j^\mu] = if_{ijk}V_k^\mu$, and $[V_i, A_j^\mu] = if_{ijk}A_k^\mu$] involving the generators (the vector and axial-vector charges V_α and A_α $\{\alpha = \pi, K, D, F, B, \ldots\}$, and their current counterparts) of the underlying badly broken symmetry groups of QCD derivable from $\mathcal{L}_{QCD}$ including quark mass terms in the infinite momentum frame (light cone frame). These ETCRs are valid even though these symmetries are broken.[1,2] In addition, the *exotic* ETCRs involving $\dot{V}_\alpha$ also hold, *i.e.*, $[\dot{V}_\alpha, V_\beta] = [\dot{V}_\alpha, A_\beta] = 0$, where $\dot{V}_\alpha = i[H, V_\alpha]$, H is the total QCD Hamiltonian and (α,β) is an *exotic* combination of flavor indices.[3] As it turns out, the judicious use of exotic ETCRs allows mass and mixing angle determination even when glueballs are considered. We also use the dynamical concept of asymptotic $SU_F(N)$ symmetry where the generator V_α, when acting on a physical Heisenberg on-mass-shell hadron state (Hilbert space state) which has momentum $\vec{k}$ ($|\vec{k}| \to \infty$), helicity λ, and physical $SU_F(N)$ index α (*i.e.* annihilation operator $a_\alpha(\vec{k}, \lambda)$), produces a quantum state which maintains its linearity (including $SU_F(N)$ particle mixings) under flavor transformations but *only* in the asymptotic infinite momentum frame limit $|\vec{k}| \to \infty$. In addition, we utilize the idea of ETCR level realization which *prescribes* how complete sets of intermediate states corresponding to observed hadronic levels schemes are to be inserted in ETCRs. Level realization allows us to avoid the severe type of problems usually associated with the arbitrary truncation of intermediate states but does sometimes lead to sets of non-linear algebraic equations (sum rules) which are often soluble only by using numerical or symbolic computational techniques. Some physical processes of keen interest that are amenable to our methods include: (1) the role of glueballs – mass and mixing angle determination – in understanding the details of hadron spectroscopy in the 1 to 3 GeV/c^2 meson mass spectrum, which unfortunately is the very same mass region where expected quarkonium excitations and even exotic mesons may play an important role; (2) radiative meson and baryon decays and the importance of the $SU_F(N)$ isoscalar singlet piece of the electromagnetic current in such decays; (3) tensor, pseudoscalar, and scalar meson two photon decays; (4) hyperon semileptonic decays; (5) ground state baryon magnetic moments; (6) single pion electroproduction processes (the relationship between the $\Delta \to N + \gamma$ transition form factors and nucleon form factors pertaining to leading-order perturbative QCD (PQCD) predictions for the quantity $R_{EM} \equiv E_{1^+}/M_{1^+}$ at high momentum transfer is of special interest); and (7) the weak axial-vector nucleon-Δ transition amplitudes important ranging from N^* production processes in isobar models to atmospheric neutrino experiments at Super-Kamiokande. We point out that our approach leads naturally to sum rules in which masses, mixing parameters, and coupling "constants" are *dynamically* interrelated very closely. For instance, quark-line rules appear naturally and automatically via this dynamic interplay in our sum rules. The technique for using the algebras is simple: one inserts the commutation relations between appropriate physical states, then one allows for appropriate intermediate (asymptotic level realization) physical states – including mixing – to occur, then one *evaluates each matrix element using the "no leakage" properties of the vector charges at infinite momentum.* We note that this procedure gives rise to sum rules which explicitly exhibit the very well-defined and close interplay among the masses (if exotic ETCRs are used), mixing angle parameters, and physical coupling constants under consideration. They in turn give rise to systems of coupled, highly nonlinear, algebraic equations, for which solutions are nontrivial. In order to make this problems tractable, one must often resort to computers for numerical or symbolic analysis. While simple in principle – in practice, the evaluation of matrix elements in the infinite momentum frame can be very tedious and complicated, especially when particles with spin greater than one are involved. Different methods may be employed for evaluation upon choosing a particular Lorentz frame of reference: (1) direct brute force evaluation; or (2) evaluation using broken symmetry group properties and the Lorentz covariance transformation properties of the matrix elements (initial state, final state, and vertex operator).[4] An extraordinary benefit of our algebraic method as opposed to other algebraic methods being used[5] is that helicity representations and group symmetry properties of the *physical* particles involved are always used, thus our results are fully relativistic, gauge invariant and can be used for particles of arbitrary spin. In addition, *the electromagnetic current is always conserved.* Thus, there is no need to impose additional group symmetries or make assumptions with regard to the detailed quark content of the particles being studied. This of course has the disadvantage of making our results rather specific to the process being studied. As in other models, the inclusion of additional intermultiplet usually results in ever increasing complexity. However, in our model, this translates to solving larger sets of non-linear algebraic equations. In these cases, we eliminate – as much as possible – those matrix elements for which we have no experimental data, solve for those matrix elements which do not require knowledge of poorly known matrix elements, or make educated approximations where possible.

2 ETCRs

Equal-time commutation relations (ETCRs) involving the generators (the vector and axial-vector charges V_α and A_α $\{\alpha = \pi, K, D, F, B, \ldots\}$) of the underlying symmetry groups of quantum chromodynamics (QCD) are derivable from $\mathcal{L}_{QCD}$ including quark mass terms. They are valid even though these symmetries are broken.[1,2] Some examples of these ETCRs or constraint algebras are: (Gell-Mann) $[V_i, V_j] = [A_i, A_j] = i f_{ijk} V_k$, and $[V_i, A_j] = i f_{ijk} A_k$. In addition the *exotic* ETCRs involving $\dot{V}_\alpha$ also hold, *i.e.*, $[\dot{V}_\alpha, V_\beta] = [\dot{V}_\alpha, A_\beta] = 0$, where $\dot{V}_\alpha = i[H, V_\alpha]$, H being the total QCD Hamiltonian and (α, β) is an *exotic* combination[3] of flavor indices. Mathematically, we have the following. We introduce quark spinor fields ${q_\alpha}^i$ each with mass m_i where $i = 1, \cdots, N$ in the flavor $SU_F(N)$ group, and $\alpha = 1, 2, 3$ are the color indices corresponding to the color $SU(3)$ group. Then, we have (suppressing color indices)

$$ {V_\mu}^a = i\bar{q}^i (\lambda^a/2)_{ij} \gamma_\mu q^j \equiv i\bar{q}(\lambda^a/2)\gamma_\mu q, \tag{1} $$

$$ V^a(t) = \int d^3x \, q^\dagger (\lambda^a/2)\gamma_\mu q, \tag{2} $$

$$ \partial^\mu {V_\mu}^a(x) = i\bar{q}^i (m_i - m_j)(\lambda^a/2)_{ij} q^j, \tag{3} $$

$$ {A_\mu}^a = i\bar{q}^i (\lambda^a/2)_{ij} \gamma_\mu \gamma_5 q^j \equiv i\bar{q}(\lambda^a/2)\gamma_\mu \gamma_5 q, \tag{4} $$

$$ A^a(t) = \int d^3x \, q^\dagger (\lambda^a/2)\gamma_\mu \gamma_5 q, \tag{5} $$

$$ \partial^\mu {A_\mu}^a(x) = i\bar{q}^i (m_i + m_j)(\lambda^a/2)_{ij} \gamma_5 q^j, \tag{6} $$

$$ [V^a(t), V^b(t)] = [A^a(t), A^b(t)] = i f^{ab}_{\;\;..c} V^c, \tag{7} $$

and

$$ [V^a(t), A^b(t)] = i f^{ab}_{\;\;..c} A^c. \tag{8} $$

In Eqs. (1-8), the λ^a, $a = 1, 2, \cdots, N^2 - 1$, satisfy the *Lie algebra*

$$[(\lambda^a/2)(t), (\lambda^b/2)(t)] = i f^{ab}_{\ldots c}(\lambda^c/2), \tag{9}$$

where the $f^{ab}_{\ldots c}$ are structure constants of the flavor group $SU_F(N)$. We take note of the following: (1) the vector charges $V^a(t)$ are not conserved and are time dependent (except when $a = 3, 8, 15, 24, 35$ when the λ^a are diagonal corresponding to isospin, strangeness, charm,...); (2) the axial-vector charges $A^a(t)$ are conserved only when *all quark masses vanish*; and (3) the algebras represented by Eqs.(1-8) *are always valid even in the presence of symmetry breaking*. This last item is what is of paramount interest to us. An avenue is open to explore – nonperturbatively – an investigation of many phenomena not now easily amenable to the usual QCD techniques.

3 Asymptotic $\mathbf{SU_F(N)}$

A key ingredient of the dynamical concept of asymptotic $SU_F(N)$ symmetry is the behaviour of the vector charge V_α when acting on a physical state which has momentum $\vec{k}$ ($|\vec{k}| \to \infty$), helicity λ, and physical $SU_F(N)$ index α: namely that the annihilation operator $a_\alpha(\vec{k}, \lambda)$ of a physical on-mass-shell hadron is assumed to maintain its linearity (including $SU_F(N)$ particle mixings) under flavor transformations generated by the charge V_α but only in the asymptotic limit $|\vec{k}| \to \infty$. Thus, the V_α are generators of $SU_F(N)$ and entertain *no* "leakage" terms in the asymptotic limit. See Eq.(13) below. In particular, we have the result that $V_\alpha|G^0\rangle = 0$, where $|G^0\rangle \equiv$ bare glueball.

In order to briefly illustrate the workings of asymptotic $SU_F(N)$ symmetry, consider the transformation of the annihilation operator $a_\alpha(\vec{k})$ under $SU_F(N)$ in broken symmetry,

$$[V_i, a_\alpha(\vec{k}, \lambda)] = i \sum_\beta u_{i\alpha\beta}(\vec{k}, \lambda) a_\beta(\vec{k}, \lambda) + \delta u_{i\alpha\lambda}(\vec{k}) \ . \tag{10}$$

In exact $SU_F(N)$, one can write

$$\begin{aligned} [V_i, a_j(\vec{k}, \lambda)] &= i f_{ijk} a_k(\vec{k}, \lambda) \ for\ j = 1, 2, \ldots, N^2 - 1 \ , \\ &= 0 \text{ for } j = 0 \ , \end{aligned} \tag{11}$$

where $a_j(\vec{k}, \lambda)$ is a $SU_F(N)$ *representation* annihilation operator. The *dynamical* assumption of asymptotic $SU_F(N)$ symmetry then states that

$$\delta u_{i\alpha\lambda}(\vec{k}) \to (|\vec{k}|)^{-(1+\epsilon)}, \ (\epsilon > 0) \text{ for } |\vec{k}| \to \infty \tag{12}$$

implying that

$$|\alpha, \vec{k}, \lambda\rangle = \sum_j C_{\alpha j} |j, \vec{k}, \lambda\rangle, \quad \vec{k} \to \infty. \tag{13}$$

The orthogonal matrix $C_{\alpha j}$ depends on $SU_F(N)$ mixing parameters and hence can be *constrained directly by the ETCRs without introducing an ad hoc mass matrix*. Here $|j, \vec{k}, \lambda\rangle$ is a $SU_F(N)$ representation state whereas $|\alpha, \vec{k}, \lambda\rangle$ is a physical state.[3] This is a key difference from conventional approaches in handling mixing problems. Our method is straightforward and has tight theoretical constraints. It is well-suited to the study of glueball-$q\bar{q}$-meson mixing. All nonlinear terms vanish like $|\vec{k}|^{-(1+\epsilon)}$, $(\epsilon > 0)$, as $\vec{k} \to \infty$. (Solvable models usually have $\epsilon = 1$ for bosons).

Thus, in the ∞-momentum frame, the *physical* annihilation operator $a_\alpha(\vec{k}, \lambda)$ is related *linearly* to the *representation* annihilation operator $a_j(\vec{k}, \lambda)$ via the orthogonal mixing matrix $C_{\alpha j}(\lambda)$. So even in the case of badly broken $SU_F(N)$ symmetry – such as $SU_F(4)$ or $SU_F(5)$ – *asymptotic* $SU_F(N)$-symmetry methods can still be successfully applied as has already been demonstrated by previous results attained such as: (1) the smallness of the isoscalar nucleon anomalous magnetic moments and the weak neutral-current axial-vector coupling constants and the prediction $k_p \cong -k_n$ for the nucleon magnetic moments;[6] (2) a correct estimate of $g_A(0)$;[7,1] (3) a new derivation of the electroweak relation $m_W = \cos\theta_W \, m_Z$;[8] (4) a derivation of the asymptotic $|\Delta I| = \frac{1}{2}$ rule and its relationship with *OZI*[9] quark-line rules;[3] (5) the *OZI* rule, such as $g_{\phi\rho\pi} \cong 0$ and $g_{\phi\pi\gamma} \cong 0$ when ϕ is almost pure $s\bar{s}$;[3] (6) a derivation of the $SU_F(3)$ Gell-Mann-Okubo mass formula[10,11,12] as an exact constraint; (7) the relation $(G_\rho/m_\rho) = (G_{K^*}/m_{K^*})$ for the $\tau \to \rho\nu_\tau$ and $\tau \to K^*\nu_\tau$ coupling constants;[13] (8) the well-known sum rule $\frac{1}{3}m_\rho\Gamma(\rho \to e^+e^-) = m_\omega\Gamma(\omega \to e^+e^-) + m_\phi\Gamma(\phi \to e^+e^-)$;[13] and (9) the derivation of experimentally well-verified mass squared relations involving hadrons composed of heavy-light quark pairs.[14,15] We note here that the mass squared derivations are *independent* of the particular details of the QCD potential and automatically incorporate relativistic corrections.

Another important implication of asymptotic $SU_F(3)$ and the well-known QCD algebra $[V_i, A_j] = i f_{ijk} A_k$ is the interesting observation that the Cabibbo sum rules for the axial-vector hyperon semi-leptonic couplings at zero four-momentum transfer squared limit are, in fact, *exact* sum rules (apart from the neglect of small inter-multiplet mixing) explicitly compatible with the Gell-Mann-Okubo hyperon mass splittings. Asymptotic symmetry thus provides theoretical understanding for the success of the Cabibbo sum rules.

4 Asymptotic Level Realization

We illustrate the asymptotic level realization hypothesis by first considering the well-known chiral $SU(2) \otimes SU(2)$ charge algebra

$$[A_{\pi^+}, A_{\pi^-}] = 2\,V_3. \tag{14}$$

Sandwich the above commutator between the identical baryon $SU_F(N)$ states $\langle B(\alpha, \lambda, \vec{s}|$ and $|B(\alpha, \lambda, \vec{t}\rangle$ with momentum $\vec{s}, \vec{t}$ ($|\vec{s}| \to \infty, |\vec{t}| \to \infty$), helicity λ, and physical $SU_F(N)$ index α. Then

$$\sum_{n_L} < B(\alpha, \lambda)|A_{\pi^+}|n_L >< n_L|A_{\pi-}|B(\alpha, \lambda >$$
$$- \sum_{n'_L} < B(\alpha, \lambda)|A_{\pi^-}|n'_L >< n'_L|A_{\pi-}|B(\alpha, \lambda) >= C(\alpha, \lambda), \tag{15}$$

where $C(\alpha, \lambda) = 2\,\langle B(\alpha, \lambda)|\,V_3\,|B(\alpha, \lambda)\rangle$ is a pure number. Among the sum over single particle hadronic intermediate states in Eq.(15), we distinguish (for given α and λ) the fractional contribution $f^L(\alpha, \lambda)$ coming from *all* the states n_L or n'_L belonging to a level $L(L = 0, 1, 2, \ldots)$, where L connotes angular momentum. Eq.(15) then becomes $C(\alpha, \lambda)(f^0(\alpha, \lambda) + f^1(\alpha, \lambda) + f^2(\alpha, \lambda) + \ldots) = C(\alpha, \lambda)$. Note that upon varying the $SU_F(N)$ index α which produces non-vanishing values of $C(\alpha, \lambda)$, the intermediate states n_L also undergo corresponding $SU_F(N)$ rotations. The asymptotic level realization hypothesis then states that $f^L(\alpha, \lambda)$ for a given level L will depend on λ but *not* on α, *i.e.*, the fractional contribution from each level L is universal and invariant under $SU_F(N)$ rotation. In other words, "block realization" of asymptotic $SU_F(N)$ in chiral algebraic constraints is postulated as an important manifestation of quark dynamics in the physical hadronic world.

We now take as external states the ground state baryons, i.e., the $1/2^+$ octet and the $3/2^+$ decuplet with helicity $\lambda = +1/2$ and apply our techniques to the fractional contribution coming from the same ground state baryons. We define relevant axial-vector matrix elements as follows,

$$\begin{aligned} \langle p, 1/2|\,A_{\pi^+}\,|n, 1/2\rangle &\equiv f \quad , \\ \langle \Delta^{++}, 1/2|\,A_{\pi^+}\,|\Delta^+, 1/2\rangle &\equiv -\sqrt{\tfrac{3}{2}}\,g \quad , \\ \langle \Delta^{++}, 1/2|\,A_{\pi^+}\,|p, 1/2\rangle &\equiv -\sqrt{6}\,h \quad . \end{aligned} \tag{16}$$

Using the commutator Eq.(14) and asymptotic level realization for the $L = 0$ ground state baryons ($J^{PC} = \frac{1}{2}^+, \frac{3}{2}^+$), the following relation, which is very well satisfied experimentally, may easily be established,

$$h^2 = \frac{4}{25} f^{\,2}, \quad g = -\frac{\sqrt{2}}{5} f \quad . \tag{17}$$

5 Interesting QCD Commutators

$$[[J^\mu_{em}(0),\, A_{\pi^+}],A_{\pi^-}] = 2\; J^\mu_3(0)\;, \tag{18}$$

$$[[J^\mu_3(0),\, A_{\pi^+}],A_{\pi^-}] = 2\; J^\mu_3(0)\;. \tag{19}$$

Here, $J^\mu_{em}(0)$ is the the electromagnetic current, and A_{π^+} and A_{π^-} are axial-vector charge operators (generators). The current $J^\mu_{em}(0)$ can be written as $J^\mu_{em} = V^\mu_3 + V^\mu_S$, where $V^\mu_3 \equiv J^\mu_3$ is the isovector part of the electromagnetic current and V^μ_S is the isoscalar part.

Insert the algebraic relations, Eq.(19), between the ground states, $\langle B(\alpha, \vec{s}, \lambda = -1/2|$ and $\left|B'(\beta, \vec{t}, \lambda = +1/2\right\rangle$ with $\vec{s} \to \infty$ and $\vec{t} \to \infty$ and use algebraic level realization. Then the following ten equations are generated,

$$\begin{aligned}\langle\Delta^{++}|\;[J^\mu_3,\,A_{\pi^+}]\,A_{\pi^-}\;|\Delta^{++}\rangle - \langle\Delta^{++}|\;A_{\pi^-}\,[J^\mu_3,\,A_{\pi^+}]\;|\Delta^{++}\rangle \\ = 2\;\langle\Delta^{++}|\,J^\mu_3\,|\Delta^{++}\rangle\end{aligned} \tag{20}$$

$$\left\langle\Delta^+\right|\,[J^\mu_3,\,A_{\pi^+}]\,A_{\pi^-}\;\left|\Delta^+\right\rangle - \left\langle\Delta^+\right|\,A_{\pi^-}\,[J^\mu_3,\,A_{\pi^+}\,]\,\left|\Delta^+\right\rangle = 2\;\left\langle\Delta^+\right|\,J^\mu_3\,\left|\Delta^+\right\rangle \tag{21}$$

$$\left\langle\Delta^0\right|\,[J^\mu_3,\,A_{\pi^+}]\,A_{\pi^-}\;\left|\Delta^0\right\rangle - \left\langle\Delta^0\right|\,A_{\pi^-}\,[J^\mu_3,\,A_{\pi^+}\,]\left|\Delta^0\right\rangle = 2\;\left\langle\Delta^0\right|J^\mu_3\left|\Delta^0\right\rangle \tag{22}$$

$$\left\langle\Delta^-\right|\,[J^\mu_3,\,A_{\pi^+}]\,A_{\pi^-}\;\left|\Delta^-\right\rangle - \left\langle\Delta^-\right|\,A_{\pi^-}\,[J^\mu_3,\,A_{\pi^+}\,]\left|\Delta^-\right\rangle = 2\;\left\langle\Delta^-\right|\,J^\mu_3\left|\Delta^-\right\rangle \tag{23}$$

$$\langle p|\;[J^\mu_3,\,A_{\pi^+}]\,A_{\pi^-}\;|p\rangle - \langle p|\;A_{\pi^-}\,[J^\mu_3,\,A_{\pi^+}\,]\,|p\rangle = 2\;\langle p|\,J^\mu_3\,|p\rangle \tag{24}$$

$$\langle n|\;[J^\mu_3,\,A_{\pi^+}]\,A_{\pi^-}\;|n\rangle - \langle n|\;A_{\pi^-}\,[J^\mu_3,\,A_{\pi^+}\,]\,|n\rangle = 2\;\langle n|\,J^\mu_3\,|n\rangle \tag{25}$$

$$\left\langle\Delta^+\right|\,[J^\mu_3,\,A_{\pi^+}]\,A_{\pi^-}\;|p\rangle - \left\langle\Delta^+\right|\,A_{\pi^-}\,[J^\mu_3,\,A_{\pi^+}\,]\,|p\rangle = 2\;\left\langle\Delta^+\right|\,J^\mu_3\,|p\rangle \tag{26}$$

$$\langle p|\;[J^\mu_3,\,A_{\pi^+}]\,A_{\pi^-}\;\left|\Delta^+\right\rangle - \langle p|\;A_{\pi^-}\,[J^\mu_3,\,A_{\pi^+}\,]\left|\Delta^+\right\rangle = 2\;\langle p|\,J^\mu_3\left|\Delta^+\right\rangle \tag{27}$$

$$\left\langle\Delta^0\right|\,[J^\mu_3,\,A_{\pi^+}]\,A_{\pi^-}\;|n\rangle - \left\langle\Delta^0\right|\,A_{\pi^-}\,[J^\mu_3,\,A_{\pi^+}\,]\,|n\rangle = 2\;\left\langle\Delta^0\right|\,J^\mu_3\,|n\rangle \tag{28}$$

$$\langle n|\;[J^\mu_3,\,A_{\pi^+}]\,A_{\pi^-}\;\left|\Delta^0\right\rangle - \langle n|\;A_{\pi^-}\,[J^\mu_3,\,A_{\pi^+}\,]\left|\Delta^0\right\rangle = 2\;\langle n|\,J^\mu_3\left|\Delta^0\right\rangle \tag{29}$$

Define (note that the Lorentz index μ has been dropped for convenience),

$$
\begin{aligned}
< p,-1/2,\vec{s}|\,J_3|\,p,1/2,\vec{t}> &\equiv b = -< n,-1/2,\vec{s}|J_3|\,n,1/2,\vec{t}>,\\
< n,-1/2,\vec{s}|\,J_3|\,\Delta^0,1/2,\vec{t}> &\equiv c = + < p,-1/2,\vec{s}|J_3|\,\Delta^+,1/2,\vec{t}>,\\
< \Delta^0,-1/2,\vec{s}|\,J_3|\,n,1/2,\vec{t}> &\equiv d = + < \Delta^+,-1/2,\vec{s}|J_3|\,p,1/2,\vec{t}>\\
< \Delta^+,-1/2,\vec{s}|\,J_3|\,\Delta^+,1/2,\vec{t}> &\equiv a\\
&= -< \Delta^0,-1/2,\vec{s}|\,J_3|\,\Delta^0,1/2,\vec{t}>,\\
< \Delta^{++},-1/2,\vec{s}|\,J_3|\,\Delta^{++},1/2,\vec{t}> &\equiv 3a\\
&= -< \Delta^-,-1/2,\vec{s}|J_3|\,\Delta^-,1/2,\vec{t}>.
\end{aligned}
\tag{30}
$$

Thus, with the definitions given in Eqs.(30), Eqs.(20) through Eqs.(29) become,

$$6ag^2 + 18ah^2 + 3dgh^2 - 3cgh - 6bh^2 = 6a \tag{31}$$

$$6ag^2 + 2ah^2 - \sqrt{2}dfh - 5cgh + 2bh^2 = 2a \tag{32}$$

$$-6ag^2 - 5dgh - 2bh^2 - 2ah^2 - \sqrt{2}cfh = -2a \tag{33}$$

$$-6ag^2 - 3dgh^2 + 3cgh + 6bh^2 - 18ah^2 = -6a \tag{34}$$

$$-2cgh + 8bh^2 - 16ah^2 - \sqrt{2}dfh - 3dgh = 2b \tag{35}$$

$$-3cgh - 8bh^2 + 16ah^2 - \sqrt{2}cfh - 2dgh = -2b \tag{36}$$

$$2cg^2 + 8ch^2 - 2bgh - 14agh + 2dh^2 - \sqrt{2}cfg = 2c \tag{37}$$

$$\frac{3}{2}cg^2 + 3bgh - 9agh + 2dh^2 - \sqrt{2}cfg + \sqrt{2}bfh + 2ch^2 + \sqrt{2}afh + cf^2 = 2c \tag{38}$$

$$9agh - \sqrt{2}afh + 2dh^2 + df^2 - \sqrt{2}dfg + 2ch^2 - \sqrt{2}bfh + \frac{3}{2}dg^2 - 3bgh = 2d \tag{39}$$

$$14agh + 8dh^2 - \sqrt{2}dfg + 2ch^2 + 2dg^2 + 2bgh = 2d \tag{40}$$

Using symbolic programs such as Mathematica, one can solve this system of ten (not all independent) simultaneous non-linear algebraic equations relatively easily – without "losing" or introducing "spurious" solutions – and one finds that the only physical solution (analytic!) is

$$a : b : c : d = 1 : -2 : \pm 2\sqrt{2}/5 : \mp 2\sqrt{2}/5, \tag{41}$$

corresponding to $h = \pm\frac{2}{5}\, f$. *It should be noted that obtaining the physical solution by "hand" is much more difficult. In addition, attempting a numerical physical solution is essentially not possible due to lack of sufficient experimental data including phases.*

One result of Eq.(41) – among many others – is the *Spin Flip Sum Rule*, with $\vec{s}, \vec{t} \to \infty$, given by,[16]

$$< p, -1/2, \vec{s}\,|\, J_3^{\mu}(0)|\; \Delta^+, 1/2, \vec{t}> = \frac{5}{8}\sqrt{2} < p, -1/2, \vec{s}\,|\, J_3^{\mu}(0)|\, p, 1/2, \vec{t}> \; . \tag{42}$$

The importance of Eq.(42) is that it directly relates the poorly known $\Delta N\gamma$ transition form factors $G_M^*(q^2)$, $G_E^*(q^2)$ (which determine the matrix element $< p, -1/2, \vec{s}\,|\, J_3^{\mu}(0)|\; \Delta^+, 1/2, \vec{t}>$) to the very well known nucleon form factors (which determine the matrix element $< p, -1/2, \vec{s}\,|\, J_3^{\mu}(0)|\, p, 1/2, \vec{t}>$) for all space-like $q^2 \leq 0$! If one takes $\vec{t}$ to lie along the z-axis and $\vec{s}$ in the $x - z$ plane such that $|\vec{t}| = t_z \to \infty\ ,\quad s_z \equiv r t_z \to \infty\ ,\quad 0 < r \leq 1\ ,\quad r, s_x$ held constant, then one obtains a relationship well-satisfied by the data.[16]

6 Conclusions

As has been shown, the potential impact of our research methods on nuclear and elementary particle physics phenomenology is quite large. First, we use ETCRs (at the very core of quantum field theory) which are valid even though the symmetries associated with the particles or processes under consideration may be badly broken. Second, the energy regime that we may probed through our techniques is not *a priori* limited to low or very high momentum transfer. Third, because our methods make the most use of any naturally occurring symmetries retained in processes under investigation, group-theoretic extrapolation techniques may often be brought to bear. Fourth, the connection between quark model theory and experiment is made manifestly explicit via matrix element dependence on momentum transfer in our sum rule approach. We look forward to the continued development and refinement of our methods as applied to the physics of hadrons and the successful use of these *nonperturbative, broken flavor symmetry* algebraic techniques by us in describing a wide variety of physical processes including even ascertaining the properties of physical glueballs and demonstrating their existence in nature.

Acknowledgments

We thank the Center for Theoretical Studies of Physical Systems (CTSPS) at Clark Atlanta University, Atlanta, Georgia (USA), for its support.

References

1. M. Gell-Mann, *The Symmetry Group of Vector and Axial Vector Currents*, *Physics* **1**, 63 (1964);
   S. Adler, *Sum Rules for the Axial-Vector Coupling-Constant Renormalization in β Decay*, *Phys. Rev.* **140**, 736 (1965);
   W. I. Weisberger, *Unsubtracted Dispersion Rules and the Renormalization of the Weak Axial-Vector Coupling*, *Phys. Rev.* **143**, 1302 (1966).
2. V. De Alfaro, S. Fubini, G. Furlan and C. Rossetti, *Currents in Particle Physics* (North-Holland Publishing, Amsterdam, 1973).
3. For a review and a literature list, see
   S. Oneda and K. Terasaki, *Algebraic Approach in Quantum Chromodynamics and Electroweak Theory*, *Prog. Theor. Phys. Suppl.* **82**, 1 (1985), especially Chapters 4 and 5;
   M. D. Slaughter and S. Oneda, *Theoretical Limit on $\iota \to \rho\gamma$ and Constraints On Glueball-$q\bar{q}$ Meson Couplings with the Pion and the Isovector Photon*, *Phys. Rev. Lett.* **59**, 1641 (1987);
   Milton D. Slaughter, *The Resonant P_{33} $S_{1+}^{3/2}(q^2)$ Electroproduction Multipole Amplitude and the $\Delta N\gamma$ Scalar Form Factor $G_C^*(q^2)$*, *Phys. Rev.* **C49**, R2894 (1994);
   M. Slaughter and S. Oneda, $1^{--} \to 0^{-+}$ *Meson Radiative and Pionic Transitions and Mass Splittings*, *Phys. Rev.* **D39**, 2062 (1989).
4. L. Durand, P. DeCelles and R. Marr, *Lorentz Invariance and the Kinematic Structure of Vertex Functions*, *Phys. Rev.* **126**, 1882 (1962);
   V. De Alfaro, S. Fubini, G. Furlan and C. Rossetti, *Currents in Particle Physics* (North-Holland Publishing, Amsterdam, 1973), Chapter 8.
5. R. Bijker, F. Iachello and A. Leviathan, *Algebraic Models of Hadron Structure. I. Nonstrange Baryons*, *Ann. Phys. (N.Y.)* **236**, 69 (1994).
6. S. Oneda, T. Tanuma and Milton D. Slaughter, *Why are the Isoscalar Neutral-Current Axial-Vector Couplings and Isoscalar Nuclear Anomalous Moments Small?*, *Phys. Lett.* **88B**, 343 (1979).
7. S. Oneda and S. Matsuda, *Asymptotic Algebraic Realization of $SU(3)$, Baryon Classification, $g_A(0)$, and D/F Ratio*, *Phys. Rev.* **D5**, 2287 (1972).
8. M. Yasuè, S. Oneda and M. D. Slaughter, *Charge Commutation Relation Approach to Electroweak Gauge Theories and the Mass of the Second Z Boson*, *Phys. Rev.* **D30**, 174 (1984).
9. S. Okubo, *Phi Meson and Unitary Symmetry Model*, *Phys. Lett.* **5**, 165 (1963);
   G. Zweig, *An $SU3$ Model for Strong Interaction Symmetry and Its Break-*

ing, CERN Report No. TH401 and TH412 (1964);
J. Iizuka, K. Okada and O. Shito, *Systematics and Phenomenology of Boson Mass Levels. 3.*, *Prog. Theor. Phys.* **35**, 1061 (1965).

10. S. Okubo, *Note on Unitary Symmetry in Strong Interactions*, *Prog. Theor. Phys.* **27**, 949 (1962);
M. Gell-Mann, *Symmetries of Baryons and Mesons*, *Phys. Rev.* **125**, 1067 (1962).
11. S. Oneda and S. Matsuda, *Fundamental Interactions in Physics*, Proc. of the 1973 Coral Gables Conf., **175** (Plenum, New York, 1973);
S. Oneda, *Group Theoretical Method in Physics*, Proc. of the Austin Conf., 1978, *Lect. Notes in Physics* **94** (Springer, Berlin, Heidelberg, New York, 1978), p. 334.
12. S. Oneda, H. Umezawa and S. Matsuda, *Asymptotic Symmetry, Particle Mixing, and K_{l3}-Decay Branching Ratios*, *Phys. Rev. Lett.* **25**, 71 (1970);
S. Oneda and S. Matsuda, *Phys. Rev.* **D2**, 324 (1970).
13. S. Oneda, *Asymptotic Flavor Symmetry and its Implication on $\tau \to \rho\nu_\tau$ and $K^*\nu_\tau$ Branching Ratio and Ground State 1^{--} Meson Multiplet*, *Phys. Rev.* **D35**, 397 (1987).
Exactly the same result has also been derived by using a version of asymptotic symmetry for the spectral functions of the currents. See
T. Das, V.S. Mathur and S. Okubo, *Symmetry, Superconvergence, and Sum Rules for Spectral Functions*, *Phys. Rev. Lett.* **18**, 761, (1967).
14. Milton Dean Slaughter and S. Oneda, *$1^{--} \to 0^{-+}$ Meson Radiative and Pionic Transitions and Mass Splittings*, *Phys. Rev.* **D39**, 2062 (1989).
15. E. Takasugi and S. Oneda, *Purely Algebraic Approach to the New Narrow Resonances*, *Phys Rev. Lett.* **34**, 1129 (1975);
H. Hallock, S. Oneda and Milton D. Slaughter, *D and D^* in the Purely Algebraic Approach to Broken $SU(4)$*, *Phys. Rev.* **D15**, 884 (1977); *Phys. Rev.* **D15**, 884 (1977).
16. Milton D. Slaughter and S. Oneda, *Phys. Rev.* **D49**, 323 (1994).

Parallel Sessions: Part II

Coherent States, Wavelets and Geometric Methods in Theoretical Physics

ON THE CHARACTERIZATION OF QUASICRYSTALLINE STRUCTURES WITH BETA-LATTICES

M. ANDRLE, A. EL KHARRAT AND JEAN-PIERRE GAZEAU

Laboratoire de Physique Théorique de la Matière Condensée
Boîte 7020, Université Paris 7 Denis Diderot
F-75251 Paris Cedex 05, France
E-mail: gazeau@ccr.jussieu.fr

We give some recent results on counting systems, called beta-integers, based on Pisot-Vijayaraghavan algebraic integers $\beta > 1$ and leading to the notion of beta-lattices in $\mathbb{R}^d$. Beta-lattices are self-similar, may be equipped with an abelian group structure, and reveal themselves as eligible labelling frames for discrete locally self-similar quasiperiodic structures with scaling factor β. As a matter of fact we present some applications of their use for mathematical models as well as for more physically derived quasiperiodic structures.

1 Introduction

Lattices, d-dimensional periodic structures involving translational symmetry, are well-known by physicists and chemists throughout the study of Crystallography and Solid State Physics. Geometrical models underlying such objects are (possibly decorated) $\mathbb{Z}$-modules: an elementary *cell* or finite pattern is infinitely repeated along precise directions in space. The variety of those directional arrangements compatible with translational symmetry has given rise to the famous crystallographic classifications: 32 point groups, and 230 space groups. The mathematical object upon which these periodic structures rest is the set of integers.

Yet, what was thought to be exclusively crystalline properties, like repetitivity and long-range order, are in fact not specific to periodic objects. Many aperiodic structures encountered in Nature or in Art also have these properties, one of the most popular recent example being afforded by quasicrystals. In many cases, these properties are characterized by the existence of an inflationary or scaling symmetry in which the scaling factor is a certain irrational algebraic integer $\beta > 1$, and it can occur that deep algebraic features of this β are the mark of (local) n-fold rotational symmetries in the structure.

The aim of this contribution is to give an overview of recent results that pertain to algebraic number theory and algorithmics of numeration systems, which lead to the notion of beta-lattices, as well as of some applications in Physics of such mathematical objects. Beta-lattices are relevant to the description of certain aperiodic structures showing inflationary symmetry possibly combined with n-fold symmetry. These structures can be quasicrystals or quasicrystalline diffraction patterns, of course. Nevertheless, they can be encountered in various other fields of Physics, like dynamical systems or hydrodynamics.[10,8] Other examples are Penrose tilings and their diffraction patterns,[12] and wavelet analysis of quasiperiodic structures.[1]

We shall restrict our presentation to one- and two-dimensional structures. In Sec.2 will be given some basic definitions concerning discrete point sets of interest for our purpose and which are nowadays known as mathematical models for quasicrystals. In Sect.3 we explain the notions of *beta-expansions* and *beta-integers*, and we give arithmetic and algebraic properties of beta-integers. We introduce the notion of *algebraic model set*, constructed upon an algebraic sieving procedure rather than a geometrical construction. The role played by integers when dealing with lattices in $\mathbb{R}^d$ is assumed by beta-integers in the definition of beta-lattices as discussed in Sect.4: cyclotomic structures in the plane are natural support for n-fold aperiodic point sets, from which the notion of *cyclotomic Pisot-Vijayaraghavan* number (or cyclotomic PV number) emerges as the fundamental object, root of the construction of beta-lattices. Sect.5 concludes this article with numerical applications in Physics, using beta-lattices as discretization scheme for simulation and analysis of quasiperiodic structures.

2 A Brief Overview of Mathematical Models for Quasicrystals

Let us start out with the notion of *Delone set* which is the obvious model of any large grouping of atoms defining a material structure. By Delone set $\Lambda \subset \mathbb{R}^d$ we mean that Λ is *uniformly discrete* (there exists $r > 0$ such that each ball of radius r contains at most one point of Λ) and *relatively dense* (there exists $R > 0$ such that each ball of radius R contains at least one point of Λ). A Delone set Λ is a *Meyer set* if $\Lambda - \Lambda$ is a Delone set or equivalently if there exists a finite set F such that $\Lambda - \Lambda \subset \Lambda + F$.

The most popular geometrical models for quasicrystals are the so-called *cut-and-project* sets. These sets actually were previously introduced by Meyer[15,16] in the context of Harmonic Analysis and Number Theory and christened by him *model sets*. First, a *cut and projection scheme* is the following

$$\begin{array}{ccc} \mathbb{R}^d \xleftarrow{\pi_1} & \mathbb{R}^d \times G & \xrightarrow{\pi_2} G \\ & \cup & \\ & D & \end{array}$$

where G is a locally compact abelian group, called the *internal space* in quasicrystallography, $\mathbb{R}^d$ is called the *physical space*, D is a lattice, *i.e.*, a discrete subgroup of $\mathbb{R}^d \times G$ such that $(\mathbb{R}^d \times G)/D$ is compact. The projection $\pi_1|_D$ is one-to-one, and $\pi_2(D)$ is dense in G.

Let $M = \pi_1(D)$ and set $* = \pi_2 \circ (\pi_1|_D)^{-1} : M \longrightarrow G$. The set $\Lambda \subset \mathbb{R}^d$ is a *model set* if there exist a cut and projection scheme and a relatively compact set $\Omega \subset G$ of non-empty interior such that $\Lambda = \{x \in M \mid x^* \in \Omega\}$. The set Ω is called a *window*.

The following was proved by Meyer: *A model set is a Meyer set. Conversely if Λ is a Meyer set, there exist a finite set F and a model set Λ_0 such that $\Lambda \subset \Lambda_0 + F$.*

We say that a Delone set Λ is self-similar of scaling factor $\theta > 1$ (*inflation factor*) if $\theta\Lambda \subset \Lambda$.

We now recall some definitions on numbers. A *PV number* is an algebraic integer > 1, solution to $X^m + c_{m-1}X^{m-1} + \cdots + c_1 X + c_0 = 0$ such that all the other roots or *Galois conjugates* have modulus strictly less than unity. A *Salem number* is an algebraic integer > 1 such that every conjugate has modulus smaller than or equal to 1, and at least one of them has modulus 1.

Another result from Meyer makes the connection between Meyer sets and those algebraic integers: *If $\Lambda \subset \mathbb{R}^d$ is a Meyer set and if $\theta > 1$ is a real number such that $\theta\Lambda \subset \Lambda$ then θ is a PV or a Salem number. Conversely for each d and for each PV or Salem number θ, there exists a Meyer set $\Lambda \subset \mathbb{R}^d$ such that $\theta\Lambda \subset \Lambda$.*

Let us illustrate this construction with one type of Fibonacci chain, well-known as a one-dimensional toy model for quasicrystals.[14] Consider the following cut and projection scheme,

$$\begin{array}{c} \mathbb{R} \xleftarrow{\pi_1} \mathbb{R} \times \mathbb{R} \xrightarrow{\pi_2} \mathbb{R} \\ \cup \\ \mathbb{Z}^2 \end{array} \tag{1}$$

Here $\pi_1(\mathbb{Z}^2) \sim \mathbb{Z}[\tau] = \{a + b\tau \mid a, b \in \mathbb{Z}\}$, with $\tau = \frac{1+\sqrt{5}}{2}$.

(One version of) the Fibonacci chain $\mathcal{F}$ is

$$\begin{aligned} \mathcal{F} &= \{x = a + b\tau \mid x' \equiv a - \frac{b}{\tau} \in \Omega = [0, 1)\} \\ &= \{\ldots, -\tau^3, -\tau, 0, \tau^2, \tau^3 + 1, \tau^4, \ldots\}. \end{aligned} \tag{2}$$

It is the set of left endpoints of a quasiperiodic tiling of $\mathbb{R}$ with two tiles L et S, of respective lengths τ^2 and τ, generated by the substitution rules $L \mapsto LLS$, $S \mapsto LS$. Starting from SL in both directions we get a bi-infinite word $\cdots LLSLS \mid LLSLLSLS \cdots$ Note that $\tau^2\mathcal{F} \subset \mathcal{F}$ and that this tiling is, by construction, *stone inflation*, which means that all the tiles when scaled by the factor τ^2 can be packed face-to-face from the original ones. In section 5 we shall develop the equivalent algebraic procedure (sketched in the above) for building the same set.

3 Beta-Integers: Arithmetic and Algebra

When a number $\beta > 1$ appears as a kind of fundamental invariant in a given structure, it is tempting to introduce into the procedure of understanding the latter a *counting system* based precisely on this β. Let us explain here what we mean by counting system.

3.1 Definition of the set of beta-integers and basic properties

Among all *beta-representations* of a real number $x \geq 0$, *i.e.*, infinite sequences $(x_i)_{i \leq k}$, such that $x = \sum_{i \leq k} x_i \beta^i$ for a certain integer k, there exists a particular one, called the *beta-expansion*, which is obtained through the "greedy algorithm".[18,17] Denote by $\lfloor y \rfloor$ and by $\{y\}$ the integer part and the fractional part of a real number y. There exists $k \in \mathbb{Z}$ such that $\beta^k \leq x < \beta^{k+1}$. Let $x_k = \lfloor x/\beta^k \rfloor$ and $r_k = \{x/\beta^k\}$. Then for $i < k$, put $x_i = \lfloor \beta r_{i+1} \rfloor$, and $r_i = \{\beta r_{i+1}\}$. We get the so-called beta-expansion $x = x_k \beta^k + x_{k-1} \beta^{k-1} + \cdots$ denoted by

$$\langle x \rangle_\beta = x_k x_{k-1} \ldots x_1 x_0 \cdot x_{-1} x_{-2} \ldots \tag{3}$$

For instance, if $\beta = \tau \approx 1 \cdot 618 \cdots$, then $x_i \in \{0, 1\}$. The tau-expansion of, say 4, is $4 = \tau^2 + 1 + 1/\tau^2$, *i.e.*, $\langle 4 \rangle_\tau = 101 \cdot 01$.

If $x < 1$ then $k < 0$, and we put $x_0 = x_{-1} = \cdots = x_{k+1} = 0$. The digits x_i obtained by this algorithm are integers from the set $A = \{0, \ldots, \lceil \beta \rceil - 1\}$, called the *canonical alphabet*, where $\lceil \beta \rceil$ denotes the closest integer to β from above. If an expansion ends in infinitely many zeros, it is said to be *finite*, and the ending zeros are omitted.

The definition of *beta-integers* is now straightforward: the set of beta-integers is the set of real numbers such that their beta-expansion is polynomial,

$$\mathbb{Z}_\beta = \{x \in \mathbb{R} \mid \langle |x| \rangle_\beta = x_k \cdots x_0\} = \mathbb{Z}_\beta^+ \cup (-\mathbb{Z}_\beta^+) , \tag{4}$$

where $\mathbb{Z}_\beta^+$ is the set of non-negative beta-integers. For instance, the first positive tau-integers are given by

$$\begin{array}{ccccccccccc} 0 & \overbrace{\qquad}^{1} & 1 & \overbrace{\qquad}^{\frac{1}{\tau}} & \tau & \overbrace{\qquad}^{1} & \tau^2 & \overbrace{\qquad}^{1} & \tau^2+1 & \overbrace{\qquad}^{\frac{1}{\tau}} & \tau^3 \\ (0) & & (1) & & (10) & & (100) & & (101) & & (1000). \end{array} \tag{5}$$

When $\beta = 2 + \sqrt{3} \equiv \theta$, the first positive theta-integers read,[a]

$$\begin{array}{ccccccccccc} 0 & \overbrace{\qquad}^{1} & 1 & \overbrace{\qquad}^{1} & 2 & \overbrace{\qquad}^{1} & 3 & \overbrace{\qquad}^{\theta-3} & \theta & \overbrace{\qquad}^{1} & \theta+1 \\ (0) & & (1) & & (2) & & (3) & & (10) & & (11). \end{array} \tag{6}$$

The set $\mathbb{Z}_\beta$ is self-similar and symmetrical with respect to the origin

$$\beta\mathbb{Z}_\beta \subset \mathbb{Z}_\beta\ ,\ \mathbb{Z}_\beta = -\mathbb{Z}_\beta .$$

3.2 Arithmetic and algebra of beta-integers

It has been shown[6] that if β is a PV number then $\mathbb{Z}_\beta$ is a Meyer set, which means that there exists a finite set F such that $\mathbb{Z}_\beta - \mathbb{Z}_\beta = \mathbb{Z}_\beta + \mathbb{Z}_\beta \subset \mathbb{Z}_\beta + F$. Hence, one can see that the set $\mathbb{Z}_\beta$ is "quasiadditive" to a certain extent. An interesting question to be addressed is how to characterize completely this beta-dependent F in order to see to what extent beta-integers differ from ordinary integers concerning additive (and multiplicative) group structures. This problem has been solved[6,7,11] for all quadratic PV units and for a few higher-degree cases.[20] The demonstrations mainly rest upon the algebraic properties of $\mathbb{Z}_\beta$, themselves mostly consequences of the bounded nature of the Galois conjugates of β: $\beta', \beta'', \ldots \in (-1, 1)$. As a matter of fact, $\mathbb{Z}_\beta$ is by construction a subset of the extension ring $\mathbb{Z}[\beta] = \sum_{k \geq 0} \mathbb{Z}\beta^k$ and has all its Galois conjugate sets $\mathbb{Z}'_\beta$, $\mathbb{Z}''_\beta, \ldots$ bounded in their respective extension rings $\mathbb{Z}[\beta']$, $\mathbb{Z}[\beta''], \ldots$ Moreover, $\mathbb{Z}_\beta$ is endowed with an internal additive law for which it has a group structure.[7,11]

3.3 Model sets based on the set of beta-integers

One of the immediate applications of beta-integers is the construction of one-dimensional model sets, such as the Fibonacci chain, based on Galois conjugation and algebraic sieving. We shall refer to such constructed sets as *algebraic model set*.

---

[a]The PV unit θ is related to twelve-fold symmetry as shall be seen later.

For quadratic PV numbers, the extension ring of β is given by $\mathbb{Z}[\beta] = \{m + n\,\beta \,|\, m, n \in \mathbb{Z}\}$. The algebraic model set is defined as,

$$\Sigma^{\Omega} = \{x = m + n\beta \in \mathbb{Z}[\beta] \mid x' = m + n\beta' \in \Omega\}, \tag{7}$$

where the closure $\overline{\Omega}$ is compact and the interior $\overset{\circ}{\Omega}$ is not empty. Then, we have the general results

- for quadratic PV unit β solution to $X^2 - aX - 1 = 0, a \geq 1$ (the Galois conjugate is $\beta' = -1/\beta$), for example $\beta = \tau$, we have

$$\mathbb{Z}^+_\beta = \Sigma^{(-1,\beta)} \cap \mathbb{R}^+ \,, \; \mathbb{Z}^-_\beta = -\mathbb{Z}^+_\beta = \Sigma^{(-\beta,1)} \cap \mathbb{R}^- \,, \tag{8}$$

$$\mathbb{Z}_\beta \subset \Sigma^{(-\beta,\beta)} \subset \mathbb{Z}_\beta/\beta \,, \tag{9}$$

and more generally the infinite increasing sequence of discretizations of $\mathbb{R}$

$$\cdots \subset \mathbb{Z}_\beta/\beta^{j-1} \subset \Sigma^{(-\beta^j,\beta^j)} \subset \mathbb{Z}_\beta/\beta^j \subset \cdots \tag{10}$$

This implies

$$\mathbb{Z}'_\beta \subset (-\beta, \beta) \;\text{ and }\; \mathbb{Z}'_\beta \cap (-1, 1) = \mathbb{Z}[\beta] \cap (-1, 1). \tag{11}$$

- for quadratic PV unit β solution to $X^2 - aX + 1 = 0, a \geq 3$ (the Galois conjugate is $\beta' = 1/\beta$), for example $\beta = \theta$,

$$\mathbb{Z}^+_\beta = \Sigma^{[0,\beta)} \cap \mathbb{R}^+ \;\text{ and }\; \mathbb{Z}^-_\beta = \Sigma^{(-\beta,0]} \cap \mathbb{R}^- \,, \tag{12}$$

$$\mathbb{Z}_\beta/\beta \subset \Sigma^{(-\beta,\beta)}/\beta \tag{13}$$

$$\mathbb{Z}_\beta \subset \Sigma^{(-\beta,\beta)} \subset \widetilde{\mathbb{Z}}_\beta \subset \Sigma^{(-2\beta,2\beta)} \,, \tag{14}$$

with $\widetilde{\mathbb{Z}}_\beta = \mathbb{Z}_\beta + \{0, \pm\frac{1}{\beta}\}$.

By using these results the Fibonacci chain constructed by *cut and projection* method in Sect.2 now reads as,

$$\mathcal{F} = \Sigma^{[0,\,1)} = \{x \in \mathbb{Z}_\tau \,|\, x' \in [0,\,1)\}\,. \tag{15}$$

This embedding allows us to put a name on each node of this chain. With this respect, the first few nodes, starting from the origin, of the Fibonacci chain are labelled by the following:

$$\ldots,\, -b_{13}\,, -b_{10}\,, -b_7\,, -b_5\,, -b_2\,,\; b_0\,,\; b_3\,,\; b_6\,,\; b_8\,,\; b_{11}\,,\; b_{14}\,, \ldots$$

where b_i denotes the ith tau-integer, with $b_0 = 0$.

4 Beta-Lattices

It is well-known that the condition $2\cos\frac{2\pi}{n} \in \mathbb{Z}$ characterizes n-fold Bravais lattices in $\mathbb{R}^2$ (and in $\mathbb{R}^3$). Let us put $\zeta = e^{\frac{2\pi i}{n}}$, $\zeta^n = 1$. If we consider the $\mathbb{Z}$-module in the plane,

$$\mathbb{Z}[\zeta] = \mathbb{Z} + \mathbb{Z}\zeta + \mathbb{Z}\zeta^2 + \cdots + \mathbb{Z}\zeta^{n-1} = \mathbb{Z}[2\cos\frac{2\pi}{n}] + \mathbb{Z}[2\cos\frac{2\pi}{n}]\zeta, \quad (16)$$

we get the cyclotomic ring of order n. This n-fold structure is generically dense in $\mathbb{C}$, except precisely for the crystallographic cases. We indeed check that $\mathbb{Z}[\zeta] = \mathbb{Z}$ for $n = 1$ or 2, $\mathbb{Z}[\zeta] = \mathbb{Z} + \mathbb{Z}i$ for $n = 4$ (square lattice), and $\mathbb{Z}[\zeta] = \mathbb{Z} + \mathbb{Z}e^{\frac{i\pi}{3}}$ for the triangular and hexagonal cases $n = 3$ and $n = 6$. If n is not crystallographic, $2\cos\frac{2\pi}{n}$ is an algebraic integer of degree $m = \varphi(n)/2 \leq \lfloor (n-1)/2 \rfloor$ where φ is the Euler function. A *cyclotomic PV number* with symmetry of order n is a PV number β such that

$$\mathbb{Z}[2\cos\frac{2\pi}{n}] = \mathbb{Z}[\beta]. \quad (17)$$

Then $\mathbb{Z}[\beta] + \mathbb{Z}[\beta]\zeta = \mathbb{Z}[\zeta]$ is a ring invariant under rotation of order n.[3] This ring is the natural framework for two-dimensional n-fold structures having β as scaling factor.

Let us give some examples of those numbers related to noncrystallographic cyclotomic structures in the plane.

$$n = 5, 10 \quad \beta = \tau = \frac{1+\sqrt{5}}{2} = 2\cos\frac{\pi}{5} \quad \text{(pentagonal and decagonal case)} \quad (18)$$

$$n = 5, 10 \quad \beta = \gamma = \frac{3+\sqrt{5}}{2} = 1 + 2\cos\frac{\pi}{5} \quad \text{(pentagonal and decagonal case)} \quad (19)$$

$$n = 8 \quad \beta = 1 + \sqrt{2} = 1 + 2\cos\frac{\pi}{4} \quad \text{(octogonal case)} \quad (20)$$

$$n = 12 \quad \beta = 2 + \sqrt{3} = 2 + 2\cos\frac{\pi}{6} \quad \text{(dodecagonal case)} \quad (21)$$

These numbers are all *quadratic PV units* (see 3.3). Another example is provided by the case $n = 7$, in which we have $\beta = 1 + 2\cos\frac{2\pi}{7}$. The latter is solution to the cubic equation $X^3 - 2X^2 - X + 1 = 0$.

At this point, we should be aware that finding a PV number such that the cyclotomic condition (17) is fulfilled for $n \geq 16$ is an open problem!

The introduction of β-integers in quasicrystal theory, playing an analogous role as integers in crystal theory, is especially relevant for reestablishing a discrete labelling frame-lattice for geometric models of quasicrystals. If we indeed replace integers by beta-integers (respectively, some completion $\widetilde{\mathbb{Z}}_\beta = \mathbb{Z}_\beta + F$ of beta-integers with finite sets, like in (14)) in Equation (16) we get the set

$$\mathbb{Z}_\beta[\zeta] = \mathbb{Z}_\beta + \mathbb{Z}_\beta\zeta + \mathbb{Z}_\beta\zeta^2 + \cdots + \mathbb{Z}_\beta\zeta^{n-1}\,, \tag{22}$$

$$(\text{resp. } \mathbb{Z}_\beta[\zeta] = \widetilde{\mathbb{Z}}_\beta + \widetilde{\mathbb{Z}}_\beta\zeta + \widetilde{\mathbb{Z}}_\beta\zeta^2 + \cdots + \widetilde{\mathbb{Z}}_\beta\zeta^{n-1}\,)\,,$$

also called n-fold cyclotomic quasiring, where n is the order of symmetry related to β. Beta-lattices in the plane are the following subsets of these quasi-rings,

$$\Gamma_q = \mathbb{Z}_\beta + \mathbb{Z}_\beta\zeta^q\,, \quad \text{if } \beta^2 - a\beta - 1 = 0 \;\; a \geq 1\,, \tag{23}$$

$$\Gamma_q = \widetilde{\mathbb{Z}}_\beta + \widetilde{\mathbb{Z}}_\beta\,\zeta^q \quad \text{if } \beta^2 - a\beta + 1 = 0 \;\; a \geq 3\,. \tag{24}$$

As an example of reciprocal, it can be shown that for $\beta = \tau$, the quasi-ring (22) is subset of suitably rescaled beta-lattice

$$\Gamma_q \subset \mathbb{Z}_\tau[\zeta] \subset \frac{\Gamma_q}{\tau^4}\,. \tag{25}$$

More items on this subject can be found in the literature.[6,7] Model sets Λ for which β plays a specific scaling role are precisely subsets of suitably deflated or inflated beta-lattices,

$$\Lambda \subset \frac{\Gamma_q}{\beta^j}\,, \quad j \in \mathbb{Z}\,. \tag{26}$$

Indeed, like for one-dimensional model sets, the embedding of two-dimensional model sets in beta-lattices is understood through an algebraic filtering procedure which is analogous to the one described in the previous section, provided we define the algebraic dual operation, denoted by *, that extends Galois conjugation to two-dimensional cases. The filtering procedure, possibly involving a phason shift Φ in $\mathbb{C}$ is,

$$\Sigma_\Phi^P = \{x \in \mathbb{Z}[\zeta] \mid x^* - \Phi \in P\} \;=\; (\mathbb{Z}[\xi] \cap (P + \Phi))^*\,, \tag{27}$$

where P is bounded in the plane. Once P is chosen, it is just a matter of determining an appropriate scale and origin in order to get $P + \Phi \subset \Gamma_q^*$. Hence we have,

$$\Sigma_\Phi^P = \{x \in \Gamma_q \mid x^* - \Phi \in P\} \;=\; \left(\Gamma_q^* \cap (P + \Phi)\right)^*\,. \tag{28}$$

Fig.1 displays the vertices of a Penrose tiling obtained though this sieving procedure in a tau-lattice.

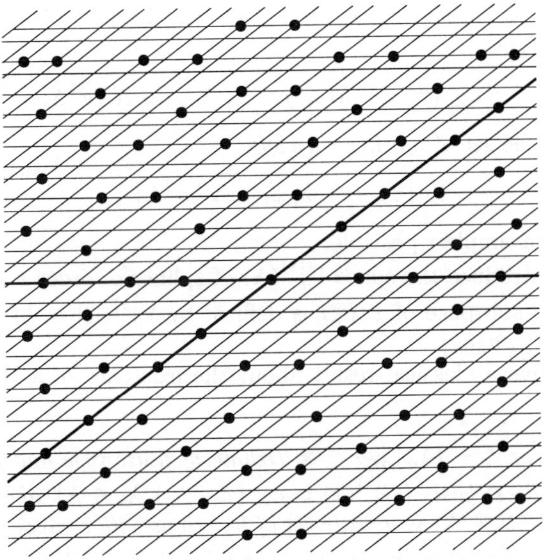

Figure 1. Embedding of the vertices of a singular Penrose tiling in a τ-lattice.

5 From Mathematics to Numerical Applications in Physics

In this section we give an overview of some results or current works in which beta-lattices are involved or used for analyzing quasiperiodic structures and phenomenons.

5.1 Spline wavelets construction for Fibonacci chains

As a first current direction of research we show how to build spline wavelets adapted to another type of Fibonacci chain, namely the model set $\Lambda = \Sigma^{[0,\tau^2)}$ (the example given in (2) is just a rescaled version of it: $\mathcal{F} = \Sigma^{[0,1)} = \frac{1}{\tau^2}\Lambda$). The necessary condition to carry out our method[4] is to have at our disposal a self-similar, locally finite Delone set. We recall here that a set Λ is self-similar if there exists a number $\theta > 1$ (*inflation factor*) such that $\theta\Lambda \subset \Lambda$, and locally finite if for all $R > 0$, the family of finite point sets $\{(\Lambda - \lambda) \cap (-R, R)\}_{\lambda \in \Lambda}$ is finite. This is true for $\Lambda = \Sigma^{[0,\tau^2)}$, $\tau^2\Lambda \subset \Lambda$. We also consider the latter Λ as an infinite sequence $\{\lambda_n\}_{n\in\mathbb{Z}}$ of real numbers $\cdots\lambda_{n-1} < \lambda_n < \lambda_{n+1}\cdots$ with $\lambda_0 = 0$. Using this notation our aim is to construct a Riesz basis of the space $L^2(\mathbb{R})$, the elements of which are of the affine wavelet type,

$$\tau^j \psi_{\lambda_n}(\tau^{2j}x - \lambda_n),\ j \in \mathbb{Z},\ \lambda_{n+1} \in \Lambda \setminus \tau^2\Lambda, \tag{29}$$

and for which the set of *mother wavelets*

$$\{\psi_{\lambda_n}(x),\ \lambda_{n+1} \in \Lambda \setminus \tau^2\Lambda\} \tag{30}$$

is finite and the simplest possible.

First we introduce the space $V_0^{(s)}(\Lambda)$ of splines of order s with nodes in Λ. This closed subspace of $L^2(\mathbb{R})$ is defined by $V_0^{(s)}(\Lambda) = \{f(x) \in L^2(\mathbb{R}) \mid \frac{\mathrm{d}^s}{\mathrm{d}x^s} f(x) = \sum_{\lambda\in\Lambda} a_\lambda \delta_\lambda\}$.

We now give a classical result about the existence of a Riesz basis for $V_0^{(s)}(\Lambda)$.[19]

Theorem 1 *For $\Lambda = \{\lambda_n\}_{n\in\mathbb{Z}} \subset \mathbb{R}$ Delone of finite type and for all $s \geq 2$, there exists a Riesz basis $\{B_{\lambda_k}^{(s)}(x) = \phi_{\lambda_k}(x - \lambda_k), \lambda_k \in \Lambda\}$ of $V_0^{(s)}(\Lambda)$ where the set $\{\phi_{\lambda_k}(x),\ \lambda_k \in \Lambda\}$ is a finite set of functions with compact support. The function $B_{\lambda_k}^{(s)}$ (called B-spline) is the unique function in $V_0^{(s)}(\Lambda)$ satisfying the following conditions,*

(i) supp $B_{\lambda_k}^{(s)} = [\lambda_k, \lambda_{k+s}]$,

(ii) $\int_{\mathbb{R}} B_{\lambda_k}^{(s)} = \frac{\lambda_{k+s} - \lambda_k}{s}$.

Note that (i) gives precise information on the (compact) support of $B_{\lambda_k}^{(s)}$ whilst (ii) is a normalization condition.

It is possible to partition the indexing set $\mathbb{Z}$ for Λ into a finite set of equivalence classes $\bar{n}$, $\mathbb{Z} = \bigcup_{\bar{n}=\bar{0}}^{\bar{q}} \bar{n}$ where class $\bar{n}$ is defined by

$$\bar{n} = \left\{n' \in \mathbb{Z} \mid B_{\lambda_{n'}}^{(s)}(x + \lambda_{n'}) = B_{\lambda_n}^{(s)}(x + \lambda_n)\ \forall x\right\}.$$

Correspondingly, for a given s, the point set Λ is partitioned into $\Lambda = \bigcup_{\bar{n}=\bar{0}}^{\bar{q}} \Lambda_{\bar{n}}$ with $\Lambda_{\bar{n}} = \{\lambda_k,\ k \in \bar{n}\}$. To each class $\bar{n}$ is biunivocally associated the function $\phi_{\bar{n}}(x) \equiv \phi_{\lambda_k}(x) = B_{\lambda_k}^{(s)}(x + \lambda_k),\ k \in \bar{n}$. In this way, the space $V_0^{(s)}(\Lambda)$ decomposes into the direct sum

$$V_0^{(s)}(\Lambda) = \bigoplus_{\bar{n}=\bar{0}}^{\bar{q}} V_{0,\bar{n}}^{(s)}, \tag{31}$$

where $V_{0,\bar{n}}^{(s)}$ is the closure of the linear span of the functions $\phi_{\bar{n}}(x-\lambda_k),\ k\in\bar{n}$.

Now changing the scale we can define, for $j\in\mathbb{Z}$, subspaces $V_j^{(s)}(\Lambda) = \{f(x)\in L^2(\mathbb{R})\ \mid\ \frac{\mathrm{d}^s}{\mathrm{d}x^s}f(x)=\sum_{\lambda\in\Lambda}a_\lambda\delta_{\tau^{-2j}\lambda}\}$.

Hence we have at our disposal an inductive chain of spaces allowing analysis at any scale. More precisely, with the above notations, we obtain a so-called *multiresolution analysis* in the following sense.

Proposition 1 *The sequence of subspaces $(V_j^{(s)}(\Lambda))_{j\in\mathbb{Z}}$ is a multiresolution analysis of $L^2(\mathbb{R})$, i.e.,*

(i) for any $j\in\mathbb{Z}$, $V_j^{(s)}(\Lambda)$ is a closed subspace of $L^2(\mathbb{R})$,

(ii) $\cdots\subset V_{-1}^{(s)}(\Lambda)\subset V_0^{(s)}(\Lambda)\subset V_1^{(s)}(\Lambda)\subset\cdots$,

(iii) $\bigcup_{j\in\mathbb{Z}}V_j^{(s)}(\Lambda)$ is dense in $L^2(\mathbb{R})$,

(iv) $\bigcap_{j\in\mathbb{Z}}V_j^{(s)}(\Lambda)=\{0\}$,

(v) $f(x)\in V_j^{(s)}(\Lambda)$ if and only if $f(\tau^{-2j}x)\in V_0^{(s)}(\Lambda)$,

(vi) there exists a finite number of functions $\phi_{\bar{n}}(x)\in V_0^{(s)}$, called scaling functions such that $\{\phi_{\bar{n}}(x-\lambda_n)\}_{n\in\bar{n},\,\bar{0}\leq\bar{n}\leq\bar{q}}$ is a Riesz basis in $V_0^{(s)}$.

As indicated by (29), the whole wavelet basis is obtained by scaling the elements of a wavelet subfamily living at a given scale of the multiresolution. This subfamily is precisely a basis of the orthogonal complement of $V_{-1}^{(s)}(\Lambda)$ in $V_0^{(s)}(\Lambda)$,

$$V_0^{(s)}(\Lambda)=V_{-1}^{(s)}(\Lambda)\oplus_\perp W_{-1}^{(s)}(\Lambda). \tag{32}$$

This decomposition can be interpreted as the building up of the "zeroth-scale" content of the multiresolution (*i.e.*, $V_0^{(s)}$) from the adding of necessary "details" (*i.e.*, $W_{-1}^{(s)}$) to the existing content at the next larger scale (*i.e.*, $V_{-1}^{(s)}$).

A preliminary characterization of $W_{-1}^{(s)}$ is given by the following proposition. Let us recall that $H^s(\mathbb{R})$ denotes the Sobolev space of functions $f\in L^2(\mathbb{R})$ such that $\frac{\mathrm{d}^\alpha}{\mathrm{d}x^\alpha}f$, with $\alpha\leq s$, is element of $L^2(\mathbb{R})$.

Proposition 2 *For $s\geq 2$, $W_{-1}^{(s)}$ is the space of sth derivatives of elements in $H^{2s}(\mathbb{R})$ which vanish on $\tau^2\Lambda$,*

$$W_{-1}^{(s)}(\Lambda)\equiv\left\{\frac{\mathrm{d}^s}{\mathrm{d}x^s}h,\ h\in H^{2s}(\mathbb{R})\ \middle|\ \forall\lambda\in\tau^2\Lambda,\ h(\lambda)=0\right\}.$$

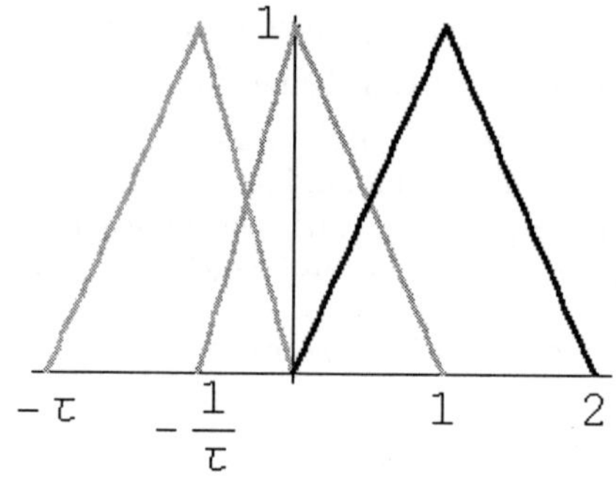

Figure 2. Linear scaling functions $\phi_{LS}(x+\tau)$, $\phi_{SL}(x+\frac{1}{\tau})$ and $\phi_{LL}(x)$.

We present the scaling functions and the corresponding wavelets for the Fibonacci chain in the case $s = 2$. The simple case *i.e.*, $s = 1$ is presented elsewhere.[2] At first we have to find the B-splines basis of $V_0^{(2)}$. From well-known properties of Fibonacci chain one knows that there are just only three possible configurations of three points. The arrangements of distances among them can be only LL, LS and SL, where $|L| = 1$ and $|S| = 1/\tau$. Thus the basis of $V_0^{(2)}$ is constructed from the following three functions shown in Fig.2 and their admissible translations (points from the same equivalence class),

$$\phi_{LL}(x) = \begin{cases} x & x \in [0,1), \\ 2-x & x \in [1,2], \end{cases} \tag{33}$$

$$\phi_{LS}(x+\tau) = \begin{cases} x+\tau & x \in [-\tau,-1/\tau), \\ \tau^2 - \tau(x+\tau) & x \in [-1/\tau,0], \end{cases} \tag{34}$$

$$\phi_{SL}(x+1/\tau) = \begin{cases} \tau(x+1/\tau) & x \in [-1/\tau,0), \\ \tau-(x+1/\tau) & x \in [0,1]. \end{cases} \tag{35}$$

The linear spline wavelet basis will be constructed from four corresponding linear spline wavelets, see Fig.3, together with their dilated and admissible translated versions.

It should also be recalled that this spline wavelet construction is applicable to either Fibonacci chains or any self-similar, locally finite Delone set.

5.2 *Quasiperiodic patterns in the Faraday experiment*

The notion of quasiperiodic structures has been mainly considered in solid state science. Yet it is possible to encounter such structures in other fields of

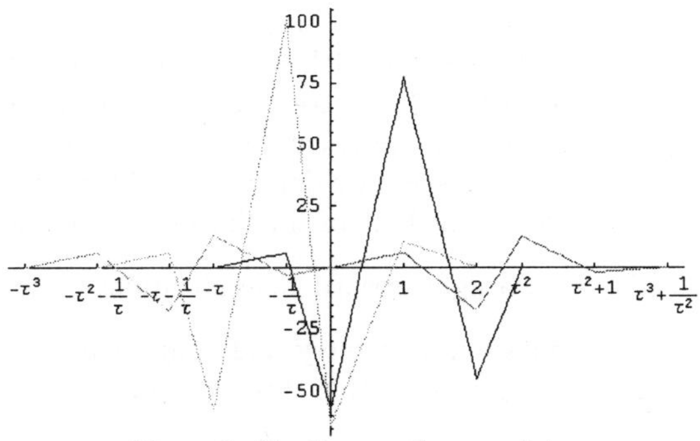

Figure 3. The linear spline wavelets.

Physics such as Hydrodynamics. The chosen example is based on the hydrodynamic Faraday instability: parametric excitation of surface waves via vertical oscillations of a flat bottomed container filled with liquid. As a result, patterns appear on the surface. Fauve and Edwards have shown experimentally[9] that this instability can display quasiperiodic patterns.

The simplest model for the Faraday instability was coined by Lifshitz and Petrich. The starting point to this model is the Swift-Hohenberg equation, in which they introduced a double diffusive term. The equation they obtained is the following,

$$\partial_t \zeta = \epsilon\, \zeta - c\left(\nabla^2 + q_1^2\right)^2 \left(\nabla^2 + q_2^2\right)^2 \zeta + \alpha\, \zeta^2 - \zeta^3\,, \tag{36}$$

the ratio q_1/q_2 being equal to $\sqrt{2+\sqrt{3}}$, the square root of the previously stated PV unit θ.

First we embed the structure formed by the peaks of the quasiperiodic pattern obtained by the Lifshitz-Petrich model into a two-dimensional theta-lattice. The Faraday surface, displayed in Fig.4, is composed of a macrostructure, made of sharp peaks, and a decoration of the macrostructure. This decoration is made of undulating circular ridges surrounding the peaks of the macrostructure. The peaks of the macrostructure form a subset of the theta-lattice. Our attempt now is to integrate the Lifshitz-Petrich equation using a theta-lattice as basic discretisation scheme for finding the quasiperiodic patterns.

5.3 Monte-Carlo Simulations of Quasicrystals

Krejcar used Metropolis Monte-Carlo method to find approximate ground-state configurations in a two-dimensional quasilattice gas model.[13] Some results of those simulations are displayed in Figs.5 and 6. A quasilattice gas model is analogous to a lattice gas model, but the periodic lattice is replaced by a quasiperiodic Delone set $\Lambda \in \mathbb{R}^n$. More precisely, quasilattice gas models are systems in which every site of Λ can be occupied by one of k different particles. Of course the model strongly depends on the choice of Λ. Krejcar used beta-lattices to perform this kind of simulation for the following reasons,

- beta-integers $\mathbb{Z}_\beta$ are straightforward generalizations of integers $\mathbb{Z}$.
- beta-lattices have inflation symmetry and support multiresolution (*i.e.*, infinite sequence of increasingly thin discretisations) of $\mathbb{R}^n$.
- beta-lattices have no pre-defined symmetry, but inflation. With this respect, they can embed a large class of mathematical models for quasicrys-

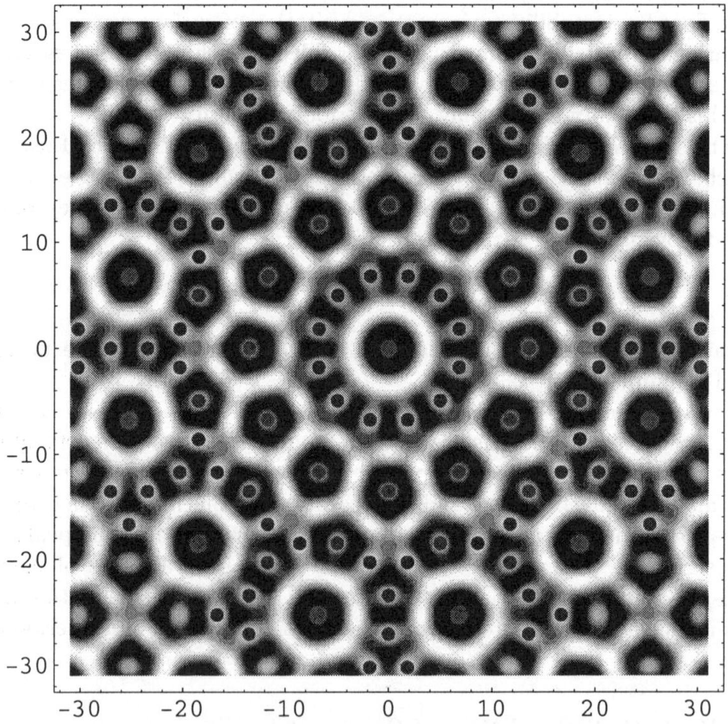

Figure 4. The macrostructure, marked here with dots, is a subset of the theta-lattice.

tals, given a certain inflation factor, such as those given in Sect.4.

- Observed quasicrystalline diffraction patterns (beyond a certain intensity) can be also embedded in a beta-lattice.

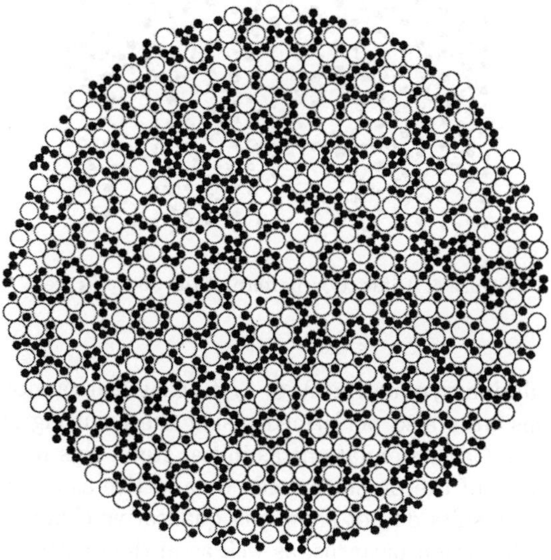

Figure 5. Monte-Carlo simulation of a quasicrystalline structure as performed by Krejcar.

The interactions considered are based on Lennard-Jones potentials, the temperature is set close to zero, and the statistical model is based on grand canonical ensemble, therefore the densities of particles appearing in the systems are not fixed by the choice of initial configurations. The densities of different particles become then some of the outcomes of the simulations.

The results of the simulations do not depend, in any appreciable manner, on the choice of initial conditions. On the other hand, the symmetry of the resulting configuration strongly depends on the choice of pair-potentials. The best results are obtained when the chosen interactions are adjusted to the underlying quasilattice, *i.e.*, when the pair-potentials reach their minima at distances close to those appearing in the quasilattice.

The stability of the stationary structures are checked by making the underlying quasilattice thinner. This action is supported by the fact that beta-

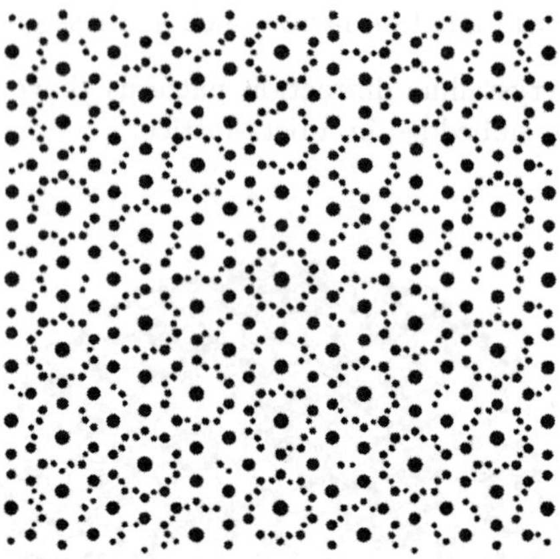

Figure 6. Diffraction pattern of the above figure.

lattices have inflation symmetry: $\Lambda \in \Lambda/\beta$. Classical random walk Monte-Carlo simulations were also performed, using the resulting structure of the quasilattice Monte-Carlo simulations as initial state, this in order to control the non-artificiality of the predetermined beta-lattice as a starting point. In the two cases the results do not show any qualitative changes in both structures and their diffraction patterns, as well as in the densities.

References

1. M. Andrle, in *Quantum Theory and Symmetries 2*, E. Kapuscik and A. Horzela, eds. (World Scientific, Singapore), to appear.
2. M. Andrle, Č. Burdík, J.P. Gazeau and R. Krejcar, *Wavelet multiresolutions for the Fibonacci chain*, *J. Phys.* **A33**, L47–L51 (2000).
3. D. Barache, B. Champagne and J.-P. Gazeau, in *Quasicrystals and Discrete Geometry*, J. Patera, ed., *Fields Institute Monograph Series*, Volume 10, Amer. Math. Soc., 1998.
4. G. Bernuau, *Propriétés spectrales et géométriques des quasicristaux. Ondelettes adaptées aux quasicristaux*, Ph. D. Thesis (Ceremade, Université Paris IX Dauphine, France, 1998) unpublished.
5. A. Bertrand, *C.R. Acad. Sc., Paris* **285**, 419–421 (1977).
6. Č. Burdík, Ch. Frougny, J.-P. Gazeau and R. Krejcar, *J. of Phys.* **A31**, 6449–6472 (1998).

7. Č. Burdík, Ch. Frougny, J.-P. Gazeau and R. Krejcar, in *Dynamical Systems: From Crystal to Chaos* (World Scientific, Singapore, 2000), pp. 125–136.
8. G. Dewel, M. Bachir, P. Borckmans and S.Metens, *C.R.A.S.P. série IIb*, 2001.
9. W.S. Edwards, S. Fauve, *J. Fluid. Mech.* **278**, 123–148 (1994).
10. A.-D. El Kharrat, J.-P. Gazeau, M. Guidy, and S. Métens, in *Quantum Theory and Symmetries 2*, E. Kapuscik and A. Horzela, eds. (World Scientific, Singapore), to appear.
11. Ch. Frougny, J.-P. Gazeau and R. Krejcar, *Theor. Comp. Sci.*, to appear (2002).
12. J.-P. Gazeau and R. Krejcar, in *From Quasicrystals to More Complex Systems*, F. Axel, F. Denoyer and J.-P. Gazeau, eds. (Springer & Les Editions de Physique, 2000).
13. J.-P. Gazeau, R. Krejcar and J. Miękisz, *Material Science and Engineering* **A294-296**, 425–428 (2002).
14. C. Janot, *Quasicrystals, a Primer* (Oxford University Press, Oxford, 1996).
15. Y. Meyer, in *Beyond Quasicrystals*, F. Axel and D. Gratias, eds. (Les Editions de Physique, Springer-Verlag, 1995).
16. R. Moody, *Model Sets: A Survey*, in *From Quasicrystals to More Complex Systems*, F. Axel, F. Denoyer and J.-P. Gazeau, eds. (Springer & Les Editions de Physique, 2000).
17. W. Parry, *Acta Math. Acad. Sci. Hungar.* **11**, 401–416 (1960).
18. A. Rényi, *Acta Math. Acad. Sci. Hungar.* **8**, 477–493 (1957).
19. L.L. Schumaker, *Spline Functions: Basic Theory*, (Wiley, New-York, 1981).
20. J.-L. Verger-Gaugry and J.-P. Gazeau, Institut Joseph Fourier, preprint (2001).

THE SCHWINGER MODEL AND THE PHYSICAL PROJECTOR: A NONPERTURBATIVE QUANTIZATION WITHOUT GAUGE FIXING

GABRIEL Y.H. AVOSSEVOU
Unité de Recherche en Physique Théorique (URPT)
Institut de Mathématiques et de Sciences Physiques (IMSP)
B.P. 2628 Porto-Novo, Republic of Benin
E-mail: avossevou@yahoo.fr

JAN GOVAERTS
Institute of Nuclear Physics, Catholic University of Louvain
2, Chemin du Cyclotron, B-1348 Louvain-la-Neuve, Belgium
E-mail: jan.govaerts@fynu.ucl.ac.be

Based on the physical projector approach, a nonpertubative quantization of the massless Schwinger model is considered which does not require any gauge fixing. The spectrum of physical states, readily identified following a diagonalization of the operator algebra, is that of a massive pseudoscalar field, namely the electric field having acquired a mass proportional to the gauge coupling constant. The physical spectrum need not be identified with confined bound fermion-antifermion pairs, an interpretation which one is otherwise led to given whatever gauge fixing procedure but which is not void of gauge fixing artefacts.

1 Introduction

Ever since its inception,[1] the massless Schwinger model, namely 1+1 dimensional massless quantum electrodynamics, has been a favourite theoretical laboratory of mathematical physics in which to test our understanding and methods of interacting relativistic quantum field theories. This model is one of the very few instances of an integrable quantum field theory possessing an exact nonperturbative solution.[1,2,3,4] It also displays most of the nonperturbative features of the theory for the strong interactions of quarks, namely quantum chromodynamics (QCD) based on the nonabelian $SU(3)_C$ colour gauge symmetry, leading to the still open issues of confinement and spontaneous dynamical chiral symmetry breaking with its cortège of characteristic consequences for hadronic interactions. Hence, it is not suprising that whenever a new nonperturbative phenomenon is suggested to occur in quantum field theory, or a new methodology towards such issues is designed, the Schwinger model is brought into the arena to test out new ideas. The present contribution is no exception. Given the recent proposal[5] of the physical projector approach to the quantization of gauge invariant systems (and more generally, constrained ones),[6] it thus seems timely to investigate the potential new insight which that approach may bring to the nonperturbative issues characteristic of the Schwinger model. The efficacy of the physical projector has already been demonstrated in the case of gauge invariant systems in 0+1 dimensions,[7] as well as topological quantum field theories in 2+1 dimensions,[8] with their own specific circumstances. It would thus be welcome to demonstrate that the physical projector approach is also of relevance with regards to the nonperturbative quantization of interacting field theories.

Within a continuum spacetime, all previous approaches[1,2,3,4] to the Schwinger model (SchM), in one way or another, are based in an essential manner on a choice of gauge fixing of the U(1) local symmetry. In effect, this leads to the reduction of the gauge boson degrees of freedom expressed in terms of the fermionic ones, except for some gauge invariant zero mode in the case of a compact space topology, namely a circle. The latter choice is made[3,4] in order to render manifest the topological features of the SchM. The conclusion is then that the spectrum of gauge invariant states is that of a massive scalar field, whose quanta are the confined states of fermion-antifermion pairs bound to one another by an electric flux line. Indeed, the one-dimensional Gauss law in vacuum implies a constant electric field, hence a linearly rising electric potential leading to a confining interaction. However, such an intuitive picture raises some difficulties of interpretation. For instance, given two point particles of opposite charges and momenta forming a scalar bound state on the circle, on which "side" of the circle should the electric field binding these two charges be set-up? As a matter of fact, other such artefacts arise as a consequence of gauge fixing, which tend to confuse the physical understanding and interpretation of the nonperturbative dynamics of the model.

Within the physical projector approach which does not require any gauge fixing, such issues are avoided altogether.[6,9] Physical states are nothing but the quanta of the electric field,[10] which is a pseudoscalar field in 1+1 dimensions, indeed the only spacetime local construct based on the original degrees of freedom which is genuinely gauge invariant. The fermionic degrees of freedom contribute in a gauge invariant manner to the physical spectrum only nonlocally through their bosonized representation and only in combination with the local gauge dependent degrees of freedom of the gauge boson sector. Even though the same physical spectrum is of course recovered as in any of the gauge fixed quantizations, we find such a physical picture much more appealing than the usual one, the more so since within the Hamiltonian formulation of the SchM, the electric field readily finds its rightful place and is obviously a gauge invariant field whose quanta must belong to the physical spectrum whatever the status of the other degrees of freedom.

A full account of the analysis is not presented,[10] restricting the discussion to its most salient features. Sect.2 introduces the model and our notations. Sect.3 describes its system of constraints and basic Hamiltonian formulation. A nonperturbative quantization is developed in Sect.4, based on the bosonization of the fermion degrees of freedom and the operator diagonalization of the model. In Sect.5, the physical spectrum is identified through the physical projector. Finally, some conclusions are presented in Sect.6.

2 The Massless Schwinger Model

The topology of the 1+1 dimensional spacetime is taken to be $\mathbb{R} \times S$, $\mathbb{R}$ standing for the time coordinate, t, and S for the spatial one, x, with the geometry of a circle of radius R and circumference $L = 2\pi R$. The Minkowski metric is $\eta_{\mu\nu} = \text{diag}\,(+-)$ $(\mu, \nu = 1, 2)$, while units such that $\hbar = 1 = c$ are used throughout. The antisymmetric tensor $\epsilon^{\mu\nu}$ is such that $\epsilon^{01} = +1$. For the Clifford-Dirac algebra $\{\gamma^\mu, \gamma^\nu\} = 2\eta^{\mu\nu}$, we shall work with the chiral representation given by

$$\gamma^0 = \sigma^1 \quad , \quad \gamma^1 = i\sigma^2 \quad , \quad \gamma_5 = \gamma^0\gamma^1 = -\sigma^3 \; , \tag{1}$$

σ^i $(i = 1, 2, 3)$ being the usual Pauli matrices.

The field degrees of freedom of the model are, on the one hand, a single massless Dirac spinor $\psi(x^\mu)$, and on the other hand, the real U(1) gauge vector field $A_\mu(x^\mu)$. The Dirac spinor decomposes into two complex Weyl spinors of opposite chiralities, $\psi = \psi_+ + \psi_-$, such that $\gamma_5\psi_\pm = \mp\psi_\pm$. The index "$\pm$" refers to the left- or right-moving character, respectively, of these two Weyl spinors in the absence of any interaction. Furthermore, the following choice of periodic and twisted boundary conditions on the circle is assumed,

$$A_\mu(t, x + L) = A_\mu(t, x) \quad , \quad \psi_\pm(t, x + L) = -e^{2i\pi\alpha_\pm}\, \psi_\pm(t, x) \; , \tag{2}$$

$\alpha_\pm$ being arbitrary real quantities defined modulo any integers, which, in fact, have to be equal modulo integers to ensure parity invariance, $\alpha_+ = \alpha = \alpha_-$ (mod $\mathbb{Z}$).

The dynamics of the model derives from the local Lagrangian density (the summation convention is implicit throughout)

$$\mathcal{L} = -\frac{1}{4}F_{\mu\nu}F^{\mu\nu} \; + \; \frac{1}{2}i\,\bar{\psi}\gamma^\mu\partial_\mu\psi \; - \; \frac{1}{2}i\,\partial_\mu\bar{\psi}\gamma^\mu\psi \; - \; e\bar{\psi}\gamma^\mu A_\mu\psi \; , \tag{3}$$

with $F_{\mu\nu} = \partial_\mu A_\nu - \partial_\nu A_\mu$ the gauge field strength, and e the U(1) gauge coupling constant. In 1+1 dimensions and in units of mass, the gauge field

A_μ is dimensionless, the spinors $\psi_\pm$ have dimension $1/2$, and e dimension unity. In covariant form, the equations of motion are

$$\gamma^\mu(i\partial_\mu - eA_\mu)\psi = 0 \quad , \quad \partial^\nu F_{\nu\mu} = e\bar{\psi}\gamma_\mu\psi = e\, J_\mu \ . \tag{4}$$

When written out in chiral components, the Weyl spinors $\psi_\pm(t \pm x)$ are seen to be left- and right-moving, respectively, in the absence of any interaction, $e = 0$.

On account of its antisymmetry, the field strength $F_{\mu\nu}$ represents a single field, F_{01}, which coincides with the electric field, indeed a pseudoscalar field in 1+1 dimensions,

$$E = -\partial_1 A^0 - \partial_0 A^1 = F_{01} = -F^{01} \ . \tag{5}$$

In particular, Gauss' law reads $\partial_1 E = e\psi^\dagger\psi$.

By construction, the system is invariant under the following global symmetries. Besides Poincaré invariance in the spacetime manifold, one also has the vector U(1) and axial $\mathrm{U}(1)_A$ symmetries whose Noether currents are, respectively,

$$J^\mu = \bar{\psi}\gamma^\mu\psi \quad , \quad J^\mu_5 = \bar{\psi}\gamma^\mu\gamma_5\psi \ , \tag{6}$$

with the duality relations $J^\mu_5 = -\epsilon^{\mu\nu} J_\nu$ and $J^\mu = -\epsilon^{\mu\nu} J_{5\nu}$. The space integral of the time component of each of these currents defines the corresponding Noether charges Q and Q_5, respectively. At the classical level, both are conserved quantities, thus generating a symmetry of the dynamics. For the quantized system however, the axial charge is no longer conserved due to quantum effects.

The U(1) vector symmetry is of course the gauge symmetry of the system, acting as

$$\psi'(t,x) = e^{-ie\epsilon(t,x)}\psi(t,x) \quad , \quad A'_\mu(t,x) = A_\mu(t,x) + \partial_\mu\epsilon(t,x) \ , \tag{7}$$

with $\epsilon(t,x)$ an arbitrary function of spacetime of the general form

$$\epsilon(t,x) = \epsilon_0(t,x) + \frac{2\pi k}{eL} x \ , \tag{8}$$

$\epsilon_0(t, x+L) = \epsilon_0(t,x)$ being an arbitrary periodic function on the spatial circle, defined modulo $2\pi/|e|$ and parametrizing small U(1) gauge transformations, while the nonperiodic contribution proportional to x represents the homotopy class of large U(1) gauge transformations of winding number k, k being an arbitrary integer which labels the corresponding element of the gauge modular group $\mathbb{Z}$. These two classes of gauge transformations are to be included into the construction of the physical projector and the identification of physical states.

3 The Basic Hamiltonian Formulation

Being straightforward enough, the Hamiltonian analysis of constraints is not detailed here. Let us only remark that the above Lagrangian density is already in Hamiltonian form in the fermionic sector,[6,11] while A^0 is necessarily the Lagrange multiplier for a first-class constraint, in fact the sole first-class constraint of the system, namely the generator of small U(1) gauge transformations which is nothing but Gauss' law.[6] Hence only A^1 needs to have its conjugate momentum included in order to identify the basic Hamiltonian formulation[6] of the SchM. In terms of the corresponding first-order action on phase space, this formulation is defined by

$$S = \int dt \int_0^L dx \left\{ \partial_0 A^1 \pi_1 + \frac{1}{2} i \psi^\dagger \partial_0 \psi - \frac{1}{2} i \partial_0 \psi^\dagger \psi - \mathcal{H} - A^0 \phi + \partial_1 (A^0 \pi_1) \right\} , \tag{9}$$

where $\mathcal{H}$ is the first-class Hamiltonian density,

$$\mathcal{H} = \frac{1}{2}\pi_1^2 - \frac{1}{2} i \psi^\dagger \gamma_5 (\partial_1 - ieA^1)\psi + \frac{1}{2} i (\partial_1 + ieA^1)\psi^\dagger \gamma_5 \psi , \tag{10}$$

while ϕ is the first-class constraint of the system,

$$\phi = \partial_1 \pi_1 + e\psi^\dagger \psi , \tag{11}$$

whose Lagrange multiplier is nothing else than the time component A^0 of the gauge field. This first-order action encodes the fundamental Grassmann graded Poisson bracket structure given by $(\alpha, \beta = +, -)$,

$$\{A^1(t,x), \pi_1(t,y)\} = \delta(x-y) \quad , \quad \{\psi_\alpha(t,x), \psi^\dagger_\beta(t,y)\} = -i\delta_{\alpha\beta}\delta(x-y) \ . \tag{12}$$

In fact, the momentum π_1 conjugate to A^1 is nothing but the electric field up to a sign, $\pi_1 = F_{10} = -E$, so that the generator ϕ of small Hamiltonian U(1) gauge transformations coincides with Gauss' law, one of the two independent equations of motion in the gauge sector, namely $\partial^\nu F_{\nu 0} = e\psi^\dagger \psi$. Manifest gauge invariance of the formulation readily follows from the brackets

$$\{\phi(t,x), \mathcal{H}(t,y)\} = 0 \quad , \quad \{\phi(t,x), \phi(t,y)\} = 0 \ . \tag{13}$$

Given the total Hamiltonian density $\mathcal{H}_T = \mathcal{H} + A^0\phi - \partial_1(A^0\pi_1)$, the Hamiltonian equations of motion read

$$\partial_0 \psi = -\gamma_5(\partial_1 - ieA^1)\psi - ieA^0\psi \ , \ \partial_0 A^1 = \pi_1 - \partial_1 A^0 \ , \ \partial_0 \pi_1 = e\psi^\dagger \gamma_5 \psi \ , \tag{14}$$

together with the constraint $\phi = \partial_1 \pi_1 + e\psi^\dagger\psi = 0$. One thus recovers the Lagrangian equation of motion for the Dirac fermion; the equation for A^1 leads back to the relation defining the conjugate momentum $\pi_1 = -E$ in terms of the field A_μ; and finally the equation for π_1 reproduces the second equation of motion in the gauge sector, namely $\partial^\nu F_{\nu 1} = e\psi^\dagger\gamma_5\psi$, Gauss' law being reproduced through the first-class constraint $\phi = 0$. The dynamics of the system is thus equally well represented through the basic Hamiltonian formulation. In particular, upon substitution of the expression for π_1 into (9), one recovers the original Lagrangian density, inclusive of the explicit surface term in (9).

Infinitesimal small U(1) gauge transformations are generated by the first-class constraint ϕ through Poisson brackets. Extended to finite transformations, one has

$$\begin{aligned}\psi'(t,x) &= e^{-ie\epsilon_0(t,x)}\psi(t,x)\ ,\\ A^{1\prime}(t,x) &= A^1(t,x) - \partial_1\epsilon_0(t,x)\ ,\\ \pi_1'(t,x) &= \pi_1(t,x)\ ,\\ A^{0\prime}(t,x) &= A^0(t,x) + \partial_0\epsilon_0(t,x)\ ,\end{aligned} \tag{15}$$

$\epsilon_0(t,x)$ being an arbitrary function of periodicity L in x and defined modulo $2\pi/|e|$. These transformations leave (9) exactly invariant, and coincide with the small gauge transformations of the Lagrangian formulation of the system.

Even though large U(1) gauge transformations $\epsilon(t,x) = (2\pi kx)/(eL)$ are not generated by the first-class constraint ϕ, nevertheless they also define an invariance of the basic Hamiltonian action (9), without any surface term being induced. The phase space transformations are then given by

$$\begin{aligned}\psi'(t,x) &= e^{-ie\epsilon(t,x)}\psi(t,x) = e^{-i\frac{2\pi k}{L}x}\psi(t,x)\ ,\\ A^{1\prime}(t,x) &= A^1(t,x) - \partial_1\epsilon(t,x) = A^1(t,x) - \tfrac{2\pi k}{eL}\ ,\\ \pi_1'(t,x) &= \pi_1(t,x)\ ,\\ A^{0\prime}(t,x) &= A^0(t,x) + \partial_0\epsilon(t,x) = A^0(t,x)\ .\end{aligned} \tag{16}$$

From these expressions, it follows that in terms of a Fourier series expansion of the field $A^1(t,x)$ on the circle, all its nonzero modes may always be gauged away entirely through an appropriate small finite gauge transformation which in any case leaves its zero mode invariant, whereas large gauge transformations affect only its zero mode which is thus defined modulo $(2\pi)/(|e|L)$. Hence only the x-independent mode of $A^1(t,x)$ is an actual dynamical gauge invariant physical degree of freedom taking its values on a circle of radius $1/(|e|L)$.[3,4] As we shall see, this topological feature translates at the quantum level in terms of θ-vacua. In contradistinction, the momentum π_1 conjugate to A^1 is totally gauge invariant and thus physical. Even though the A^1 zero mode is not exatly gauge invariant except modulo $(2\pi)/(|e|L)$ whereas the π_1 zero mode is genuinely gauge invariant, one may suspect that these zero modes could be conjugate to one another. However, since the nonzero modes of A^1

are pure gauge, which are the gauge invariant physical degrees of freedom of the system that are conjugate to the gauge invariant nonzero modes of π_1?

4 Nonperturbative Canonical Quantization

The operator quantization of the model proceeds from its basic Hamiltonian formulation. Thus, the space of quantum states must provide a representation space for the commutation and anticommutation relations (only the nonvanishing (anti)commutators are given throughout)

$$[A^1(x), \pi_1(y)] = i\delta(x-y) \quad , \quad \{\psi_\alpha(x), \psi^\dagger_\beta(y)\} = \delta_{\alpha\beta}\delta(x-y) \ , \qquad (17)$$

where it is understood that one is working within the Schrödinger picture considered, say, at $t = 0$, an argument which henceforth is no longer displayed. Quantum dynamics of the system is generated through the first-class Hamiltonian density (10) and constraint (11). However, being composite operators, these first-class quantities require a choice of operator ordering which must remain anomaly free and gauge invariant, namely such that the gauge commutation relations

$$[\phi(x), \mathcal{H}(y)] = 0 \quad , \quad [\phi(x), \phi(y)] = 0 \qquad (18)$$

are still obtained at the operator level. In the case of the SchM, it is possible to meet all these requirements and construct an explicit exact diagonalization of the quantum Hamiltonian. This is achieved through a bosonization of the fermionic sector of degrees of freedom.[3,4]

4.1 Bosonization of the fermionic degrees of freedom

By lack of space, we do not provide details of this construction, but only its basic ingredients.[10] A quantization of the system amounts to the specification of a quantum representation space for the commutation and anticommutation relations (17). We shall obtain the above field degrees of freedom $A^1(x)$, $\pi_1(x)$ and $\psi_\pm(x)$ as composite operators of an alternative set of degrees of freedom whose commutation relations define the space of quantum states. For this purpose, let us introduce the following mode algebras,

$$[\varphi_n, \varphi^\dagger_m] = \delta_{n,m} \quad , \quad [Q_\pm, p_\pm] = i \quad , \quad [A_{\pm,n}, A^\dagger_{\pm,m}] = \delta_{n,m} \ , \qquad (19)$$

where for the φ_n and $\varphi^\dagger_m$ modes the indices n and m take all integer values, while for the operators $A_{\pm,n}$ and $A^\dagger_{\pm,m}$ they only take strictly positive integer values. We also have $Q^\dagger_\pm = Q_\pm$ and $p^\dagger_\pm = p_\pm$. Furthermore, the eigenvalue spectrum of the $Q_\pm$ operators is constrained to lie within the interval $[0, 2\pi]$, namely $Q_\pm$ stands for the coordinate of a circle of radius unity. Consequently, the eigenvalue spectra of their conjugate operators $p_\pm$ for the Heisenberg algebras $[Q_\pm, p_\pm] = i$ are quantized as $(n_\pm + \lambda_\pm)$, $n_\pm$ being any integers and $\lambda_\pm$ (mod $\mathbb{Z}$) being the U(1) holonomies labelling all inequivalent representations of the Heisenberg algebra on the circle of radius unity.[12] It is clear that the complete representation space of the commutation relations (19) is the tensor product of these two Heisenberg algebra representations characterized by $\lambda_\pm$ together with Fock space representations for each of the independent modes φ_n and $A_{\pm,n}$ and their adjoint operators $\varphi_n^\dagger$ and $A^\dagger_{\pm,n}$, the former being annihilation and the latter creation operators.

Given the fundamental algebra (19), let us now introduce the following operators, for all strictly positive integer values $n \geq 1$,

$$\begin{aligned} q_\pm &= Q_\pm + N_0(\varphi_0 + \varphi_0^\dagger) \ , \\ a_{+,n} &= A_{+,n} + N_n\sqrt{n}\left(\varphi_n^\dagger + \varphi_{-n}\right) \ , \\ a^\dagger_{+,n} &= A^\dagger_{+,n} + N_n\sqrt{n}\left(\varphi_n + \varphi^\dagger_{-n}\right) \ , \\ a_{-,n} &= A_{-,n} + N_n\sqrt{n}\left(\varphi_n + \varphi^\dagger_{-n}\right) \ , \\ a^\dagger_{-,n} &= A^\dagger_{-,n} + N_n\sqrt{n}\left(\varphi_n^\dagger + \varphi_{-n}\right) \ , \end{aligned} \tag{20}$$

N_n being some normalization factor to be specified presently, as well as the bosonic fields (in the Schrödinger picture),

$$\begin{aligned} \varphi(x) &= A\sum_{n=-\infty}^{+\infty} N_n \left[\varphi_n\, e^{\frac{2i\pi}{L}nx} + \varphi_n^\dagger\, e^{-\frac{2i\pi}{L}nx}\right] \ , \\ \pi_\varphi(x) &= \frac{-i}{2AL}\sum_{n=-\infty}^{+\infty} \frac{1}{N_n}\left[\varphi_n\, e^{\frac{2i\pi}{L}nx} - \varphi_n^\dagger\, e^{-\frac{2i\pi}{L}nx}\right] \ , \\ \Phi_\pm(x) &= Q_\pm \pm \frac{2\pi}{L}p_\pm x + \sum_{n=1}^{\infty}\frac{1}{\sqrt{n}}\left[A^\dagger_{\pm,n}\, e^{\pm\frac{2i\pi}{L}nx} + A_{\pm,n}\, e^{\mp\frac{2i\pi}{L}nx}\right] \ , \\ \phi_\pm(x) &= q_\pm \pm \frac{2\pi}{L}p_\pm x + \sum_{n=1}^{\infty}\frac{1}{\sqrt{n}}\left[a^\dagger_{\pm,n}\, e^{\pm\frac{2i\pi}{L}nx} + a_{\pm,n}\, e^{\mp\frac{2i\pi}{L}nx}\right] \ , \end{aligned} \tag{21}$$

the factor A being yet another normalization constant in terms of which N_n is expressed as

$$N_n = \frac{1}{|A|}\,\frac{1}{\sqrt{L}\sqrt{2\omega_n}} \ , \quad \omega_n = \sqrt{\mu^2 + \left(\frac{2\pi}{L}n\right)^2} \ , \tag{22}$$

where $\mu > 0$ is a given mass parameter. When applied to the SchM, the quantities A and μ are related to the gauge coupling constant e. Note that we also have $\phi_\pm(x) - \Phi_\pm(x) = \varphi(x)/A$. Only the fields $\varphi(x)$, $\pi_\varphi(x)$ and $\Phi_\pm(x)$ define the independent set of basic phase space degrees of freedom.

A straightforward analysis then establishes that the only nonvanishing commutation relations for these fields are

$$\begin{array}{c}[\varphi(x),\pi_\varphi(y)]=i\delta(x-y)\ ,\\ [\Phi_\pm(x),\pi_\varphi(y)]=0\quad,\quad[\Phi_\pm(x),\partial_1\Phi_\pm(y)]=\pm2i\pi\delta(x-y)\ ,\\ [\phi_\pm(x),\pi_\varphi(y)]=\frac{i}{A}\delta(x-y)\quad,\quad[\phi_\pm(x),\partial_1\phi_\pm(y)]=\pm2i\pi\delta(x-y)\ .\end{array}\tag{23}$$

In other words, $\varphi(x)$ and its conjugate momentum $\pi_\varphi(x)$ define the degrees of freedom of a scalar field which is periodic in x on the circle of radius R and takes its values in the real line, while $\Phi_\pm(x)$ are two independent chiral bosons each taking their values in the circle of radius unity and obeying quantum boundary conditions such that $\Phi_\pm(x+L)=\Phi_\pm(x)+2\pi(n_\pm+\lambda_\pm)$, $(n_\pm+\lambda_\pm)$ being the eigenvalues of the momentum zero modes $p_\pm$, namely corresponding to twisted chiral bosons on the circle of radius unity when $\lambda_\pm\neq0$ (mod $\mathbb{Z}$). These nontrivial topology properties of the chiral boson zero modes turn out to be crucial for the correct representations of the gauge invariance properties of the system under both small and large U(1) gauge transformations.

In order to make contact with the fermionic degrees of freedom, let us now introduce the following field operators,

$$\begin{array}{l}\psi_\pm(x)=\frac{1}{\sqrt{L}}\,e^{\pm\frac{i\pi}{L}x}\,e^{i\frac{\pi}{2}\eta p_\mp}\,e^{\pm i\lambda q_\pm}\,e^{\frac{2i\pi\lambda}{L}p_\pm x}\\ \qquad\times\prod_{n=1}^{\infty}e^{\pm\frac{i\lambda}{\sqrt{n}}a^\dagger_{\pm,n}e^{\pm\frac{2i\pi}{L}nx}}\prod_{n=1}^{\infty}e^{\pm\frac{i\lambda}{\sqrt{n}}a_{\pm,n}e^{\mp\frac{2i\pi}{L}nx}}\ ,\end{array}\tag{24}$$

where η and λ are two real constants such that $\eta^2=1=\lambda^2$. Except for the first three factors, this operator is nothing but the exponential $e^{\pm i\lambda\phi_\pm(x)}$ with an operator ordering such that all the $q_\pm$ and $a^\dagger_{\pm,n}$ are to the left of all the $p_\pm$ and $a_{\pm,n}$ operators. These fields possess the following holonomy properties

$$\psi_\pm(x+L)=-e^{2i\pi\lambda\lambda_\pm}\,\psi_\pm(x)\ ,\tag{25}$$

on account of the topology properties of the $Q_\pm$ and $p_\pm$ zero modes of the chiral fields $\Phi_\pm(x)$ and $\phi_\pm(x)$. That the fields $\psi_\pm(x)$ are indeed fermions in 1+1 dimensions follows from their anticommutation relations obtained from the fundamental algebra in (19),

$$\{\psi_\alpha(x),\psi^\dagger_\beta(y)\}=\delta_{\alpha\beta}\delta(x-y)\quad,\quad\alpha,\beta=+,-\ .\tag{26}$$

In other words, the bosonic fields $\phi_\pm(x)$ and their quantum states also provide a representation of the fermionic algebra in (17), irrespective of the choice of sign factors $\eta=\pm1$ and $\lambda=\pm1$.

Hence, the fermionic matter sector of the SchM may be bosonized in this manner in terms of the above bosonic field degrees of freedom, provided the U(1) holonomies $\lambda_\pm$ are chosen such that $\lambda_\pm=\lambda\alpha_\pm$ (mod $\mathbb{Z}$). Note that in order to obtain the anticommutation relations in (17) on their own, so far it is only through the combinations $\phi_\pm(x) = \Phi_\pm(x) + \varphi(x)/A$ that the fields $\varphi(x)$ and $\Phi_\pm(x)$ are involved in this construction. However, when considered together with the bosonic commutations relations for the gauge field in (17), the whole set of degrees of freedom associated to each of the independent fields $\varphi(x)$, $\pi_\varphi(x)$ and $\Phi_\pm(x)$ separately is essential to establish the complete operator diagonalization of the model.

4.2 Gauge invariant point-splitting regularization

Given the above bosonization of the fermionic fields, one must now define the composite operators associated to the first-class Hamiltonian density (10) and constraint (11). Since gauge invariance must be preserved at all steps, a gauge invariant point-splitting regularization of short-distance singularities is a relevant choice. Namely, associated for example to the product $\psi^\dagger_\pm(x)\psi_\pm(x)$ which is ill-defined as such as a composite operator, one considers the gauge invariant quantity

$$\psi^\dagger_\pm(y)\, e^{ie\int_x^y du A^1(u)}\, \psi_\pm(x) \tag{27}$$

in the limit where y goes to x, and subtracts any divergent term that arises. Since the Wilson line phase factor is inserted in this point-split product of the two fields, the result is manifestly gauge invariant. This example applies directly to the definition of the vector and axial current operators. A similar approach is used for the products of fermionic fields, now including the U(1) covariant derivative, that contribute to the first-class Hamiltonian.

In terms of the bosonized representation of the fermion fields, one then obtains the following definitions for composite quantum operators,

$$\psi^\dagger_\pm\psi_\pm = -\frac{\lambda}{2\pi}\left(\partial_1\phi_\pm \mp e\lambda A^1\right) , \tag{28}$$

so that the first-class constraint reads,

$$\phi = \partial_1\pi_1 + e\psi^\dagger\psi = \partial_1\pi_1 - \frac{e\lambda}{2\pi}\partial_1(\phi_+ + \phi_-) , \tag{29}$$

while the fermion contributions to the first-class Hamiltonian density $\mathcal{H}$ are

$$\frac{1}{2}i\psi^\dagger_\pm\left(\partial_1 - ieA^1\right)\psi_\pm - \frac{1}{2}i\left(\partial_1 + ieA^1\right)\psi^\dagger_\pm\psi_\pm = \pm\frac{1}{4\pi}\left(\partial_1\phi_\pm \mp e\lambda A^1\right)^2 \mp \frac{\pi}{12L^2} . \tag{30}$$

Note that the vector and axial currents as well as the first-class constraint, which are composite quantities in terms of the fermionic fields, are local non-composite operators in terms of the bosonic degrees of freedom, a property not shared, though, by the first-class Hamiltonian whose definition thus still requires the specification of some choice of operator ordering.

4.3 The diagonalized operator basis

Let us now turn to the definition of the first-class Hamiltonian operator in correspondence with (10). Given the above bosonization and gauge invariant point-splitting regularization, one finds

$$\mathcal{H}=\frac{1}{2}\pi_1^2+\frac{1}{4\pi}\left(\partial_1\phi_+-e\lambda A^1\right)^2+\frac{1}{4\pi}\left(\partial_1\phi_-+e\lambda A^1\right)^2\ , \tag{31}$$

where the Casimir energy contribution $-\pi/(6L^2)$ has been ignored since the energy is defined up to an arbitrary constant anyway. However, completing the sum of squares as follows

$$\begin{aligned}\mathcal{H}=\tfrac{1}{2}\pi_1^2+&\tfrac{1}{8\pi}\left(\partial_1\phi_++\partial_1\phi_--\tfrac{2\pi}{e\lambda}\partial_1\pi_1+\tfrac{2\pi}{e\lambda}\partial_1\pi_1\right)^2\\ +&\tfrac{1}{8\pi}\left(\partial_1\phi_+-\partial_1\phi_--2e\lambda A^1\right)^2\ ,\end{aligned} \tag{32}$$

one recognizes the explicit appearance of the first-class constraint ϕ in the form (29), so that one may also write

$$\begin{aligned}\mathcal{H}=\tfrac{1}{2}\pi_1^2&+\tfrac{\pi}{2e^2}\left(\partial_1\pi_1\right)^2+\tfrac{e^2}{2\pi}\left(A^1-\tfrac{1}{2e\lambda}\partial_1(\phi_+-\phi_-)\right)^2\\ &+\tfrac{\pi}{2e^2}\phi^2-\tfrac{\pi}{e^2}\phi(\partial_1\pi_1)\ .\end{aligned} \tag{33}$$

Given this expression, let us then introduction the following definitions,

$$\varphi=-\frac{1}{\mu}\pi_1\ \ ,\ \ \pi_\varphi=\mu\left[A^1-\frac{1}{2e\lambda}\partial_1(\phi_+-\phi_-)\right]\ , \tag{34}$$

μ being the mass parameter

$$\mu=\frac{|e|}{\sqrt{\pi}}\ . \tag{35}$$

In particular, note that except for the mass factor μ, the field φ is nothing else than the electric field, $\varphi=E/\mu$. The first-class Hamiltonian then reads

$$\mathcal{H}=\frac{1}{2}\pi_\varphi^2+\frac{1}{2}\left(\partial_1\varphi\right)^2+\frac{1}{2}\mu^2\varphi^2+\frac{1}{2}\left(\frac{\phi}{\mu}\right)^2+\left(\frac{\phi}{\mu}\right)\left(\partial_1\varphi\right)\ , \tag{36}$$

a most suggestive result indeed!

To confirm the fact that the fields φ and π_φ are conjugate to one another, one need only compute their algebra given the quantization rules (17), leading to the canonical commutation relations,

$$[\varphi(x),\pi_\varphi(y)]=i\delta(x-y)\ . \tag{37}$$

Furthermore, the remaining commutation relations are

$$[\phi_\pm(x), \pi_\varphi(y)] = -\frac{i\pi\mu}{e\lambda}\delta(x-y) \quad , \quad [\phi_\pm(x), \partial_1\phi_\pm(y)] = \pm 2i\pi\delta(x-y) \ . \quad (38)$$

This set of commutations relations is thus equivalent, upon bosonization, to the original set in (17). Furthermore, it is seen to coincide exactly with the relations in (23) provided only one sets $A = -e\lambda/(\pi\mu)$ and identifies the mass scale μ with the quantity $\mu = |e|/\sqrt{\pi}$, so that $N_n = \sqrt{\pi}/(\sqrt{L}\sqrt{2\omega_n})$.

Consequently, all quantities must now be expressed in terms of the mode degrees of freedom $Q_\pm$, $p_\pm$, φ_n, $A_{\pm,n}$ and their adjoint operators, or equivalently in terms of the fields $\varphi(x)$, $\pi_\varphi(x)$ and $\Phi_\pm(x)$. Thus, the first-class constraint is finally given by

$$\phi = -\frac{e\lambda}{2\pi}\partial_1\left[\Phi_+ + \Phi_-\right] \ , \quad (39)$$

namely,

$$\begin{aligned}\phi(x) = -\tfrac{e\lambda}{L}\Big\{&(p_+ - p_-) + \\ &+ i\textstyle\sum_{n=1}^{\infty}\sqrt{n}\left((A^\dagger_{+,n} + A_{-,n})\,e^{\frac{2i\pi}{L}nx} - (A^\dagger_{-,n} + A_{+,n})\,e^{-\frac{2i\pi}{L}nx}\right)\Big\} \ ,\end{aligned} \quad (40)$$

while the first-class Hamiltonian reads

$$H = \sum_{n=-\infty}^{+\infty}\omega_n\varphi^\dagger_n\varphi_n + \int_0^L dx\left[\frac{1}{2}\left(\frac{\phi}{\mu}\right)\left(\frac{\phi}{\mu}\right) + \left(\frac{\phi}{\mu}\right)(\partial_1\varphi)\right] \ , \quad (41)$$

where the contribution of the (φ, π_φ) degrees of freedom is displayed explicitly in terms of their creation and annihilation operators following the usual normal ordering prescription for that sector. For the gauge degrees of freedom contribution to H though, since the operators $\phi(x)$ and $\varphi(x)$ commute with themselves and one another, these operators are multiplied as shown, without any normal ordering prescription in terms of the modes $Q_\pm$, $p_\pm$, $A_{\pm,n}$ and $A^\dagger_{\pm,n}$ being applied. The reason for this choice is that these contributions to the total Hamiltonian then vanish identically for all physical states, while otherwise some infinite normal ordering constant would arise. Finally, the vector and axial charge operators are given by

$$\begin{aligned} Q &= \textstyle\int_0^L dx\,\psi^\dagger\psi = -\lambda(p_+ - p_-) \ , \\ Q_5 &= \textstyle\int_0^L dx\psi^\dagger\gamma_5\psi = -\frac{i\lambda\sqrt{2\mu}}{2\sqrt{\pi}\sqrt{L}}\left[\varphi_0 - \varphi_0^\dagger\right] \ . \end{aligned} \quad (42)$$

In conclusion, the SchM and its dynamics has been quantized in terms of the decoupled set of basic bosonic fields, $\varphi(x)$ and $\pi_\varphi(x)$ on the one hand,

and $\Phi_\pm(x)$ on the other hand, whose commutation relations (19) are those of a periodic scalar field taking its values in the real line and two twisted chiral bosons taking their values in a circle of radius unity. These two sectors of operators, which commute with one another, see (23), diagonalize the commutator and anticommutator algebra (17) defining the quantum system as well as its first-class Hamiltonian H and constraint $\phi(x)$ operators, (40) and (41). In particular, the field $\varphi(x)$ coincides, up to the mass scale μ, with the electric field, whose nonperturbative quantum dynamics is that of a pseudoscalar field whose quanta have a mass μ proportional to the U(1) gauge coupling e of the original gauge field A_μ to the original fermionic degrees of freedom. The appearance of this screening length $1/\mu$ for the electric field is a purely quantum effect, and is one of the consequences within the physical sector of the system of its underlying fermionic one even though the latter degrees of freedom are not gauge invariant except through specific combinations. The structure of the physical ground state is also dependent on gauge invariant topology features of the fermionic matter excitations. Furthermore, the physical spectrum of U(1) gauge invariant states consists only of superpositions of the massive free quanta of the electric field. These results are described next.

5 The Physical Projector and Spectrum

Having defined a quantization of the model, different issues may be addressed. First, U(1) gauge invariance remains a symmetry of the quantum dynamics, namely the first-class algebra $[\phi(x),\phi(y)]=0$ and $[H,\phi(x)]=0$ is preserved for the quantum operators. Likewise, the Poincaré algebra remains satisfied,[10] namely manifest spacetime covariance is maintained through quantization on the space of quantum states, as well as on the subspace of physical states to be constructed hereafter since the Poincaré algebra commutes with the U(1) generator ϕ.

Switching to the Heisenberg picture (as indicated by the subscript "H" in the relations hereafter), it is possible to establish the operator equations of motion generated by the total Hamiltonian $H_T = H + \int_0^L dx\,\left[A^0\phi + \partial_1(A^0\varphi)/\mu\right]$. For the operators of interest, one finds

$$\begin{gathered}
[\partial_0^2-\partial_1^2+\mu^2]\varphi_H=\partial_1\left(\frac{\phi_H}{\mu}\right)\ ,\\
\partial_0\left(\frac{\phi_H}{\mu}\right)=0\ ,\\
\partial_\mu\, J^\mu_H=0\quad,\quad \frac{d}{dt}Q_H=0\ ,\\
\partial_\mu\, J^\mu_{5\,H}=\frac{\mu^3}{e}\varphi_H=\frac{e}{\pi}E_H\quad,\quad \frac{d}{dt}Q_{5H}=\frac{e}{\pi}\int_0^L dx\,E_H\ .
\end{gathered}\tag{43}$$

Note that the axial current $J^{\mu}_{5\,H}$ and charge $Q_{5\,H}$ are no longer conserved at the quantum level as they are at the classical one, showing that this classical global symmetry is explicitly broken by quantum properties of the system. This chiral anomaly is directly proportional to the electric field, namely the actual physical configuration space of the system. In contradistinction, the local U(1) vector symmetry is explicitly preserved in the quantized system on account of the gauge invariant regularization through point splitting.

Finally, let us turn to the issue of gauge invariance. It is clear that among the basic fields, the electric field sector (φ, π_φ) is explicitly gauge invariant, while the chiral bosons $\Phi_\pm(x)$ are not. Given any L-periodic function $\epsilon_0(x)$ and an operator $\mathcal{O}(x)$, the corresponding finite small gauge transformation generated by the first-class constraint $\phi(x)$ is given by the adjoint action of the following unitary operator

$$\mathcal{O}'(x) = U(\epsilon_0)\,\mathcal{O}(x)\,U^{-1}(\epsilon_0) \quad , \quad U(\epsilon_0) = e^{i\int_0^L dx\,\epsilon_0(x)\phi(x)} \; . \tag{44}$$

For the basic fields, one finds

$$\varphi'(x) = \varphi(x) \quad , \quad \pi'_\varphi(x) = \pi_\varphi(x) \quad , \quad \Phi'_\pm(x) = \Phi_\pm(x) \mp e\lambda\epsilon_0(x) \; . \tag{45}$$

Since the latter relation for $\Phi_\pm(x)$ implies that the fields $\phi_\pm(x)$ also transform with the same rules, the correct phase transformation of the fermionic fields (24) is reproduced. Given a mode expansion of the gauge parameter,

$$\epsilon_0(x) = \sum_{n=-\infty}^{+\infty} \epsilon_n^{(0)}\, e^{\frac{2i\pi}{L}nx} \; , \tag{46}$$

the above rules translate into the gauge transformation for chiral modes,

$$Q'_\pm = Q_\pm \mp e\lambda\epsilon_0^{(0)} \quad , \quad A'_{\pm,n} = A_{\pm,n} \mp e\lambda\sqrt{n}\,\epsilon^{(0)}_{\mp n} \quad , \quad {A^\dagger}'_{\pm,n} = A^\dagger_{\pm,n} \pm e\lambda\sqrt{n}\,\epsilon^{(0)}_{\pm n} \; , \tag{47}$$

showing that the chiral momentum zero modes $p_\pm$ are invariant under small gauge transformations, as are of course also all the modes φ_n and $\varphi_n^\dagger$ of the fields $\varphi(x)$ and $\pi_\varphi(x)$. Furthermore, recall that the gauge parameter $\epsilon_0(x)$ is defined modulo $2\pi/|e|$, hence so is its zero mode $\epsilon_0^{(0)}$. Consequently, the actual reason for our previous assumption that the chiral coordinate zero modes $Q_\pm$ take their values on a circle of radius unity, namely that they are defined modulo 2π, stems directly from their properties under small U(1) gauge transformations.

Consider now a large gauge transformation of homotopy class k, thus associated to the parameter $\epsilon(x) = (2\pi kx/eL)$. Since in this case the fermion fields $\psi_\pm$ transform with the phase factor $e^{-2i\pi kx/L}$ while the gauge field $A^1(x)$ is shifted by the quantity $-\partial_1\epsilon(x) = -2\pi k/(eL)$, it follows that among all the modes of the basic fields $\varphi(x)$, $\pi_\varphi(x)$ and $\Phi_\pm(x)$, only the chiral momentum zero modes $p_\pm$ must transform under large gauge transformations according to the following rule

$$p'_\pm = p_\pm - \lambda k \ . \tag{48}$$

This transformation is generated by a unitary operator of the form

$$U_k = C_k \, e^{ik\lambda(Q_+ + Q_-)} \quad , \quad p'_\pm = U_k \, p_\pm \, U_k^{-1} \ , \tag{49}$$

where C_k is a cocycle factor introduced in order to obtain the following group composition law for any two large gauge transformations of U(1) homotopy classes k and ℓ,

$$U_k \, U_\ell = U_{k+\ell} \ . \tag{50}$$

The general solution to this condition is $C_k = e^{i\theta k\lambda}$ where θ is an arbitrary real constant defined modulo 2π. Hence, the operator representation of large U(1) gauge transformations of homotopy class k is simply $U_k = e^{i\theta k\lambda} \, e^{ik\lambda(Q_+ + Q_-)}$, up to any transformation $U(\epsilon_0)$.

These gauge transformations also answer a question raised previously. The gauge invariant degrees of freedom which are conjugate to the gauge invariant electric field are those of the momentum $\pi_\varphi(x)$, which is indeed a specific combination of the original gauge dependent degrees of freedom which is gauge invariant under both small and large gauge transformations. Namely, $\pi_\varphi(x)$ is essentially the difference of the original gauge field $A^1(x)$ with the gradient $\partial_1(\Phi_+ - \Phi_-)(x)$ of the difference of the chiral bosons. In particular, because of this gradient contribution, the zero mode of the electric field, which is proportional to $(\varphi_0 + \varphi_0^\dagger)$, is conjugate to the combination $(A_0^1 - \pi(p_+ + p_-)/(e\lambda L))$, which is proportional to $(\varphi_0 - \varphi_0^\dagger)$, rather than to the $A^1(x)$ zero mode A_0^1 on its own as might have been anticipated. This situation is in contradistinction with results obtained within a gauge fixed formulation.[3,4] Moreover, the fact that this zero mode A_0^1 is defined modulo $(2\pi)/(|e|L)$ because of its properties under large gauge transformations, is now seen to correspond to the analogous property for the chiral momentum zero modes $p_\pm$ which are defined modulo integers for the same reason, since the combination $(\varphi_0 - \varphi_0^\dagger)$ is invariant under both small and large gauge transformations. Note that the actual phase space topology of the twisted chiral zero mode sector $(Q_\pm, p_\pm)$ is that of a two-torus for each chiral boson.

Having identified the operators $U(\epsilon_0)$ and U_k which generate all small and large gauge transformations on the space of quantum states, the physical projector is readily constructed as the sum of all these transformations,[5] in order to retain only the totally gauge invariant components of any state,[6] namely the actual physical states of the system. Hence, the physical projector is given by

$$\mathbb{E}=\sum_{k=-\infty}^{+\infty}\int_0^{\frac{2\pi}{|e|}}d\epsilon_0^{(0)}\prod_{n=1}^{\infty}\int_{-\infty}^{\infty}d(\mathrm{Re}\,\epsilon_n^{(0)})\int_{-\infty}^{\infty}d(\mathrm{Im}\,\epsilon_n^{(0)})\,U_k\,U(\epsilon_0)\;, \tag{51}$$

where the absolute normalizations of these integrals are not specified at this stage, the operator $\mathbb{E}$ being a non-normalizable projector density.

In order to describe the physical spectrum, let us now introduce a basis for the representation space of the algebra (19). In the electric field sector, we have the normalized Fock vacuum $|0>_\varphi$ which is annihilated by the modes φ_n and which is acted on with the creation operators $\varphi_n^\dagger$ to generate the basis of normalized states,

$$|\{k_p\}>_\varphi=\prod_{p=-\infty}^{+\infty}\frac{1}{\sqrt{k_p!}}\left(\varphi_p^\dagger\right)^{k_p}|0>_\varphi\;, \tag{52}$$

in which $k_p\geq 0$ quanta of integer momentum $-\infty<p<+\infty$ are excited. In the chiral boson sector, the creation operators $A_{\pm,n}^\dagger$ act on a normalized Fock vacuum $|0>_\Phi$ which is annihilated by the operators $A_{\pm,n}$, to also generate the full Fock basis of quantum excitations of the chiral fields. Finally, the chiral zero mode sector $(Q_\pm,p_\pm)$ is represented in the orthonormalized momentum eigenbasis $|n_+,n_->$, $<n'_+,n'_-|n_+,n_->=\delta_{n'_+,n_+}\delta_{n'_-,n_-}$, such that,

$$p_\pm|n_+,n_->=(n_\pm+\lambda_\pm)|n_+,n_->\;. \tag{53}$$

From the explicit expression for the physical projector, it is then possible to work out its action on this basis of the quantum space of states.[10] One then finds

$$\mathbb{E}=\mathbb{1}_\varphi\otimes|\Omega_\theta><\Omega_\theta|\;, \tag{54}$$

where the tensor product refers to that between the electric and chiral sectors of degrees of freedom, while the gauge invariant vacuum $|\Omega_\theta>$ in the sector of gauge dependent degrees of freedom is given by the non-normalizable state

$$|\Omega_\theta>=\sum_{\ell=-\infty}^{+\infty}e^{i\ell\theta}\,e^{-\sum_{n=1}^{\infty}A_{+,n}^\dagger A_{-,n}^\dagger}\,|n_+=\ell,n_-=\ell;0>_\Phi\;, \tag{55}$$

which is thus seen to correspond to a θ-vacuum associated to a coherent superposition, characterized by the parameter θ, of the chiral momentum eigenstates. In fact, gauge invariance enforces the constraint $p_+=p_-$, which not only requires identical eigenvalues $n_+=\ell=n_-$ but also identical values modulo integers for the U(1) holonomies $\lambda_+ = \lambda_-$ (mod $\mathbb{Z}$), namely the parity invariant condition $\alpha_+ = \alpha = \alpha_-$ (mod $\mathbb{Z}$) since $\alpha_\pm = \lambda\lambda_\pm$ (mod $\mathbb{Z}$).

Consequently, all physical states of the system are spanned by the free massive quanta of the electric field, namely the multiparticle states

$$|\{k_p\} >_\varphi \otimes |\Omega_\theta > \ , \tag{56}$$

which are all invariant under the action of both small and large U(1) gauge transformations, $U(\epsilon_0)$ and U_k. In particular, the first-class constraint operator $\phi(x)$ annihilates any physical state, since $\phi(x)|\Omega_\theta >= 0$. This basis of states also diagonalizes the total Hamiltonian H_T of the system, which thus coincides with the first-class one H on the subspace of physical states, leading to the physical energy spectrum

$$E(\{k_p\}) = \sum_{p=-\infty}^{+\infty} k_p\,\omega_p \ , \ \ \omega_p = \sqrt{\left(\frac{2\pi}{L}p\right)^2 + \mu^2} \ , \ \ \mu = \frac{|e|}{\sqrt{\pi}} \ , \tag{57}$$

whose spacetime momentum value is also

$$P^1(\{k_p\}) = \sum_{p=-\infty}^{+\infty} k_p \left(\frac{2\pi p}{L}\right) \ . \tag{58}$$

Finally, note that the non-normalizable character of physical states resides only within the gauge invariant contribution of the gauge dependent degrees of freedom, namely within the factorized single gauge invariant θ-vacuum $|\Omega_\theta >$.

For the sake of illustration, let us also consider the expectation values of some quantities of interest. Given any multiparticle state, the expectation value of the electric field is readily seen to vanish identically at all times, with nonvanishing fluctuations around that mean value however,

$$< E(x) >= 0 \ \ , \ \ <: E^2(x) :>= \frac{1}{L}\sum_{p=-\infty}^{+\infty} \frac{\mu\, k_p}{\sqrt{(\frac{2\pi}{L}p)^2 + \mu^2}} \ . \tag{59}$$

On the other hand, the expectation value of the gauge potential $A^1(x)$ within the same multiparticle states reads,

$$< A^1(x) >= \frac{2\pi}{e\lambda L}\lambda_\pm = \frac{2\pi}{eL}\alpha_\pm \ . \tag{60}$$

Even though the operator $A^1(x)$ is not gauge invariant, its expectation value for gauge invariant states should be; nonetheless, it need not necessarily vanish for that reason. Indeed, first note that on account of translation invariance in x, the quantity $< A^1(x) >$ measures only the expectation value of the zero mode of $A^1(x)$, which is thus certainly invariant under small U(1) gauge transformations. Furthermore, because of large U(1) gauge transformations, this zero mode is actually defined only modulo $(2\pi)/(|e|L)$. This is indeed also the property shared by the above result for $< A^1(x) >$, given that the U(1) holonomies $\lambda_\pm$, which are directly related to the choice of boundary conditions of the fermionic fields $\psi_\pm(x)$ in x, are defined modulo integers. Consequently, even though nonvanishing, the expectation value $< A^1(x) >$ is perfectly gauge invariant, a possibility which arises since the zero mode of that operator is defined on a circle of radius $1/(|e|L)$ rather than the real line. Note that consistency of this result requires once again the two U(1) holonomies $\lambda_\pm$, or the two factors $\alpha_\pm$, to be equal to one another modulo integers. It is interesting to remark that the expectation value $< A^1(x) >$ is also a measure of the fermionic boundary conditions. In contradistinction, the expectation value of the other original gauge dependent fields, namely the Weyl spinors $\psi_\pm(x)$, always vanishes for multiparticle states, $< \psi_\pm(x) >= 0$, which is again a result perfectly consistent with gauge invariance. Since these degrees of freedom vary continuously for small as well as large U(1) gauge transformations in contradistinction to the zero mode of $A^1(x)$, being gauge invariant necessarily their expectation value for gauge invariant states much vanish identically.

Note that the U(1) charge operator Q, see (42), vanishes identically for any physical state, on account of gauge invariance under small gauge transformations of these states, $Q|\{k_p\} >_\varphi \otimes|\Omega_\theta >= 0$. None of the physical states thus carries any electric charge. This conclusion includes the physical ground state $|0 >_\varphi \otimes|\Omega_\theta >$, showing that the exact U(1) vector gauge symmetry of the quantized system is not spontaneously broken. Had the global axial symmetry $\mathrm{U}(1)_A$ not be broken explicitly by a quantum anomaly, the issue of the Wigner or Goldstone mode realization of that symmetry could have been raised as well. However, the quantum axial charge Q_5, see (42), does not commute with the quantum Hamiltonian H, so that its action does not conserve energy even though it maps physical states into physical states since it commutes with the first-class constraint operator $\phi(x)$. For instance, acting on the physical ground state, one finds

$$Q_5|0 >_\varphi \otimes|\Omega_\theta >= i\frac{\lambda\sqrt{2\mu}}{2\sqrt{\pi}\sqrt{L}}\,|k_0 = 1 >_\varphi \otimes|\Omega_\theta > , \qquad (61)$$

resulting in a zero momentum 1-particle excitation of the electric field. For the same reason, the expectation value of the chiral charge vanishes identically for any multiparticle state, including the physical ground state, $< Q_5 >= 0$.

Furthermore, any finite axial transformation acts as follows on the physical ground state,

$$e^{i\alpha Q_5}\,|0>_\varphi\otimes|\Omega_\theta>=|z_0>_\varphi\otimes|\Omega_\theta>\qquad\text{with}\qquad z_0=-\frac{\alpha\lambda\sqrt{2\mu}}{2\sqrt{\pi}\sqrt{L}}\,,\tag{62}$$

where $|z_0>_\varphi$ stands for the holomorphic zero momentum coherent state in the $(\varphi(x),\pi_\varphi(x))$ sector defined by

$$|z_0>_\varphi=e^{-\frac{1}{2}|z_0|^2}\,e^{z_0\varphi_0^\dagger}\,|0>_\varphi\,,\tag{63}$$

z_0 being an arbitrary complex variable. Consequently, axial transformations act by producing coherent zero momentum excitations of the electric field. However, they do not induce a change in the θ parameter of the θ-vacuum $|\Omega_\theta>$ in the sector of gauge dependent degrees of freedom. This conclusion, as well as some of the above results, should be contrasted with those available in the literature following any gauge fixing procedure,[2,3,4] some of which are thus seen to be actually artefacts due to gauge fixing.

6 Concluding Remarks

Through quantization but without any gauge fixing procedure whatsoever, the original gauge and fermionic degrees of freedom of the massless Schwinger model have been reorganized into a diagonalized set of basic gauge invariant and gauge dependent operators. The gauge invariant ones are those of the electric field whose conjugate momentum variable is the single gauge invariant combination of the vector field A^1 with the bosonized Weyl spinors. The gauge dependent ones are those of the specific combinations of the bosonized Weyl spinors with the electric field which are decoupled from the gauge invariant sector. The gauge invariant dynamics and physical content of the system is that of a free pseudoscalar field of mass $\mu=|e|/\sqrt{\pi}$, namely the electric field, which may be excited in any multiparticle state. The gauge dependent degrees of freedom contribute only through a uniquely defined gauge invariant combination which is the θ-vacuum $|\Omega_\theta>$. This state is comprised, see (55), on the one hand, of a condensate which is the coherent superposition of pairs of nonzero mode excitations of the chiral bosons $\Phi_+(x)$ and $\Phi_-(x)$ of vanishing total momentum, and on the other hand, of a θ-vacuum coherent superposition of the chiral boson zero mode momentum eigenstates sharing a common eigenvalue.

The latter feature stems from the gauge invariant topological properties of the zero modes of the gauge dependent sector under both small and large U(1) gauge transformations. The θ variable, defined modulo 2π, parametrizes the sole freedom left at the quantum level in the gauge invariant physical contribution of the gauge dependent sector. Up to that parameter, which labels all possible distinct quantum representations of the modular group $\mathbb{Z}$ of large U(1) gauge transformations, there exists only a single gauge invariant quantum state within the sector of gauge dependent degrees of freedom, namely the θ-vacuum $|\Omega_\theta>$. All the gauge invariant topological features of the system thus reside in that θ-vacuum for the gauge invariant combination of its zero modes.

The quantization and solution of the model is nonperturbative in that taking the limit of a vanishing gauge coupling constant e leads to a singular behaviour. There is no analytic deformation possible of the set of field operators that diagonalize the interaction-free system into that which diagonalizes the interacting system, thus precluding any perturbation expansion. Note also that the compactification of the spatial coordinate into a circle is essential in bringing to the fore all the topology features of the system within its zero mode sector which are implied by gauge invariance under both small and large U(1) gauge transformations. The radius R or circumference L of that circle provides a regularization of the system which allows for a discrete spectrum of quantum states free of infrared divergences in 1+1 dimensions, and thus in particular a clear separation of the zero mode contributions and their topological properties.

The physical projector, free of the necessity of any gauge fixing, is thus able to unambiguously identify, through a nonperturbative quantization which diagonalizes the dynamics of the massless Schwinger model, and in agreement with results established following whatever gauge fixing procedure at least in as far as the physical spectrum is concerned, the actual physical content of the model, without suffering, however, from artefacts resulting from gauge fixing which render the physical interpretation sometimes confusing. The physical projector has also achieved the quantization of a compact phase space, namely the product of the two two-tori corresponding to the zero mode sectors of the two gauge dependent chiral bosons, without having recourse to any geometric quantization methods beyond the usual rules of canonical quantization.

Acknowledgements

G.Y.H.A. acknowledges the financial support of the "Coopération Universitaire au Développement, Commission Interuniversitaire Francophone" (CUD-CIUF) of the Belgian French Speaking Community, and wishes to thank the Institute of Nuclear Physics (Catholic University of Louvain, Belgium) for its hospitality while this work was being pursued.

References

1. J. Schwinger, *Phys. Rev.* **125**, 397 (1962); *ibid* **128**, 2425 (1962).
2. The literature on the Schwinger model being so vast, see for example, J.H. Lowenstein and J.A. Swieca, *Ann. Phys.* **68**, 172 (1971); A. Casher, J. Kogut and L. Susskind, *Phys. Rev.* **D10**, 732 (1974); S. Coleman, R. Jackiw and L. Susskind, *Ann. Phys.* **93**, 267 (1975); S. Coleman, *Ann. Phys.* **101**, 239 (1976); D. Wolf and J. Zittartz, *Z. Phys.* **B51**, 65 (1983); *ibid* **B59**, 117 (1985); R. Link, *Phys. Rev.* **D42**, 2103 (1990).
3. N.S. Manton, *Ann. Phys.* **159**, 220 (1985), and references therein.
4. J.E. Hetrick and Y. Hosotani, *Phys. Rev.* **D38**, 2621 (1988), and references therein.
5. J.R. Klauder, *Ann. Phys.* **254**, 419 (1997); J.R. Klauder, *Nucl. Phys.* **B547**, 397 (1999); J.R. Klauder, *Quantization of Constrained Systems, Lect. Notes Phys.* **572**, 143 (2001), `hep-th/0003297`.
6. For a detailed discussion and references to the original literature, see, J. Govaerts, *Hamiltonian Quantisation and Constrained Dynamics* (Leuven University Press, Leuven, 1991); J. Govaerts, *The Quantum Geometer's Universe: Particles, Interactions and Topology*, contribution to these Proceedings.
7. J. Govaerts and J.R. Klauder, *Ann. Phys.* **274**, 251 (1999); V.M. Villanueva, J. Govaerts and J.-L. Lucio-Martinez, *J. Phys.* **A33**, 4183 (2000); J. Govaerts, *The Cosmological Constant of One-Dimensional Matter Coupled Quantum Gravity is Quantized*, preprint STIAS-02-002, `hep-th/0202134`.
8. J. Govaerts and B. Deschepper, *J. Phys.* **A33**, 1031 (2000).
9. J. Govaerts, *J. Phys.* **A30**, 603 (1997).
10. G.Y.H. Avossevou, Ph.D. Thesis (Institut de Mathématiques et de Sciences Physiques, Republic of Benin, April 2002), unpublished.
11. L. Faddeev and R. Jackiw, *Phys. Rev. Lett.* **60**, 1692 (1988); J. Govaerts, *Int. J. Mod. Phys.* **A5**, 3625 (1990).
12. J. Govaerts and V.M. Villanueva, *Int. J. Mod. Phys.* **A15**, 4903 (2000); J. Govaerts, Proc. First International Workshop on Contemporary Problems in Mathematical Physics, 31 October-5 November 1999, eds. J. Govaerts, M.N. Hounkonnou and W.A. Lester, Jr. (World Scientific, Singapore, 2000), pp. 244-259.

PHASE SPACE GEOMETRY IN CLASSICAL AND QUANTUM MECHANICS

JOHN R. KLAUDER

Departments of Physics and Mathematics
University of Florida
Gainesville, FL 32611
E-mail: klauder@phys.ufl.edu

Phase space is the state space of classical mechanics, and this manifold is normally endowed only with a symplectic form. The geometry of quantum mechanics is necessarily more complicated. Arguments will be given to show that augmenting the symplectic manifold of classical phase space with a Riemannian metric is sufficient for describing quantum mechanics. In particular, using such spaces, a fully satisfactory geometric version of quantization will be developed and described.

1 Introduction

What is the difference between classical mechanics and quantum mechanics?

From a certain perspective, it is surely true that the differences appear to be vast. One theory is deterministic, the other is stochastic. One theory involves point particles and their trajectories, the other involves wave functions generally spread out over space. One theory involves commuting algebraic expressions, the other involves generally noncommuting algebraic expressions corresponding, in each case, to observables. These differences are indeed vast and well known, and of course they are all true. In brief, if one wants to find big differences, then it is not too difficult to do so. However, let us take a different perspective.

Instead of focusing on vast distinctions, let us see how close we can bring the formulations of classical and quantum mechanics to each other. If we can bring them close to each other, then we may be able to shed comparative light on one side from the vantage point of the other side, a procedure which may be rather useful in gaining further understanding of both theories.

Prior to the discovery of quantum mechanics, there was a wealth of significant developments in the arena of classical mechanics. The range of applicability of classical mechanics is enormous, and, for many sorts of problems, a classical description is frequently all that is needed. With the advent of quantum mechanics, a wider family of problems may be successfully addressed. Sometimes one reads that is is necessary to let $\hbar$, Planck's constant/2π, go to zero, i.e., $\hbar \to 0$, if one wishes to describe classical mechanics. Our viewpoint is quite the opposite in that $\hbar$ is, after all, not zero in the real world, and it should be accepted for the nonzero value that it has, $\hbar \simeq 10^{-27}$ erg·sec. Consequently, it should not be a case of classical *or* quantum descriptions, but, instead, classical *and* quantum descriptions. In other words, *classical and quantum formulations should coexist.*

The first part of the present article is devoted to a picture of classical and quantum physics fully consistent with the view that both formulations coexist. The second part of this article discusses the meaning of the variables used in phase space path integrals of different types, and shows how quantum mechanics impacts on the meaning of phase space variables that are used in classical mechanics. In the third part of this article, we show how the addition of a metric to the phase space of classical mechanics can be considered the key concept in defining the basic ingredients in quantum mechanics, and in particular the impact that the addition of the metric to the classical phase space has on the phase space of quantum mechanics.

Throughout this article we will rely on the concept of coherent states and their important application in building a bridge between the classical and the quantum world views. It is perhaps appropriate, therefore, that at this moment we present a mini review of the definition and some properties of coherent states before we put them to use in what follows.

1.1 Coherent states: definitions and properties

Quantum mechanics deals with vectors, let us call them $|\psi\rangle$ following the notation of Dirac, in a Hilbert space $\mathfrak{H}$, $|\psi\rangle \in \mathfrak{H}$, with an inner product denoted by $\langle\phi|\psi\rangle$ taken to be linear in the right hand vector and antilinear in the left hand vector. Linear operators act on vectors and yield new vectors, and in the quantum mechanics of a single canonical system, two basic operators are P and Q, an irreducible pair, which satisfy the Heisenberg commutation relation, $[P, Q] = -i\hbar 1\!\!1$, where $1\!\!1$ denotes the unit operator. These operators are to be taken as self adjoint, which means that they can be used to generate unitary groups of transformations acting in the Hilbert space. In particular, we introduce the family of unitary operators given by

$$U[p, q] \equiv e^{-iqP/\hbar}\, e^{ipQ/\hbar} \,, \tag{1}$$

defined for all $(p, q) \in \mathbb{R}^2$, and which, we state, have the following multiplication rule

$$U[p, q]\, U[p', q'] = e^{i(pq'-qp')/2\hbar}\, U[p + p', q + q'] \,. \tag{2}$$

Next, we choose a distinguished vector $|\eta\rangle$, called the *fiducial vector*, which is normalized such that $\| |\eta\rangle \| \equiv \sqrt{\langle\eta|\eta\rangle} = 1$, and consider the set of vectors each of which is defined as

$$|p,q\rangle \equiv U[p,q]\,|\eta\rangle \tag{3}$$

for all $(p,q) \in \mathbb{R}^2$. As defined, every vector in this set has a unit norm, $\|\,|p,q\rangle\,\| = 1$. The set of states defined in this manner constitutes a *set of coherent states*.[1]

Any set of coherent states as defined above satisfies two fundamentally important properties. The first property is that *the coherent state vectors are continuously parameterized*. This means that if the parameters $(p',q') \to (p,q)$ in the sense of convergence in $\mathbb{R}^2$, i.e., $|p'-p|+|q'-q| \to 0$, then it follows that the vectors $|p',q'\rangle \to |p,q\rangle$ in the strong sense, that is $\|\,|p',q'\rangle - |p,q\rangle\,\| \to 0$. The second property is that the vectors not only span the full Hilbert space, but that *they admit a resolution of unity as an integral over one dimensional projection operators* given by

$$\mathbb{1} = \int |p,q\rangle\langle p,q|\, dp\, dq/(2\pi\hbar)\ . \tag{4}$$

Here the integral runs over all points $(p,q) \in \mathbb{R}^2$, and the integral converges in the weak sense (as well as in the strong sense).

As minimum requirements in the choice of $|\eta\rangle$ it is generally convenient to require that

$$\langle\eta|P|\eta\rangle = 0\ , \qquad \langle\eta|Q|\eta\rangle = 0\ , \tag{5}$$

which has the virtue that

$$\langle p,q|P|p,q\rangle = p\ , \qquad \langle p,q|Q|p,q\rangle = q\ , \tag{6}$$

leading to the physical interpretation of the variables p and q as *expectation values* in the coherent states. It is furthermore common to choose $|\eta\rangle = |0\rangle$, the ground state of an harmonic oscillator, typically satisfying the relation $(Q+iP)\,|0\rangle = 0$.

2 Equations of Motion, and Boundary Data

Classical mechanics is described by time dependent phase space paths, $q(t)$ and $p(t)$, where q denotes position and p denotes momentum. The equations that govern the time dependence may be determined as the extremal equations – also known as the Euler-Lagrange equations – that arise from stationary variation of the action functional

$$I = \textstyle\int [p\dot{q} - H(p,q)]\, dt \tag{7}$$

holding q fixed at both the initial and final times, say, $t = 0$ and $t = T$, respectively. In this expression, $H(p,q)$ denotes the all-important Hamiltonian function, which, for convenience, we have assumed to be time independent. Variation of the action leads to

$$\delta I = \int[(\dot{q} - \partial H/\partial p)\,\delta p - (\dot{p} + \partial H/\partial q)\,\delta q]\,dt \tag{8}$$

since the surface term, $p\delta q|$, vanishes. Setting $\delta I = 0$ for general variations $\delta p(t)$ and $\delta q(t)$, we arrive at Hamilton's equations of motion

$$\dot{q} = \frac{\partial H}{\partial p}\,, \qquad \dot{p} = -\frac{\partial H}{\partial q}\,. \tag{9}$$

How these equations are derived and whether or not they have solutions are *two fundamentally different issues.* In other words, holding q fixed at two different times, as was used in this derivation, does *not* imply there is necessarily one and only one solution to these equations. Sometimes there may be a unique solution, but there also may be *no* solution, or even *many* solutions. As examples, consider $\dot{q} = p$ and $\dot{p} = -q$, $q(0) = 0$ and $q(2\pi) = 1$ (*no* solution) or $q(2\pi) = 0$ (*many* solutions). In higher dimensions, there may be no solution for a path that needs to pass through a wall, etc.

Accepting the difference between the derivation and the solution of the equations of motion allows us to consider more general situations. Suppose the action functional had the form

$$I' = \int[\tfrac{1}{2}(p\dot{q} - q\dot{p}) - H(p,q)]\,dt \tag{10}$$

which differs by a total derivative from the previous expression, specifically

$$I' = I - \tfrac{1}{2}\int[d(pq)/dt]\,dt\,. \tag{11}$$

We now propose to derive the same basic equations of motion from I' that we obtained from I. To do so requires that we hold both q *and* p fixed at the beginning and end of the time interval in order that, after integration by parts, we again find that

$$\delta I' = \int[(\dot{q} - \partial H/\partial p)\,\delta p - (\dot{p} + \partial H/\partial q)\,\delta q]\,dt\,. \tag{12}$$

Insisting that $\delta I' = 0$ leads again to the equations of motion

$$\dot{q} = \frac{\partial H}{\partial p}\,, \qquad \dot{p} = -\frac{\partial H}{\partial q}\,. \tag{13}$$

These equations are the same as before, but now we are asked to seek solutions based on the boundary conditions $p(0), q(0) = p', q'$ and $p(T), q(T) = p'', q''$. Clearly, in the general case the proposed solution is over specified and does not exist. However, in the present case – and unlike the previous case – we know how to limit the specified data so as to ensure a *unique* solution. One rule to obtain a unique solution would be to choose only the initial data $p(0), q(0)$, while another choice would be to choose only the final data $p(T), q(T)$.

3 On the Meaning of the Variables p and q

In the phase space path integral formulation of quantum mechanics, Feynman[2] states that we may compute the propagator according to the formula

$$\langle q''|e^{-i\mathcal{H}T/\hbar}|q'\rangle \equiv \mathcal{M}\int e^{(i/\hbar)\int[p\dot{q}-H(p,q)]\,dt}\,\mathcal{D}p\,\mathcal{D}q\,. \tag{14}$$

Here one is instructed to integrate over all functions $p(t)$ and $q(t)$, $0<t<T$, subject only to the conditions that $q(T) = q''$ and $q(0) = q'$. For present purposes it is not too important just how the integral on the right side is to be defined; rather we are more interested in the simpler question of what is the *meaning* of q. The meaning of the variable q can be determined from the eigenvalue equation $Q|q''\rangle = q''|q''\rangle$, which asserts that q has the meaning of the sharp value of position associated with an eigenvalue. We can determine the meaning of the variable p by using a related phase space path integral given by

$$\langle p''|e^{-i\mathcal{H}T/\hbar}|p'\rangle \equiv \mathcal{M}\int e^{(i/\hbar)\int[-q\dot{p}-H(p,q)]\,dt}\,\mathcal{D}p\,\mathcal{D}q\,. \tag{15}$$

In this case the integrals run over all functions $q(t)$ and $p(t)$, $0<t<T$, subject only to the conditions that $p(T) = p''$ and $p(0) = p'$. Consequently, the meaning of p is that of a sharp eigenvalue associated with the equation $P|p''\rangle = p''|p''\rangle$.

It may seem reasonable that both p and q have the meaning of sharp values until one realizes that this situation refers to *quantum mechanics* and not *classical mechanics* since $\hbar \neq 0$. In fact, this interpretation asserts that we can specify both *p and q simultaneously* for *all* intermediate t values, while the uncertainty relation asserts that this is impossible for *any* t value.

It is interesting to observe that there is another interpretation of such phase space path integrals,[3,1] namely, that

$$\langle p'',q''|e^{-i\mathcal{H}T/\hbar}|p',q'\rangle \equiv \mathcal{M}\int e^{(i/\hbar)\int[p\dot{q}-H(p,q)]\,dt}\,\mathcal{D}p\,\mathcal{D}q\,. \tag{16}$$

Here, unlike the previous cases, one integrates over all $p(t)$ and all $q(t)$, with $0<t<T$, subject to the conditions $p(T), q(T) = p'', q''$ and $p(0), q(0) = p', q'$. The initial and final eigenstates in this case are *not* sharp eigenstates but

rather are the *coherent states* $|p,q\rangle$ discussed above for a general $|\eta\rangle$ which satisfies (5). For coherent states, the meaning of p and q is *not* that of sharp eigenvalues but rather is that of *mean* values, as already noted previously, namely,

$$\langle p,q|P|p,q\rangle = p\,, \qquad \langle p,q|Q|p,q\rangle = q\,. \tag{17}$$

As mean values, it is perfectly acceptable to specify values of $p(t)$ and $q(t)$ simultaneously for all t, $0 < t < T$, and thus there is absolutely no contradiction with the uncertainty relation.

One should wonder how it is that the same formal path integral has two different evaluations; cf. Eqs.(14) and (16). The key word in the previous sentence is "formal", which implies that the so-called path integrals are in fact undefined as they stand and these formal expressions need to be *defined*. That the results given in (14) and (16) are different, means that the individual formal expressions have received their definition by different rules. These matters are well spelled out elsewhere[4] and need not be repeated here.

The conclusion of the discussion in this section is that the only reasonable interpretation of the variables p and q is as mean values rather than truly sharp values since we live in a world where $\hbar \neq 0$. This interpretation has important implications and is not changed by any attempt to define (14) or (15) by lattice limits as are customarily used to give some level of proper definition to such expressions.[5]

4 Shadow Metric

Classical mechanics takes place on a phase space manifold which is equipped with a *symplectic form* ω leading to a *symplectic manifold*. The symplectic form takes as its argument two vectors and returns a number. For example,

$$\omega(dp,dq) = \omega_{ab}\,dp^a\,dq^b\,, \tag{18}$$

where $\omega_{ba} = -\omega_{ab}$ and $\det[\omega_{ab}] \neq 0$. A symplectic manifold has what is called a *symplectic geometry* induced by the symplectic form. Such a geometry is rather loose, like that of a rubber sheet which may be stretched by different amounts in different directions and still retain its "geometry". This kind of geometry is quite distinct from the more familiar Riemannian geometry determined by a Riemannian metric, and it is basic that a symplectic manifold appropriate for classical mechanics should not be assumed to be a metric space endowed with a Riemannian metric.

Our point of view is that one nevertheless *needs* a metric in order to give physics to the usual mathematical expressions that appear in classical mechanics.[6] For example, one can make canonical coordinate transformations such that the mathematical expression for the Hamiltonian assumes the simple form p for essentially any system. How is one to read out of the universal expression $H = p$ that the given expression actually refers to, say, an oscillator, or perhaps an anharmonic oscillator, etc? Clearly, one needs more information to make that choice correctly. And make no mistake, one definitely needs to make that choice because in quantum mechanics one solves for the spectrum of an oscillator or an anharmonic oscillator, etc., and no ambiguity in that situation can be admitted. To quantize a given classical theory one must not only know its mathematical formulation in terms of the phase space variables, but, at the same time, one must know to what physical system it refers. Unfortunately, this kind of information is simply *not* part of a traditional classical theory.

We must then seriously consider augmenting traditional classical mechanics with another structure, the purpose of which is to keep track of the physics of the various mathematical expressions that enter into the theory. The structure we propose is that of a Riemannian metric, generally having the form

$$d\sigma^2 = A(p,q)\,dp^2 + 2B(p,q)\,dp\,dq + C(p,q)\,dq^2 \ , \tag{19}$$

with $A > 0$, $C > 0$, and $AC > B^2$. For the sake of illustration, let us choose the simplest example given by

$$d\sigma^2 = dp^2 + dq^2 \ , \tag{20}$$

which by its very form describes a two dimensional flat space expressed in Cartesian coordinates. This metric does not enter into the analysis of classical mechanics; specifically, it does not enter at all into the classical equations of motion. Its sole purpose is "to stand on the sidelines" and to determine the physics of any given expression. How do we intend to do this? Let us adopt the expression $\frac{1}{2}(p^2 + q^2) + \lambda q^4$ to correspond to the physics of an anharmonic oscillator provided that in the same coordinates the metric reads $d\sigma^2 = dp^2 + dq^2$. As the coordinates change, the expression for the Hamiltonian of this system changes form, but so does the expression for the metric! To understand the physical meaning of the new expression for the Hamiltonian, one need only find the transformation that restores the metric to Cartesian form and – presto! – the Hamiltonian in these coordinates assumes its usual physical form. In short, *the metric encodes the physical meaning of the mathematical expression that represents the Hamiltonian.* For the classical theory, this metric is separate from the theory itself, and for that reason it has been called a "shadow metric", one that is not part of the necessary mathematical formalism. However, as we have just argued, it is needed so as to complete the whole theory by ensuring that the physical meaning of the given expressions are not lost. In point of fact – and perhaps without even being aware of it – one intuitively introduces a "shadow metric", or its equivalent, so one can maintain a proper physical interpretation of the expressions that appear; otherwise the mathematics would be disconnected from the physics, a situation that would be untenable.

5 Continuous Time Regularization

As remarked earlier, path integrals in general, and phase space path integrals in particular, are formal expressions that need some form of regularization in order for them to become defined. The most common regularization is a so-called lattice regularization which replaces the action integral by a Riemann sum and integrates over the path values at the discrete time steps that are thereby introduced. The result of interest arises in the limit that the lattice spacing goes to zero, a limit which is taken as the final step in the calculation. Of course, there are many different possible choices that one can make for the discrete form of the classical action all of which yield the usual classical action for continuous and differentiable paths in the limit that the lattice spacing vanishes. However, and this is the important point, there is no guarantee that the limit of the integral exists as the lattice spacing vanishes, and even when it exists, there is no guarantee that the result is physically acceptable for any of the various different choices of lattice action. This is an issue of great importance that needs to be analyzed whenever one is in any doubt about such problems.

On the other hand, a lattice form of regularization is not the only choice that can be used, and in this section we wish to discuss a form of regularization that maintains the classical action as an integral over time, and which is called a *continuous time regularization* to distinguish it from the lattice form previously discussed. The key idea here is the introduction of an additional factor that serves as a regularization device. This regularization factor – and its removal – takes the form[7]

$$\lim_{\nu\to\infty} \mathcal{M}\int e^{(i/\hbar)\int[p\dot{q}-H(p,q)]\,dt}\, e^{-(1/2\nu)\int[\dot{p}^2+\dot{q}^2]\,dt}\, \mathcal{D}p\,\mathcal{D}q\,. \tag{21}$$

Observe what has been done: we have inserted a real damping factor in the integrand that formally tends to unity as $\nu\to\infty$. But, that limit is reserved until the integral over paths has actually been performed. It is important to observe that the regularized expression (21) can actually be given an unambiguous mathematical version as

$$\lim_{\nu\to\infty} 2\pi e^{\nu T/2\hbar} \int e^{(i/\hbar)\int[p\,dq-H(p,q)\,dt]}\, d\mu_W^\nu(p,q)\;, \tag{22}$$

where μ_W^ν denotes a Wiener measure on a flat two-dimensional phase space expressed in Cartesian coordinates and for which the parameter ν denotes the diffusion constant. Note well that this expression has *no* formal prefactor, and with $\int p\,dq$ interpreted as a (Stratonovich) stochastic integral,[8] (22) defines, for each $\nu < \infty$, a completely unambiguous path integral over continuous phase space trajectories. Moreover, it can be shown[7] for a wide class of Hamiltonians that the limit exists and that the limit actually provides a solution to Schrödinger's equation.

6 Existence of a Flat Space Metric in Any Canonical Quantum Theory

There are several standard rules of quantization – such as those of Heisenberg and Schrödinger – that do not explicitly use a metric on a flat phase space in their construction. On the other hand, the fact that such rules do not use such a flat space metric does not mean that there is no such metric. In fact, perhaps surprisingly, there is *always* such a metric implicitly present, as we now proceed to demonstrate. Everyone would agree that any canonical quantization procedure leads to vectors in a Hilbert space and to canonical self adjoint operators Q and P that satisfy the Heisenberg commutation relation, $[Q,P] = i\hbar\mathbb{1}$. With the aid of these basic elements, we can always construct vectors of the form

$$|p,q;\psi\rangle \equiv e^{-iqP/\hbar}\, e^{ipQ/\hbar}\,|\psi\rangle\;, \tag{23}$$

for a general vector $|\psi\rangle$, in whatever representation is involved. In terms of $d|p,q;\psi\rangle \equiv |p+dp,q+dq;\psi\rangle - |p,q;\psi\rangle$, we next form the expression

$$\|\,\hbar d|p,q;\psi\rangle\,\|^2 - |\langle p,q;\psi|\,\hbar d|p,q;\psi\rangle|^2\;, \tag{24}$$

which is evidently quadratic in the differentials dp and dq. Granting minimal domain requirements, it readily follows that (24) becomes

$$\langle(\Delta Q)^2\rangle\, dp^2 + \langle(\Delta P\Delta Q + \Delta Q\Delta P)\rangle\, dp\,dq + \langle(\Delta P)^2\rangle\, dq^2\;, \tag{25}$$

where in this expression, we have used the notation $\langle(\cdot)\rangle \equiv \langle p,q;\psi|(\cdot)|p,q;\psi\rangle$, and $\Delta Q \equiv Q - \langle Q\rangle$, etc. It is clear that the given expression generates a metric on phase space, and since the metric coefficients $\langle(\Delta Q)^2\rangle$, etc., are constants, then the metric characterizes a *flat* two dimensional phase space. This feature is embedded into any version of canonical quantization one can imagine. A flat space metric may not be explicitly *used* in arriving at your favorite quantization, but it is nevertheless present in the multitude of consequences that follow from the quantization itself.

Moreover, for many states of physical interest, e.g., a typical harmonic oscillator ground state, it follows that the coefficients appearing in (25) are all proportional to $\hbar$. Thus, as seems entirely natural, the phase space metric induced by any quantization procedure is fully of a quantum character itself.

7 Metrical Quantization

It is clear that one may sometimes choose a different set of axioms to derive a common body of knowledge, and that dictum certainly applies to the process of quantization, as exemplified by the approaches of Heisenberg, Schrödinger, and Feynman. We have observed that traditional canonical quantization procedures do not make use of a flat phase space metric, but that such a metric nevertheless arises in a natural way within any quantization scheme. Suppose we took that metric, normally a secondary feature, and promoted it to primary status, namely postulating its existence as one of our axioms of quantization, indeed, as our *primary* axiom of quantization. The existence of a flat phase space metric and its elevation to prominence has another advantage for, as we have observed above, a metric we have called the "shadow metric" is an all but essential ingredient in order to keep track of the physics of the mathematical expressions under consideration.

Thus as our first axiom in the program of metrical quantization for canonical systems, we postulate the existence of a flat space metric $d\sigma(p,q)^2$ on our classical phase space which for the sake of the present discussion we take in the form

$$d\sigma(p,q)^2 = \hbar(dp^2 + dq^2) \ , \qquad (26)$$

a relation that evidently characterizes a flat two dimensional phase space expressed in Cartesian coordinates. Observe that this is a quantum object since it is proportional to $\hbar$. In the formal classical limit in which $\hbar \to 0$, it follows that the metric vanishes; this property ensures that the metric is strictly quantum in character and not part of the usual classical theory as conventionally interpreted. On the other hand, since $\hbar > 0$ in the real world, we can still use (26) as our shadow metric to give physics to our expressions.

Quantum mechanics *needs to know* what system is under discussion, it needs to know so that it can determine the energy spectrum, for example, of one specific system and not that of any other system. Thus it seems absolutely natural that we must combine the metric directly with the quantization formalism so that in fact the quantization will "know" what system is under consideration.

In an earlier section we discussed a continuous time regularization as a mathematical device to provide a regularization to an otherwise ill defined formal phase space path integral. Now, as we look at that expression once again, we see that the regularization itself involved the very same flat phase space metric we have in mind. That is, the given regularization accomplished the further goal of adding to the formalism an appropriate metric as the "keeper of the physics". In short, *not only does the continuous time regularization of the phase space path integral presented in (21) or (22) solve the issue of giving a proper mathematical meaning to the path integral, it* **simultaneously** *solves the problem of maintaining a proper physical interpretation throughout the quantization procedure.*

With this discussion as background, we are in a position to formalize the two axioms of Metrical Quantization.[5] First we adopt the phase space metric (26). The second, and final step, is to say how we use this metric and that is to postulate that the propagator is defined by

$$\langle p'', q''|e^{-i\mathcal{H}T/\hbar}|p', q'\rangle \equiv \lim_{\nu\to\infty} 2\pi e^{\nu T/2\hbar} \int e^{(i/\hbar)\int[p\,dq - H(p,q)\,dt]}\, d\mu_W^\nu(p,q)\,, \quad (27)$$

where the Brownian motion paths are carried by the metric (26) in the sense of (21). This is all that is needed since, according to the Gel'fand-Naimark-Segal reconstruction theorem,[9] everything quantum in character about (27) is a consequence of the functional form of the right hand side. In particular, and as the notation suggests, it follows that the right hand side generates the propagator in a canonical coherent state representation based on self adjoint Heisenberg operators Q and P that satisfy $[Q, P] = i\hbar\mathbb{1}$. Morevover, in the present case it follows that the fiducial vector necessarily satisfies the relation $(Q + iP)|0\rangle = 0$. Additionally, it also follows that the Hamiltonian operator $\mathcal{H}$ is related to the classical Hamiltonian $H(p, q)$ by *antinormal ordering*; that is, the very act of introducing the regularization and making the formal phase space path integral well defined has removed all ambiguity including the usual factor ordering ambiguity. In short, the very form of the regularization has already implicitly *chosen* an ordering prescription for the quantization!

8 Covariance under Canonical Coordinate Transformations

The fact that (27) [or more loosely, (21)] is well defined as an integral means that we are free to change the variables of integration as usual. Let us discuss some of the changes we might make and see how they effect the form of the integrand.

From a classical point of view, recall that a classical canonical transformation associates the old and new canonical coordinates via a one-form such as

$$\overline{p}\,d\overline{q} = pdq + dF(\overline{q}, q) \ , \tag{28}$$

where $\overline{p}, \overline{q}$ denote the new canonical coordinates while p, q denote the old canonical coordinates. Furthermore, the Hamiltonian transforms as a scalar in classical mechanics, which means that

$$\overline{H}(\overline{p}, \overline{q}) \equiv H(p(\overline{p}, \overline{q}), q(\overline{p}, \overline{q})) = H(p, q) \ . \tag{29}$$

In classical mechanics, such expressions make sense because one is dealing with smooth functions of time that are both continuous and have continuous derivatives.

However, in the well defined path integral of (27) we are dealing with Brownian motion paths $p(t), q(t)$ which are continuous paths but which are *nowhere* differentiable! This fact means that expressions like $\int pdq$ cannot be defined as ordinary integrals, but rather they should be interpreted as *stochastic integrals*.[8] For the case at hand, we may adopt a Riemann sum prescription based on the mid-point rule of definition, namely

$$\int pdq \equiv \lim \Sigma \, \tfrac{1}{2}(p_{l+1} + p_l)(q_{l+1} - q_l) \ , \tag{30}$$

in the limit that the lattice spacing goes to zero. This is the so-called Stratonovich prescription, which has the virtue that it satisfies the rules of the ordinary calculus[8] despite the fact that the paths $p(t)$ and $q(t)$ are nowhere differentiable![a]

This correspondence with the ordinary rules of calculus means that in the continuous time regularized path integral the classical action will transform just as it does in the classical theory. The Wiener measure will also undergo a coordinate transformation, but it will still describe Brownian motion on a *flat* two dimensional phase space, although in general it will now do so in curvilinear coordinates rather than the initial Cartesian coordinates. Therefore under a canonical coordinate transformation (27) becomes[7,10]

$$\langle \overline{p}'', \overline{q}'' | e^{-i\mathcal{H}T/\hbar} | \overline{p}', \overline{q}' \rangle = \lim_{\nu \to \infty} \mathcal{M} \int e^{(i/\hbar)\int [\overline{p}\,d\overline{q} + d\overline{G}(\overline{p}, \overline{q}) - \overline{H}(\overline{p}, \overline{q})\,dt]} \, d\overline{\mu}_W^{\nu}(\overline{p}, \overline{q}) \tag{31}$$

[a]The commonly used alternative rule, which is the so-called Itô prescription, and given by $\lim \Sigma \, p_l(q_{l+1} - q_l)$, has other virtues but it generally does not satisfy the ordinary rules of calculus. Due to our initial choice of coordinates, we are free to choose either the Stratonovich or the Itô rule with which to make our coordinate transformations since both rules lead to the same result in Cartesian coordinates for the integral in question.

In this expression, the function $\overline{G}(\overline{p},\overline{q})$ [which arises from $F(\overline{q},q)$] appears as a total derivative and so amounts, in effect, to the addition of a phase factor to each of the coherent states.

Here, at last, we can see the virtue of the present formulation very clearly. Although the coordinate form of the Hamiltonian may well have changed, and thus the physical meaning of the Hamiltonian cannot be read directly from its functional form in the non-Cartesian coordinates, the phase space path integral nevertheless still refers to the original physical system and what ensures this is the fact that the metric on flat space needed to carry the Brownian motion paths acts as a shadow metric maintaining control over the proper physical interpretation. Moreover, we can also see that the quantum Hamiltonian $\mathcal{H}$ has in no way changed so that indeed the original physical system is still under discussion. Observe also that the result is still expressed in terms of the original coherent states – apart from a trivial phase factor – the only difference being the way in which they are parametrized. In particular,

$$|\overline{p},\overline{q}\rangle \equiv e^{-i\overline{G}(\overline{p},\overline{q})/\hbar}\, e^{-iq(\overline{p},\overline{q})P/\hbar}\, e^{ip(\overline{p},\overline{q})Q/\hbar}\, |0\rangle\ . \tag{32}$$

These states still enjoy a resolution of unity, but which is now expressed in the form

$$\mathbb{1} = \int |\overline{p},\overline{q}\rangle\langle\overline{p},\overline{q}|\, d\overline{p}d\overline{q}/(2\pi\hbar)\ . \tag{33}$$

In summary, and in a manner of speaking, our main point is that quantum phase space may be said to arise from classical phase space simply by the introduction of a metric. Thus, viewed in the right way, perhaps classical and quantum mechanics are closer to each other than is commonly believed!

References

1. J.R. Klauder and B.-S. Skagerstam, *Coherent States* (World Scientific, Singapore, 1985).
2. R.P. Feynman, *Phys. Rev.* **84**, 108 (1951).
3. J.R. Klauder, *Ann. Phys. (NY)* **11**, 123 (1960).
4. J.R. Klauder, *Acta Physica Austriaca, Suppl.* **XXII**, 3 (1980).
5. J.R. Klauder, in *Quantum Future*, P. Blanchard and A. Jadczyk, eds. (Springer-Verlag, Berlin, 1999), p. 129.
6. J.R. Klauder, *Foundations of Physics* **27**, 1467 (1997).
7. I. Daubechies and J.R. Klauder, *J. Math. Phys.* **26**, 2239 (1985).
8. N. Ikeda and S. Watanabe, *Stochastic Differential Equations and Diffusion Processes* (North-Holland, Amsterdam, 1981).

9. G.G. Emch, *Algebraic Methods in Statistical Mechanics and Quantum Field Theory* (Wiley-Interscience, New York, 1972).
10. J.R. Klauder, *Ann. Phys. (NY)* **188**, 120 (1988).

YANG-MILLS-VLASOV SYSTEM ON A CURVED SPACETIME

N. NOUTCHEGUEME, P. NOUNDJEU,
S.B. TCHAPNDA AND D. TEGANKONG
University of Yaounde I, Cameroon
E-mail: nnoutch@uycdc.uninet.cm

We prove a local in time existence theorem for the Cauchy problem for the Yang-Mills system in temporal gauge, with current generated both by a distribution function that satisfies the Vlasov equation, and an unknown Yang-Mills charge density, subject to a conservation equation. We prove a global in time existence theorem in the case of massless particles.

1 Introduction

In kinetic theory of relativistic plasmas, the Yang-Mills-Vlasov system governs the collisionless evolution of particles with a non-abelian charge: the Yang-Mills charge. A classical example of such a plasma, also called "quagma", given in chromodynamics is that of the "quark-gluon" plasma of primordial particles that seems to have existed in the early universe. In this paper, we suppose that the Yang-Mills field, that is subject to the Yang-Mills equations, derives from a potential we take in temporal gauge, and that the current that is the source of the Yang-Mills field is generated both by a distribution function, subject to the Vlasov equation, and by an unknown charge density, subject to a conservation equation, which is a consequence of the conservation of the current.

We prove for the coupled system, that is a non-linear integro-differential system, the existence of local in time solutions, in appropriate weighted functional spaces, that impose no decay at spatial infinity. We apply this result in the case where the curved spacetime is the Einstein cylinder, to prove, using the conformal transformations, a global in time existence theorem on the Minkowski spacetime, in the case of massless particles such as photons or primordial particles. Our results generalize previous ones,[2,3] that study the case of a zero charge density.

2 Notations, Hypotheses and Equations

A greek index ranges from 0 to 3 and a latin index from 1 to 3 unless otherwise specified. We adopt the Einstein summation convention.

- Background manifold: a C^∞, 4-dimensional, time-oriented spacetime (V, g), endowed with a metric g of hyperbolic signature (+,-,-,-). We have: $V = \mathbb{R} \times S$, S a 3-dimensional manifold; the local coordinates are (x^α) with x^0 or t on $\mathbb{R}$ (time), (x^i) on S (space); $S_t := \{t\} \times S$, spatial hypersurface; $\mathbb{R} \times \{x^i\}$ temporal lines. We take: $g = \alpha^2 dt^2 - \widetilde{g_{ij}} dx^i dx^j$ where $\widetilde{g_{ij}} := -g_{ij}$, (g_{ij}) negative definite. The metrics $g_t := (\widetilde{g_{ij}})$ induced by g on S_t are properly riemannian, and uniformly equivalent to a metric $\overline{g}$ on S, with injectivity radius $\delta > 0$. We suppose that there exists 4 constants, $A_1, A_2, B_1, B_2 > 0$, such that, in every geodesic ball of radius δ of (S_t, g_t) one has,

$$A_1 \sum_{i=1}^{3} (\xi^i)^2 \leq \tilde{g}_{ij} \xi^i \xi^j \leq A_2 \sum_{i=1}^{3} (\xi^i)^2; \; B_1 \leq \alpha \leq B_2.$$

- To define norms of tensors, we associate to g the elliptic metric, $\gamma = \alpha^2 dt^2 + \widetilde{g_{ij}} dx^i dx^j$

- We suppose the covariant derivatives of the curvature tensors of g and $\overline{g}$ bounded.

- $A : V \to \mathcal{G}$, denotes the Yang-Mills potential, a 1-form on V with values in $\mathcal{G}$, Lie algebra of a Lie group G. $\mathcal{G}$ is a N-dimensional vector space on $\mathbb{R}$, with fixed basis (ε_a), and that is endowed with: the Lie brackets [,]; and an Ad-invariant scalar product denoted "."; locally $A = (A_\mu) = (A^a_\mu)$.

- $F : V \to \mathcal{G}$ denotes the curvature of A, that is the Yang-Mills field associated to A. F is an antisymmetric 2-form; locally,

$$F = (F_{\lambda\mu}) = (F^a_{\lambda\mu}) \; ; \; F^a_{\lambda\mu} = \nabla_\lambda A^a_\mu - \nabla_\mu A^a_\lambda + [A_\lambda, A_\mu], \tag{1}$$

where ∇ is the covariant derivative in g and $[A_\lambda, A_\mu]^a = C^a_{bc} A^b_\lambda A^c_\mu$ with C^a_{bc} the structure constants of G. We take A in temporal gauge, which means that $A_0 = 0$. Hence, (1) gives,

$$\partial_0 A_i = F_{0i} \; . \tag{2}$$

- $f : \mathbb{P} \equiv T(V) \times \mathcal{G} \to \mathbb{R}^+$ denotes the distribution function of the particles; $\mathbb{P}$ has local coordinates $(x^\alpha, p^\alpha, q^a)$. The particles have a rest mass $m \geq 0$; a Yang-Mills charge $q : V \to \mathcal{G}$, C^∞ function on V with constant norm e in $\mathcal{G}$; and an unknown charge density that is a scalar funtion $\rho : V \to \mathbb{R}^+$. We also have $g_{\alpha\beta} p^\alpha p^\beta = m^2$ and then,

$$p^0 = \alpha^{-1}(\widetilde{g_{ij}} p^i p^j + m^2) \; ; \; O : q.q = \mid q \mid^2 = e^2 \; . \tag{3}$$

The trajectories of the particles are in the subfiber $\mathbb{P}_{m,e} := \mathbb{R} \times TS \times O$ of $\mathbb{P}$ defined by (3), and whose local coordinates are (x^0, x^i, p^i, q^A) with the index

$A = 1, 2, \ldots, N-1$. We have, on the trajectories that are paths in $\mathbb{P}$,

$$\frac{dx^\alpha}{ds} = p^\alpha \ ; \ \frac{dp^\alpha}{ds} = -\Gamma^\alpha_{\lambda\mu} p^\lambda p^\mu + p^\beta q.F_\beta{}^\alpha \ ; \ \frac{dq^a}{ds} = -C^a_{bc} p^\alpha A^b_\alpha q^c \tag{4}$$

(Γ connection of g). The trajectories then have tangent vector

$$Y = Y(F, A) := (p, P, Q) \tag{5}$$

where

$$P^\alpha := -\Gamma^\alpha_{\lambda\mu} p^\lambda p^\mu + p^\beta q.F_\beta{}^\alpha \quad ; \quad Q^a := -C^a_{bc} p^\alpha A^b_\alpha q^c .$$

The Vlasov equation in f expresses the conservation of the number of the collisionless particles and writes as $L_Y f = 0$ namely, on $\mathbb{P}_{m,e}$,

$$L_Y f := p^\alpha \frac{\partial f}{\partial x^\alpha} + P^i \frac{\partial f}{\partial p^i} + Q^A \frac{\partial f}{\partial q^A} = 0 \tag{6}$$

- $J : V \to \mathcal{G}$ denotes the Yang-Mills current. We have $J = J(f, \rho) = (J^\beta) = (J^{\beta,A})$, where we take

$$J^{\beta,A}(x) = \int_{\mathbb{R}^3 \times O} p^\beta q^A f(x, p, q) \omega_p \omega_q - \rho(x) U^\beta q^A(x), \tag{7}$$

with $\omega_p = \frac{|g|^{\frac{1}{2}}}{p_0} dp^1 dp^2 dp^3$, ω_q being the canonical volume element of the sphere O of $\mathcal{G}$, and U a given vector field on V satisfying $U^\alpha \nabla_\alpha U^\beta = 0$; $U^\alpha U_\alpha = 1$; $U^0 > 0$. The Yang-Mills equations in F write,

$$\widehat{\nabla_\alpha} F^{\alpha\beta} = J^\beta, \tag{8}$$

where $\widehat{\nabla_\alpha} = \nabla_\alpha + [A_\alpha, .]$ denotes the gauge covariant derivative. The identity $\widehat{\nabla_\alpha}\widehat{\nabla_\beta} F^{\alpha\beta} = 0$ implies, given (8), that the current J must satisfy the conservation law $\widehat{\nabla_\alpha} J^\alpha = 0$. But one shows that, if f satisfies the Vlasov equation (6), then the integral term in the r.h.s of (7) is gauge covariant divergence free. Then, ρ must satisfy,

$$\nabla_\alpha (\rho U^\alpha) = 0 \ . \tag{9}$$

We study the Cauchy problem for the system (2),(8),(9),(6) in the variables $u = (A, F, \rho, f)$.

To determine the six unknowns F^{0i}, F_{ij} that stand for the electric and the magnetic parts of F, we add to (8) a part of the Bianchi identities in F,

$$\widehat{\nabla_\alpha} F_{\beta\gamma} + \widehat{\nabla_\beta} F_{\gamma\alpha} + \widehat{\nabla_\gamma} F_{\alpha\beta} = 0. \tag{10}$$

We take Eq.(8) for $\beta = i$, choose (10) for $\alpha = 0$, $\beta = j$, $\gamma = k$, and have for F,

$$\begin{cases} LF = d, \text{with } LF = (\nabla_\alpha F^{\alpha i}; \nabla_0 F_{ij} + \nabla_i F_{j0} + \nabla_j F_{0i}); \\ d = d(A, F, \rho, f) := (-[A_j, F^{ij}] + J^i(f, \rho); [A_i, F_{0i}] - [A_j, F_{0i}]). \end{cases} \tag{11}$$

Denote by $(\sum)$ the coupled system (2),(11),(9), (6) in u. The Cauchy data on S and $\widehat{S} := S \times \mathbb{R}^3 \times O$, are: 1) for A, a 1-form $a : S \to \mathcal{G}$; 2) for F, a couple $\Phi = (E_0, B_0)$ where $E_0 : S \to \mathcal{G}$ is a vector and $B_0 : S \to \mathcal{G}$ a 2- form; 3) for ρ a function $\rho_0 : S \to \mathbb{R}^+$; 4) for f a function $\varphi : \widehat{S} \to \mathbb{R}^+$. A solution (F^{0i}, F_{ij}) of (11) will solve the remaining equations (10) for $\alpha = i, \beta = j, \gamma = k$ and (8) for $\beta = 0$ if we take $B_0 = \overline{d}a$ and if the Cauchy data satisfy the constraint

$$\widehat{\overline{\nabla}_i} E_0^i + \int_{\mathbb{R}^3 \times O} p^0 q \varphi \omega_p \omega_q - \rho_0 U^0 q = 0. \tag{12}$$

Here, $\overline{d}$ and $\widehat{\overline{\nabla}}$ are the exterior differential and the gauge covariant derivative on $(S, \overline{g})$; (12) can be solved as p.d.e in (E_0^i) for φ and ρ_0 given.[1]

3 Linear System Associated to Σ

Put $\sim$ on fixed quantities and deduce from (Σ) the **linear** system: $(\widetilde{\Sigma}) : LF = \widetilde{d}; \partial_0 A_i = \widetilde{F_{0i}};\ \nabla_\alpha(\rho U^\alpha) = 0;\ L_{\widetilde{Y}} f = 0$ with $\widetilde{d} = d(\widetilde{A}, \widetilde{F}, \widetilde{\rho}, \widetilde{f})$; $\widetilde{Y} = Y(\widetilde{F}, \widetilde{A})$.

We solve $(\widetilde{\Sigma})$ on $V_T := [0, T] \times S$ and $\widehat{V_T} := V_T \times \mathbb{R}^3 \times O$, $T > 0$. If $X \subset Z$, $C_0^\infty(X)$ denotes the space of restrictions to X of functions or tensors with compact support in Z, set $C_0^\infty(T) = [C_0^\infty(V_T)]^3 \times C_0^\infty(\widehat{V_T})$; $\widetilde{u} = (\widetilde{A}, \widetilde{F}, \widetilde{\rho}, \widetilde{f})$; $u_0 = (a, \Phi, \rho_0, \varphi)$; $\mathcal{C}_0 = \{\overline{g}_{ij} p^i p^j > c > 0\}$.

Proposition 1 *Let $u_0 \in [C_0^\infty(S)]^3 \times C_0^\infty(\widehat{S})$. Then if $\widetilde{u} \in C_0^\infty(T)$, the linear system $(\widetilde{\Sigma})$ has a unique solution $u \in C_0^\infty(T)$ taking the Cauchy data u_0 on S, for every $T > 0$ if $m > 0$, and, if $m = 0$, for $T > 0$ small, if $supp\, \varphi \subset \mathcal{C}_0$.*

Proof: Use classical results on linear hyperbolic systems for the equations in A, F, ρ. Now $L_{\widetilde{Y}} f = 0$ writes $\frac{df}{d\tau} = 0$ on the trajectories of $\widetilde{Y}$, given by (4), and that integrate, using classical results on differential systems; f then expresses directly in terms of φ.

4 Functional Spaces and A Priori Estimates (Localization)

Let $B = B_\delta$ be a geodesic ball of radius $\delta > 0$ of $(S, \overline{g})$. Define $\Omega_t \subset \mathbb{R} \times B_\delta$ by

$$\Omega_t = \left\{ (x^0, x^i), 0 \leq x^0 \leq t; K(T - x^0) \geq [\sum_{i=1}^{3} (x^i)^2]^{\frac{1}{2}} \right\}$$

where $K > 0$ and $0 \leq t \leq \frac{T}{2} \leq \frac{\delta}{2K}$. Set $\omega_\tau = \Omega_t \cap S_\tau$; $\widehat{\omega_\tau} = \omega_\tau \times \mathbb{R}^3 \times O$; $\widehat{\Omega_t} = \Omega_t \times \mathbb{R}^3 \times O$; $\widehat{B} = B \times \mathbb{R}^3 \times O$; $0 \leq \tau \leq t$. Let

- $E^s_k(\widehat{\Omega_t})$, $k \in \mathbb{R}^+$, $s \in \mathbb{N}$, be the completion of $C^\infty_0(\widehat{\Omega_t})$ in the norm $\| f \|_{E^s_k(t)} = sup_{0 \leq \tau \leq t} \| f \|^\tau_{s,k}$, where

$$\| f \|^\tau_{s,k} = [\sum_{|I| \leq s} y_{I,f}(\tau)]^{\frac{1}{2}} \tag{13}$$

with

$$y_{I,f}(\tau) = \int_{\widehat{\omega_\tau}} (p^0)^{2k+2(\hat{I}+\check{I})+1} \mid D^I f \mid^2 \theta_\tau \ .$$

- $E^s(\Omega_t)$ be the completion of $C^\infty_0(\Omega_t)$ in the norm $\| U \|_{E^s(t)} = sup_{0 \leq \tau \leq t} \| U \|^\tau_s$, where $\| U \|^\tau_s = [\sum_{l \leq s} z_{l,U}(\tau)]^{\frac{1}{2}}$, with $z_{l,U}(\tau) = \int_{\omega_\tau} \mid \nabla^l U \mid^2 \mu_\tau$ with $U = A$, F, ρ and where

$$\mid \nabla^l U \mid^2 = \gamma_{\lambda_1 \mu_1} \cdots \gamma_{\lambda_l \mu_l} \gamma_{\alpha_1 \beta_1} \cdots \gamma_{\alpha_k \beta_k} \nabla^{\lambda_1} \ldots \nabla^{\lambda_l} U^{\alpha_1 \ldots \alpha_k} . \nabla^{\mu_1} \ldots \nabla^{\mu_l} U^{\beta_1 \ldots \beta_k} \tag{14}$$

$$\begin{aligned} &\widetilde{E^s_k}(\widehat{\Omega_t}) := \{ f \in E^{s-1}_k(\widehat{\Omega_t}) \ ; \\ &\| f \|_{\widetilde{E^s_k}(t)} := \| f \|_{E^{s-1}_k(t)} + Esssup_{0 \leq \tau \leq t} [\sum_{|I| = s} y_{I,f}(\tau)]^{\frac{1}{2}} < +\infty \} \end{aligned} \tag{15}$$

$$\begin{aligned} &\widetilde{E^s}(\Omega_t) := \{ U \in E^{s-1}(\Omega_t) \ ; \\ &\| U \|_{\widetilde{E^s}(t)} := \| U \|_{E^{s-1}(t)} + Esssup_{0 \leq \tau \leq t} [z_{s,U}(\tau)]^{\frac{1}{2}} < +\infty \} \end{aligned} \tag{16}$$

$$\begin{aligned} \mathcal{E}^s_k(\Omega_t) &:= [E^s(\Omega_t)]^3 \times E^s_k(\widehat{\Omega_t}) \ ; \\ \widetilde{\mathcal{E}^s_k}(\Omega_t) &:= [\widetilde{E^s}(\Omega_t)]^3 \times \widetilde{E^s_k}(\widehat{\Omega_t}) \ ; \\ \mathcal{H}^s_k(\omega_0) &:= [H^s(\omega_0)]^3 \times H^s_k(\widehat{\omega_0}) \ , \end{aligned}$$

where $H^s_k(\widehat{\omega_0})$ is the completion of $C^\infty_0(\widehat{\omega_0})$ in the norm

$$\| \varphi \|_{H^s_k} := [\sum_{|I|\leq s} \int_{\widehat{\omega_0}} (p^0)^{2k+2(\hat{I}+\check{I})+1} \mid D^I \varphi \mid^2 \theta_0]^{\frac{1}{2}} \ . \tag{17}$$

If $\mathcal{B}_\delta$ [resp. $\widehat{\mathcal{B}_\delta}$] is a cover of S [resp. $\widehat{S}$] by the balls B (resp. $\widehat{B}$), set

$$H^{l.u}_s(S) = \{v, \| v \|_{H^{l.u}_s(S)} := sup_{B\in\mathcal{B}_\delta}[\sum_{l\leq s} \int_B \mid \overline{\nabla}^l v \mid^2 \mu_0]^{\frac{1}{2}} < +\infty\} \tag{18}$$

$$\begin{aligned} H^{l.u}_{k,s}(\widehat{S}) &= \{\varphi, \| \varphi \|_{H^{l.u}_{k,s}(\widehat{S})} \\ &:= sup_{\widehat{B}\in\widehat{\mathcal{B}_\delta}}[\sum_{l\leq s} \int_{\widehat{B}} (p^0)^{2k+2(\hat{l}+\check{l})+1} \mid \widehat{\overline{\nabla}}^l \varphi \mid^2 \theta_0]^{\frac{1}{2}} < +\infty\} \end{aligned} \tag{19}$$

$\mathcal{H}^{l,u}_{k,s}(\widehat{S}) := [H^{l.u}_s(S)]^3 \times H^{l.u}_{k,s}(\widehat{S})$; $\widetilde{\mathcal{E}}^s_{loc}(V_t) := [\widetilde{E}^s_{loc}(V_t)]^3 \times \widetilde{E}^s_{k,loc}(\widehat{V}_t)$ where, by definition,

$$U \in \widetilde{E}^s_{loc}(V_t) \Leftrightarrow U \in \widetilde{E^s}(\Omega^j_t \ \ ; \ \ f \in \widetilde{E}^s_{k,loc}(\widehat{V}_t) \Leftrightarrow f \in \widetilde{E^s_k}(\widehat{\Omega^j_t}) \ \ , \ \ \forall j \in J \ ,$$

if $V_t = \bigcup_{j\in J} \Omega^j_t$; $\widehat{V}_t = \bigcup_{j\in J} \widehat{\Omega^j_t}$.

In the above definitions, omit the dot "." in (14) if $U = \rho : V_T \to \mathbb{R}$; in (13), (17), (19), $\hat{I}$, $\check{I}$ and $\hat{l}$, $\check{l}$ are the number of derivatives in p^i and q^A; $\theta_\tau = \mu_\tau \wedge \omega_p \wedge \omega_q$, μ_τ induced by $\mu(g)$ on ω_τ; in (15), (16), $s \geq 1$ and the derivatives of order s are in the sense of distributions. $\widetilde{\mathcal{E}}^s_k(\Omega_t)$ is the dual of a Banach space; in (18), (19), $\overline{\nabla}$ and $\widehat{\overline{\nabla}}$ are the covariant derivatives in the sense of distributions on $(S, \overline{g})$, $(\widehat{S}, \widehat{\overline{g}})$ where $\widehat{\overline{g}} = \overline{g}_{ij}dx^i dx^j + (p^0)^{-2}\overline{g}_{ij}\overline{D}p^i\overline{D}p^j + dS^2_0$ where $\overline{D}p^i = dp^i + \overline{\Gamma}^i_{jk}p^i dx^k$, $\overline{\Gamma}$ connection of $\overline{g}$, and ds^2_0 the canonical metric of O.

Proposition 2 *(A priori estimates) Let $A, F, \rho, d \in C^\infty_0(\Omega_t)$; $f, b \in C^\infty_0(\widehat{\Omega_t})$; $s \in \mathbb{N}; \tau \in [0,t]; C > 0$. If $LF = d; \partial_0 A_i = F_{0i}; \nabla_\alpha(\rho U^\alpha) = 0; L_Y f = b$, then :*
$\| F \|^\tau_s \leq C[\| F \|^0_s + \int_0^\tau \| d \|^{\tau'}_s d\tau'];\| A \|^\tau_s \leq C[\| A \|^0_s + \int_0^\tau \| F \|^{\tau'}_s d\tau']$;
$\| \rho \|^\tau_s \leq C \| \rho \|^0_s$;
$[\| f \|^\tau_{s,k} \leq C[\| f \|^0_{s,k} + \int_0^\tau \| b \|^{\tau'}_{s,k} d\tau']exp[C \int_0^\tau (1+ \| A \|^{\tau'}_s + \| F \|^{\tau'}_s)d\tau']$,
$s \geq 4$
$\| J \|^\tau_s \leq C[\| f \|^\tau_{s,k} + \| \rho \|^\tau_s]$, $k > \frac{3}{2}$
If $m = 0$, suppose in addition that $suppf \subset \{p^0 \geq k_0 > 0\}$.

Proof Multiply each equation by a suitable function in order to have by Leibnitz's formula, a divergence term on the l.h.s. and the products at the r.h.s.; integrate over Ω_τ, $\widehat{\Omega}_\tau$, taking K large; use Stokes theorem, Schwarz inequality, Sobolev embedding and multiplication theorems, and obtain the desired inequalities for $s = 0$, by Gronwall lemma. Conclude by induction on $s \in \mathbb{N}$.

5 The Local in Time Existence Theorem

Theorem 1 *(Existence and Uniqueness) Let $u_0 \in \mathcal{H}^{l.u}_{k,s}(S)$, where the parameters are such that $s > \frac{N}{2} + 5, k > \frac{3}{2}$, be the Cauchy data satisfying the constraints. Then there exists a number $t_0 > 0$ depending only on $\| u_0 \|_{\mathcal{H}^{l.u}_{k,s}(S)}$, such that, the Cauchy problem for the non-linear system (Σ) has a unique solution $u \in \widetilde{\mathcal{E}}^s_{k,loc}(V_{t_0})$, taking the Cauchy data u_0 on S.*

If $m = 0$, suppose in addition that $supp\varphi \subset \mathcal{C}_0$.

Proof First take $u_0 \in C^\infty_0(0) := [C^\infty_0(\omega_0)]^3 \times C^\infty_0(\widehat{\omega}_0)$; construct a solution u in $\mathcal{E}^s_k(\Omega_t)$, where t is a decreasing function of $\| u_0 \|_{\mathcal{H}^s_k(\omega_0)}$, by iterating the C^∞ solutions of $(\widetilde{\Sigma})$ given by Proposition 1, and using a priori estimates (Proposition 2). Next take $u_0 \in \mathcal{H}^s_k(\omega_0)$ and approach u_0 in $\mathcal{H}^s_k(\omega_0)$ by a sequence $(u_n) \subset C^\infty_0(0)$. Use this convergence to obtain the unique solution $u \in \widetilde{\mathcal{E}^s_k}(\Omega_t)$, by iterating this time the solutions of the Cauchy problem with data u_n on ω_0. Finally take $u_0 \in \mathcal{H}^{l.u}_{k,s}(S)$. Consider the cover $\mathcal{B}_\delta$ of S by the balls $B_j, j \in J$. Each B_j is the basis on S of the truncated cone $\Omega^j_{t_j} \subset [0, t_j] \times B_j, t_j > 0$, that is the existence domain of the unique solution $u_j \in \widetilde{\mathcal{E}}^s_k(\Omega^j_{t_j})$ of the Cauchy problem whose data are the restriction of u_0 on B_j. The family (t_j) is bounded from below by a number $t_0 > 0$, that depends only on $\| u_0 \|_{\mathcal{H}^{l.u}_{k,s}(S)}$ and the family $(\Omega^j_{t_0}), j \in J$, is a cover of V_{t_0}. The desired solution u is then defined on V_{t_0} by $u = u_j$ on $\Omega^j_{t_0}, j \in J$.

6 The Global in Time Existence Theorem in the Case of Massless Particles

Recall that if $\widetilde{g}$ is another metric on V, then the two metrics g and $\widetilde{g}$ are said to be conformal each other if there exists a function $\Omega > 0, C^1$ on V such that $\widetilde{g}_{\alpha\beta} = \Omega^{-2} g_{\alpha\beta}$; hence; $\widetilde{g}^{\alpha\beta} = \Omega^2 g^{\alpha\beta}$.

A direct calculation shows that the corresponding connections Γ and $\widetilde{\Gamma}$

are related by

$$\Gamma^{\alpha}_{\lambda\mu} = \widetilde{\Gamma}^{\alpha}_{\lambda\mu} + \Omega^{-1}(\delta^{\alpha}_{\lambda}\partial_{\mu}\Omega + \delta^{\alpha}_{\mu}\partial_{\lambda}\Omega - g^{\alpha\beta}g_{\lambda\mu}\partial_{\beta}\Omega) \tag{20}$$

We make the following identifications between covariant tensors on (V, g) that is 4-dimensional, on the one hand, and the corresponding tensors on $(V, \widetilde{g})$, on the other hand, that we denote with $\sim$:

$$\widetilde{A}_{\alpha} = A_{\alpha} \ ; \ \widetilde{F}_{\alpha\beta} = F_{\alpha\beta} \ ; \ \widetilde{p_{\alpha}} = p_{\alpha} \ ; \widetilde{U}_{\alpha} = U_{\alpha} \ ,$$

hence,

$$\widetilde{A}^{\alpha} = \Omega^2 A^{\alpha}; \widetilde{F}^{\alpha\beta} = \Omega^4 F^{\alpha\beta}; \widetilde{p}^{\alpha} = \Omega^2 p^{\alpha}; \widetilde{U}^{\alpha} = \Omega^2 U^{\alpha}; \widetilde{\omega}_{\widetilde{p}} = \Omega^2 \omega_p \ . \tag{21}$$

Now $\widetilde{f}$, $\widetilde{\rho}$ being a distribution function and a charge density over $(V, \widetilde{g})$ define the corresponding quantities on (V, g) by,

$$f(x^{\lambda}, p^{\alpha}, q^{a}) = \widetilde{f}(x^{\lambda}, \widetilde{p}^{\alpha}, q^{a}); \ \rho = \Omega^{-2}\widetilde{\rho} \ . \tag{22}$$

Then a straightforward calculation gives, using (20), (21), (22), (7),

$$\widetilde{\widehat{\nabla}}_{\alpha}\widetilde{F}^{\alpha\beta} = \Omega^4 \widehat{\nabla}_{\alpha}F^{\alpha\beta} \ ; \ \widetilde{\nabla}_{\alpha}(\widetilde{\rho}\widetilde{U}^{\alpha}) = \Omega^4 \nabla_{\alpha}(\rho U^{\alpha}) \ ; \ \widetilde{J}(\widetilde{f}, \widetilde{\rho}) = \Omega^4 J(f, \rho) \ ,$$

$$L_{\widetilde{Y}}\widetilde{f} = \Omega^2 L_Y f - p_{\lambda}p^{\lambda}g^{\alpha\beta}\partial_{\beta}\Omega\frac{\partial f}{\partial p^{\alpha}}; \partial_0\widetilde{A_i} = \partial_0 A_i \ , \tag{23}$$

where $\widetilde{\widehat{\nabla}}$ denotes the gauge covariant derivative in $(V, \widetilde{g})$ and $\widetilde{Y}$ the vector on $(V, \widetilde{g})$ analogous to Y on (V, g) and defined by (5).

Now it is well known that the Minkowski spacetime $(\mathbb{R}^4, \eta)$ transforms conformally into a bounded open subset of the Einstein cylinder $(\mathbb{R} \times S^3; g_0)$.

Theorem 2 *Let $m = 0$ and let $u_0 = (a, \Phi, \rho_0, \varphi)$ be Cauchy data on $\mathbb{R}^3$ and $\widehat{\mathbb{R}^3} \equiv \mathbb{R}^6$ that can identify to $\overline{u_0} = (\overline{a}, \overline{\Phi}, \overline{\rho_0}, \overline{\varphi}) \in \mathcal{H}^{l.u}_{k,s}(S^3)$, $s > \frac{N}{2} + 5$; $k > \frac{3}{2}$ with $supp\overline{\varphi} \subset \mathcal{C}_0$, $\overline{u}_0$ sufficiently small in norm and satisfying the constraints. Then the Cauchy problem for the Yang-Mills-Vlasov system has a global solution on the Minkowski spacetime taking the Cauchy data u_0 on $\mathbb{R}^3$ and $\widehat{\mathbb{R}}^3$.*

Proof Take $(V, g) = (\mathbb{R} \times S^3, g_0)$, the Einstein cylinder; consider the quantities with $\sim$ in (23) as defined on $\mathbb{R}^4$ through the canonical diffeomorphism that transforms conformally the Minkowski spacetime into a bounded subset $]-T, T[\times S^3$ of $(\mathbb{R} \times S^3, g_0)$, $T > 0$. Apply Theorem 1 that gives the existence of a solution for the Cauchy problem on $V_{t_0} = [0, t_0] \times S^3 \subset]-T, T[\times S^3$, use (23) and the fact that since $m = 0$, $p_{\lambda}p^{\lambda} = 0$, and hence $L_{\widetilde{Y}}\widetilde{f} = \Omega^2 L_Y f$, to obtain the desired result.

References

1. Y. Choquet-Bruhat, D. Christodoulou, *Existence of global solution of Yang-Mills-Higgs and spinor fields*, *Ann. Ec. Norm. Sup.* **14, 4**, 481–500 (1981).
2. Y. Choquet-Bruhat, N. Noutchegueme, *Système de Yang-Mills-Vlasov en jauge temporelle*, *Ann. Inst. Henri Poincare* **55**, 759–787 (1991).
3. Y. Choquet-Bruhat, N. Noutchegueme, *Solutions globales du système de Yang-Mills-Vlasov (masse nulle)*, *Comptes Rendus Acad. Sci. Paris* **311**, 785–787 (1990).

Parallel Sessions: Part III

Functional Analysis

Special Functions and Orthogonal Polynomials

ON GENERALIZED CONTINUOUS D SEMI-CLASSICAL HERMITE AND CHEBYCHEV ORTHOGONAL POLYNOMIALS OF CLASS ONE

E. AZATASSOU[1] AND M.N. HOUNKONNOU[1,2]

[1] *Unité de Recherche en Physique Théorique (URPT)*
Institut de Mathématiques et de Sciences Physiques (IMSP)
B.P. 613 Porto-Novo, Benin

[2] *Laboratoire de Physique Mathématique et de Mathématique Appliquée,*
Institut International de Recherche et de Formation (INIREF)
B.P. 2740 Cotonou, Benin
E-mail: hounkonnou@yahoo.fr

In this contribution, starting from the system of equations for recurrence coefficients generated by continuous D semi-classical Laguerre-Freud equations of class 1, we deduce the β constant recurrence relation coefficient γ_n leading to the generalized D semi-classical Hermite and Chebychev orthogonal polynomials of class 1. Various interesting cases are pointed out.

1 Introduction

The semi-classical D Laguerre-Freud equations of class 1 are given by[1,2,3,4,5],

$$\begin{cases} (a_2 - 2nc_3)\,(\gamma_n + \gamma_{n+1}) = \\ \qquad = 2\sum_{k=0}^{n-1}(c_3(\beta_k^2 + \beta_k\beta_n + \beta_n^2) + c_2(\beta_k + \beta_n) + c_1) \\ \qquad +4c_3\sum_{k=1}^{n-1}\gamma_k - (a_2\beta_n^2 + a_1\beta_n + a_0)\ , n \geq 2, \\ \\ [a_2 - (2n+1)c_3]\gamma_{n+1}\beta_{n+1} = \\ \qquad = \sum_{k=0}^{n}(c_3\beta_k^3 + c_2\beta_k^2 + c_1\beta_k + c_0) \\ \qquad +c_3\left[2\gamma_{n+1}n\beta_n + \sum_{j=0}^{n}\beta_j + 3\sum_{k=1}^{n}\gamma_k(\beta_k + \beta_{k-1})\right] \\ \qquad +c_2\left[(2n+1)\gamma_{n+1} + 2\sum_{k=1}^{n}\gamma_k\right] - (a_2\beta_n + a_1)\gamma_{n+1} \quad , n \geq 1, \end{cases} \tag{1}$$

as deduced from the Pearson differential equation

$$D(\rho\Phi) - \rho\Psi = 0, \tag{2}$$

where $\rho(x)$ is a positive weight function, $\Phi(x)$ and $\Psi(x)$ taking the form,

$$\Phi(x) = \sum_{k=0}^{3} c_k x^k, \tag{3}$$

$$\Psi(x) = \sum_{k=0}^{2} a_k x^k. \tag{4}$$

The coefficients γ_n and β_n in Eq.(1) verify the polynomial three term recurrence relation

$$P_{n+1}(x) = (x - \beta_n)P_n(x) - \gamma_n P_{n-1}, \quad P_0 = 1; \quad P_1 = x - \beta_0; \quad \gamma_1 \neq 0, \tag{5}$$

and satisfy the following initial conditions deduced from Eq.(1),

$$\begin{cases} (a_2 - 2c_3)(\gamma_1 + \gamma_2) & = 6c_3\beta_0^2 + 4c_2\beta_0 + 2c_1 \\ & \quad -(a_2\beta_1^2 + a_1\beta_1 + a_0), \\ a_2\gamma_1 & = -(a_2\beta_0^2 + a_1\beta_0 + a_0), \\ (a_2 - c_3)\gamma_1\beta_1 & = c_3\beta_0^3 + c_2\beta_0^2 + c_1\beta_0 + c_0 \\ & \quad +c_3(\beta_0 + \beta_1) + c_2\gamma_1 - (a_2\beta_0 + a_1)\gamma_1, \\ \beta_0 & = \frac{M_1}{M_0}. \end{cases} \tag{6}$$

where M_0 and M_1 are the moments of zero- and first-order, respectively.

Let us examine the particular situation when the constant c_3 in relation (3) giving $\Phi(x)$ is set to zero. In this case, one can provide the expression of the recurrence coefficients γ_{n+1} and β_{n+1} as follows, valid for $n \geq 1$,

$$\gamma_{n+1} = \tfrac{2nc_1 + 2nc_2\beta_n}{a_2} + \tfrac{2c_2}{a_2}(\textstyle\sum_{k=0}^{n-1}\beta_k) - \beta_n^2 - \tfrac{a_1}{a_2}\beta_n - \tfrac{a_0}{a_2} - \gamma_n, \tag{7}$$

$$\beta_{n+1} = \tfrac{\sum_{k=0}^{n}(c_2\beta_k^2 + c_1\beta_k + c_0) + 2c_2(\sum_{k=1}^{n}\gamma_k)}{a_2\gamma_{n+1}} + \tfrac{c_2}{a_2}(2n+1) - \beta_n - \tfrac{a_1}{a_2}. \tag{8}$$

To characterize the semi-classical Hermite or Chebychev orthogonal polynomials, let us assume that, in Eqs.(7) and (8), the coefficients β_n take a same constant value, $\beta_n = \beta$, for all positive integer $n \geq 1$. Eq.(7) then becomes

$$\gamma_{n+1} = \frac{(2c_1 + 4c_2\beta)}{a_2}n + \gamma_1 - \gamma_n, \quad n \geq 1, \tag{9}$$

$$\gamma_1 = -(\beta^2 + \frac{a_1}{a_2}\beta + \frac{a_0}{a_2}). \tag{10}$$

Applying the change of variable,

$$\delta = \frac{(2c_1 + 4c_2\beta)}{a_2}, \tag{11}$$

in (9) leads to the two following relations, obtained recursively,

$$\gamma_{2p} = p\delta, \tag{12}$$

$$\gamma_{2p+1} = p\delta + \gamma_1. \tag{13}$$

Let us now come back to the Eq.(8). Eq.(9) rewrites as

$$\gamma_{k+1} + \gamma_k = \delta k + \gamma_1 \quad , \forall k \geq 1, \tag{14}$$

so that

$$\sum_{k=1}^{2p} \gamma_k = p^2\delta + p\gamma_1, \tag{15}$$

$$\sum_{k=1}^{2p+1} \gamma_k = \delta(p^2 + p) + (p+1)\gamma_1. \tag{16}$$

Thus, Eq.(8), which gives the following relations, for all $p \geq 1$,

$$2a_2\beta = \frac{(2p+1)\alpha + 2c_2(p^2\delta + p\gamma_1)}{p\delta + \gamma_1} + c_2(4p+1) - a_1, \tag{17}$$

$$2a_2\beta = \frac{(2p+2)\alpha + 2c_2[(p^2+p)\delta + (p+1)\gamma_1]}{(p+1)\delta} + c_2(4p+3) - a_1, \tag{18}$$

where

$$\alpha = c_2\beta^2 + c_1\beta + c_0 \ , \tag{19}$$

leads to the coupled system of six equations

$$c_2\delta = 0, \tag{20}$$

$$2a_2\beta\delta = 2\alpha - 2c_2\gamma_1^2 + 4c_2\gamma_1 + (c_2 - a_1)\delta, \tag{21}$$

$$2a_2\beta\gamma_1 = \alpha + \gamma_1(c_2 - a_1), \tag{22}$$

$$2c_2\delta(\delta + \gamma_1) = 0, \tag{23}$$

$$\alpha(\delta - 2\gamma_1) - 2c_2\left[\gamma_1(\gamma_1 + 2\delta) + \delta^2\right] = 0, \tag{24}$$

$$(\delta + \gamma_1)(\alpha - 2c_2\gamma_1) - 3\alpha\gamma_1 = 0. \tag{25}$$

Different cases have to be examined.

2 The Case $c_2 = 0$

This corresponds to a solution of (20) and $\delta = \frac{2c_1}{a_2}$ such that we get

$$\alpha = 0 \ \text{ or } \ \delta - 2\gamma_1 = 0, \tag{26}$$

as a consequence from Eqs.(24) and (25), together with the following condition $\beta = (a_2c_0 - a_1c_1)/a_2c_1, \ c_1 \neq 0$, deduced from Eq.(21).

2.1 The case $\alpha = 0$

From Eq.(19), we deduce $\beta = -\frac{c_0}{c_1}$, $c_1 \neq 0$ so that $p \geq 1$,

$$\gamma_{2p+1} = \frac{2c_1}{a_2} p + \gamma_1, \tag{27}$$

$$\gamma_{2p} = \frac{2c_1}{a_2} p, \tag{28}$$

with the conditions $c_0/c_1 = a_1/2a_2$. As the relations (27) and (28) show, when all coefficients take a common constant value, $\beta_n = \beta$, the recurrence coefficients γ_n behave as a linear (affine) function of n. This is characteristic for the Hermite orthogonal polynomials. Hence, a family of generalised semi-classical D Hermite polynomials of class 1 is thus characterized by relations (27), (28) and for all real x by

$$\Phi(x) = x - \beta, \tag{29}$$

$$\Psi(x) = a_2 x^2 - 2a_2\beta x + a_0, \tag{30}$$

$$\rho(x) = k \exp\left[a_2(\frac{x^2}{2} - \beta x)\right] \mid x - \beta \mid^{(-a_2\beta^2 + a_0 - 1)}, \tag{31}$$

and with $a_2 < 0$, $a_0 - a_2\beta^2 \neq 1$, by

$$\gamma_{2p+1} = \frac{2}{a_2} p + \gamma_1; \tag{32}$$

$$\gamma_{2p} = \frac{4}{a_2} p - \beta^2 - 2\beta - \frac{a_0}{2} - \gamma_1. \tag{33}$$

The moments of zero and first order are given by, respectively,

$$M_0 = 2k \exp[\frac{-a_2}{2}\beta^2] \int_0^{+\infty} \exp[\frac{a^2}{2}u^2] u^{a_0 - a_2\beta^2 - 1} du \tag{34}$$

$$M_1 = \beta M_0. \tag{35}$$

2.2 The case $\alpha \neq 0$.

The only situation to be examined is

$$\delta - 2\gamma_1 = 0, \tag{36}$$

giving $\delta = 2\gamma_1$, which leads to the initial value

$$\gamma_1 = -\left(\beta^2 + \frac{a_1}{a_2}\beta + \frac{a_0}{a_2}\right), \tag{37}$$

with the condition $c_1 = -(a_2\beta^2 + a_1\beta + a_0)$. Therefore, the coefficients γ_n read, ($\Psi(x)$ is normalized choosing $a_2 = 1$),

$$\gamma_{2p} = \frac{2c_1}{a_2}p = -2(\beta^2 + a_1\beta + a_0)p, \tag{38}$$

$$\gamma_{2p+1} = \frac{2c_1}{a_2}p - \beta^2 - \frac{a_1}{a_2}\beta - \frac{a_0}{a_2}. \tag{39}$$

We thus get, for all real x,

$$\Phi(x) = -(\beta^2 + a_1\beta + a_0)x + c_0, \tag{40}$$

$$\Psi(x) = x^2 + a_1 x + a_0, \tag{41}$$

$$\rho(x) = C\exp\left[\frac{-x^2}{2m} - (\frac{c_0}{m^2} + \frac{a_1}{m})x\right] \mid x - \frac{c_0}{m} \mid^{-k}, \quad m > 0, \tag{42}$$

$$k = \frac{c_0}{m^2}\left(\frac{c_0}{m} + a_1\right) + \frac{a_0}{m} + 1, \quad k < 0, \tag{43}$$

$$\beta = \frac{-a_1}{2}, \quad m\beta = c_0, \tag{44}$$

$$m = \beta^2 + a_1\beta + a_0. \tag{45}$$

3 The Case $c_2 \neq 0$

This leads to $\beta = -\frac{c_1}{2c_2}$ and

$$\gamma_{2p} = 0, \quad \gamma_{2p+1} = \gamma_1, \tag{46}$$

with the conditions (20)-(25) reduced to

$$\begin{aligned} 2\alpha - 2c_2\gamma_1^2 + 4c_2\gamma_1 &= 0, \\ 2a_2\beta\gamma_1 - \alpha - \gamma_1(c_2 - a_1) &= 0, \\ \alpha + c_2\gamma_1 &= 0, \\ \left(\frac{c_1}{c_2} - \frac{a_1}{a_2}\right)\beta &= \frac{a_0}{a_2} - \frac{c_0}{c_2}. \end{aligned} \tag{47}$$

Eq.(46) with the conditions (47) (with $\beta = -\frac{c_1}{2c_2}$ and $c_2 \neq 0$) are determined by constant coefficients; hence these relations may characterize various families of generalized D semi-classical Chebychev orthogonal polynomials of class 1.

4 Conclusion

To conclude this contribution, let us point out that the recurrence coefficients β_n and γ_n related to classical Hermite or Chebychev orthogonal polynomials[6] can be easily recovered from the system of equations (1) in which the corresponding coefficients c_i and a_j defining the functions $\Phi(x)$ and $\Psi(x)$ have to be replaced. For example, by substitution in (1), one finds $c_i = 0$ $(i = 1, 2, 3)$, $c_0 = 1$, $a_0 = a_2 = 0$, $a_1 = -2$, corresponding to classical monic orthogonal polynomials $\Phi(x) = 1$ and $\Psi(x) = -2x$, leading to $\beta_n = 0$ and $\gamma_n = -(1/2)n$. In contradistinction, taking $\Psi(x) = -x$ and $\Phi(x) = 1 - x^2$, (namely, $\rho(x) = (1 - x^2)^{-1/2}$), one finds $\beta_n = 0$ and $\gamma_n = 1/4$ which characterize the classical monic Chebychev orthogonal polynomials of first kind.

Acknowledgement

The authors wish to thank Prof. A. Ronveaux for discussions, and for having called their attention to the topic of this work.

References

1. E. Azatassou, M.N. Hounkonnou and A. Ronveaux, *Laguerre-Freud Equations for the Operator* $D_{q,\omega}$, in *Proceedings of the First International Workshop on Contemporary Problems in Mathematical Physics*, J. Govaerts, M.N. Hounkonnou and W. A. Lester, Jr., eds. (World Scientific, Singapore, 2000) pp. 336–346.
2. S. Belmedhi, *Formes linéaires et polynômes orthogonaux semi-classiques de classe* $s = 1$. *Description et Classification*, Thèse d'Etat (Université Pierre et Marie Curie, Paris VI, France, 1990), unpublished.
3. M. Foupouagnigni, Ph.D. Thesis (Institut de Mathématiques et de Sciences Physiques, Republic of Benin, 1998), unpublished.
4. M. Foupouagnigni, M.N. Hounkonnou and A. Ronveaux, *Laguerre-Freud Equations for the Recurrence Coefficients of* D_ω *-Semi-classical Orthogonal Polynomials of Class One*, *J. Comput. App. Math.* **99**, 143–154 (1998).
5. L. Salto Díaz, *Polinomios* $D_\omega-$ *semi classicos*, Tesis Doctoral (Universidad de Alcalá de Henares, 1995), unpublished.
6. A.F. Nikiforov and V.B. Uvarov, *Special Functions of Mathematical Physics* (Birkhauser, Basel, 1988).

THE TRANSFORMATION OF POLYNOMIAL EIGENFUNCTIONS OF LINEAR SECOND-ORDER Q-DIFFERENCE OPERATORS: A SPECIAL CASE OF Q-JACOBI POLYNOMIALS

G. BANGEREZAKO[1] AND M.N. HOUNKONNOU[2]

[1] *Université du Burundi, Faculté des Sciences, Département de Mathématique*
B.P. 2700 Bujumbura, Burundi
E-mail: gbang@avu.org

[2] *Unité de Recherche en Physique Théorique (URPT)*
Institut de Mathématiques et de Sciences Physiques (IMSP)
B.P. 613 Porto-Novo, Republic of Benin
E-mail: hounkonnou@yahoo.fr

A systematic discussion of the Mielnik-Ovcharov-Samsonov factorization method is given for the general linear second order q-difference equation. A special approach for the factorization of some classes of such equations is also described. As an application, the set of the Little q-Jacobi $p_n(qx; c, q|q)$ or equivalently Big q-Jacobi $P_n(q^3x; q, c, 0; q)$ polynomials is transformed into a non-classical one.

1 Introduction

Consider the factorization chain

$$\begin{aligned} H_j(x) - \mu_j &= H_j^+(x)H_j^-(x), \\ H_{j+1}(x) - \mu_j &= H_j^-(x)H_j^+(x), \quad x \in \mathbb{R}, \quad j \in \mathbb{Z}, \end{aligned} \tag{1}$$

where H is a differential or a difference operator. Including H in such a factorization chain may help either to solve it and to derive important properties for its solutions, or to produce from it other solvable operators that do not belong to the same family as H itself. In the first case one may impose to the chain some self-similarity, demanding for example that the dressing variable j acts not just as an index but as a full independent variable. In the second case, one needs to have at one's disposal a set (at least one) of transformation eigenfunctions for the transformable operator H, not belonging to the set of transformable eigenfunctions. In other words, a certain "bispectrality" is required for the operator H.

The first method here referred to as the Infeld-Hull-Miller (IHM) factorization method (FM),[5,11,12] has already been applied by Schrödinger in 1940 for solving the harmonic oscillator.[19] Nowadays, most of the known exactly solvable second-order differential or difference operators can be solved by that technique. The second method, here referred to as the Mielnik-Samsonov-Ovcharov (MSO) factorization methods (FM)[13,18] appeared as complementary to the first method by allowing to generate new exactly solvable Hamiltonians from already known ones (namely, exactly solved by the first method).

In this work, we study the MSO factorization method for the general linear second-order q-difference equation

$$[u(x)E_q + v(x) + w(x)E_q^{-1}]y(x) = \lambda y(x), \tag{2}$$

where $c(x) = -(a(x) + b(x)), E_q^i f(x) = f(q^i x)$, $i \in \mathbb{Z}$. Here and in all that follows, $y(x) = S(h(x))$, for some function S, and $h(x)$ a certain first degree Laurent polynomial in x: $h(x) = k_1 x + k_2 x^{-1} + k_3$. We then apply the results to a special case of the hypergeometric q-difference equation[15] (with $h(x) = x$),

$$[\sigma(x)D^q D_q + \tau(x)D_q]y(x) = \lambda y(x), \tag{3}$$

where $D_q f(x) = (E_q - 1)f(x)/(qx - x)$, $D^q f(x) = (1 - E_q^{-1})f(x)/(x - x/q)$, with $\sigma(x) = \sigma_0 x^2 + \sigma_1 x + \sigma_2$, $\tau(x) = \tau_0 x + \tau_1$, corresponding to a special case of the Little q-Jacobi $p_n(qx; c, q|q)$ or equivalently Big q-Jacobi $P_n(q^3 x; q, c, 0; q)$ polynomials.[9] Non-hypergeometric type functions (associated to polynomial sequences) resulting from the transformation of the latter are given as particular cases.

This work has some points of overlap with similar results in the literature.[1,6,16,17] The MSO FM is obviously also useful in the issues of quasi-exactly solvability.[2,3,8,7,10]

2 MSO FM for the Second Order q-Difference Equation

2.1 The general case

Consider the general second-order q-difference operator

$$H(x) = u(x)E_q + v(x) + w(x)E_q^{-1}. \tag{4}$$

Assume next that it is "bispectral" (not in the specific sense of other work[4]), in the sense that it admits two sequences of different systems of eigen-elements, say (λ_n, y_n) and (γ_n, z_n),

$$\begin{aligned} Hy_n(x) &= \lambda_n y_n(x), \\ Hz_n(x) &= \gamma_n z_n(x), \quad n = 0, 1, \ldots \end{aligned} \tag{5}$$

In that case, one can use one of the two eigen-elements, say for instance (γ_n, z_n), to transform H into another solvable operator $\tilde{H}$ in the following manner. Factorize H and define $\tilde{H}$ as follows,

$$H - \gamma_m = L_m R_m,$$
$$\tilde{H} - \gamma_m = R_m L_m, \quad m = 0, 1, \ldots, \tag{6}$$

where

$$R_m = 1 + f(x,m) E_q^{-1}, \qquad L_m = u(x) E_q + g(x,m),$$
$$f(x,m) = -\frac{z_m(x)}{z_m(x/q)}, \qquad g(x,m) = -w(x)\frac{z_m(x/q)}{z_m(x)}. \tag{7}$$

It follows from (6) that the functions $\tilde{y}_n(x,m)$, $m, n = 0, 1, \ldots$ defined by

$$[u(x)E_q + g(x,m)]\tilde{y}_0(x,m) = 0,$$
$$\tilde{y}_n(x,m) = [1 + f(x,m)E_q^{-1}]y_{n-1}(x), \quad m = 0, 1, \ldots, n = 1, 2 \ldots, \tag{8}$$

are eigenfunctions of $\tilde{H}(x,m)$ corresponding to the eigenvalues γ_m and λ_n, for $m = 0, 1, \ldots$, $n = 0, 1, \ldots$, respectively. Here, we shall refer to H and y_n as the *transformable* operator and functions, respectively, z_n as the *transformation* functions and finally $\tilde{H}$ and $\tilde{y}_n$ as the *transformed* operator and functions, respectively. The point here is that if

$$\frac{y_n(x)}{y_n(x/q)} \neq \frac{z_n(x)}{z_n(x/q)}, \tag{9}$$

then, for a fixed m, the transformed functions $\tilde{y}_n(x,m)$, $n = 0, 1, \ldots$ are non-trivial solutions of the transformed operator $\tilde{H}$ (in polynomial theory, when transforming polynomials into polynomials, one requires more conditions than (9), demanding for example that the functions z_n be the usual product of polynomials and an exponential type function while the y_n are polynomials). Moreover under some additional conditions, the transformed functions $\tilde{y}_n$ admit most of the mathematical properties of the transformable y_n, such as difference eigenvalue equations, closure and orthogonality, difference and recurrence relations, duality, IHM and MSO factorization, etc...

Difference equations

Clearly, the functions $\tilde{y}_n(x,m)$ satisfy the eigenvalue equation

$$\tilde{H}(x,m)\tilde{y}_0(x,m) = \gamma_m \tilde{y}_0(x,m),$$
$$\tilde{H}(x,m)\tilde{y}_n(x,m) = \lambda_{n-1}\tilde{y}_n(x,m), n = 1, 2, \ldots, \tag{10}$$

for

$$\tilde{H}(x,m) = u(x)E_q + \tilde{v}(x,m) + \tilde{w}(x,m)E_q^{-1}, \tag{11}$$

where

$$\tilde{v}(x,m) = g(x,m) + f(x,m)u(x/q) + \gamma_m$$
$$= v(x) + f(x,m)u(x/q) - u(x)f(qx,m),$$
$$\tilde{w}(x,m) = f(x,m)g(x/q,m) = w(x)\frac{g(x/q,m)}{g(x,m)}. \tag{12}$$

<u>Orthogonality and closure</u>

Consider the functions $\rho(x)$ and $\tilde{\rho}(x)$ defined by

$$\frac{\rho^2(qx)}{\rho^2(x)} = \frac{u(x)}{f(qx,m)g(qx,m)}\frac{h(xq^{-\frac{1}{2}})-h(xq^{\frac{1}{2}})}{h(xq^{\frac{1}{2}})-h(xq^{\frac{3}{2}})} = \frac{u(x)}{w(qx)}\frac{h(xq^{-\frac{1}{2}})-h(xq^{\frac{1}{2}})}{h(xq^{\frac{1}{2}})-h(xq^{\frac{3}{2}})},$$
$$\frac{\tilde{\rho}^2(qx,m)}{\tilde{\rho}^2(x,m)} = \frac{u(x)}{f(qx,m)g(x,m)}\frac{h(xq^{-\frac{1}{2}})-h(xq^{\frac{1}{2}})}{h(xq^{\frac{1}{2}})-h(xq^{\frac{3}{2}})} = \frac{u(x)g(qx,m)}{w(qx)g(x,m)}\frac{h(xq^{-\frac{1}{2}})-h(xq^{\frac{1}{2}})}{h(xq^{\frac{1}{2}})-h(xq^{\frac{3}{2}})}, \tag{13}$$

where h is a first-order Laurent polynomial in x.

Interesting relations exist between $\rho(x)$, $\tilde{\rho}(x,m)$ and $\tilde{y}_0(x,m)$. One has

$$\tilde{\rho}^2(x,m) = \rho^2(x)g(x,m),$$
$$\tilde{y}_0(x,m) = \frac{1}{\rho^2(x)g(x,m)z_m(x)(h(xq^{-\frac{1}{2}}) - h(xq^{\frac{1}{2}}))}$$
$$= \frac{1}{\tilde{\rho}^2(x,m)z_m(x)(h(xq^{-\frac{1}{2}}) - h(xq^{\frac{1}{2}}))}. \tag{14}$$

Next, as is easily seen, the similarity reductions $\rho H \rho^{-1}$ and $\tilde{\rho}\tilde{H}\tilde{\rho}^{-1}$ send H and $\tilde{H}$, respectively, into their formal symmetric form, which looks like,

$$c(qx)\frac{h(xq^{\frac{1}{2}})-h(xq^{\frac{3}{2}})}{h(xq^{-\frac{1}{2}})-h(xq^{\frac{1}{2}})}E_q + b(x) + c(x)E_q^{-1}, \tag{15}$$

or

$$a(x)E_q + b(x) + a(x/q)\frac{h(xq^{-\frac{3}{2}})-h(xq^{-\frac{1}{2}})}{h(xq^{-\frac{1}{2}})-h(xq^{\frac{1}{2}})}E_q^{-1}. \tag{16}$$

Denoting by $\ell^2(q^\beta, q^\alpha; \rho^2)$ the linear space of q-discrete functions

$$\psi(x),\ x = q^\beta, q^{\beta-1}, \ldots, q^\alpha;\ \alpha, \beta \in \mathbb{Z} \cup \{-\infty, \infty\}, \tag{17}$$

in which a discrete-weighted inner product is defined by

$$(\psi, \phi)_{\rho^2} = \int_{q^\beta}^{q^\alpha} \psi(x)\phi(x)\rho^2(x)d_h x$$
$$= \sum_{\alpha}^{\beta} \psi(q^{i+\frac{1}{2}})\phi(q^{i+\frac{1}{2}})\rho^2(q^{i+\frac{1}{2}})(h(q^i) - h(q^{i+1})). \tag{18}$$

The analogous space for $\tilde{\rho}^2$ is denoted as $\tilde{\ell}^2(q^{\tilde{\beta}}, q^{\tilde{\alpha}}; \tilde{\rho}^2)$. Now let ℓ_1^2 and $\tilde{\ell}_1^2$ be the respective subspaces of ℓ^2 and $\tilde{\ell}^2$ containing those elements for which

$$u(xq^{-\frac{1}{2}})\varrho(xq^{-\frac{1}{2}})(h(x/q) - h(x)) \\ \times[\psi(xq^{\frac{1}{2}})\phi(xq^{-\frac{1}{2}}) - \psi(xq^{-\frac{1}{2}})\phi(xq^{\frac{1}{2}})] \,|_{q^\alpha}^{q^{\beta+1}} = 0 \tag{19}$$

when ϱ^2 equals ρ^2 and $\tilde{\rho}^2$, respectively (if one of the boundary point α or β is at ∞, the cancellation of the expression in (19) is required only at the other boundary point). Using summation by parts, one easily verifies that in ℓ_1^2 and $\tilde{\ell}_1^2$, respectively, the operators H and $\tilde{H}$ are symmetric, that is

$$(H\psi, \phi)_{\rho^2} = (\psi, H\phi)_{\rho^2} \quad , \quad (\tilde{H}\psi, \phi)_{\tilde{\rho}^2} = (\psi, \tilde{H}\phi)_{\tilde{\rho}^2}. \tag{20}$$

Next, let ℓ_2^2 be the preimage of the image of ℓ^2 in $\tilde{\ell}^2$ by R_m, and $\tilde{\ell}_2^2$ be the preimage of the image of $\tilde{\ell}^2$ in ℓ^2 by L_m, both constrained to the conditions that

$$u(xq^{-\frac{1}{2}})\rho(xq^{-\frac{1}{2}})(h(x/q) - h(x)) \\ \times\phi(xq^{\frac{1}{2}})\psi(xq^{-\frac{1}{2}}) \,|_{q^\alpha}^{q^{\beta+1}} = 0, \tag{21}$$

for $\psi \in \ell^2$ and $\phi \in \tilde{\ell}^2$ (here also, if one of the boundary points α or β is at ∞, the cancellation of the expression in (21) is required only at the other boundary point). Also, using summation by parts, one proves that, defined in ℓ_2^2 and $\tilde{\ell}_2^2$, respectively, the operators R_m and L_m are $(\rho^2, \tilde{\rho}^2)$-mutually adjoint in the sense that

$$(\phi, R_m\psi)_{\tilde{\rho}^2} = (L_m\phi, \psi)_{\rho^2}. \tag{22}$$

Let us already note that $\ell_2^2 \subset \ell_1^2 \subset \ell^2$ while $\tilde{\ell}_2^2 \subset \tilde{\ell}_1^2 \subset \tilde{\ell}^2$ (taking into account in particular, the first equation in (14)). We are led to the following

Proposition 2.1 *Suppose that*
(c1) $y_n(x) \in \ell_2^2(q^\beta, q^\alpha; \rho^2)$, $n = 0, 1, 2, \ldots;$
(c2) For m fixed, $\tilde{y}_n(x, m) \in \tilde{\ell}_2^2(q^{\tilde{\beta}}, q^{\tilde{\alpha}}; \tilde{\rho}^2)$, $n = 0, 1, 2, \ldots,$
without the constraints of quadratic integrability for them;
(c3) The system $y_n(x)$ *is orthogonal and closed in* $\ell^2(q^\beta, q^\alpha; \rho^2)$.
Then, for m fixed, the system $\tilde{y}_n(x, m)$, $n = 0, 1, \ldots$ *or* $\tilde{y}_n(x, m)$, $n = 1, 2, \ldots$ *is orthogonal and closed in* $\tilde{\ell}_2^2(q^{\tilde{\beta}}, q^{\tilde{\alpha}}; \tilde{\rho}^2)$.

Proof The proof of the orthogonality of the new system is straightforward, given the above considerations: it can be deduced from (20) or (22) under the condition (19) or (21), respectively. To prove its closure (i.e., no non-vanishing element in the space, that may be orthogonal to the system in its totality), one uses (22) and the closure of the system $y_n(x)$ in $\ell^2(q^\beta, q^\alpha; \rho^2)$, to derive that $\tilde{y}_0(x,m)$ is the unique non-vanishing element orthogonal to the system $\tilde{y}_n(x,m) = R_m y_n(x)$. This means that the system $\tilde{y}_n(x,m)$, $n = 0,1,2,\ldots$ (or $\tilde{y}_n(x,m)$, $n = 1,2,\ldots$ if $\tilde{y}_0(x,m)$ is not quadratically integrable) is closed in the space.

Difference and recurrence relations

Assume that the transformable functions satisfy the difference relations

$$\begin{aligned} \alpha_n y_{n+1} &= H_n^- y_n, \\ \beta_n y_n &= H_n^+ y_{n+1}, \quad n = 0,1,\ldots \end{aligned} \tag{23}$$

On the other hand, from (6), one has

$$\begin{aligned} \tilde{y}_{n+1} &= R_m y_n, \\ (\lambda_n - \gamma_m) y_n &= L_m \tilde{y}_{n+1}, \quad n = 0,1,\ldots \end{aligned} \tag{24}$$

A combination of (23) and (24) leads to the following three-term difference relations for $\tilde{y}_n$, $n = 1,2,\ldots,$

$$\begin{aligned} \alpha_n(\lambda_n - \gamma_m)\tilde{y}_{n+2} &= R_m H_n^- L_m \tilde{y}_{n+1}, \\ \beta_n(\lambda_{n+1} - \gamma_m)\tilde{y}_{n+1} &= R_m H_n^+ L_m \tilde{y}_{n+2}. \end{aligned} \tag{25}$$

Using the difference eigenvalue equation satisfied by the $\tilde{y}_n$ (see (10)) and the preceding relations, one can naturally reach first order difference relations connecting $\tilde{y}_n$, $n = 1,2,\ldots$ (in this connection, the concluding remark of other work on this topic[13] needs some clarification).

Assume now that the transformable functions y_n satisfy a three-term recurrence relation

$$y_{n+1} + (b_n - h(x))y_n + a_n^2 y_{n-1} = 0, \tag{26}$$

so that, using the first relation in (24), one shows that the transformed $\tilde{y}_n$, $n = 1,2,\ldots$, satisfy the following five-term recurrence relation

$$\begin{aligned} &\tilde{y}_{n+4} + [b_{n+2} + b_{n+1} - h(x) - h(x/q)]\tilde{y}_{n+3} \\ &+[(b_{n+1} - h(x))(b_{n+1} - h(x/q)) + a_{n+1}^2 + a_{n+2}^2]\tilde{y}_{n+2} \\ &+a_{n+1}^2[b_{n+1} + b_n - h(x) - h(x/q)]\tilde{y}_{n+1} + a_{n+1}^2 a_n^2 \tilde{y}_n = 0. \end{aligned} \tag{27}$$

However, we remark that the preceding relations do not include $\tilde{y}_0$. If for a_n^2 in (26), one has $a_0^2 = 0$, or if one assumes that $y_{-1} = 0$, using the second relation in (24), one establishes the following difference recurrence relations for the system of transformed functions $\tilde{y}_n$, $n = 0, 1, 2, \ldots$,

$$\begin{aligned}(\lambda_{n-1} - \gamma_m)(\lambda_n - \gamma_m)L_m\tilde{y}_{n+2}\\ +(\lambda_{n-1} - \gamma_m)(\lambda_{n+1} - \gamma_m)(b_n - h(x))L_m\tilde{y}_{n+1}\\ +(\lambda_{n+1} - \gamma_m)(\lambda_n - \gamma_m)L_m\tilde{y}_n = 0.\end{aligned} \qquad (28)$$

Duality

Assume that the transformable functions $y_n(x) = y_n(q^s)$ are also explicit functions of n. In that case, one can consider the functions $\theta_s(n) = \tilde{y}_n(q^s, m)$ dual to the transformed $\tilde{y}_n(x, m)$, $n = 0, 1, \ldots$ defining

$$\begin{aligned}&\theta_s(0) = \tilde{y}_0(q^s, m),\\ &\theta_s(n) = R_m y_n(q^s) = y_n(q^s) + f(q^s, m)y_n(q^{s-1}),\ n = 1, 2, \ldots\end{aligned} \qquad (29)$$

From (6), one finds that the functions $\theta_s(n)$ satisfy the three-term recurrence relation

$$\begin{aligned}\theta_{s+1}(n) + (\tilde{v}(q^s) - \delta_n)\theta_s(n) + \tilde{w}(q^s)u(q^{s-1})\theta_{s-1}(n) = 0,\\ \delta_0 = \gamma_m,\ \delta_n = \lambda_{n-1},\ n \geq 1.\end{aligned} \qquad (30)$$

If $\tilde{w}(1)u(q^{-1}) = 0$, then the functions in (30) are, up to a multiplication by $\theta_0(n)$, polynomials in δ_n of degree s. If $\tilde{v}(q^s)$ is real for $s \geq 0$ and $\tilde{w}(q^s)u(q^{s-1}) > 0$, $s > 0$, so that the polynomials are naturally orthogonal with positive discrete weight (Favard's theorem).

IHM factorization.

Considering the formulas (25), and the second difference eigenvalue equation in (10)), one finds that for a fixed m, the operator

$$\tilde{\mathcal{H}}_n(x, m) = \tilde{H}(x, m) - \lambda_{n-1},\quad n = 1, 2, \ldots \qquad (31)$$

admits the following factorization of IHM type

$$\begin{aligned}r(x, n)\tilde{\mathcal{H}}_n - \mu_n = B_n A_n\\ t(x, n)\tilde{\mathcal{H}}_{n+1} - \mu_n = A_n B_n,\end{aligned} \qquad (32)$$

for some first order difference operators A_n and B_n and some functions r and t, while $\mu_n = -\alpha_{n-1}\beta_{n-1}(\lambda_{n-1} - \gamma_m)(\lambda_n - \gamma_m)$. We will not evaluate the unknown expressions as we are only interested in the principle of the approach here.

MSO factorization

Here also, the operator $\tilde{H}$ admits a factorization of MSO type. To demonstrate this fact, we need only to ensure that for fixed m, the operator $\tilde{H}(x,m)$ is also "bispectral". Indeed, we have

$$\tilde{H}(x,m)\tilde{y}_n(x,m) = \tilde{\lambda}_n\tilde{y}_n(x,m), \; n = 0,1,\ldots,$$
$$\tilde{\lambda}_0 = \gamma_m; \;\; \tilde{\lambda}_n = \lambda_{n-1}, \; n = 1,2,\ldots, \tag{33}$$

and

$$\tilde{H}(x,m)\tilde{z}_k(x,m) = \gamma_k\tilde{z}_k(x,m), k = 0,1,\ldots, \; k \neq m, \tag{34}$$

where $\tilde{z}_k(x,m) = R_m z_k(x)$. Here, one can generalize (6) to obtain a chain similar to that in (1),

$$H^j(x,m_0,\ldots,m_{j-1}) - \gamma_{m_j} = L_{m_j}R_{m_j},$$
$$H^{j+1}(x,m_0,\ldots,m_j) - \gamma_{m_j} = R_{m_j}L_{m_j}, \tag{35}$$

with

$$H^0(x,m_0) = H(x),$$
$$z_n^0(x,m_0) = z_n(x),$$
$$y_n^0(x,m_0) = y_n(x),$$
$$R_{m_j} = 1 - \frac{\tilde{z}^j_{m_j}(x,m_0,\ldots,m_{j-1})}{\tilde{z}^j_{m_j}(x/q,m_0,\ldots,m_{j-1})}E_q^{-1},$$
$$L_{m_j}\tilde{y}_0^{j+1}(x,m_0,\ldots,m_j) = 0,$$
$$\tilde{y}_n^{j+1}(x,m_0,\ldots,m_j) = R_{m_j}\tilde{y}_n^j(x,m_0,\ldots,m_{j-1}), n = 1,2,\ldots,$$
$$\tilde{z}_n^{j+1}(x,m_0,\ldots,m_j) = R_{m_j}\tilde{z}_n^j(x,m_0,\ldots,m_{j-1}), n = 0,1\ldots,(n \neq m_j) \tag{36}$$

2.2 A special approach

Consider again the eigenvalue equation

$$[U(x)E_q + V(x) + W(x)E_q^{-1}]Q_n(x) = \Upsilon_n Q_n(x). \tag{37}$$

As already noted, it is not difficult to rewrite (37) in the form

$$[A(x)E_q + B(x) + C(x)E_q^{-1}]Y_n(x) = \Lambda_n Y_n(x), \tag{38}$$

where $B(x) = -(A(x) + C(x))$. Hence we will here consider (38) as a general starting equation.

Consider the situation when $C(x)$ does not depend explicitly on q and $A(x) = cC(x)$, c being a constant (a similar reasoning should be developed considering that $A(x)$ does not depend explicitly on q and $C(x) = cA(x)$). In that case (38) reads

$$[cC(x)E_q - (cC(x) + C(x)) + C(x)E_q^{-1}]Y_n(x,q) = \Lambda_n(q)Y_n(x,q). \quad (39)$$

Substituting q by $1/q$ in (39), and performing a similarity reduction on the obtained operator in the left hand side, one gets

$$\begin{aligned}[cC(x)E_q - (cC(x) + C(x)) + C(x)E_q^{-1}]\pi(x)Y_n(x,1/q)\\ = \Lambda_n(1/q)\pi(x)Y_n(x,q),\end{aligned} \quad (40)$$

where

$$\pi(qx)/\pi(x) = 1/c. \quad (41)$$

This means that the operator in the left hand side of (39) and (40) is "bispectral" with two distinguished systems of eigen-elements $(\lambda_n(q), Y_n(x,q))$ and $(\Lambda_n(1/q), Z_n(x,q))$ where $Z_n(x,q) = \pi(x)Y_n(x,1/q)$. Hence it can be transformed according to the scheme studied in the first subsection. But as one can see, if $A(x)$ is for example a polynomial, the functions $Y_n(x,q)$ are not in general orthogonal. This is why we rewrite (39) and (40) in a more convenient form for the transformation. For that, assuming that $\lambda_n(q) \neq 0$, for $n \geq 1$ (this is generally the case for solutions of polynomial type), we define the functions $y_n(x,q)$ by, for $n = 0, 1, 2, \cdots$,

$$y_n(x,q) = \tfrac{1}{\Lambda_{n+1}(q)}[cE_q - (c+1) + E_q^{-1}]Y_{n+1}(x,q). \quad (42)$$

As one can verify, the functions $y_n(x,q)$ are given by

$$y_n(x,q) = \tfrac{Y_{n+1}(x,q)}{C(x)}, \; n = 0, 1 \ldots, \quad (43)$$

and satisfy the eigenvalue equation

$$[u(x)E_q + v(x) + w(x)E_q^{-1}]y_n(x,q) = \lambda_n(q)y_n(x,q), \quad (44)$$

where

$$\begin{aligned}u(x) = cC(qx);\; v(x) = -(c+1)C(x) - \Lambda_1(q),\\ w(x) = C(x/q);\; \lambda_n(q) = \Lambda_{n+1}(q) - \Lambda_1(q).\end{aligned} \quad (45)$$

In particular, if $y_0(x,q) \equiv const$, then $v(x) = -(u(x) + w(x))$. Similarly, the functions

$$z_n(x,q) = \pi(x)y_n(x,1/q) \quad (46)$$

satisfy the equation

$$[u(x)E_q + v(x) + w(x)E_q^{-1}]z_n(x,q) = \gamma_n(q)z_n(x,q), \quad (47)$$

where $\gamma_n(q) = \Lambda_{n+1}(1/q) - \Lambda_1(q)$. Thus, the operator in the left hand side of (44) and (47) is "bispectral" and, under additional boundary constraints, the functions $y_n(x,q)$ are orthogonal with the weight

$$\rho(x) = \frac{w(qx)}{x\pi(x)}. \tag{48}$$

Hence the considerations from the first subsection can be repeated here.

An example

The transformable and transformation functions. Applying the preceding considerations to the q-hypergeometric case (see (3)),

$$[A(x)E_q + B(x) + C(x)E_q^{-1}]Y_n(x) = \Lambda_n Y_n(x), \tag{49}$$

with

$$\begin{aligned} A(x) &= [(\sigma_0 + (1-1/q)\tau_0)x^2 + (\sigma_1 + (1-1/q)\tau_1)x + \sigma_2]/x^2, \\ C(x) &= [q(\sigma_0 x^2 + \sigma_1 x + \sigma_2)]/x^2, \; B(x) = -(A(x) + C(x)), \end{aligned} \tag{50}$$

one is led to the following simple "bispectral" situation

$$\begin{aligned} [u(x)E_q + v(x) + w(x)E_q^{-1}]y_n(x,q) &= \lambda_n(q)y_n(x,q), \\ [u(x)E_q + v(x) + w(x)E_q^{-1}]z_n(x,q) &= \gamma_n(q)z_n(x,q), \end{aligned} \tag{51}$$

where

$$u(x) = -c(q^3x-1)/x; \; w(x) = -(xq-1)/x; \; v(x) = -(u(x)+w(x)), \tag{52}$$

$$\lambda_n = q^{1-n}(1-q^n)(cq^{2+n}-1); \; \gamma_n = q^{1-n}(q^{2+n}-1)(c-q^n), \tag{53}$$

and the functions $y_n(x,q)$ are a special case of the Little q-Jacobi $p_n(qx;c,q|q)$ or equivalently Big q-Jacobi $P_n(q^3x;q,c,0;q)$ polynomials.[9] The transformation functions $z_n(x,q)$ being on the other hand defined as in (41) and (46). For their use in expressions (25) and (27), here we give the difference relations for the polynomials $y_n(x,q)$ (they are not given in this form in the literature) as well as the recurrence ones. We have

$$\begin{aligned} c_1(n)y_{n+1}(x,q) &= [r(x)E_q + f_n(x)]y_n(x,q) \\ -a_{n+1}^2 c_1(n+1)c_1(n)y_n(x,q) &= [r(x)E_q + g_n(x)]y_{n+1}(x,q), \end{aligned} \tag{54}$$

$$y_{n+1}(x,q) + (b_n - x)y_n(x,q) + a_n^2 y_{n-1}(x,q) = 0 \tag{55}$$

where ($Q = q^n$)

$$c_1(n) = -\frac{cQ^2q^3 + cQ^2q^2 + Q^6c^3q^7 - c^2Q^4q^6 - c^2Q^4q^4 - q - q^5c^2Q^4 + q^4Q^2c}{Q(cQ^2q^3-1)(cQ^2q-1)}, \tag{56}$$

$$b_n = \frac{(Q^2c^2q^2 - Qcq^2 + cQ^2q^2 - 2Qcq + c - Qc + 1)Q}{(cQ^2q - 1)(cQ^2q^3 - 1)q}, \tag{57}$$

$$a_n^2 = \frac{Q^2(Qc-1)(Qcq-1)(-1+Q)(-1+qQ)c}{(-1+cQ^2)(cQ^2q-1)^2(-1+cQ^2q^2)q^3}, \quad r(x) = -c(q^3x - 1), \tag{58}$$

$$f_n(x) = \frac{qx}{Q} - \frac{-1-c+Qcq^2+Qcq}{cQ^2q^3-1}, \quad g_n(x) = xcq^4Q - cqQ\frac{Qq^2+Qcq^2-q-1}{cQ^2q^3-1}. \tag{59}$$

Note finally that the polynomials $y_n(x)$ are orthogonal on the interval $[0, q^{-\frac{5}{2}}]$ with respect to the weight $\rho(x)$ given by (48) where $w(x)$ is given by (52) and $\pi(x)$ by (41). Since the interval of orthogonality is finite, they are also closed in the corresponding inner product space.

The transformed functions. For a given m, the properties of the transformed functions $\tilde{y}_n(x, m)$, $n = 0, 1, \ldots$ are those derived in the first subsection of the current section: they satisfy type (10) difference equations, and difference and recurrence relations of type (25), (27) and (28). And since the conditions of orthogonality of proposition (2.1) are satisfied, these functions are orthogonal in the inner product space $\tilde{\ell}_2^2(0, q^{-\frac{5}{2}}; \tilde{\rho}^2)$ where $\tilde{\rho}^2$ is given in (14). For the issue of closure, we have that the system $\tilde{y}_n(x, m)$, $n = 0, 1, \ldots$ is closed in that space since the unique element $\tilde{y}_0(x, m)$ orthogonal to it in its totality is not quadratically integrable. On the other hand, the transformed operator $\tilde{H}(x, m)$ admits IHM and MSO factorizations according to the scheme given in the first subsection. However, there are no interesting duality relations.

How do the transformed objects look like? Let us note that, using simple procedures in Maple V,[14] allows one to evaluate explicitly any one of these functions at least for not too large values of m and n (depending on the software and computer capabilities).

The case $m = 1$ illustrates the first non-classical situation for the transformed objects. This is the only case for which the volume required to display date is manageable. The main data are ($m = 1$, $n = 0, 1, 2$),

$$f(x, 1) = -\frac{(cq - q^4)x - c + q}{((c - q^3)x - c + q)c}; \quad g(x, 1) = \frac{(xq-1)c(xc - q^3x - c + q)}{(xcq - c - xq^4 + q)x}, \tag{60}$$

$$\begin{aligned}\tilde{v}(x, 1) = {} & [(c^3q^4 - 2q^7c^2 + c^2q^2 + cq^{10} - 2cq^5 + q^8)x^3 + (-c^3q^4 - c^3q^3 \\ & -qc^3 + q^7c^2 + q^6c^2 + 2q^5c^2 + c^2q^4 - 2qc^2 - 2cq^8 + cq^5 + 2cq^4 + q^3c \\ & +cq^2 - q^8 - q^6 - q^5)x^2 + (c^3q^3 + qc^3 + c^3 - 3c^2q^4 - c^2q^3 + qc^2 + cq^5 \\ & -q^3c - 3cq^2 + q^6 + q^3 + q^5)x - c^3 + qc^2 + cq^2 - q^3]/ \\ & [((cq - q^4)x - c + q)((c - q^3)x - c + q)x],\end{aligned} \tag{61}$$

$$\tilde{w}(x,1) = -\frac{(xcq-c-xq^4+q)(x-1)(xc-q^3x-cq+q^2)}{(xc-q^3x-c+q)^2x}, \tag{62}$$

$$\tilde{\rho}^2(x,1) = c_1\rho^2(x)g(x,1) = c_2\frac{q^2x-1}{x^3\pi(x)}\frac{(xq-1)(xc-q^3x-c+q)}{(xcq-c-xq^4+q)}, \tag{63}$$

$$\begin{aligned}\tilde{y}_0(x,1) &= \frac{c_3}{\tilde{\rho}^2(x,1)\pi(x)y_1(x,1/q)(h(xq^{-\frac{1}{2}})-h(xq^{\frac{1}{2}}))} \\ &= c_4\frac{(xq-1)(xc-q^3x-c+q)}{(q^2x-1)(xcq-c-xq^4+q)^2},\end{aligned} \tag{64}$$

where c_i, $i = 1, \ldots, 4$ are constants of integration,

$$\tilde{y}_1(x,1) = \frac{(-q+c)(xc-c-q^3x+1)}{(xc-q^3x-c+q)c}, \tag{65}$$

$$\begin{aligned}\tilde{y}_2(x,1) = [&q(c^2q^2x + c - c^2q^2 - c^2x^2q - q + xc^2q^4 - xc^3q \\ &+q^3c^2x + cq^7x^2 - c^2q^4x^2 + c^2q^5x + c^3q^4x^2 - c^2q^7x^2 - xc^3q^4 \\ &+x^2cq - q^5cx - xcq - q^3cx - xcq^4 + cq^2 + qc^3 - qc^2 + c^2x + xq^4 \\ &-q^4x^2 + xq + xc^2q + cq^4x^2 - c^2 + cq - cq^2x - xc)] \\ &/[(q^3c-1)(xc-q^3x-c+q)c].\end{aligned} \tag{66}$$

Note that if $w(x) \sim x^\alpha$ while $x \to \infty$, then $\tilde{y}_0(x,m) \sim \frac{1}{x^{m+\alpha}}$ as well as $\tilde{y}_n(x,m) \sim x^{m+n-1}$, $n = 1, 2, \ldots$ (in our particular case, $\alpha = 0$ and $m = 1$).

Finally, let us remark that similar transformation formulas for the special Meixner polynomials $M_n^{(2,c)}(x)$ are also available in the literature.[1]

References

1. G. Bangerezako and M.N. Hounkonnou, *The transformation of polynomial eigenfunctions of linear second-order difference operators: a special case of Meixner polynomials*, *J. Phys.* **A34**, 5653–5666 (2001).
2. C.M. Bender and G.V. Dunne, *Quasi-exactly solvable systems and orthogonal polynomials*, *J. Math. Phys.* **37**, 6–11 (1996).
3. A. Gangopadhyaya, A. Khare and U.V. Sukhatme, *Methods for generating quasi-exactly solvable potentials*, *Phys. Lett.* **A208**, 261–268 (1995).
4. J. Duistermaat and F. A. Grunbaum, *Differential equations in the spectral parameter*, *Comm. Math. Phys.* **103**, 177-240 (1986).
5. L. Infeld and T. Hull, *The factorization method*, *Rev. Mod. Phys.* **23**, 21–68 (1951).
6. E.G. Kalnins and W. Miller, *Symmetry techniques for q-series: Askey-Wilson polynomials*, *Rocky Mountain J. Math.* **19**, 223-230 (1989).

7. A. Khare and B.P. Mandal, *Do quasi-exactly solvable systems always correspond to orthogonal polynomials?*, *Phys. Lett.* **A239**, 197–200 (1998).
8. A. Khare and U. Sukhatme, *New solvable and quasiexactly solvable periodic potentials*, *J. Math. Phys.* **40**, 5473–5494 (1999).
9. R. Koekoek and R. Swarttouw, *The Askey-scheme of hypergeometric orthogonal polynomials and its q-analogue*, Report 98-17, Tech. Univ. Delft, Fac. Tech. Math. (Delft, 1998).
10. G. Krylov and M. Robnik, *Polynomial families and Schrödinger equation: one example for nonhypergeometric type of correspondence*, *J. Phys.* **A34**, 5403–5415 (2001).
11. W. Miller, *Lie theory and difference equations I*, *J. Math. An. Appl.* **28**, 383–399 (1969).
12. W. Miller, *Lie theory and q-difference equations*, *SIAM J. Math. Anal.* **1**, 171–188 (1970).
13. B. Mielnik, *Factorization method and new potentials with the oscillator spectrum*, *J. Math. Phys.* **25**, 3387–3389 (1984).
14. M.B. Monagan, K.O. Geddes, G. Labahn and S. Vorkoetter, *Maple V Programming Guide* (Springer-Verlag New York, Berlin, Heidelberg, 1996).
15. A. Nikiforov, S. Suslov and V. Uvarov, *Classical Orthogonal Polynomials of a Discrete Variable* (Spring-Verlag, Berlin-Heidelberg-New York, 1991).
16. J. Oscar Rosas-Ortiz, *Exactly solvable hydrogen-like potentials and factorization method*, *J. Phys.* **A31**, 10163–10179 (1998).
17. J. Oscar Rosas-Ortiz, *On the factorization method in quantum mechanics*, in Proc. First Inter. Worskshop on Symmetries in Quantum Mechanics and Quantum Optics (Burgos, Spain, 1998).
18. B. Samsonov and I. Ovcharov, *Darboux tranformation and non-classical orthogonal polynomials*, *Izv. vuzov (Fizika)* **4**, 58–65 (1995), in russian.
19. E. Schrödinger, *A method for determining quantum mechanical eigenvalues and eigenfunctions*, *Proc. Roy. Irish. Acad.* **A46**, 9–16 (1940).

ON A GENERALIZATION OF THE METHOD BY BARBAROUX *ET AL.* FOR THE IMPROVEMENT ON THE RATE OF DECAY OF AN OPERATOR RESOLVENT

G. HONNOUVO[1] AND M.N. HOUNKONNOU[2]

[1]*International Chair in Mathematical Physics and Applications (ICMPA)*
B.P. 2628 Porto-Novo, Republic of Benin
E-mail: g_honnouvo@yahoo.fr

[2]*Unité de Recherche en Physique Théorique (URPT)*
Institut de Mathématiques et de Sciences Physiques (IMSP)
B.P. 2628 Porto-Novo, Republic of Benin
E-mail: hounkonnou@yahoo.fr

In this paper, we provide a generalization of a method developed by Barbaroux *et al.* [See **Lemma 3.1** and the sequel in *Helv. Phys. Act.* **70**, 16–43 (1997)], to study the localization near band edges for random Schrödinger operators. We also give a relevant application for illustration.

1 Introduction

Let us consider the unperturbed Schrödinger operator H_0,[1]

$$H_0 = (-i\nabla - A)^2 + V_0, \tag{1}$$

where A is a vector potential and V_0 a background electrostatic potential. Let V be the perturbation potential of H_0. We consider the following assumptions.

(H1) The operator H_0 is essentially self-adjoint on $C_0^\infty(\mathbb{R}^d)$.

(H2) The spectrum of H_0, $\sigma(H_0)$, is semibounded and contains an open gap, that is, there exist finite constants $D_0 \geq 0$ and $-D_0 \leq B_- < B_+ \leq \infty$ such that

$$\sigma(H_0) \subset] - D_0, B_-] \cup [B_+, \infty[. \tag{2}$$

(H3) Let $M_0 = \sup_x ||V(x)|| < \infty, \ C_0 > 1/ \ M_0 < B + C_0$.
If we take $A = \frac{B}{2}(x, -y)$ and $V_0 = 0$, then $H_A \equiv H_0$ is the Landau Hamiltonian.

(H4) Let $\rho(x) \equiv (1 + ||x||^2)^{-1}$. The operator $H_0(\alpha) \equiv e^{i\alpha\rho} H_0 e^{-i\alpha\rho}$, defined for $\alpha \in \mathbb{R}$, admits an analytic continuation as a type-A analytic family to a strip

$$S(\alpha_0) \equiv \{x + iy \in \mathbb{C}/|y| < \alpha_0\}$$

for some $\alpha_0 > 0$.

The discussion of the physical interpretation of this system may be found elsewhere.[1]

2 Main results

In this section, we give a generalization of the basic technical result obtained by Barbaroux *et al.*,[1] based on an alternative form of the Combes-Thomas method[3] which allows an improvement on the rate of decay of an operator resolvent.

Theorem 2.1 *Let A and B be two self-adjoint operators such that $d\pm \equiv dist(\sigma(A) \cap \mathbb{R}^{\pm}, 0) > 0$ and $||B|| < 1$. Then*

i) For $\beta \in \mathbb{R}$ $/|\beta| < \sqrt{d_+ d_-}$, one has $0 \in \rho(A + i\beta B)$.

ii) For $\beta \in \mathbb{R}$ as in (i), one has $||(A + i\beta B)^{-1}|| \leq \frac{1}{1-||B||} sup(d_+^{-1}, d_-^{-1})$.

Proof Let $P\pm$ be the spectral projectors for A corresponding to the sets $\sigma(A) \cap \mathbb{R}^{\pm}$, respectively, and define $u_\pm = P_\pm u$. By Schwarz's inequality, one has

$$\begin{aligned} ||u|| \cdot ||(A + i\beta B)u|| &\geq Re < (u_+ - u_-), (A + i\beta B)(u_+ + u_-) > \\ &\geq [< u_+, Au_+ > - < u_-, Au_- > -2\beta Im < u_+, Bu_- >] \\ &\geq d_+||u_+||^2 + d_-||u_-||^2 - 2\beta Im < u_+, Bu_- > \\ &\geq d_+||u_+||^2 + d_-||u_-||^2 - ||B||(d_+||u_+||^2 + d_-||u_-||^2) \\ &\geq (1 - ||B||)(d_+||u_+||^2 + d_-||u_-||^2) \\ &\geq (1 - ||B||)\, inf\{d_+, d_-\}||u||^2, \end{aligned} \tag{3}$$

which implies that

$$0 \in \rho(A + i\beta B) \text{ and } ||(A + i\beta B)^{-1}|| \leq \frac{1}{1 - ||B||} sup(d_+^{-1}, d_-^{-1}). \quad \blacksquare$$

This result has been proved previously[1], however under the assumption $|\beta| < \frac{1}{2}\sqrt{d_+ d_-}$, thus restricting the range of applications.

Theorem 2.2 *Let $\tilde{H}$ be a semibouned self-adjoint operator with a spectral gap $G \equiv]E_-, E_+[\subset \rho(\tilde{H})$. Let W be a symmetric operator such that*

$$D\left[(\tilde{H} + C_0)^{\frac{1}{2}}\right] \subset D(W),$$

and

$$||(\tilde{H} + C_0)^{-\frac{1}{2}} W (\tilde{H} + C_0)^{-\frac{1}{2}}|| < 1,$$

for some C_0 such that $\tilde{H} + C_0 > 1$. For any $E \in G$, let $\Delta_\pm \equiv dist(E_\pm, E)$. Then

i) The energy $E \in \rho(\tilde{H} + i\beta W)$, for all real β satisfying

$$|\beta| < \left\{ \frac{\Delta_- \quad \Delta_+}{(E_- + C_0)(E_+ + C_0)} \right\}^{\frac{1}{2}}.$$

ii) For any real β and any energy E as in i), one has

$$||(\tilde{H} + i\beta W - E))^{-1}|| \leq \frac{1}{1 - ||B||} sup\left(\frac{E_- + C_0}{\Delta_-}, \frac{E_+ + C_0}{\Delta_+} \right).$$

Proof Let $E \in G$ and C_0 be as above. Define a self-adjoint operator $A = (\tilde{H} + C_0)^{-1}(\tilde{H} - E)$ and $B = (\tilde{H} + C_0)^{-\frac{1}{2}} W (\tilde{H} + C_0)^{-\frac{1}{2}}$. Since $\tilde{H}$ is self-adjoint, so is A and $B^* = (\tilde{H} + C_0)^{-\frac{1}{2}} W^* (\tilde{H} + C_0)^{-\frac{1}{2}}$. Since $D\left[(\tilde{H} + C_0)^{\frac{1}{2}}\right] \subset D(W)$, B is also self-adjoint. Note that $0 \in \rho(A)$ and $d\pm \equiv dist\Big(\sigma(A) \cap \mathbb{R}^\pm, 0\Big) = \frac{\Delta_\pm}{(E_\pm + C_0)} > 0$.

Applying Theorem 2.1 to these operators A and B, we see that for β as in i), $0 \in \rho(A + i\beta B)$ and that $||(A + i\beta B)^{-1}|| \leq \frac{1}{1 - ||B||} \sup\left(\frac{E_- + C_0}{\Delta_-}, \frac{E_+ + C_0}{\Delta_+}\right)$. Let $P_\pm$ be as in proof of Theorem 2.1. For any $w \in D(\tilde{H})$,

$$\begin{aligned} ||(\tilde{H} + i\beta W - E)w|| &= ||(\tilde{H} + C_0)^{\frac{1}{2}}(A + i\beta B)(\tilde{H} + C_0)^{\frac{1}{2}} w|| \\ &\geq ||(A + i\beta B)(\tilde{H} + C_0)^{\frac{1}{2}} w||, \end{aligned} \tag{4}$$

since $(\tilde{H}+C_0)\geq 1$. We now repeat estimate (2.1) taking $u\equiv(\tilde{H}+C_0)^{\frac{1}{2}}w$. This gives

$$\begin{aligned}||(\tilde{H}+i\beta W-E)w|| &\geq (1-||B||)||(\tilde{H}+C_0)^{\frac{1}{2}}u||^{-1}\\ &\times\left(d_+||P_+((\tilde{H}+C_0)^{\frac{1}{2}}w)||^2+d_-||P_-((\tilde{H}+C_0)^{\frac{1}{2}}w)||^2\right)\\ &\geq (1-||B||).\inf(d_-,d_+).||(\tilde{H}+C_0)^{\frac{1}{2}}w||, \qquad (5)\end{aligned}$$

where $u=(\tilde{H}+C_0)^{\frac{1}{2}}w$. Since $||(\tilde{H}+C_0)^{\frac{1}{2}}w||\geq||w||$, the result ii) follows from (5) and Theorem 2.1. ■

Let us now consider the case of a finite number of gaps for the operator $\tilde{H}$.

Theorem 2.3 *Let $\tilde{H}$ be a semibounded self-adjoint operator and let $G_n\equiv\,]E_-^n,E_+^n[\subset\rho(\tilde{H})$ $(n=1,..,N)$ be a finite family of gaps for $\tilde{H}$ such that $E_+^n<E_-^{n+1}$ with $1\leq n\leq N-1$. Let W be a symmetric operator such that*

$$D\left[(\tilde{H}+C_0)^{\frac{1}{2}}\right]\subset D(W),$$

and

$$||(\tilde{H}+C_0)^{-\frac{1}{2}}W(\tilde{H}+C_0)^{-\frac{1}{2}}||<1,$$

for some C_0 such that $\tilde{H}+C_0>1$. For any $(E_1,...,E_N)$ such that $E_n\in G_n$, let $\Delta_\pm\equiv\inf\{dist(E_\pm^n,E_n)\}$. Then

i) The energy $E_n\in\rho(\tilde{H}+i\beta W)$ for all real β satisfying

$$|\beta|<\left\{\frac{\Delta_-\quad\Delta_+}{(E_-+C_0)(E_++C_0)}\right\}^{\frac{1}{2}}.$$

iii) For any real β and any energy E as in i), one has

$$||(\tilde{H}+i\beta W-E))^{-1}||\leq\frac{1}{1-||B||}sup\left\{\frac{E_-+C_0}{\Delta_-},\frac{E_++C_0}{\Delta_+}\right\}.$$

Proof For the proof, let us remark that for any E_n and any n,

$$sup\left\{\frac{E_-^n+C_0}{\Delta_-^n},\frac{E_+^n+C_0}{\Delta_+^n}\right\}\leq sup\left\{\frac{E_-+C_0}{\Delta_-},\frac{E_++C_0}{\Delta_+}\right\},$$

with $\Delta_\pm^n \equiv dist(E_\pm^n, E_n)$ and

$$\left(\frac{\Delta_- \quad \Delta_+}{(E_- + C_0)(E_+ + C_0)}\right)^{\frac{1}{2}} \leq \left(\frac{\Delta_-^n \quad \Delta_+^n}{(E_-^n + C_0)(E_+^n + C_0)}\right)^{\frac{1}{2}}.$$

From Theorem 2.2, it results that $E_n \in \rho(\tilde{H} + i\beta W)$ for $n = 1, 2, \cdots, N$ and

$$||(\tilde{H} + i\beta W - E_n))^{-1}|| \leq \frac{1}{1 - ||B||} sup\left\{\frac{E_- + C_0}{\Delta_-}, \frac{E_+ + C_0}{\Delta_+}\right\}. \blacksquare$$

3 Application

Let us consider a one-particle Hamiltonian which describes an electron in two dimensions (x, y), subjected to a constant magnetic field of strength $B > 0$ in the perpendicular z-direction. The Hamiltonian is H_A, and $B = \nabla \times A$ is the magnetic field. The spectrum of $\sigma(H_A)$ consists of an increasing sequences $E_n(B)$ of eigenvalues, each of infinite multiplicity given by[2]

$$E_n(B) = (2n + 1)B, \quad n = 0, 1, 2, \cdots$$

Let $G_{n_i}(B) =](2n_i + 1)B, \ (2n_i + 3)B[$ $(i = 1, 2, \cdots, N)$ be N gaps in $\sigma(H_A)$; $n_i < n_{i+1}$. Let $E_i \in G_{n_i}(B)$ for $i = 1, 2, \cdots, N$ as well as $\Delta_+ = \inf_i\{dist((2n_i + 3)B, \ E_i)\}$, $\Delta_- = \inf_i\{dist((2n_i + 1)B, \ E_i)\}$, $E_-(B) = (2n_N + 1)B, \ E_+(B) = (2n_N + 3)B$.

Theorem 3.1 *Under the assumption (H3), one has*

i) The energy $E_i \in \rho(H_A + i\beta V)$ for all real β satisfying

$$|\beta| < \left\{\frac{\Delta_- \quad \Delta_+}{(E(B)_- + C_0)(E(B)_+ + C_0)}\right\}^{\frac{1}{2}}.$$

ii) For any real β and any energy E as in i), one has

$$||(H_A + i\beta V - E))^{-1}|| \leq \frac{1}{1 - ||B||} sup\left(\frac{E(B)_- + C_0}{\Delta_-}, \frac{E(B)_+ + C_0}{\Delta_+}\right).$$

Here $B = (H_A + C_0)^{-\frac{1}{2}} V (H_A + C_0)^{-\frac{1}{2}}$.

Proof The proof is the same as that of Theorem 2.3. ■

Acknowledgments

G.H. thanks the Abdus Salam International Center for Theoretical Physics for financial support through a grant. M.N.H acknowledge the Belgian Cooperation CUD-CIUF-UNB for financial support.

References

1. J.M. Barbaroux, J.M. Combes and P.D. Hislop, *Localization near band edges for random Schrödinger operators*, *Helv. Phys. Acta* **70**, 16–43 (1997), as references therein.
2. J.M. Combes and P.D. Hislop, *Landau Hamiltonians with random potentials: Localization and the density of state*, Preprint CPT-Marseille (July 2000).
3. J.M. Combes and L. Thomas, *Asymptotic behavior of eigenfunctions for multiparticle Schrödinger operators*, *Comm. Math. Phys.* **34**, 254–270 (1973).

EXPANSION OF HYPERGEOMETRIC POLYNOMIALS: SEVERAL APPROACHES

A. RONVEAUX

Physique Mathématique, Facultés Universitaires N. D. de la Paix
61, Rue de Bruxelles, B-5000 Namur, Belgium
E-mail: Andre.Ronveaux@fundp.ac.be

A. ZARZO

Instituto Carlos I de Física Teórica y Computacional, Facultad de Ciencias
Universidad de Granada, Spain
and
Depto. Matemática Aplicada, E.T.S. Ingenieros Industriales
Universidad Politécnica de Madrid, Spain
E-mail: azarzo@etsii.upm.es

I. AREA

Departamento de Matemática Aplicada, E.T.S.E. Telecomunicación
Universidade de Vigo, Campus Lagoas-Marcosende
E-36200 Vigo, Spain
E-mail: area@dma.uvigo.es

E. GODOY

Departamento de Matemática Aplicada, E.T.S.I. Industriales
Universidade de Vigo, Campus Lagoas-Marcosende
E-36200 Vigo, Spain
E-mail: egodoy@dma.uvigo.es

1 Introduction

This short survey tries to explain the power and the limitations of several techniques giving the expansions of hypergeometric polynomials in terms of other polynomial families given *a priori* (connection problem). This presentation avoids technical details and heavy computations and shows only very few examples, referring to many works presented in several publications.[15,9,3,6,7,8,23,25] Among the techniques proposed, we shall first discuss the Fields-Wimp formula and next the two-step approaches mixing direct and inverse representations. In case of existence of orthogonality for the expanding family, we shall insist on the powerful recurrent approach that can also be applied to generalized linearization problems, to asymptotic results inside classical orthogonal polynomials, and can also be extended for the expansion of multivariable polynomials.

By lack of space, we restrict ourselves to hypergeometric polynomials but all what we say can be extended to the case of expansion of basic (q-) hypergeometric polynomials.[16,30,20]

2 Generalities

A typical hypergeometric polynomial can be written as,[4]

$$ {}_{r+1}F_s\left(\begin{matrix} -n\,,\,a_1\,,\,\ldots\,,\,a_r \\ b_1\,,\,\ldots\,,\,b_s \end{matrix}\,\middle|\,x\right) = \sum_{k=0}^{n}(-n)_k\frac{(a_1)_k\ \cdots\ (a_r)_k}{(b_1)_k\ \cdots\ (b_s)_k}\frac{x^k}{k!}, \tag{1} $$

with $(a)_k := a(a+1)\cdots(a+k-1)$, and $(b)_j \neq -\ell$, ℓ positive integer.

This function of $r+s+1$ variables, denoted now as

$$ {}_{r+1}F_s(-n, a_R; b_S; x) \qquad \text{with} \quad a_R = (a_1, \ldots, a_r)\,, \quad b_S = (b_1, \ldots, b_s)\,, \tag{2} $$

is obviously a polynomial of degree n in the variable x but also in anyone of the variables a_i.

As a function of x,

$$ y(x) = {}_{r+1}F_s(-n, a_R; b_S; x) \tag{3} $$

is solution of a differential equation which can be written as

$$ \left[\theta\prod_{j=1}^{s}(\theta+b_j-1) - x(\theta-n)\prod_{i=1}^{r}(\theta+a_i)\right]\ y(x) = 0, \tag{4} $$

where the operator θ is defined by $\theta := x\ \mathrm{d}/\mathrm{d}x$.

Many situations ask to expand $P_n \equiv {}_{r+1}F_s(-n; a_R; b_S; x)$ in another polynomial family Q_m, of degree m in x or in a_i, defining the connection coefficients $\mathbf{C_m(n)}$ by

$$ P_n = \sum_{m=0}^{n} \mathbf{C_m(n)}\ Q_m\ . \tag{5} $$

In this connection problem,[9] it appears very often that the family Q_m is orthogonal, which means that there exists a real scalar product denoted $\langle\cdot,\cdot\rangle$ such that $\langle Q_n, Q_m\rangle = 0$ if $n \neq m$.

In that case, we get obviously,

$$ \mathbf{C_m(n)} = \frac{\langle P_n, Q_m\rangle}{\langle Q_m, Q_m\rangle}\ . \tag{6} $$

This formula asks, however, to compute the integrals, sums or q-sums hidden in the scalar product which is not always easy or may even be impossible. See for instance.[3,15,16]

When both families P_n and Q_m are hypergeometric, it is sometimes convenient to apply the so-called "FIELDS-WIMP" formula.[12,16,7,24,37,39]

3 FIELDS–WIMP Formula Approach

With obvious up and down notations emphasizing the role of the parameters in numerator or denominator, the Fields-Wimp formula reads as:

$$ {}_{r+t}F_{s+u}\left(\begin{matrix} a_R\,,\, c_T \\ b_S\,,\, d_U \end{matrix}\middle| xw\right) = \sum_{n=0}^{\infty} \frac{(a_R)_n(\alpha)_n(\beta)_n}{(b_S)_n(\gamma+n)_n}\,\frac{(-x)^n}{n!} \times {}_{r+2}F_{s+1}\left(\begin{matrix} n+\alpha\,,\, n+\beta\,,\, n+a_R \\ 1+2n+\gamma\,,\, n+b_S \end{matrix}\middle| x\right) \times {}_{t+2}F_{u+2}\left(\begin{matrix} -n\,,\, n+\gamma\,,\, c_T \\ \alpha\,,\, \beta\,,\, d_U \end{matrix}\middle| w\right). \tag{7} $$

The structure of this relation decoupling in some way x and w, is particularly suited to solve connection problems of the form[24]

$$ P_n(cx) = \sum_{n=0}^{n} \mathbf{C_m(n)}\, Q_m(x)\,. \tag{8} $$

An appropriate choice of parameters in Eq.(7) restricted to a finite sum gives explicit coefficients $\mathbf{C_m(n)}$ where both families P_n and Q_m are Jacobi polynomials (as discussed for example elsewhere[24])

$$ P_n^{(\alpha,\beta)}(x) = {}_2F_1\left(\begin{matrix} -n\,,\, n+\alpha+\beta+1 \\ \alpha+1 \end{matrix}\middle| \frac{1-x}{2}\right), \tag{9} $$

or also if P_n and Q_m are two arbitrary Sister Celine polynomials[37]

$$ f_n\,(a_R;\, b_S\,;x) = {}_{r+2}F_{s+2}\left(\begin{matrix} -n\,,\, n+1\,,\, a_R \\ 1\,,\, \frac{1}{2}\,,\, b_S \end{matrix}\middle| x\right). \tag{10} $$

Many other situations fit within the Fields-Wimp formula and explain why connection coefficients are also hypergeometric functions evaluated at some point ($\omega = 1$ for instance). However, if this approach is powerful, mainly from the arbitrary degree of freedom in the parameters, many expansions like $P_n(x^\ell)$ (ℓ positive integer), $P_n(ax+b)$ or $P_n(x)Q_m(x)$ (linearization problem[9]) cannot be approached this way. Different techniques have been developed.

Let us mention that a q-generalization of the Fields-Wimp formula, called Verma's extension,[39] exists with the same degree of generality (concerning the number of parameters) but also with the same limitations shown before.

4 Connection Via Inversion

The idea consists in splitting the connection problem into two parts.[28,7,5,40] If the coefficients D_k or I_r are known in both expansions

$$P_n(x) = \sum_{k=0}^{n} D_k(n)\, x^k\,, \quad x^k = \sum_{r=0}^{k} I_r(k)\, Q_r(x)\,, \tag{11}$$

the connection coefficients $\mathbf{C_m(n)}$ are given by the "convolution"

$$\mathbf{C_m(n)} = \sum_{k=m}^{n} D_k(n)\, I_m(k)\,. \tag{12}$$

More complicated situations appear if for instance a more natural basis forces $P_n(x)$ to be developed in terms of falling factorial $x^{[k]}$,

$$x^{[k]} = x(x-1)\cdots(x-k+1) = (-1)^k\,(-x)_k\,. \tag{13}$$

Of course this link between $x^{[k]}$ and power x^r is given by the Stirling number of first kind[1] $\mathcal{S}_i(j)$ and the connection coefficients read now as,

$$\mathbf{C_m(n)} = \sum_{i=m}^{n} I_m(i) \sum_{j=1}^{n} \mathcal{S}_i(j)\, D_j(n) \tag{14}$$

where

$$x^{[j]} = \sum_{i=0}^{j} \mathcal{S}_i(j)\, x^i\,. \tag{15}$$

Stirling numbers of second kind[1] appear in the same way in the opposite situation: expansion of $P_n(x)$ in x^k and inversion of $Q_t(x)$ in terms of $x^{[k]}$.

The interesting thing is that the Fields-Wimp formula contains both types of inversion formula. Extracting in the left hand side of (7) coefficients of x^n, we get up to a convenient choice of parameters,[27,7]

$$\frac{\prod_{j=2}^{r+1}(a_j)_n}{\prod_{j=1}^{s}(b_j)_n}\, x^n = \sum_{k=0}^{n}(-1)^k \binom{r}{k}\; {}_{r+1}F_s\left(\left.\begin{matrix}-k\,, a_2\,, \ldots\,, a_{r+1}\\ b_1\,, \ldots\,, b_s\end{matrix}\right| x\right)\,. \tag{16}$$

The above formula can be applied when none of the numerator and denominator parameters depends on k. For instance, it can be applied to get the inversion coefficients in case of Laguerre, Hermite, Kravchuk, Meixner or Charlier polynomials. In some situations, for instance in case of polynomials orthogonal on quadratic lattices[26] or even in case of Hahn, Jacobi or Bessel polynomials, one of the numerator parameters depends also on k. For solving the inversion problem we need to use the following formula,[27]

$$\frac{\prod_{j=3}^{p+1}(a_j)_n}{\prod_{j=1}^{s}(b_j)_n} x^n = \sum_{k=0}^{n} \binom{n}{k} \frac{(-1)^k}{(a_2+k)_k(a_2+2k+1)_{n-k}} \times {}_{p+1}F_s\left(\left. \begin{matrix} -k,\, a_2+k,\, a_3,\, \ldots, a_p,\, a_{p+1} \\ b_1, b_2, \ldots, b_s \end{matrix} \right| x \right). \tag{17}$$

A typical use of this formula gives[7] the inversion of $(-z)_n\,(z+\gamma+\delta+1)_n$ in terms of Racah polynomials[20]

$$R_n(x(z);\alpha,\beta,\gamma,\delta) = {}_4F_3\left(\left. \begin{matrix} -n\,,\, n+\alpha+\beta+1\,,\, -z\,,\, z+\gamma+\delta+1 \\ \alpha+1\,,\, \beta+\delta+1\,,\, \gamma+1 \end{matrix} \right| 1 \right). \tag{18}$$

If this technique of connection via inversion is quite general, the price to pay comes from the fact that very often both coefficients D_k and I_r are hypergeometric quantities.

A completely different approach consists in computing $\mathbf{C_m(n)}$ as solution of a recurrence relation easily obtained in many cases.

5 Recurrent Approach

Recurrence relation for $C_m(n)$ appears naturally when each family P_n and Q_m are orthogonal with respect to a scalar product such that the operator multiplication by x is self-adjoint. If we denote the scalar product in the Q family by $\langle\cdot,\cdot\rangle$ ($\langle Q_r(x), Q_s(x)\rangle = 0$, $r \neq s$) the condition

$$\langle xQ_r(x), Q_s(x)\rangle = \langle Q_r(x), xQ_s(x)\rangle \tag{19}$$

generates a three-term recurrence relation inside the Q_n family,

$$xQ_n(x) = A_nQ_{n+1}(x) + B_nQ_n(x) + C_nQ_{n-1}(x)\,. \tag{20}$$

The orthogonality of the P_n family generates a similar recurrence relation for $P_n(x)$, with coefficients denoted by $\bar{A}_n$, $\bar{B}_n$ and $\bar{C}_n$.

Now multiplication by x in both sides of the connection formulae (5) and re-expansion of $P_{n\pm1}(x)$ in Q_m generates easily a "cross rule" linking $C_m(n)$ to their four neighbours $C_m(n \pm 1)$ and $C_{m\pm1}(n)$.[34,21]

This two indices recurrence relation is not very useful from a computational point of view. But if we restrict the orthogonality of Q_m to a classical orthogonality, we can relax the orthogonality condition on the P_n family asking only $P_n(x)$ to be solution of a linear differential, difference or q-difference equation $\mathcal{L}[P_n(x)] = 0$ of order L with polynomial coefficients. The `Navima`[a] algorithm[6,17,34] works with these assumptions.

From (5)

$$\sum_{m=0}^{n} C_m(n)\ \mathcal{L}[Q_m(x)] = \mathcal{L}[P_n(x)] = 0\,, \tag{21}$$

it is easy to see that each term in $\mathcal{L}[Q_m(x)] = 0$ like $x^s\ Q_m^{(r)}(x)$ is a linear combination of constant coefficient terms $Q_k^{(L)}$:

$$x^s Q_m^{(r)}(x) = \sum_k L_{s,r,k}\ Q_k^{(L)}(x)\ . \tag{22}$$

Two properties are used in this expansion, taking into account that the family $Q_m(x)$ is "classical" which means solution of second order equation,[26,40,30]

$$\sigma(x)\ D^2\ Q_m(x) + \tau(x)\ D\ Q_m(x) + \lambda_m\ Q_m(x) = 0\ , \tag{23}$$

where $D = \frac{\mathrm{d}}{\mathrm{d}x}$, $\sigma(x)$ and $\tau(x)$ are polynomials of degree at most 2 and 1, and λ_m is constant. The two properties are,

(i) the family $\{Q_m^{(r)}\}$ is orthogonal for each r;

(ii) each $Q_m^{(r)}$ is a linear combination of $Q_{m+1}^{(r+1)}$, $Q_m^{(r+1)}$, $Q_{m-1}^{(r+1)}$.

With the above assumptions, any hypergeometric polynomial like in (1) can be expanded in any classical orthogonal family solution of (23) (Jacobi, Bessel, Laguerre, Hermite) using the operador $\mathcal{L}$ defined by the differential equation (4).

Mutadis mutandis this algorithm works in the same way if the operator $D = \frac{d}{dx}$ in $\mathcal{L}$ is replaced by the difference operator Δ ($\Delta f(x) = f(x+1) - f(x)$) and also by the q-derivative operator D_q ($D_q f(x) = \frac{f(qx)-f(x)}{x(q-1)}$).

[a] http://www.uvigo.es/webs/t10/navima

The assumption $\mathcal{L}[P_n(x)] = 0$, weaker than the orthogonality assumption, allows to use this technique in order to solve linearization problems like

$$P_i(x)\bar{P}_j(x) = \sum_{k=0}^{i+j} L_{ijk}\, Q_k(x) \tag{24}$$

from the operator $\bar{\bar{\mathcal{L}}}$ such that $\bar{\bar{\mathcal{L}}}[P_i\bar{P}_j] = 0$ and easily built from $\mathcal{L}[P_i(x)] = 0$ and $\bar{\mathcal{L}}[\bar{P}_j(x)] = 0$.

The recurrent approach applied to linearization problems like (24) and also to more general products like $\prod_{r=1}^{k} P_{i_r}(x)$ built possibly with several different families, raises the problem to obtain the differential equation for these products from the differential equation of each classical family P_i, $\bar{P}_j$. These equations are given explicitly for $[P_n(x)]^N$, $P_i(x)P_j(x)P_k(x)$, $P_i(x)\bar{P}_j(x)$ and an algorithmic procedure is described in order to manage more complicated situations.[10,19,32]

In case $\mathcal{L}$ is a differential operator, expansion of $P_m(x^\ell)$ (ℓ integer) is also possible building first $\tilde{\mathcal{L}}[P_m(x^\ell)] = 0$.

The benefit of the assumptions used in the `Navima` algorithm allows to replace the previous "cross rule" by a "row rule" of type:

$$\sum_{j=-J}^{K} A_j(m,n)\ \mathbf{C_{m+j}(n)} = 0 \tag{25}$$

which can be solved in many cases.

Let us mention that from both "cross rules" generated by the starting classical family $\{P_n\}$ and the family of its derivatives $\{P_n'\}$ (still classical and orthogonal), it is possible to obtain by elimination[21] a "column rule" of type

$$\sum_{j=0}^{R} B_j(m,n)\mathbf{C_m(n-j)} = 0\,.$$

Next, we present an example where monic Hermite polynomials are involved. The coefficients $\mathbf{C_m(n)}$ in the expansion,[17]

$$\boxed{H_n(x^2) = \sum_{m=0}^{2n} \mathbf{C_m(n)}\ H_m(x)} \tag{26}$$

satisfy the **five term** recurrence relation

$$16(5-m+2n)\mathbf{C_{m-5}(n)}+16(-15+11m-2m^2-6n+3mn)\mathbf{C_{m-3}(n)}$$
$$+4(-1+m)(-18+19m-6m^2-6n+6mn)\mathbf{C_{m-1}(n)}$$
$$+2(-1+m)m(1+m)(3-4m+2n)\mathbf{C_{m+1}(n)}$$
$$+(1-m)m\,(1+m)(2+m)(3+m)\mathbf{C_{m+3}(n)}=0$$

valid for $5 \leq m \leq 2n-3$, with the initial conditions

$$\begin{cases} \mathbf{C_{2n}(n)} & = 1\,, \\ \mathbf{C_{2n-2}(n)} & = \dfrac{n(2n-1)}{2}\,, \\ \mathbf{C_{2n-4}(n)} & = \dfrac{n(n-1)(1-8n+4n^2)}{8}\,, \\ \mathbf{C_{2n-6}(n)} & = \dfrac{n(n-1)(n-2)(2n-5)(4n^2-8n-3)}{48}\,. \end{cases} \tag{27}$$

Of course, some of the polynomial $P_n(x)$ met before can be out of the class of hypergeometric polynomials, like Bernstein polynomials.[36] The ℓ-associated of orthogonal polynomials $\{\bar{P}_n^{[\ell]}(x)\}$ defined from the recurrence relation (20) (written in monic form)

$$P_{n+1}^{[\ell]}(x)=(x-\bar{B}_{n+\ell})P_n^{[\ell]}(x)-\bar{C}_n P_{n-1}^{[\ell]}(x)\,, \qquad n>1\,,$$
$$P_0^{[\ell]}(x):=1\,, \quad P_1^{[\ell]}(x)=x-\bar{B}_\ell\,,$$

play an important role in the theory of perturbation of the recurrence coefficients $\{\bar{B}_n,\bar{C}_n\}$ of the given monic family $\{P_n(x)\}$.

It can be shown that when the family $\{P_n(x)\}$ is classical, then $\{P_n^{[\ell]}(x)\}$ is no more hypergeometric but it is a 2-sum of product of 2 hypergeometric polynomials. However, $P_n^{[\ell]}(x)$ is solution of a fourth order differential equation[13,14,31,22,41,11,29] which allows, using the recurrent approach, to expand in the starting family $\{P_n(x)\}$ or in any other classical family.

Moreover, the recurrent approach gives also asymptotic properties[38] between two families linked by a limit. For instance, from

$$\lim_{N\to\infty} k_n^{(a/N)}(x;N)=C_n^{(a)}(x)\,, \tag{28}$$

with $k_n^{(a)}(x;N)$ and $C_n^{(a)}(x)$ being respectively monic Kravchuk and Charlier polynomials[26] (both hypergeometric), we expand first

$$k_n^{(a/N)}(x;N) = \sum_{m=0}^{n} \mathbf{C_m(n)}\; C_m^{(a)}(x) \tag{29}$$

and the `Navima` algorithm gives[18]

$$a^2m(m+1)\; \mathbf{C_{m+1}(n)} + a(m-1)m\; \mathbf{C_m(n)} + (m-n-1)N\; \mathbf{C_{m-1}(n)} = 0\,, \tag{30}$$

valid for $1 \le m \le n-1$ with the initial conditions $\mathbf{C_n(n)} = \mathbf{1}$ and $\mathbf{C_{n-1}(n)} = an(n-1)/N$, since monic polynomials have been considered. Solving the recurrence up to $1/N$ gives a contribution of only $\mathbf{C_{n-1}(n)}$ and $\mathbf{C_{n-2}(n)}$ which allows to write

$$k_n^{(a/N)}(x;N) = C_n^{(a)}(x) + \frac{an(n-1)}{2N}\left[2C_{n-1}^{(a)}(x) + aC_{n-2}^{(a)}(x)\right] + \mathrm{O}\left(\frac{1}{N^2}\right)\,. \tag{31}$$

A systematic study of all limit transition, vertical or transverse, in the Askey tableau is already available.[18,35]

6 Multivariable Polynomials

In order to develop a multivariable polynomial $Q(x_1,\cdots,x_k)$ in products of classical orthogonal polynomials $P_{r,n_r}(x_r)$ of degree n_r in one variable x_r,

$$Q(x_1,\ldots,x_k) = \sum_{(n_1,\cdots,n_k)\in\mathbb{N}^k} \mathbf{u_{n_1\cdots n_k}} P_1(x_1)\;\cdots\; P_{k,n_k}(x_k)\,, \tag{32}$$

an extension of the recurrent technique works now if $Q(x_1,\ldots,x_k)$ is solution of a linear Partial Differential Equation of any order with polynomial coefficients.[33]

We can compute for each partial derivative $\partial/\partial x_r$ and for each x_r multiplication, the action on the polynomial in x_r and iterate as often as necessary. This becomes quickly cumbersome so it is more convenient to define appropriate operator acting on **indices**. For instance we write,[33,23]

$$x_r \sum_{n=0}^{\infty} \mathbf{u_n}\; P_n(x_r) = \sum_{n=0}^{\infty} \mathcal{X}(\mathbf{u_n})\; P_n(x_r) \tag{33}$$

where the operator $\mathcal{X}(\mathbf{u_n})$ acts on the indices n as

$$\mathcal{X} = A_{n-1}\mathcal{E}_n^{-1} + B_n I + C_{n+1}\mathcal{E}_n\,, \tag{34}$$

and where the shift operator $\mathcal{E}_n^{\pm 1}\{\mathbf{u_n}\}$ transforms $\mathbf{u_n}$ into $\mathbf{u_{n\pm 1}}$.

A package called ORTHOSERIES, written in `Maple`, and developed by L. Rebillard, makes all manipulations for a given polynomial Q, which satisfies a linear partial differential equation, choosing the expanding family for each variable. Typical results are shown in the following examples.[33]

i) $Q(x_1,x_2,x_3) = 8x_3^3 - 24x_1^2x_3 - 4x_1^2x_2 + \frac{4}{3}\,x_2^3$ is an harmonic homogeneous polynomial solution of the Laplace equation,

$$\nabla^2 Q = \frac{\partial^2}{\partial x_1^2}Q(x_1,x_2,x_3) + \frac{\partial^2}{\partial x_2^2}Q(x_1,x_2,x_3) + \frac{\partial^2}{\partial x_3^2}Q(x_1,x_2,x_3) = 0$$

Choosing the Legendre, Laguerre and Hermite family[26] for the variable x_1, respectively x_2, x_3, the program gives

$$\begin{aligned} Q(x_1,x_2,x_3) = & \left(7 + \frac{41}{3}\,\alpha + 8\alpha^2 + \frac{4}{3}\,\alpha^3\right)\, P_0(x_1)L_0^{(\alpha)}(x_2)H_0(x_3) \\ & - (23 + 20\alpha + 4\alpha^2)P_0(x_1)L_1^{(\alpha)}(x_2)H_0(x_3) \\ & - (1+\alpha)P_2(x_1)L_0^{(\alpha)}(x_2)H_0(x_3) + P_2(x_1)L_1^{(\alpha)}(x_2)H_0(x_3) \\ & + (24 + 8\alpha)P_0(x_1)L_2^{(\alpha)}(x_2)H_0(x_3) - 8P_0(x_1)L_3^{(\alpha)}(x_2)H_0(x_3) \\ & + 3P_0(x_1)L_0^{(\alpha)}(x_2)H_1(x_3) - 3P_2(x_1)L_0^{(\alpha)}(x_2)H_1(x_3) \\ & + P_0(x_1)L_0^{(\alpha)}(x_2)H_3(x_3)\,. \end{aligned} \tag{35}$$

ii) The connection coefficients in

$$L_m^{(\alpha+\beta+1)}(x_1 + x_2) = \sum_{(n_1,n_2)\in\mathbb{N}^2} \mathbf{u_{n_1,n_2}} L_{n_1}^{(\alpha)}(x_1)L_{n_2}^{(\beta)}(x_2)\,, \tag{36}$$

using the operator

$$L = (x_1 + x_2)\partial_1^2 + (2 + \alpha + \beta - x_1 - x_2)\partial_1 + m \tag{37}$$

annihilating $L_m^{(\alpha+\beta+1)}(x_1 + x_2)$, satisfy the recurrence relation

$$\begin{aligned} & - n_2\mathbf{u_{n_1+1,n_2-1}} - (n_2 + \beta + 1)\mathbf{u_{n_1+1,n_2+1}} + (-n_1 + m)\mathbf{u_{n_1,n_2}} \\ & + 2(n_1 + 1 + n_2 - m)\mathbf{u_{n_1+1,n_2}} + (m - 1 + \beta - n_1)\mathbf{u_{n_1+2,n_2}} = 0\,. \end{aligned} \tag{38}$$

The initial condition allows us to find that $\mathbf{u_{n_1,n_2}} = 1$ if $n_1 + n_2 = m$ and $\mathbf{u_{n_1,n_2}} = 0$ if $n_1 + n_2 \neq m$ and therefore we recover the well-known result,[2]

$$L_m^{(\alpha+\beta+1)}(x_1+x_2) = \sum_{n=0}^{m} L_n^{(\alpha)}(x_1) L_{m-n}^{(\beta)}(x_2)\,. \tag{39}$$

Remarks

We are aware of the (little bit) partial overview presented in this short review of limited length and we apologize for many missing references. We gave some priority to the recurrent approach for which the `Navima` group has some partial experience, and we hope however that the positioning of the several techniques are sufficiently clarified in order to help the reader.

Acknowledgments

This work was partially financial supported by the "XXII Comisión Mixta Permanente del acuerdo cultural entre España y Bélgica (Comunidad Francesa)". The financial support of A. Ronveaux from C.U.D. and University of Cotonou in order to take part in COPROMAPH 2 (Benin-October/November 2001), is also greatly appreciated. The work of I. Area and E. Godoy was partially supported by Ministerio de Ciencia y Tecnología of Spain under grant BFM2001-3423. A. Zarzo acknowledges financial support from DGES (Ministerio de Educación y Cultura of Spain) under contract PG95-1205 and also from Junta de Andalucía (FQM-0207).

References

1. M. Abramowitz and I. Stegun, *Handbook of Mathematical Functions* (Dover, New York, 1965).
2. W.A. Al-Salam, *Characterization theorems for orthogonal polynomials*, in *Orthogonal Polynomials: Theory and Practice*, P. Nevai, ed., *NATO ASI Series C: Math. and Phys. Sciences* **294**, 1–24 (Kluwer, Dordrecht, 1990).
3. R. Álvarez–Nodarse, R.J. Yáñez and J.S. Dehesa, *Modified Clebsch-Gordan-type expansions for products of discrete hypergeometric polynomials*, *J. Comput. Appl. Math.* **89**, 171–197 (1998).
4. G.E. Andrews, R. Askey and R. Roy, *Special Functions*, *Encyclopedia of Mathematics and its Applications* **71** (Cambridge University Press, Cambridge, 1999).

5. I. Area, E. Godoy, A. Ronveaux and A. Zarzo, *Inversion problems in the q-Hahn tableau*, *J. Symbolic Comput.* **28**, 767–776 (1999).
6. I. Area, E. Godoy, A. Ronveaux and A. Zarzo, *Minimal recurrence relations for connection coefficients between classical orthogonal polynomials: Discrete case*, *J. Comput. Appl. Math.* **89**, 309–325 (1998).
7. I. Area, E. Godoy, A. Ronveaux and A. Zarzo, *Solving connection and linearization problems within the Askey scheme and its q-analogue via inversion formulas*, *J. Symbolic Comput.* **133**, 151–162 (2001).
8. P.L. Artés, J.S. Dehesa, A. Martínez-Finkelshtein and J. Sánchez–Ruiz, *Linearization and connection coefficients for hypergeometric-type polynomials*, *J. Comput. Appl. Math.* **99**, 15–26 (1998).
9. R. Askey, *Orthogonal Polynomials and Special Functions*, *CBMS-NSF Reg. Conferences Series in Applied Mathematics* **21** (SIAM, Philadelphia, Pennsylvania, 1975).
10. S. Belmehdi, S. Lewanowicz and A. Ronveaux, *Linearization of the product of orthogonal polynomials of a discrete variable*, *Appl. Math.* **24**, 445–455 (1997).
11. S. Belmehdi and A. Ronveaux, *The fourth-order differential equation satisfied by the associated orthogonal polynomials*, *Rend. Mat. Appl.* **VII. Ser. 11(2)**, 313–326 (1991).
12. J.L. Fields and J. Wimp, *Expansions of hypergeometric functions in hypergeometric functions*, *Math. Comp.* **15**, 390–395 (1961).
13. M. Foupouagnigni, A. Ronveaux and W. Koepf, *Fourth order q–difference equation for the first associated of the q–classical orthogonal polynomials*, *J. Comput. Appl. Math.* **101**, 231–236 (1999).
14. M. Foupouagnigni, M.H. Hounkonnou and A. Ronveaux, *The fourth-order difference equation satisfied by the associated orthogonal polynomials of the Δ-Laguerre-Hahn class*, *J. Symb. Comput.* **28**, 801–818 (1999).
15. G. Gasper, *Projection formulas for orthogonal polynomials of a discrete variable*, *J. Math. Anal. Appl.* **45**, 176–198 (1974).
16. G. Gasper and M. Rahman, *Basic Hypergeometric Series*, *Encyclopedia of Mathematics and its Applications* **35** (Cambridge University Press, Cambridge, 1990).
17. E. Godoy, A. Ronveaux, A. Zarzo and I. Area, *Minimal recurrence relations for connection coefficients between classical orthogonal polynomials: Continuous case*, *J. Comput. Appl. Math.* **84**, 257–275 (1997).
18. E. Godoy, A. Ronveaux, A. Zarzo and I. Area, *On the limit relations between classical continuous and discrete orthogonal polynomials*, *J. Comput. Appl. Math.* **91**, 97–105 (1998).

19. M.H. Hounkonnou, S. Belmehdi and A. Ronveaux, *Linearization of arbitrary products of classical orthogonal polynomials*, *Appl. Math.* **27**, 187–196 (2000).
20. R. Koekoek and R.F. Swarttouw, *The Askey-scheme of hypergeometric orthogonal polynomials and its q-analogue*, Report 98-17, Fac. Techn. Math. Informatics, Delft University of Technology (Delft, 1998).
21. W. Koepf and D. Schmersau, *Representations of orthogonal polynomials*, *J. Comput. Appl. Math.* **90**, 57–94 (1998).
22. J. Letessier, A. Ronveaux and V. Valent, *Fourth-order difference equation for the associated Meixner and Charlier polynomials*, *J. Comput. Appl. Math.* **71**, 331–341 (1996).
23. S. Lewanowicz, *Recurrence relations for the connection coefficients of orthogonal polynomials of a discrete variable*, *J. Comput. Appl. Math.* **76**, 213–229 (1996).
24. S. Lewanowicz, *The hypergeometric functions approach to the connection problem for the classical orthogonal polynomials*, internal report (1997).
25. Y.L. Luke, *The Special Functions and their Approximations* (Academic Press, New York, 1969), 2 Volumes.
26. A.F. Nikiforov, S.K. Suslov and V.B. Uvarov, *Classical Orthogonal Polynomials of a Discrete Variable* (Springer, Berlin, 1991).
27. A.P. Prudnikov, Yu.A. Brychkov and O.I. Marichev, *Integrals and Series* (Gordon and Breach, New York, 1986), 3 Volumes.
28. E.D. Rainville, *Special Functions* (MacMillan, New York, 1960).
29. A. Ronveaux, *Fourth-order differential equations for numerator polynomials*, *J. Phys.* **A21**, L749–L753 (1988).
30. A. Ronveaux, *Three lectures on classical orthogonal polynomials*, in Proc. First International Workshop on Contemporary Problems in Mathematical Physics, J. Govaerts, M.N. Hounkonnou and W.A. Lester, Jr., eds. (World Scientific, Singapore, 2000), pp. 283–306.
31. A. Ronveaux, E. Godoy, A. Zarzo and I. Area, *Fourth-order difference equation for the first associated of classical discrete orthogonal polynomial*, *J. Comput. Appl. Math.* **90**, 45–50 (1998).
32. A. Ronveaux, M.N. Hounkonnou and S. Belmehdi, *Generalized linearization problems*, *J. Phys.* **A28**, 4423–4430 (1995).
33. A. Ronveaux and L. Rebillard, *Expansion of multivariable polynomials in sum of products of orthogonal polynomials*, to be published in *Appl. Math. and Computation* (special issue, in print).
34. A. Ronveaux, A. Zarzo and E. Godoy, *Recurrence relation for connection coefficients between two families of orthogonal polynomials*, *J. Comput. Appl. Math.* **62**, 67–73 (1995).

35. A. Ronveaux, A. Zarzo, I. Area and E. Godoy, *Transverse limits in the Askey tableau*, *J. Comput. Appl. Math.* **99**, 335-337 (1998).
36. A. Ronveaux, A. Zarzo, I. Area and E. Godoy, *Bernstein bases and Hahn-Eberlein orthogonal polynomials*, *Integral Transforms and Special Functions* **7**, 87–96 (1998).
37. J. Sánchez–Ruiz and J.S. Dehesa, *Some connection and linearization problems for polynomials in and beyond the Askey scheme*, *J. Comput. Appl. Math.* **133**, 579–591 (2001).
38. N.M. Temme and J.L. López, *The role of Hermite polynomials in asymptotic analysis*, CWI Report MAS–R9926 (Amsterdam, 1999), *Methods Appl. Anal.*, to appear.
39. A. Verma, *Certain expansions of the basic hypergeometric functions*, *Math. Comp.* **20**, 151–157 (1966).
40. A. Zarzo, I. Area, E. Godoy and A. Ronveaux, *Results for some inversion problems for classical continuous and discrete orthogonal polynomials*, *J. Phys.* **A30**, L35–L40 (1997).
41. A. Zarzo, A. Ronveaux and E. Godoy, *Fourth-order differential equation satisfied by the associated of any order of all classical orthogonal polynomials. A study of their distribution of zeros*, *J. Comput. Appl. Math.* **49**, 349–359 (1993).

List of Participants

Prof. A. Afouda
Université d'Abomey-Calavi
B.P. 526, Cotonou, Republic of Benin

Dr. L. Agbohounto
Université d'Abomey-Calavi
Republic of Benin

Mrs. Y. Ahanzo Glèlè
Faculté des Sciences de la Santé, Assistance Médicale
Université d'Abomey-Calavi, Republic of Benin

Mr. Y.K.M. Ahipo
UFR, Département de Mathématique et d'Informatique
Université de Ouagadougou, Burkina-Faso

Dr. B. Ahounou
Institut de Mathématiques et de Sciences Physiques (IMSP)
Université d'Abomey-Calavi, Porto-Novo, Republic of Benin

Dr A. Akpo
Département de Physique
Faculté des Sciences et Techniques
Université d'Abomey-Calavi
B.P. 526, Cotonou, Republic of Benin

Prof. S.T. Ali
Department of Mathematics and Statistics
Concordia University, Montréal, Québec, Canada

Dr. R. Aminou
Université d'Abomey-Calavi, Republic of Benin

Prof. N. G. Andjiga
Département de Mathématique
Université de Yaoundé, Cameroon

Prof. K. Assiamoua
Département de Mathématique
Université de Lome, Togo

Mr. H. Atakoun
Institut de Mathématiques et de Sciences Physiques (IMSP)
Université d'Abomey-Calavi, Porto-Novo, Republic of Benin

Mr. C. Atindogbe
Institut de Mathématiques et de Sciences Physiques (IMSP)
Université d'Abomey-Calavi, Porto-Novo, Republic of Benin

Mr. G.Y.H. Avossevou
Institut de Mathématiques et de Sciences Physiques (IMSP)
Université d'Abomey-Calavi, Porto-Novo, Republic of Benin

Prof. E. Azatassou
Département de Mathématique
Université d'Abomey-Calavi, Republic of Benin

Prof. D. Bagayoko
Department of Physics
Southern University and A&M College
P. O. Box 11776, Baton Rouge, LA 70813, USA

Mr. E. Baloïtcha
Unité de Recherche en Physique Théorique (URPT)
Institut de Mathématiques et de Sciences Physiques (IMSP)
Université d'Abomey-Calavi, Porto-Novo, Republic of Benin

Prof. Y. Birregah
Département de Mathématique
Université de Lome, Togo

Prof. B.C. Capo Hounkpati
The Gbe International Laboratory
Université d'Abomey-Calavi, Republic of Benin

Mrs. L. da Matha
Institut de Mathématiques et de Sciences Physiques (IMSP)
Université d'Abomey-Calavi, Porto-Novo, Republic of Benin

Mr. O. da Matha
Institut de Mathématiques et de Sciences Physiques (IMSP)
Université d'Abomey-Calavi, Porto-Novo, Republic of Benin

Mr. A. K. Dade
Université d'Abomey-Calavi, Republic of Benin

Prof. G. Debiais
Groupe de Physique Fondamentale
Université de Perpignan
52 avenue de Villeneuve, F-66860 Perpignan, France

Prof. C.S. Diatta
Institut de Technologies Nucléaires Appliquées (ITNA)
Université Check Anta Diop de Dakar, Senegal

Dr. I. Diedhou
Institut de Technologies Nucléaires Appliquées (ITNA)
Université Check Anta Diop de Dakar, Senegal

Mr. G. Edah
Institut de Mathématiques et de Sciences Physiques (IMSP)
Université d'Abomey-Calavi, Porto-Novo, Republic of Benin

Mr. H. Enjieu
Institut de Mathématiques et de Sciences Physiques (IMSP)
Université d'Abomey-Calavi, Porto-Novo, Republic of Benin

Prof. J.-P. Ezin
Institut de Mathématiques et de Sciences Physiques (IMSP)
Université d'Abomey-Calavi, Porto-Novo, Republic of Benin

Prof. K. Fadimba
Département de Mathématique
Université de Lome, Togo

Prof. C. Gnanvo
Département de Mathématique
Université d'Abomey-Calavi, Republic of Benin

Prof. G.A. Goldin
Science and Mathematics Partnerships SERC
Rutgers – The State University of New Jersey
118 Frelinghuysen Road, Piscataway, NJ 08854-8019, USA

Mr. D.S. Gossou
Institut de Mathématiques et de Sciences Physiques (IMSP)
Université d'Abomey-Calavi, Porto-Novo, Republic of Benin

Ms. L. Gouba
Institut de Mathématiques et de Sciences Physiques (IMSP)
Université d'Abomey-Calavi, Porto-Novo, Republic of Benin

Dr. A. Goudjo
Département de Mathématique
Université d'Abomey-Calavi, Republic of Benin

Prof. J. Govaerts
Institut de Physique Nucléaire
Université catholique de Louvain
2, Chemin du cyclotron, B-1348 Louvain-la-Neuve, Belgium

Mr. K. Guèdjè
Institut de Mathématiques et de Sciences Physiques (IMSP)
Université d'Abomey-Calavi, Porto-Novo, Republic of Benin

Mr. A. Hassan
Institut de Mathématiques et de Sciences Physiques (IMSP)
Université d'Abomey-Calavi, Porto-Novo, Republic of Benin

Mr. G. Honnouvo
Institut de Mathématiques et de Sciences Physiques (IMSP)
Université d'Abomey-Calavi, Porto-Novo, Republic of Benin

Mr. B. Hounga
Unité de Recherche en Physique Théorique (URPT/INIREF)
Institut de Mathématiques et de Sciences Physiques (IMSP)
Université d'Abomey-Calavi, Porto-Novo, Republic of Benin

Dr. E. Houngninou
Université d'Abomey-Calavi, Republic of Benin

Prof. M.N. Hounkonnou
Institut de Mathématiques et de Sciences Physiques (IMSP)
Université d'Abomey-Calavi, Porto-Novo, Republic of Benin

Dr. J. Hounsou
Institut de Mathématiques et de Sciences Physiques (IMSP)
Université d'Abomey-Calavi, Porto-Novo, Republic of Benin

Dr. A.N. Issa
Département de Mathématique
Université. d'Abomey-Calavi, Republic of Benin

Dr. Y. Justum
Laboratoire de Chimie Physique
Université de Paris-Sud (Paris XI), Orsay, France

Prof. A. Kebede
Department of Physics
NC A& T State University
Greensboro, NC 27411, USA

Prof. J. Klauder
Departments of Physics and Mathematics
University of Florida
Gainesville, FL 32611, USA

Mr. B. Kone
UFR, Département de Mathématique et d'Informatique
Université de Ouagadougou, Burkina-Faso

Dr. Y.B. Kouagou
Institut de Mathématiques et de Sciences Physiques (IMSP)
Université d'Abomey-Calavi, Porto-Novo, Republic of Benin

Dr. B. Kpamègan
Université d'Abomey-Calavi, Republic of Benin

Mr. L. Leadi
Institut de Mathématiques et de Sciences Physiques (IMSP)
Université d'Abomey-Calavi, Porto-Novo, Republic of Benin

Prof. K. Maazou
Université de Niamey, Republic of Niger

Dr. A. Marcos
Département de Mathématique
Université d'Abomey-Calavi, Republic of Benin

Mr. F. Massamba
Institut de Mathématiques et de Sciences Physiques (IMSP)
Université d'Abomey-Calavi, Porto-Novo, Republic of Benin

Dr. B. Mbodji
UFR de Mathématiques Appliquées et Informatique
Université Gaston Berger, St. Louis, Senegal

Prof. N.G. Medhin
Center for Theoretical Studies of Physical Systems (CTSPS)
Clark Atlanta University, Atlanta, GA 30314, USA

Mr. J.E.B. Mendy
Institut de Mathématiques et de Sciences Physiques (IMSP)
Université d'Abomey-Calavi, Porto-Novo, Republic of Benin

Mr. S. Mora
Université d'Abomey Calavi, Republic of Benin

Mr. H. Mouhamadou
Institut de Mathématiques et de Sciences Physiques (IMSP)
Université d'Abomey-Calavi, Porto-Novo, Republic of Benin

Prof. S. Moussiliou
Université d'Abomey-Calavi, Republic of Benin

Dr. M.B. M'Passi
Département de Physique
Université Marien Ngouabi de Brazzaville
B.P. 69, Brazzaville, Congo

Prof. A.Z. Msezane
Center for Theoretical Studies of Physical systems (CTSPS)
Clark Atlanta University, Atlanta, GA 30314, USA

Mr. A. Musesa Landa
Institut de Mathématiques et de Sciences Physiques (IMSP)
Université d'Abomey-Calavi, Porto-Novo, Republic of Benin

Mr. L.J. Mutuba
Institut de Mathématiques et de Sciences Physiques (IMSP)
Université d'Abomey-Calavi, Porto-Novo, Republic of Benin

Dr. H. Mweene
Department of Physics
Zambia University, Lusaka, Zambia

Ms. U.T. Ndongmouo
Institut de Mathématiques et de Sciences Physiques (IMSP)
Université d'Abomey-Calavi, Porto-Novo, Republic of Benin

Mr F. Ngakeu
Institut de Mathématiques et de Sciences Physiques (IMSP)
Université d'Abomey-Calavi, Porto-Novo, Republic of Benin

Mr. N. Ngarhasta
UFR, Département de Mathématique et d'Informatique
Université de Ouagadougou, Burkina-Faso

Prof. N. Noutchéguémè
Département de Mathématique
Université de Yaoundé, Cameroon

Prof. A. Odzijewicz
Faculty of Mathematical and Physics
University of Bialystok, Sosnowa State, Poland

Dr. C. Ogouyandjou
Institut de Mathématiques et de Sciences Physiques (IMSP)
Université d'Abomey-Calavi, Porto-Novo, Republic of Benin

Mr. B. Olory
Université d'Abomey-Calavi, Republic of Benin

Mr. H. Onibon
LTHE/IRD - 1025 B.P. 53 Domaine Universtaire
F-38041 Grenoble Cedex 9, France

Dr. M.-F. Ouédraogo
UFR, Département de Mathématique et d'Informatique
Université de Ouagadougou, Burkina-Faso

Dr. H. M. Quadjovie
Département de Mathématique
Université de Lome, Togo

Dr. N.P. Rapapa
Department. of Physics
Faculty of Science and Technology
University of Lesotho, Lesotho

Prof. T. Ratiu
Département de Mathématique
Ecole Polytechnique Fédérale de Lausanne, Switzerland

Prof. K. Reed
Lawrence Livermore National Laboratory
Berkeley, USA

Prof. A. Ronveaux
Facultés Universitaire Notre Dame de la Paix
Namur, Belgium

Mr. G. Rubomboza
Université du Burundi, Burundi

Dr. D. Sangare
UFR de Mathématiques Appliquées et Informatique
Université Gaston Berger, St. Louis, Senegal

Mr. C. Sassaï
UFR, Département de Mathématique et d'Informatique
Université de Ouagadougou, Burkina-Faso

Mr. S. Sawadogo
UFR, Département de Mathématique et d'Informatique
Université de Ouagadougou, Burkina-Faso

Mr. D.K. Senam
Institut de Mathématiques et de Sciences Physiques (IMSP)
Université d'Abomey-Calavi, Porto-Novo, Republic of Benin

Dr. A. Sene
UFR, de Mathématiques Appliquées et Informatique
Université Gaston Berger, St. Louis, Senegal

Dr. M. D. Slaughter
Department of Physics
University of New Orleans
New Orleans LA 70148, USA

Mr. O. So
UFR, Département de Mathématique et d'Informatique
Université de Ouagadougou, Burkina-Faso

Mr. A. Sodoga
Institut de Mathématiques et de Sciences Physiques (IMSP)
Université d'Abomey-Calavi, Porto-Novo, Republic of Benin

Prof. B. Some
UFR, Département de Mathématique et d'Informatique
Université de Ouagadougou, Burkina-Faso

Prof. H. Sosso
Université de Ndjamena, Tchad

Mr. J. de D. Tapsoba
UFR, Département de Mathématique et d'Informatique
Université de Ouagadougou, Burkina-Faso

Prof. K. Tcharie
Département de Mathématique
Université de Lome, Togo

Mrs. C. Tchohoungbo
Republic of Benin

Dr. J. Tossa
Institut de Mathématiques et de Sciences Physiques (IMSP)
Université d'Abomey-Calavi, Porto-Novo, Republic of Benin

Dr. B. Toukourou
Lycée National Léon M'Ba
Libreville, Gabon

Mr. R. Yamapi
Institut de Mathématiques et de Sciences Physiques (IMSP)
Université d'Abomey-Calavi, Porto-Novo, Republic of Benin

Dr. C.B. Yêhouénou
Université d'Abomey-Calavi, Republic of Benin

Dr. T. Zossou
Université d'Abomey-Calavi, Republic of Benin

Mrs. P. Zoungrana
UFR, Département de Mathématique et d'Informatique
Université de Ouagadougou, Burkina-Faso